Einstein Studies

Editors: Don Howard   John Stachel

Published under the sponsorship
of the Center for Einstein Studies,
Boston University

Volume 1: Einstein and the History of General Relativity
        Don Howard and John Stachel, editors

Volume 2 : Conceptual Problems of Quantum Gravity
        Abhay Ashtekar and John Stachel, editors

IN PRESS :

Volume 3 : Studies in the History of General Relativity
        Jean Eisenstaedt and A. J. Kox, editors

Volume 4 : Recent Advances in General Relativity
        Allen I. Janis and John R. Porter, editors

Abhay Ashtekar   John Stachel

Editors

Conceptual Problems of Quantum Gravity

Based on the
Proceedings of the 1988 Osgood Hill Conference,
North Andover, Massachusetts
15–19 May 1988

Birkhäuser

Boston · Basel · Berlin

Abhay Ashtekar
Department of Physics
Syracuse University
Syracuse, New York 13244
USA

John Stachel
Center for Einstein Studies
and
Department of Physics
Boston University
Boston, Massachusetts 02215
USA

Library of Congress Cataloging-in-Publication Data
Conceptual problems of quantum gravity : based on the proceedings of
   the 1988 Osgood Hill Conference, North Andover, Massachusetts, 15--19
   May 1988 / Abhay Ashtekar, John Stachel, editors.
         p.   cm. -- (Einstein studies : v. 2)
      "Sponsored by the Boston University Center for Einstein Studies" - Pref.
      Includes bibliographical references and index.
      ISBN 0-8176-3443-6 (alk. paper)
      1. Quantum gravity--Congresses. 2. Astrophysics--Congresses.
   I. Ashtekar, Abhay. II. Stachel, John J., 1928-   . III. Boston
   University. Center for Einstein Studies. IV. Series.
   QC178.C63 1991   91-9536
   530.1 ' 42--dc20   CIP

Printed on acid-free paper.

© The Center for Einstein Studies. 1991
The Einstein Studies series is published under the sponsorship of the Center for Einstein
Studies, Boston University.

ISBN 0-8176-3443-6
ISBN 3-7643-3443-6

Typeset in TeX by ARK Publications, Inc., Newton Centre, MA
Printed and bound by Edwards Brothers, Inc., Ann Arbor, Michigan.
Printed in the United States of America.

9 8 7 6 5 4 3 2 1

This volume, like the conference, is dedicated to the memory of
Paul Dirac (1902–1984)

CONTENTS

Preface to Volume Two

The conference on which this volume is based was actually the second one on quantum gravity held at Osgood Hill. The first, sponsored by the Boston University Institute of Relativity Studies, an earlier avatar of the Center for Einstein Studies, took place October 31–November 3, 1972.

A number of participants took part in the earlier meeting. Chris Isham gave the first of his magisterial survey talks on quantum gravity, which have since become a standard ingredient in the recipe for a good conference on the subject. (He pointedly reminded me that his opening talk was then allotted two hours). Others who attended the earlier meeting are Bryce De Witt, Jimmy York, Karel Kuchař, and myself.

Sadly, one participant we shall never again see in this world: Paul Dirac, who presented his ideas on the possibility of a distinction between macroscopic and microscopic metric tensors. At the recent meeting, Karel and I recalled with pleasure the wonderful exchange between Dirac and John Wheeler on the nature of "the quantum principle," a discussion that —like the rest of the conference—is recorded on tape, awaiting the ears of future historians. With the unanimous agreement of the participants, the proceedings of this meeting were dedicated to Dirac's memory, as is this volume.

Another highlight of that first conference was John Wheeler's dramatic renunciation, under the influence of Andrei Sakharov's ideas about gravity, of his geometrodynamical program—a remarkable early confirmation of Sakharov's thesis about the convergence of American and Soviet societies.

Pursuing this political metaphor, one might say that the first half century of relations betweeen general relativity and quantum mechanics was dominated by the struggle for hegemony between two rival imperialisms. I refer, of course, to unified field-theoretical imperialism which grew out of general relativity and maintains that some sort of classical nonlinear field theory will ultimately be able to account for all quantum phenomena; and to quantum field-theoretical imperialism, which grew out of quantum mechanics and looks upon the field equations of general relativity as no more than a particularly messy challenge to the well-established formalisms and interpretative schema of special-relativistic quantum field theory.

The first Osgood Hill conference on quantum gravity was dedicated—in my mind at least—to advocacy of the doctrine of peaceful coexistence and competition between the two imperialisms, and to the exploration of third paths of development, not based on the assumption that either imperialism would ultimately triumph. As Abhay Ashtekar indicates in the *Introduction*, this viewpoint is much more widely accepted today than it had been in the seventies when it seemed that field-theoretical imperialism was on the verge of global triumph. In keeping with the then prevalent climate of opinion, our 1972 session on non-conventional approaches to quantum gravity was dubbed "the crackpot session" by Roger Penrose, who contributed an early exposition of twistor theory to it.

The only record in print of the first meeting is an article about it in *Nature* (Volume 240, pp. 382–383). The proceedings were never published, a mistake that is obviously not being repeated.

The Organizing Committee for this second conference on quantum gravity, sponsored by the Boston University Center for Einstein Studies, consisted of Abhay Ashtekar, Jim Hartle, Chris Isham, Roger Penrose (who unfortunately could not attend because because he was awarded the Wolf prize that week), Abner Shimony, and myself. The meeting was held May 15–19, 1988 at the University's Osgood Hill conference center in North Andover, Massachusetts. (A list of the participants will be found at the end of this volume). A most convivial, tolerant, and non-imperialistic spirit prevailed at the meeting—although it must be admitted that the discussion of some topics put the good will of many participants to a severe test. We have included edited versions of the discussons after many of the talks so that the reader can get an impression of the often sharp (in both senses of the word), but always good-humored exchanges at the sessions. As the title indicates, the central focus of the meeting was on the serious issues of principle raised by any attempt to create a single framework that can encompass the widely divergent conceptual schemata of general relativity and of quantum field theory. The reader will find a survey of these issues and how they arose from earlier work on quantum gravity in Ashtekar's *Introduction*.

The papers have been organized by topics, and the order does not completely coincide with the order in which they were presented at the sessions. Thus, discussion comments sometimes refer forward to papers that appear later in the book. In addition, comments in the general discussion at the end of the meeting have been included together with the initial discussion at the end of the appropriate talk.

The reader will find occasional references to several papers that were not submitted for publication in this volume. Murray Gell-Mann spoke on "Quantum Mechanics of This Specific Universe,"[1] David Boulware on "Status of Higher Derivative Theories," and Sidney Coleman on "Why There is Nothing Rather Something: Worm Holes and the Cosmological Constant."[2]

Progress has been made on a number of technical problems discussed at the meeting, and Abhay and I look foward confidently to the third Osgood Hill conference on quantum gravity in a few years. But many of the conceptual problems are likely to be still around for a long time, and it is our hope that many physicists, particularly students, will find this volume a useful introduction to the important and fascinating task of struggling with them.

John Stachel

NOTES

[1] For the contents of Gell-Mann's paper see Gell-Mann, M. and Hartle, J.B. "Quantum Mechanics in the Light of Quantum Cosmology." In *Complexity, Entropy and the Physics of Information.* Santa Fe Institute Studies in the Sciences of Complexity, vol. III, Wojciech H. Zurek, ed. Reading, MA: Addison Wesley (1990).

[2] For the contents of Coleman's paper see *Nuclear Physics B* 310: 643 (1988).

Acknowledgments

The editors thank Boston University for its support of the Center for Einstein Studies, as well as the Dean Edmonds Foundation, which also helped to finance the Osgood Hill Conference on Conceptual Problems of Quantum Gravity.

They thank the Organizing Committee (see the *Preface*) for their work on the scientific aspects of the meeting, and Ann Lehar of the Einstein Papers for her help with the technical aspects of its organization. The work of four participants who were then graduate students at Boston and Syracuse Universities, Joy Christian (BU), Joseph Romano (SU), Mohammed Samiullah(BU), and Ranjeet Tate (SU) contributed much to the smooth running of the meeting, as did the work of Elaine Lord, Diane O'Brien, and the rest of the staff of the Osgood Hill Conference Center.

Ranjeet Tate worked hard on the transcription of the taped discussions, and Syracuse University contributed the services of Carol Roth, whose efficient work helped to turn the raw transcriptions into polished versions. The Boston University Center for Einstein Studies and NSF Grants PHY86-12412 and PHY 16733 to Syracuse University supported editorial work on this volume.

George Adelman, Editorial Director of Birkhäuser Boston, and Lauren Cowles, its former Physics Editor, contributed valuable advice and invaluable encouragement to the editorial work. It was a pleasure to work with Ann Kostant who helped with the design, layout and typesetting of the book.

Introduction: The Winding Road to Quantum Gravity

Abhay Ashtekar

> *Traveler, there are no paths;*
> *Paths are made by walking.*
> *–Antonio Machado*

General relativity and quantum theory are undoubtedly among the greatest intellectual achievements of this century. Each of them has profoundly altered the conceptual fabric that underlies our understanding of the physical world. Furthermore, each has been successful in describing the physical phenomena in its own domain to an astonishing degree of accuracy. And yet, they offer us *strikingly* different pictures of physical reality. Indeed, at first one is surprised that physics could keep progressing blissfully in the face of so deep a conflict. The reason of course is the "accidental" fact that the values of fundamental constants in our universe conspire to make the Planck length so small and Planck energy so high compared to laboratory scales. It is because of this that we can happily maintain a schizophrenic attitude, and use the precise, geometric picture of reality offered by general relativity while dealing with cosmological and astrophysical phenomena, and the quantum-mechanical world of chance and intrinsic uncertainties while dealing with atomic and subatomic particles. Although this strategy is quite appropriate as a practical stand, it is of course highly unsatisfactory from a conceptual viewpoint. Everything in our past experience in physics tells us that the two pictures we currently use must be approximations, special cases that arise as appropriate limits of a single, universal theory. That theory must therefore represent a synthesis of general relativity and quantum mechanics. This would be the quantum theory of gravity. Not only should it correctly describe all the known physical phenomena, but it should also adequately handle the Planck regime. This is the future theory that we invoke when faced with phenomena, such as the big bang and the final state of black holes, where the worlds of general relativity and quantum mechanics must unavoidably meet.

Research in quantum gravity dates back at least to the thirties when the issue of the necessity of quantizing the gravitational field was first analyzed in some detail. However, it is probably fair to say that the field really "took off" only in the late fifties. The general developments since then loosely represent three stages, each spanning roughly a decade. The purpose of this chapter is to present a sketch of these developments in broad strokes.

First, there was the begining: exploration. The goal was to do unto gravity as one would do unto any other physical field. The electromagnetic field had been successfully quantized using two approaches: canonical and covariant. In the canonical approach, electric and magnetic fields obeying Heisenberg's uncertainty principle are at the forefront, and quantum states naturally arise as gauge-invariant functionals of the vector potential on a spatial three-slice. In the covariant approach on the other hand, one first isolates and then quantizes the two radiative modes of the Maxwell field in space-time, without carrying out a (3+1)-decomposition, and the quantum states naturally arise as elements of the Fock space of photons. Attempts were made to extend these techniques to general relativity. In the electromagnetic case the two methods are completely equivalent. Only the emphasis changes in going from one to another. In the gravitational case, however, the difference is *profound*. This is not accidental. The reason is deeply rooted in one of the essential features of general relativity, namely the dual role of the space-time metric.

To appreciate this point, let us begin with field theories in Minkowski space-time, say Maxwell's theory to be specific. Here, the basic dynamical field is represented by a tensor field $F_{\mu\nu}$ on Minkowski space. The space-time geometry provides the kinematical arena on which the field propagates. The background, Minkowskian metric provides the light cones and the notion of causality. We can foliate this space-time by a one-parameter family of space-like three-planes, and analyze how the values of the electric and magnetic fields on one of these surfaces determine those on any other surface. The isometries of the Minkowski metric let us construct physical quantities such as fluxes of energy, momentum, and angular momentum carried by electromagnetic waves.

In general relativity, by contrast, there is no background geometry. The space-time metric itself is the fundamental dynamical variable. On the one hand, it is analogous to the Minkowski metric in Maxwell's theory; it determines space-time geometry, provides light cones, defines causality, and dictates the propagation of all physical fields (including itself). On the other hand, it is the analog of the Newtonian gravitational potential, and therefore the basic dynamical entity of the theory, similar, in this respect, to the $F_{\mu\nu}$ of the Maxwell theory. This dual role of the metric is in effect a precise statement of the equivalence principle that is at the heart of general relativity. It is this feature that is largely responsible for the powerful conceptual economy of general relativity, its elegance and its aesthetic beauty. However, this feature also brings with it a host of problems. We see already in the classical theory several manifestations of these difficulties. It is because there is no background geometry, for example, that it is so difficult to analyze singularities of the theory and to define the energy and momentum carried by gravitational waves. Since there is no a priori space-time, to introduce notions as basic as causality, time, and evolution, one must first solve the dynamical equations and *construct* a space-time. As an extreme example, consider black holes, whose definition requires the knowledge of the causal structure of the entire space-time. To find if the given initial conditions lead to the formation of a black hole, one must first obtain their maximal evolution, and, using the causal structure determined by *that* solution, ask if its future infinity has a past boundary. Thus, because there is no longer a clean separation between the kinematical arena and dynamics, in the classical theory substantial care and effort is already needed even in the formulation of basic physical questions.

In quantum theory the problems become significantly more serious. To see this, recall first that, because of the uncertainty principle, already in nonrelativistic quantum mechanics, particles do not have well-defined trajectories; time-evolution only produces a probability amplitude, $\Psi(\vec{x}, t)$, rather than a specific trajectory, $\vec{x}(t)$. Similarly, in quantum gravity, even after evolving an initial state, one would not be left with a specific space-time. In the absence of a space-time geometry, how is one to introduce even habitual physical notions such as causality, time, scattering states, and black holes?

The canonical and the covariant approaches have adopted dramatically different attitudes to face these problems. In the canonical approach, one notices that, in spite of the conceptual difficulties mentioned above, the Hamiltonian formulation of general relativity is well-defined, and attempts to use it as a stepping stone to canonical quantization. One wants to construct the mathematical framework first and then use it to suggest a physical interpretation. The fundamental canonical commutation relations are to lead us to the physical uncertainty principle. The motion generated by the Hamiltonian is to be thought of as time evolution. The fact that certain operators on the fixed ("spatial") three-manifold commute is supposed to capture the appropriate notion of causality. The emphasis is on preserving the geometrical character of general relativity, on retaining the beautiful fusion of gravity and geometry that Einstein created. In the sixties, the Hamiltonian formulation of the classical theory was worked out in detail, and a precise quantization program was laid down. However, this path, which had been so successful in particle physics, meandered away from the mainstream developments of quantum field theory. In the canonical program, it is hard to see how gravitons are to emerge, how scattering matrices are to be computed, how Feynman diagrams are to dictate dynamics and virtual processes are to give radiative corrections. To use a well-known phrase, the emphasis on geometry of the canonical program "drove a wedge between general relativity and the theory of elementary particles." More importantly, due to difficult technical problems, many of the internal goals of the program itself remained unattained. The quantum version of Einstein's equations turned out to be too difficult to solve exactly. Since no perturbation expansion preserving the spirit of the program was available, much of the discussion remained rather formal, and most of the conceptual problems of quantum gravity remained unsettled.

In the covariant approach the emphasis is just the opposite. Field-theoretic techniques are to be applied even if it means sacrificing the geometric beauty of general relativity. Indeed, as we shall see a little later, even the dynamics given by Einstein's equations are not to be regarded as sacrosanct. The first step in this program is to split the space-time metric $g_{\mu\nu}$ in two parts, $g_{\mu\nu} = \eta_{\mu\nu} + Gh_{\mu\nu}$, where $\eta_{\mu\nu}$ is to be a background, kinematical metric, often chosen to be flat, G is Newton's constant, and $h_{\mu\nu}$, the deviation of the physical metric from the chosen background, the dynamical field.[†] The two roles of the metric tensor are now split. With this splitting most of the conceptual problems discussed above seem to melt away. Thus, in the transition to the quantum theory it is only $h_{\mu\nu}$ that is quantized. The quanta of this field propagate on the classical background space-time with metric $\eta_{\mu\nu}$. If the background is in fact chosen to be flat, one can use the Casimir operators of the Poincaré group and show that the quanta have spin two and rest mass zero. These are the gravitons. The Einstein-Hilbert Lagrangian tells us how they interact with one another. Thus, in this program, quantum general relativity was first reduced to another quantum field theory in Minkowski space. One could apply to it all the machinery of perturbation theory that had been so successful in particle physics. One now had a definite program to compute amplitudes for various scattering processes. Unruly gravity appeared to be tamed and forced to fit into the mold created to describe other forces of nature. It was in the sixties that this procedure was first spelled out in detail. The quantum effects of gravity could now be visualized vividly in terms of exchanges of virtual gravitons. As in other field theories, calculations of physical amplitudes could be carried out using Feynman diagrams. Consistent Feynman rules for computing graviton-graviton interactions were discovered. Differences between abelian and nonabelian couplings were noted. Covariant quantum gravity was shown to be perturbatively unitary. Thus, by the early seventies, the covariant approach had led to several concrete results.

We now come to the second stage of development. This was marked by enthusiasm, hope. The motto was: Go forth, perturb, and expand. The enthusiasm was first generated by the discovery that Yang-Mills theory coupled to fermions is renormalizable if the masses of gauge particles arise from a spontaneous symmetry-breaking mechanism. This led to a successful theory of electroweak interactions. Particle physics witnessed a renaissance of quantum field theory. The enthusiasm spilled over to gravity; it reenforced the hope that perturbative methods would also be applicable to quantum general relativity. Consequently, much of the work in quantum gravity in the seventies was carried out within the covariant approach. Courageous calculations were performed to estimate radiative corrections. Unfortunately, however, this research soon ran into its first road block. Detailed calculations revealed that quantum general relativity is perturbatively non-renormalizable at two loops. To appreciate the significance of this result, let us return to the quantum theory of photons and electrons. This theory is perturbatively renormalizable. This means that, although individual terms in the perturbation expansion of a physical amplitude may diverge due to radiative corrections involving closed loops of virtual particles, these infinities are of a specific type; they can be systematically absorbed in the values of free parameters of the theory, the fine structure constant and the electron mass. Thus, by renormalizing these parameters, individual terms in the perturbation series can be systematically rendered finite. (However, the series when summed almost certainly diverges, a point that will become relevant later.) In quantum general relativity, such a systematic procedure is not available; infinities that arise due to radiative corrections are genuinely troublesome. Put differently, when one includes Feynman diagrams with two or more loops of virtual gravitons and attempts to extract finite answers, the quantum theory acquires an infinite number of undetermined parameters. It therefore has no predictive power at all!

Buoyed, however, by the success of perturbative methods in electro-weak interactions, the community was reluctant to give them up in the gravitational case. In the case of weak interactions, it was known for some time that the observed low energy phenomena could be explained using Fermi's simple four-point interaction. The problem was that this Fermi model led to a non-renormalizable theory. The new, renormalizable theory of weak interactions referred to above does reduce to the Fermi model in the low-energy regime, but differs from it at a fundamental level. It marshals new processes that are significant only at high energies, but which cure the problem of non-renormalizability. It was natural to look for similar processes in the gravitational case. This led to higher-derivative theories, theories in which the Lagrangian contains, in addition to the Einstein-Hilbert term, curvature-squared terms with appropriate coupling constants. The new terms play a significant role only at high energies, and leave the low-energy or long-distance behavior of the theory unaltered. Thus, in the classical, macroscopic regime, the new theory would essentially reproduce general relativity. For appropriate choices of coupling constants, the theory does indeed become renormalizable. In fact the theory is asymptotically free; at high energies, it behaves as if it were a free field theory. Unfortunately, however, it fails to be unitary! In the perturbative treatment, the Hamiltonian is unbounded from below, signaling a dramatic instability. Thus, although promising at first, this idea of mimicking the successful theory of weak interactions finally fails.

In these approaches the gravitational field was generally considered in isolation. The unification of electroweak interactions led to another suggestion. Perhaps a viable theory would result *only* when gravity is coupled to suitably chosen matter fields. The most striking example of this attitude is probably supergravity. The hope was that the infinities of the bosonic fields including gravity would be cancelled by those of the fermionic fields, giving us a renormalizable theory of gravity interacting with matter. Much effort went into analyzing the possibility that the most sophisticated version of this theory—$N = 8$ supergravity— can be employed as a genuine "grand unified theory." It turns out that some cancellations of infinities do occur, and that supergravity is indeed better behaved than general relativity coupled to matter in the standard way. Furthermore, the Hamiltonian is manifestly positive and the theory is unitary. Unfortunately, however, the theory again turns out to be non-renormalizable. Thus, as far as full quantum gravity—rather than quantum field theory in curved spaces—is concerned, the results in the seventies and early eighties, while important, were mainly negative in character.

Perturbative methods did revive once again in the eighties, however, due to another renaissance; that of string theory. In string theory, one abandons point particles and the corresponding local fields and considers extended objects—strings—as fundamental. It is remarkable, however, that the familiar physical particles—including the spin two, rest mass zero graviton—do in fact arise naturally in the theory as various components of string excitations. What is more, the theory has only one free parameter. All couplings are therefore fixed automatically; very little is fed in from outside. Consequently, hope was expressed that this theory would incorporate all the information about the physical world; it would be the "theory of everything." Technical successes of the theory generated a great deal of enthusiasm. The theory is unitary and, although a conclusive proof is yet to be published, there is general consensus among experts that the theory is *finite*—not just renormalizable—order by order in perturbation theory. Unfortunately, however, when summed, the series diverges and does so uncontrollably. One might wonder at first if this should be regarded as a serious failure of the theory. After all even in quantum electrodynamics the series is believed to diverge. Recall, however, that quantum electrodynamics is an inherently incomplete theory. It ignores many processes that must come into play at high energies or short distances. In particular, it completely ignores the microstructure of space-time, and simply assumes that space-time can be approximated by the smooth continuum of Minkowski space at arbitrarily small distances. Therefore, it can blame the divergence of its perturbation series on all these effects that lie outside its realm. A "theory of everything," on the other hand, cannot escape so easily. It must face the high energy regime—including the complications at the Planck scale—squarely. It must incorporate all these effects. It cannot plead ignorance and shift the burden of infinities to yet another theory.

Thus, we have now entered the third stage of development. The mood is markedly different. The heady enthusiasm has given way to deeper reflection. The conceptual questions of the sixties are again a focal point. The key lesson we have learned from the second stage is that perturbative, field-theoretic methods are not viable in quantum gravity. The assumption that space-time can be represented by a smooth continuum at arbitrarily small length scales simply leads to inconsistencies. We can neither ignore the microstructure of space-time nor presuppose its nature. We must let quantum gravity unravel it. Irrespective of whether we prefer strings or local fields, general relativity or higher derivative theories, we must face the problem of quantization nonperturbatively. As a well-known poetic expression goes, if we begin by "steam-rollering" space-time into flatness and linearity, we will learn nothing by "attempting to wave the magic wand of quantum theory over the resulting corpse."

Once again the orientation of research has shifted. Nonperturbative ideas are being explored in string theory. For, in spite of divergence of the perturbation series, it may well exist as an exact theory. Furthermore, the nonperturbative effects may be crucial in removing the current ambiguities, particularly in the reduction of the theory to four space-time dimensions. In quantum general relativity itself, the canonical program has been revived using new technical and conceptual tools. Some exact solutions to the quantum Einstein equations have been obtained. These have given us an inkling as to how different the true microstructure of space-time is likely to be from a smooth continuum. Furthermore, gravity is no longer treated separately, divorced from particle physics. A unified mathematical framework seems to bind it to other basic interactions. At the end of the seventies, the path integral—or, sum-over-histories— program was launched using Riemannian (i.e., positive-definite) metrics. The program is being vigorously applied to minisuperspaces that arise in cosmology. Conceptual questions of quantum cosmology are being analysed using fresh ideas. Finally, new numerical initiatives have been launched to test if quantum general relativity can exist nonperturbatively as a field theory.

Had perturbative methods worked, we would have been left with a rather simple picture of space-time, and many of the conceptual problems that arise due to the dual role of the metric would have simply disappeared, or at least been simplified enormously. Now we *have* to face those questions seriously. Is it appropriate in quantum gravity to work with a fixed space-time manifold? Shouldn't the topology also be subject to quantum fluctuations? Is there then a quantum theory of topology? Can the topological fluctuations representing worm holes be responsible for the laboratory values of coupling constants? In particular, do they force the cosmological constant to vanish? Another set of questions concerns the issue of constructing a quantum theory of the universe as a whole. Is such a program viable or even meaningful? These questions have revived interest in quantum measurement theory. Is it possible that it is in fact some aspect of quantum gravity that is responsible for the "collapse of the wave function," the key ingredient of the Copenhagen interpretation? A related question is: Why does the world around us look so classical? Is it because of some special initial conditions? Normally, in physics, initial conditions are simply supplied externally. However, we are now attempting to construct a "complete" theory. Should there be a place in it for a theory of initial conditions? Are there laws that govern the initial state of the universe? The final state of an evaporating black hole? What is the role of time in quantum gravity? Is the unitarity of evolution that we take for granted in quantum mechanics exact, or can small deviations occur due to quantum gravity effects? If so, how small? Finally, most of the work to date assumes that the rules of quantum mechanics are to remain intact in the process of its unification with general relativity. Why should this be? Isn't there a need to generalize the basic framework of quantum mechanics itself to accommodate the absence of a background space-time?

One cannot help but be struck by how qualitatively different these questions are compared to those that were pursued in the seventies. In retrospect, it appears that a complacent attitude towards such issues had then developed in the false security of perturbative methods. Indeed, in most circles it wasn't even respectable to show one's unease with quantum measurement theory, or suggest that rules of quantum mechanics are not absolutely sacrosanct. The eighties brought a growing realization that one cannot escape from such issues indefinitely.

The goal of this volume is to summarize the current status of these conceptual problems of quantum gravity.

January 1990

NOTE

† The term "covariant" is thus somewhat misleading. It may nonetheless be used to distinguish this approach from the canonical method since it does not involve a $(3 + 1)$-decomposition of space-time.

PART I

Quantum Mechanics, Measurement, and the Universe

Einstein and Quantum Mechanics

John Stachel

This paper is divided into five sections. Four of them are largely historical: The first ("A Lost Leader") discusses the charge that Einstein deserted his own earlier approach to the role of probability in physics when he refused to accept the fundamentally indeterministic element in quantum mechanics. The second section discusses Einstein's lifelong search for what he called a "constructive theory" of the quantum, and the third tries to highlight the essential elements in Einstein's critique of quantum mechanics. The fourth section ("The Other Einstein") recounts Einstein's persistent doubts about the entire field program as a foundation for theoretical physics, and his fragmentary suggestions for an alternative program.

The fifth section discusses three lessons that Einstein drew from his successful search for a generally covariant theory of gravitation: the indivisibility of gravitation and inertia; the lack of any prior physical individuation of points in a generally covariant theory before a solution to the field equations is given (no metric—no events); and the radically local character of general relativity in which even the global topology of the manifold depends on the solution to the field equations (no metric– no anything). It attempts to apply these lessons to a few contemporary problems of classical and quantum gravity.

1. "A Lost Leader"[1]

Max Born prefaced an essay on "Einstein's Statistical Theories" (Born 1949) with a few intensely personal words about Einstein:

> He has seen more clearly than anyone before him the statistical background of the laws of physics, and he was a pioneer in the struggle for conquering the wilderness of quantum phenomena. Yet later, when out of his own work a synthesis of statistical and quantum physics emerged

which seemed to be acceptable to almost all physicists, he kept himself aloof and sceptical. Many of us regard this as a tragedy—for him, as he gropes his way in loneliness, and for us who miss our leader and standard bearer (Born 1949, pp. 163–164).

In an unpublished reply to Born's essay, Einstein wrote:

This article is a moving hymn to a beloved friend, who in his old age (shall we say) unfortunately has succumbed to occultism, one who will not believe in spite of all the evidence that God plays dice. In one point, however, Born does me an injustice, namely, when he thinks that I have been untrue to myself in this respect since earlier I often availed myself of statistical methods. In truth, I never believed that the foundation [*Grundlage*] of physics could consist of laws of a statistical nature (cited from Stachel 1986a, p. 374).

Einstein was quite correct in his rejection of Born's charge of apostasy. From his earliest statistical papers on (see Einstein 1989), Einstein's primary definition of the probability of any state of a system was always based on consideration of the temporal evolution of that system. Consider a (bounded) system that passes through a number of states in the course of that evolution. If, during the total time interval T, the system spends the time interval t in some state, then Einstein defines the probability of that state as:

$$\lim_{T \to \infty} t/T.$$

In practice, Einstein then generally assumed something like an ergodic hypothesis so that he could replace time averages with ensemble averages; but this was a secondary step following the primary, time-average definition of probability. Aside from the familiar problems with infinite-time limits, this definition makes the probability observable in principle. Even more important, the definition does not require a knowledge of the form of the dynamical laws that govern the temporal evolution of the system. Early on, Einstein had objected to Planck's definition of probabilities, in particular for states of an oscillator in equilibrium with a radiation field in his derivation of the black-body law, by counting possible complexions of the system. In the absence of a dynamical theory of such a system, Einstein asked, how does one justify the assumption that all complexions are equiprobable?

In order to explain the absorption and emission of light by a quantum system, Einstein later postulated transition probabilities between the states of such a system by analogy with similar probabilities that could be derived for a classical charged oscillator in equilibrium with a radiation field (Einstein 1916). He was well aware of the problems that this step raised for the causality principle, asking in 1920:

> Can the quantum absorption and emission of light ever be understood in the sense of the complete causality requirement, or would a statistical residue remain? I must admit that there the courage of certainty fails me. But I would abandon *complete* causality very, very reluctantly (Einstein to Max Born, 27 January 1920; cited from Born 1971, p. 23, translation modified).

Einstein thus saw no contradiction between employing probabilistic concepts in theoretical physics and maintaining a belief in "complete causality." In some cases, for example in the theory of Brownian motion, it was already possible to see the connection between the probabilistic and the causal viewpoints. In other cases, for example in the quantum theory, the nature of this connection might still be quite obscure. But this did not prevent him from continuing to search for such a connection, a search that continued even after the advent of quantum mechanics (see Section 3).

2. Einstein's Search for a Constructive Theory of the Quantum

In order to appreciate the reasons for Einstein's discontent with quantum mechanics, it is important first to understand the nature of Einstein's concern with the foundations of theoretical physics, to understand what he regarded as a fully satisfactory physical explanation. We can best begin by considering the distinction he enunciated in 1919 between principle-theories and constructive theories:

> We can distinguish between various kinds of theories in physics. Most of them are constructive. They attempt to build up a picture of the more complex phenomena out of the materials of a relatively simple formal scheme from which they start out. Thus the kinetic theory of gases seeks to reduce mechanical, thermal, and diffusional processes to movements of molecules—i.e., to build them up out of the hypothesis of molecular motion. When we say that we have succeeded in *understanding* a group of natural processes, we invariably mean that a constructive theory has been found that covers the processes in question.
>
> Along with this most important class of theories there exists a second, which I will call "principle-theories." These employ the analytic, not the synthetic method. The elements which form their basis and starting point are not hypothetically constructed but empirically discovered

ones, general characteristics of natural processes, principles that give rise to mathematically formulated criteria which the separate processes or the theoretical representations of them have to satisfy. Thus, the science of thermodynamics seeks by analytical means to deduce necessary conditions, which separate events have to satisfy, from the universally experienced fact that perpetual motion is impossible.

The advantages of the constructive theory are completeness, adaptability, and clearness, those of the principle theory are logical perfection and security of foundations.

The theory of relativity belongs to the latter class. (Einstein 1919, p. 228).

Since this quotation dates from 1919, one might be inclined to wonder whether it reflects Einstein's outlook a decade and more earlier, during the years in which he was developing his ideas on relativity and the quantum. A recently discovered letter, at any rate, shows that he held quite similar views by the beginning of 1908. He wrote that he considered a physical theory to be satisfactory only

if it builds its structures out of *elementary* foundations. . . we are still far from having satisfactory elementary foundations for electrical and mechanical processes. . . The theory of relativity is just as little ultimately satisfactory as, for example, classical thermodynamics was before Boltzmann had interpreted entropy as probability. If the Michelson–Morley experiment had not put us in the most awkward position, no one would have accepted the theory of relativity as a (partial) salvation (Einstein to Arnold Sommerfeld, 14 January 1908; German text in Eckert and Pricha 1984, translation cited from Einstein 1989, Introduction).

Einstein's distinction between principle and constructive theories is sometimes interpreted as implying that he thought that principle-theories were *better*. We see from the quotations above that Einstein never held such a view. Principle-theories are more secure, but they are less satisfying. For Einstein, to understand was to have a constructive theory, a *Weltbild* (picture of the world).

It is important to appreciate the emotional significance the achievement of such understanding had for him:

Man tries to make for himself in the fashion that suits him best a simplified and intelligible picture of the world; he then tries to some extent to substitute this cosmos of his for the world of experience, and thus to overcome it. This is what the painter, the poet, the speculative philosopher and the natural scientist do, each in his own fashion. Each makes a cosmos and its construction the pivot of his emotional life in order to find in this way the peace and security which he cannot find in the narrow whirlpool of personal experience (Einstein 1918, p. 225).

The large emotional charge associated with his views on quantum mechanics, in particular, can perhaps be glimpsed in the following quotation:

> I find the idea that there should not be laws for being [*das Seiende*], but only laws for probabilities, simply monstrous [*scheusslich*] (a nauseatingly indirect description) (Einstein to Tatiana Ehrenfest, 12 October 1938, cited from Stachel 1986a, p. 356).

As you see, this is a very emotional topic for Einstein. To grasp, to comprehend the world is not just an intellectual experience, it is also an emotional experience—the drive for which he regards as necessary for really great accomplishments in science. And who are we to quarrel with Einstein's judgement in such matters? You also see that for Einstein, the theory of relativity was in a sense a last resort. It was the failure of his constructive attempts that led him to set up the special theory of relativity, as he explained much later:

> All my attempts ... to adapt the theoretical foundations of physics to this [new] knowledge failed completely. It was as if the ground had been pulled out from under one, with no firm foundation to be seen anywhere upon which one could have built (Einstein 1949, cited from Einstein 1979, p. 43).

Again, I call your attention to the emotional charge attached to his words. He goes on:

> Reflections of this type made it clear to me as long ago as shortly after 1900 ... that neither mechanics nor electrodynamics could (except in limiting cases) claim exact validity. Gradually I despaired of the possibility of discovering the true laws by means of constructive efforts based on known facts. The longer and more desperately I tried, the more I came to the conviction that only the discovery of a universal formal principle could lead us to assured results (Einstein 1949, cited from Einstein 1979, p. 49).

It could lead to assured results; but by itself, a theory of principle could not produce an ultimately satisfactory theory—that is, a constructive theory. It was a tool to help progress by providing new guidelines in the further search for constructive theories of matter and light.

I shall not discuss Einstein's work on the theory of relativity any further here. But I maintain that Einstein approached the quantum theory in a very similar spirit. He never regarded the work that he did in 1905 or in subsequent years on what he first called "the quantum hypothesis" (see Einstein 1989, Doc. 14) as an explanation, in his sense, since it did not constitute a constructive theory. Rather, it was a fruitful way of attempting to describe—at least partially—the nature of certain quantum phenomena, and thus to pose the quantum problem in a form amenable to further constructive efforts. In his 1911 report to the first Solvay Congress he says:

> We are all agreed that the so-called quantum theory of today is indeed a useful tool, but no theory in the ordinary meaning of the word, at any rate not a theory that could now be developed in a coherent manner. On the other hand, it has been proven that classical mechanics . . . no longer can be regarded as a usable system for the theoretical representation of all physical phenomena . . . So the question arises: on the validity of which general principles of physics may we hope to rely in the field of concern to us [i.e., quantum phenomena] (cited from Stachel 1986a, p. 358).

He goes on to discuss reliable principles such as the conservation of energy and Boltzmann's principle, but I just want to indicate the flavor of Einstein's approach to these problems. Einstein's strategy was, thus, first to formulate the quantum hypothesis and apply it to as many phenomena as possible. But this step was only a way station in his search for a constructive theory. It is, if you like, a useful formulation of the problem, rather than a solution. In that sense it is analogous to the way he looked upon the relativity principle as a guidepost, a tool to help in the search for constructive theories.

Now I come to his first abortive attempt at a constructive theory in 1909, which took the form of a unified field theory of matter and radiation. This is noteworthy since earlier he had inclined to the idea of entirely replacing fields by particles; or perhaps it is better to say the idea of replacing systems having an infinite number of degrees of freedom by systems with only a finite number of degrees of freedom; for, to Einstein, this was the essential distinction between fields and particles. In fact, the earliest extant account of his views on electromagnetic theory, a recently discovered letter of 1899, reads very much like an adumbration of the later Wheeler–Feynman program: get rid of fields, which are just a way of summing up the direct interactions between particles (see Einstein 1987, Doc. 52). And, you may remember, he opens his 1905 quantum paper by pointing out the formal distinction between the way Maxwell's theory treats an electromagnetic system, using infinitely many degrees of freedom, and the way statistical mechanics treats material systems, using a finite number of degrees of freedom. He goes on to suggest that this use of infinitely many degrees of freedom may be responsible for the difficulties with electromagnetic theory that he points out in the paper, in particular what we now call the ultraviolet catastrophe. The first thing he does in the paper is to show that if you take classical statistical mechanics and Maxwell's theory seriously you get a perfectly definite result for the black-body radiation distribution law, a result that was later called the Rayleigh–Jeans law. But this result is obviously incorrect since it results in an infinite total electromagnetic energy in a finite volume of space (see Einstein 1989, Doc. 14).

In more high-falutin' philosophical terms, Einstein's first inclination was to accept an exclusively particle ontology: particles or systems with a finite number of degrees of freedom are the only reality, and one has to reduce the apparent fields in the world to direct interactions between particles. By 1909 Einstein had completely reversed field (to make a bad pun) by adopting a field ontology, and was trying to explain particles (both electrons and light quanta) as some sort of singularities or other cohesive structures in a (nonlinear) field. He did not publish a full theory of this type, but did give a few hints about the sort of theory he was trying to develop. The basic problem he felt was that the quantum of electric charge remained a "stranger" (*Fremdling*) in Maxwell's theory, put in by hand. How could one explain it? As we have seen, he did not accept the idea of simply postulating such structures; one had to construct them in his sense. He opined in 1909 that

> the next phase in the development of theoretical physics will bring us a theory of light that may be regarded as a sort of fusion of the wave and emission theories of light (Einstein 1989, Doc. 60, translation cited from the Introduction).

This is the first hint of what later came to be called the wave-particle duality. But the evidence he brought forth in support of this claim seems to have convinced Einstein by 1909 that what was really needed was some sort of modification of the Maxwell field theory that would allow structures within the field that could be interpreted as explaining the particle-like properties of both radiation—the light quantum, and matter— essentially the electron. Einstein was impressed by the existence of a dimensionless constant formed out of e, h and c—what we now call the fine structure constant. He was a little worried by the fact that its numerical value was not close to one, but much less so than, for example, Lorentz. On the basis of the existence of this constant, he anticipated that

The same theoretical modification that leads to the elementary quantum [of charge] will lead to the quantum spectrum of radiation as a consequence.

In other words, Einstein expected that the quantum of charge and the quantum of radiation would both emerge from the same theory. Of course, he did not succeed in finding such a theory in 1909, and all attempts have failed to this day. Einstein tried to find (special-) relativistically invariant, nonlinear modifications of Maxwell's equations that would lead to the existence of stable structures in the field: either singularities, or finite, nondissipative concentrations of energy. Although he did not succeed, he wrote encouragingly about the project in 1909:

I have not yet succeeded ... in finding a system of equations that I could see was suited to the construction of the elementary quantum of electricity and the light quantum. The manifold of possibilities does not seem to be so large, however, that one should shrink from the task (Einstein 1989, Doc. 56; translation cited from the Introduction).

He does not refer to this project again in his published papers, but in his correspondence during the next few years he oscillates back and forth between hope and despair about its prospects. Philipp Frank tells about his first visit to Einstein in 1912, while the latter was working in Prague. Einstein's room in the Physics Institute looked out over a park that belonged to a neighboring mental institution. Einstein remarked to Frank, "There you see that portion of the lunatics who are *not* working on the quantum theory" (Frank 1979, pp. 141–143). About this time, he turned his attention more and more toward the search for a relativistic theory of gravitation, which eventually resulted in the general theory of relativity, and published nothing further on quantum problems until 1916.

In later years, after he developed the general theory of relativity, Einstein came to doubt whether synthetic, constructive efforts, starting from contexts closely linked to empirical evidence, could ever lead to the kind of unified theory that he had in mind. He shifted to the search for a formal scheme that, starting from a small number of highly abstract concepts, would lead through a long deductive chain to an explanation of the quantum effects. In a 1953 letter he wrote:

I am firmly convinced that every attempt to arrive at a rational theory by *synthetic* construction will have an unsatisfactory result. Only a new basis for all of physics, from which all possible processes can be deduced with logical necessity (as for example is the case with thermodynamics), can bring a convincing solution (Einstein to M. Renninger, 11 June 1953; cited from Stachel 1986a, pp. 359–360).

In other letters from about this time he wrote:

> Now you will understand why I lapsed into my apparently Don Quixotic attempts to generalize the gravitational equations. If one cannot trust Maxwell's equations, and a representation [*Darstellung*] by means of fields and differential equations is indicated, on account of the principle of general relativity, and one has come to despair of arriving at a deeper basis [*Tieferlegung*] of the theory by intuitive [*anschaulich*]-constructive means; then no other sort of effort seems open (Einstein to Max von Laue, 17 January 1952; cited from Stachel 1986a, p. 360).

> I do not believe in macro- and micro-laws, but only in (structure) laws of general rigorous validity. And I believe that these laws are *logically simple,* and that reliance on their logical simplicity is our best guide. Thus, it would not be necessary to start with more than a relatively small number of empirical facts. If nature is not arranged in correspondence with this belief then we have altogether very little hope of understanding it more deeply... This is not an attempt to convince you in any way. I just wanted to show you how I came to my attitude. I was especially strongly impressed with the realization that, by using a semiempirical method, one would never have arrived at the gravitational equations of empty space (Einstein to David Bohm, 24 November 1954; cited from Stachel 1986a, p. 360).

These quotations indicate that the sharp line that Einstein had earlier drawn between theories of principle and constructive theories has become blurred. But the goal of *deducing* quantum phenomena from some unified theory, rather than positing their existence from the outset, remains constant in his approach.

3. Einstein's Critique of Quantum Mechanics

Very often Einstein's critique is presented as if what worried Einstein most about quantum mechanics is its inherent radically-probabilistic element. It is true that he stated he could more easily conceive of a completely chaotic universe than one governed by probabilistic laws.

> I still do not believe that the Lord God plays dice. If he had wanted to do this then he would have done it quite thoroughly, and not stopped with a plan for gambling. In for a penny, in for a pound [*Wenn schon, denn schon*] (Einstein to Mr. and Mrs. Fritz Reiche, 15 August 1942; cited from Stachel 1986a, p. 374.)

In other words if everything is going to have a random element then why not have *everything* random? Why bother with laws of randomness? In that case we would not have to search for laws at all, he adds. I think he remarked somewhere that the only rational occupation for a person who believes in this fundamentally probabilistic element in the universe would be that of croupier.

But in spite of this evidence, I maintain that the probability issue was not the deepest source of his dissatisfaction with the prevailing interpretation of quantum mechanics.

> The sore point [*Der wunde Punkt*] lies less in the renunciation of causality than in the renunciation of a reality thought of as independent of observation (Einstein to Georg Jaffe, 19 January 1954; cited from Stachel 1986a, p. 374).

It is important to emphasize that he did not see this renunciation as a defect of quantum mechanics as such, but as a defect of the prevailing interpretation of that theory which regards it as the most complete possible description of an individual physical system. He accepted the ensemble point of view, which he called the Born interpretation—although I am not sure whether Born saw it that way—as a perfectly reasonable interpretation of quantum mechanics—which does not lead to such problems as that raised in the so-called Einstein–Podolsky–Rosen (EPR) argument. Let me cite a short summary by Einstein of this argument:

> It is not my opinion that there is a logical inconsistency in the quantum-theory itself and the "paradoxon" does not try to show it. The intention is to show that the statistical quantum theory is not compatible with certain principles, the convincing power of which is independent of the present quantum theory.
>
> There is the question: Does it make sense to say that two parts A and B of a system do exist independently of each other if they are (in ordinary language) located in different parts of space at a certain time, if there are no considerable interactions between those parts (expressed in terms of potential energy) at the considered time? ... I mean by "independent of each other" that an action on A has no immediate influence on the part B. In this sense I express a principle (a)
>
> (a) independent existence of the spatially separated.
> This has to be considered with the other thesis (b)
>
> (b) The ψ-function is the complete description of an individual physical situation.
>
> My thesis is that (a) and (b) cannot be true together, for if they would hold together the special kind of measurement concerning A could not influence resulting ψ-function for B (after measurement of A).

> The majority of quantum theorists discard (a) tacitly to be able to conserve (b); I, however, have strong confidence in (a) so I feel compelled to relinquish (b) (Einstein to Leon Cooper, 31 October 1949; cited from Stachel 1986a, p. 375).

I think this is a very good summary of the conclusion that Einstein drew from the EPR argument. He was not always well understood, in particular by Born, who badly misunderstood Einstein—so badly that they reached an impasse in their correspondence on the subject. Wolfgang Pauli, who was then at the Institute, intervened in a letter to Born that gives a beautiful example of objective critical evaluation of a position with which one does not agree. Although Pauli disagreed with Einstein and was closer to Born's point of view, he understood that Born was just not tuned in on Einstein's wave length. So Pauli undertook the task of explaining to Born what Einstein was talking about.

> ... I was unable to recognize Einstein whenever you talked about him in either your letter or your manuscript. It seems as if you had erected some dummy Einstein for yourself, which you then knocked down with great pomp (Pauli to Born, 31 March 1954; cited from Born 1971, p. 221).

I believe that this occupation or hobby still has many practitioners. It is rather important to understand what Einstein meant before one refutes him. But let me continue with the quotation from Pauli:

> In particular Einstein does not consider the concept of "determinism" to be as fundamental as it is frequently held to be ... he *disputes* that he uses as criterion for the admissibility of a theory the question: "Is it rigorously deterministic?". . . he was *not at all* annoyed with you, but only said that you were a person who will not listen (ibid.).

There are still many people in the world who fit that description—persons who will not listen—but one might try to follow Einstein's example and not get annoyed with them.

In this case the situation did not improve with time in spite of Pauli's letter. John Bell recently characterized Born's summary of his discussion with Einstein, written more than twenty years later when Born edited the letters for publication, in the following words: "Misunderstanding could hardly be more complete" (Bell 1987, p. 144).

To return to the main thread of my discussion, as Pauli understood it, the real stumbling point for Einstein was what has come to be called the nonlocality of quantum mechanics, not its indeterminism. He held that, if one adopted the ensemble or statistical interpretation of the theory, there would be no problem.

> It is ... to be expected that behind quantum mechanics there lies a
> lawfulness and a description that refer to the individual system. That
> it is not attainable within the bounds of concepts [taken] from classical
> mechanics is clear; the latter however is in any case outmoded as the
> foundation [*Fundament*] of physics (Einstein to Gregory Breit, 2 August
> 1935; cited from Stachel 1986a, p. 375).

So Einstein felt that quantum mechanics started out with a set of basic
concepts that were outmoded. If, then, one could not advance further
than quantum mechanics did on the basis of these concepts, this was not
really so surprising to him. On the contrary, as we will see in a moment,
what he found surprising is that one could get so far with a set of concepts
that are known to be inadequate. He elaborated on this inadequacy of
the classical-mechanical starting point of quantum mechanics.

> I do not at all doubt that the contemporary quantum theory (more ex-
> actly "quantum mechanics") is the most complete theory compatible
> with experience, as long as one bases the description on the concepts
> of material point and potential energy as fundamental concepts (Einstein
> 1953; p. 6; translation cited from Stachel 1986a, pp. 375–376).

Arthur Fine has noted that while the old Einstein often is considered as
a reactionary in physics and the quantum mechanicians are considered
to be radicals, from Einstein's point of view, the situation was just the
opposite (Fine 1986, Chapter 2). The adherents of quantum mechanics
were clinging to the classical-mechanical concepts, and trying to find
some way to continue to do physics using only this outmoded conceptual
basis, whereas he was searching for a *new* conceptual basis for physics,
which he regarded as the radical step. To continue with the quotation
from Einstein:

> [The difficulties of quantum mechanics] are connected with the fact
> that one retains the classical concept of force or potential energy and
> only replaces the law of motion by something entirely new... To me
> it seems, however, that one will finally recognize that something must
> take the place of forces acting or potential energy... something that has
> an atomistic structure in the same sense as the electron itself. "Weak
> fields" or forces as active causes will not then occur at all, just as
> little as mixed states (Einstein 1953, pp. 12, 14; translation cited from
> Stachel 1986a, p. 376).

And in one of his last letters, he again touched on this point:

> I believe, however, that the renunciation of the objective description
> of "reality" is based upon the fact that one operates with fundamental
> concepts that are untenable in the long run (like f[or] i[instance] clas-
> sical thermodynamics). Quite understandably most physicists resist the

idea that we are still very far from a deeper insight into the structure of reality (Einstein to André Lamouche, 20 March 1955; cited from Stachel 1986a, p. 376).

Each of us likes to feel that he or she was born into the generation that finally has achieved the great illumination about the nature of reality.

Did Einstein believe in some sort of hidden variable program to underpin quantum mechanics? Bernard D'Espagnat and John Bell, for example, believe that he did, as do John Clauser and my dear colleague Abner Shimony (see Stachel 1986a, p. 376, for citations and references); but I don't think so. While some statements by Einstein seem to allow such an interpretation, in commenting on Bohm's hidden variable theory in 1953, he stated flatly:

> I think that it is not possible to get rid of the statistical character of the present quantum theory by merely adding something to the latter without changing the fundamental concepts about the whole structure (Einstein to Aron Kupperman, 14 November 1953, Item 8-036 in the Einstein archive).

Einstein hoped that such new fundamental concepts would emerge from his search for a unified field theory, which

> owes its origin in part to the conjecture that a rational generally relativistic field theory might offer the key to a more complete quantum theory (Einstein 1953, p. 14).

Let me cite a few more quotations from Einstein that indicate in somewhat more detail what he had in mind. The first is from a letter written in 1950:

> In a consistent field theory there is no real definition of the field... *A priori* no bridge to the empirical is given. There is not, for example, any "particle" in the strict sense of the word, since it does not fit into the program of representing reality by everywhere continuous, indeed even analytic functions ... The upshot is that a comparison with the empirically known can only be expected to come from finding exact solutions of the system, in which empirically "known" structures and their interactions are "reflected." Since this is immensely difficult, the sceptical attitude of contemporary physicists is quite understandable (Einstein to Michele Besso, 15 April 1950, Einstein and Besso, 1972, pp. 438–439; translation cited from Stachel 1986a, p. 376).

Thus, for Einstein, once one had a unified field theory the path to the empirically observable was the great problem: how could one connect the theory with observation? It would certainly not be a matter of simply integrating over certain "hidden variables" in order to recover the concepts of classical mechanics. Here is an excerpt from a letter that explains the problem in the context of his last attempt at a unified theory, the theory of the asymmetrical metric tensor field:

> Our situation is this. We stand before a closed box that we cannot open, and try to discuss what is inside and what is not. The similarity to Maxwell's theory is only external, so that we cannot transfer the concept of "force" from this theory to the asymmetric field theory. If this theory is at all useful, then one cannot assume any separation between particles and field of interaction. In addition, there is no concept at all of the *motion* of something more or less rigid. The question here is exclusively: are there singularity-free solutions? Is their energy in particular localized in such a way as demanded by our knowledge of the atomic and quantum character of reality? The answer to this question is not really attainable with contemporary mathematical methods. Thus I do not see how one can guess whether any sort of action-at-a-distance or any type of object, insofar as we have attained a semiempirical knowledge of them, can be represented by the theory (Einstein to John Moffat, 24 August 1953; cited from Stachel 1986a, p. 377).

Now, if the concept of a hidden-variable theory, not to speak of the concept of a *local* hidden-variable theory, is to be given any precise meaning—and not just used as a description of *anything* that is non-quantum mechanical—then I don't think that we can characterize what Einstein is talking about in these quotations as a hidden-variable theory. Max Jammer has expressed similar views on this question (Jammer 1979, pp. 160–161, Jammer 1982, pp. 72–73).

To summarize and conclude this section, after 1930 Einstein never denied the great explanatory power of quantum mechanics, nor challenged its validity; but he did not agree that this success required acceptance of its underlying conceptual structure as the basis for all further progress in theoretical physics. He wrote to Schrödinger:

> The wonder [about quantum mechanics] is only that one can represent so much with it, although the most important source of knowledge, group invariance, finds such incomplete application there . . . It is the case that a logically coherent theory that is connected appropriately to the real state of affairs usually has great extrapolatory power, even if it is little related to the deeper truth [*der Wahrheit in der Tiefe*] (Einstein to Schrödinger, 16 July 1946; cited from Stachel 1986a, p. 377).

I will later return to the question of what Einstein meant by his remark about group invariance as the most important source of theoretical knowledge; but first I shall discuss another topic.

4. The Other Einstein

As I mentioned earlier, Einstein initially felt that the use of systems with only a finite number of degrees of freedom might prove the ultimate answer to the problems posed by the existence of the quantum. By 1909 he had shifted to the "unified field theory" point of view, to which, according to the conventional wisdom, he held more or less continuously until the end of his life.

But there is another Einstein who, as early as 1916, once more expressed the conviction that the entire continuum approach ultimately might have to be abandoned, and continued to consider this possibility until the end of his life. You may feel that there is a contradiction here, but as Whitman said:

> Do I contradict myself?
> Very well then I contradict myself.
> (I am large, I contain multitudes.)
> (*Song of Myself,* Whitman 1949, p. 142).

Einstein was certainly large, and "the other Einstein" is one of his aspects, even if one that is not very well known.

Let me cite some quotations that reveal "the other Einstein," and his comments on some of the very problems we have been discussing at this conference. In 1916, he wrote to one of his former students:

> . . . you have correctly grasped the drawback that the continuum brings. If the molecular view of matter is the correct (appropriate) one; i.e., if a part of the universe is to be represented by a finite number of moving points, then the continuum of the present theory contains too great a manifold of possibilities. I also believe that this too great is responsible for the fact that our present means of description miscarry with the quantum theory. The problem seems to me how one can formulate statements about a discontinuum without calling upon a continuum (space-time) as an aid; the latter should be banned from the theory as a supplementary construction not justified by the essence of the problem, [a construction] which corresponds to nothing "real." But we still lack the mathematical structure unfortunately. How much have I already plagued myself in this way!
>
> Yet I see difficulties of principle here too. The electrons (as points) would be the ultimate entities in such a system (building blocks). *Are there* indeed such building blocks? Why are they all of equal magnitude? Is it satisfactory to say: God in his wisdom made them all equally big, each like every other, because he wanted it that way; if

it had pleased him, he could also have created them different? With the continuum viewpoint one is better off in this respect because one doesn't have to prescribe elementary building blocks from the beginning. Further, the old question of the vacuum! But these considerations must pale beside the overwhelming fact: The continuum is more ample than the things to be described (Einstein to Walter Dällenbach, November 1916; cited from Stachel 1986a, p. 379).

He continued to discuss this possibility over the years. Here is an excerpt from a letter of 1935:

In spite of the successes of quantum mechanics, I do not believe that this method can offer a usable *foundation* [*Fundament*] of physics. I see in it something analogous to classical statistical mechanics, only with the difference that here we have not found the equations corresponding to those of classical mechanics.

In any case one does not have the right today to maintain that the foundation must consist in a *field theory* in the sense of Maxwell. The other possibility, however, leads in my opinion to a renunciation of the time-space continuum and to a purely algebraic physics. Logically this is quite possible (the system is described by a number of integers; "time" is only a possible standpoint [*Gesichtspunkt*], from which the other "observables" can be considered—an observable logically coordinated to all the others. Such a theory doesn't have to be based upon the probability concept. For the present, however, instinct rebels against such a theory (Einstein to Paul Langevin, 3 October 1935; cited from Stachel 1986a, pp. 379–380).

Of course, even more serious than his instinctive rebellion was the fact that he had no idea how to create such a theory. Judging from his letters, he evidently became more and more pessimistic about the continuum point of view. In 1952, for example, he wrote:

In present-day physics there is manifested a kind of battle between the particle-concept and the field-concept for leadership that will probably not be decided for a long time. It is even doubtful if one of the two rivals finally will be able to maintain itself as a fundamental concept (Einstein to Herbert Kondo, 11 August 1952; cited from Stachel 1986a, p. 380).

He wrote to Besso in 1954:

I consider it entirely possible that physics cannot be based upon the field concept, that is, on continuous structures. Then *nothing* will remain of my whole castle in the air including the theory of gravitation, but also nothing of the rest of contemporary physics (Einstein to Basso, 10 August 1954, Einstein and Besso 1972, p. 527; cited from Stachel 1986a, p. 380).

In the same year, he wrote to André Lichnerowicz:

> Your objections regarding the existence of singularity-free solutions which could represent the field together with the particles I find most justified. I also share this doubt. If it should finally turn out to be the case, then I doubt in general the existence of a rational physically useful *field* theory. But what then? Heine's classical line comes to mind: "And a fool waits for the answer" (Einstein to André Lichnerowicz, 25 February 1954; cited from Stachel 1986a, p. 380).

He wrote to Bohm:

> I must confess that I was not able to find a way to explain the atomistic character of nature. My opinion is that if the objective description through the field as an elementary concept is not possible, then one has to find a possibility to avoid the continuum (together with space and time) altogether. But I have not the slightest idea what kind of elementary concepts could be used in such a theory (Einstein to David Bohm, 28 October 1954; cited from Stachel 1986a, p. 380).

The reason Einstein didn't extensively publicize these views is that he felt that unless one has a real theory it is useless to publish such speculative remarks. But contemporary physicists may find them fascinating, since they speak to so many of our current concerns.

I have found only one place—but I must admit that I have not looked everywhere—where he speculated in somewhat more detail on what such a non-continuum theory might involve.

> The alternative continuum-discontinuum seems to me to be a real alternative; i.e., there is here no compromise. By discontinuum theory I understand one in which there are no differential quotients. In such a theory space and time cannot occur, but only numbers and number-fields and rules for the formation of such on the basis of algebraic rules with exclusion of limiting processes. Which way will prove itself, only success can teach us.

> Physics up to now is naturally in its essence a continuum physics, in spite of the use of the material point, which looks like a discontinuous conceptual element, and has no more right of existence in field description. Its strength lies in the fact that it posits parts which exist quasi-independently, *beside* one another. Upon this rests the fact that there are reasonable laws, that is rules which can be formulated and tested for the individual parts. Its weakness lies in the fact that it has not been possible up to now to see how that atomistic aspect including quantum relations can result as a consequence. On the other hand dimensionality (as four-dimensionality) lies at the foundation of the theory.

An algebraic theory of physics is affected with just the inverted advantages and weaknesses, aside from the fact that no one has been able to propose a possible logical schema for such a theory. It would be especially difficult to derive something like a spatio-temporal quasi-order from such a schema. I cannot imagine how the axiomatic framework of such a physics would appear, and I don't like it when one talks about it in dark apostrophes [*Anredungen*]. But I hold it entirely possible that the development will lead there; for it seems that the state of any finite spatially limited system may be fully characterized by a finite number of numbers. This speaks against the continuum with its infinitely many degrees of freedom. The objection is not decisive only because one doesn't know, in the contemporary state of mathematics, in what way the demand for freedom from singularity (in the continuum theory) limits the manifold of solutions (Einstein to H.S. Joachim, 24 August 1954; cited from Einstein 1986a, p. 581).

I was interested to learn that even near the end of his life Einstein was still on the lookout for new mathematical tools that might turn speculations that he thought it best to keep private into the basis of a real theory. Abraham Fraenkel, the mathematician, reports an interview he had with Einstein in 1951. With the background I have reported, you will see at once why Einstein was so excited by what Fraenkel told him:

In December 1951 I had the privilege of talking to Professor Einstein and describing the recent controversies between the (neo-) intuitionists and their "formalistic" and "logicistic" antagonists; I pointed out that the first attitude would mean a kind of atomistic theory of functions, comparable to the atomistic structure of matter and energy. Einstein showed a lively interest in the subject and pointed out that to the physicist such a theory would seem by far preferable to the classical theory of continuity. I objected by stressing the main difficulty, namely, the fact that the procedures of mathematical analysis, e.g., of differential equations, are based on the assumption of mathematical continuity, while a modification sufficient to cover an intuitionistic-discrete medium cannot easily be imagined. Einstein did not share this pessimism and urged mathematicians to try to develop suitable new methods not based on continuity (Fraenkel 1954).

Now I have arrived at the point where I can discuss Einstein's remark, cited in the last section, that the most important source of theoretical knowledge in physics, group invariance, finds incomplete application in quantum theory. This brings up one of the most profound sources of Einstein's skepticism about the ultimate status of quantum mechanics. He attached primary significance to the concept of general covariance writing, for example, in 1954:

You consider the transition to special relativity as the most essential thought of relativity, not the transition to general relativity. I consider the reverse to be correct. I see the most essential thing in the overcoming of the inertial systems, a thing that acts upon all processes but undergoes no reaction. This concept is in principle no better than that of the center of the universe in Aristotelian physics (Einstein to Georg Jaffe, January 19, 1954; cited from Stachel 1986a, p. 377).

That is, Einstein puts the a priori postulation of inertial systems on a level with the postulation of a center of the universe in the Aristotelian system. Therefore, he felt a quantum theory constructed on the basis of Galilean or even special-relativistic space-times could not constitute a satisfactory foundation of physics.

Contemporary physicists do not see that it is hopeless to take a theory that is based on an independent rigid space (Lorentz-invariance) and later hope to make it general-relativistic (in some natural way) (Einstein to Max von Laue, September 1950; cited from Stachel 1986a, p. 378).

He justified himself against criticism that he had neglected the development of relativistic quantum field theory, in similar terms:

I have not really studied quantum field theory. This is because I cannot believe that special relativity theory suffices as the basis of a theory of matter, and that one can afterwards make a nongenerally relativistic theory into a generally relativistic one. But I am well aware of the possibility that this opinion may be erroneous (Einstein to K. Roberts, September 6, 1954; cited from Stachel 1986a, p. 378).

In spite of his statement that he had failed to understand Bohr's concept of complementarity, Einstein did share one element of Bohr's viewpoint: Einstein didn't really believe that the wave aspect of the electron is as fundamental as the particle aspect; and conversely, that the particle aspect of the light quantum is as fundamental as the wave aspect. As you may know, Bohr did not believe in a full equality between the wave and the particle aspects of matter and radiation, respectively. He believed that the classical limit of the quantum theory led to the more important aspect; therefore the particle aspect of the electron was physically more fundamental while the wave aspect was more a mathematical construct. Conversely, for the photon the classical limit was given by the wave theory, and the particle aspect was more a mathematical construct. Einstein once said something very similar:

I do not believe that the light-quanta have reality in the same immediate sense as the corpuscles of electricity. Likewise I do not believe that the particle-waves have reality in the same sense as the particles themselves.

The wave-character of particles and the particle-character of light will—in my opinion—be understood in a more indirect way, not as immediate physical reality (Einstein to Paul Bonofield, 18 September 1939; cited from Stachel 1986a, pp. 373–374)

5. Some Lessons From General Relativity

If one looks at how Einstein developed the general theory of relativity, one sees that he drew a certain number of lessons from the way that he did so, lessons that he subsequently incorporated in his approach to the foundations of theoretical physics. We can consider these lessons, and try to apply them to our problems today. However, those who have not studied the history of the development of the subject are not always aware of these lessons (the converse of this statement is unfortunately not true). I am not claiming that these views on the foundations of physics must be accepted uncritically just because Einstein held them, but only that they are worth considering—that they are not automatically irrelevant either, just because Einstein held them.

The first lesson at which Einstein arrived is the fundamental importance to any theory of gravitation of the equivalence principle, which he very soon expressed by the phrase that inertia and gravitation are *wesensgleich,* the same in their essence or essentially the same. The distinction between gravitation and inertia is not absolute, but it depends on the frame of reference employed. Once the appropriate mathematical tool had been developed, this lesson could be expressed by the statement that the affine connection of space-time cannot be split up uniquely into two parts, one of which would be an inertial connection, the other a gravitational field tensor (remember, the difference between two symmetric connection fields is a tensor field). Until the end of his life, Einstein placed great stress on the significance of the affine connection as the correct mathematical representation of the gravitational-cum-inertial field. Replying to von Laue, who, in preparing the post-World War II revision of his relativity textbook (von Laue 1956), espoused the point of view that only a nonvanishing Riemann tensor represented a real gravitational field, Einstein wrote:

What characterizes the existence of a gravitational field from the empirical standpoint is the nonvanishing of the Γ^l_{ik} [the components of the affine connection—JS], not the nonvanishing of the R_{ikmn} [the components of the Riemann tensor—JS]. If one does not think in such intuitive [*anschaulich*] ways, one cannot comprehend why something like curvature should have anything to do with gravitation at all. In any case, no reasonable person would have hit upon something in that way. The key to the understanding of the equality of inertial and gravitational mass would have been missing (Einstein to von Laue, September 1950; cited from Stachel 1989, p. 89).

A couple of comments are in order. First of all, one may note Einstein's physically intuitive (*anschaulich*) way of thinking. There are several quotations from Einstein about his mode of thought that are quite well-known (see especially Einstein 1945), in which he noted that his mode of thought was dependent on visual and even kinesthetic imagery. So I will quote an account that has not been previously cited for this purpose as far as I know. A student who was taking Einstein's 1917 relativity course at the University of Berlin reported in his diary that in private conversations Einstein said:

> he is a poor calculator, he does not work with abstract ideas ... He visualizes [gravitational] waves with the help of an elastic body that represents an elastically oscillating system. He also applied motions of his fingers for this elastic system to the 3 last types [of real, transverse-transverse gravitational waves, as opposed to the purely coordinate waves], thus in this way one has to make them plausible to oneself ... (excerpts from the diary of Rudolf Jakob Humm, Zentralbibliothek Zürich).[2]

You see that Einstein had adumbrated the Weber bar!

Secondly, I believe that Einstein's emphasis on the fundamental role of the affine connection is much more palatable to physicists today, including many relativists, than it was some years ago. This is because of the current emphasis on gauge theories, in which the fundamental role of the connection is clear. And the theoretical discovery of the thermal ambience of accelerating particle detectors has also emphasized that one cannot make a distinction between quantum and thermal fluctuations in a frame-independent way (see, e.g., Sciama, Candelas, and Deutsch 1981). So I think the viewpoint that the Riemann tensor isn't *everything* in general relativity is more accepted today than it was when I was a student, when a sort of monotheistic belief in the Riemann tensor was accepted gospel among many relativists.

This approach implies that no division of the connection into an inertial and a gravitational part has any absolute validity since such a division is entirely frame-dependent. Obviously, if one *could* make such an absolute division, that is, if one had a tensor of first differential order to represent the gravitational field, one could solve with ease many outstanding problems in general relativity. At the classical level, for example, you would have a locally-defined gravitational energy density. Formal quantization of gravity would become more or less trivial, since you could quantize this tensor field on the background inertial connection.

The first person who seems to have realized how useful such a division would be is Nathan Rosen. In 1939, in order to produce it, he introduced a second, flat metric into the theory, undoing what Einstein had done in 1915. Thus, Einstein's approach can be expressed in the injunction: What God has joined together (inertia and gravitation), let no man put asunder. Einstein gave Rosen a piece of advice about his theory that I think was very good, but which Rosen did not follow: "Throw it in the waste basket."[3]

What I want to emphasize here is that people who attempt to quantize general relativity with the help of a background metric, flat or not, are violating this injunction. They are essentially introducing, whether they realize it or not, an inertial connection—the unique one associated with the background metric— and there with a gravitational tensor of the first order. As indicated above, if you give me such a gravitational tensor, I too can do wonderful things. But the trouble is that one *doesn't* have such a tensor in general relativity. In addition to the obvious question of what the conformal (i.e., light-cone) structure of the nonquantized background metric has to do with the actual propagation of the quantized gravitational field, such a bimetric approach violates the equivalence principle since it implies the existence of an invariant decomposition of the connection, treats one part as an inertial connection that forms part of the nonquantized background, and treats the other part as a gravitational tensor field that is to be quantized. Those quantum field theorists who approach general relativity as if it were no more than a nonlinear Lorentz-invariant field theory with a particularly nasty gauge group are especially prone to this sin of splitting asunder what belongs together.

Such approaches to quantum gravity can lead to truly comical conclusions. Some time ago I had a discussion with a French physicist who told me that it was simple to quantize general relativity: all one has to do is to take a conformally-flat metric tensor, and quantize the conformal factor. When I asked what made him think that one must employ a conformally flat metric, he replied: "Oh, because if you don't, I have no idea how to quantize general relativity." Such approaches to the quantization problem remind me of Einstein's comment that he didn't think much of a carpenter who, when he had to bore a hole in a piece of wood, looked for the thinnest spot. One should bore the hole where it is needed, no matter how hard the job.

The next lesson involves the meaning of the concept of general covariance. Einstein's interpretation of this concept is closely connected with the question of why almost three years elapsed between the time when he decided that the metric tensor was the correct mathematical representation of gravitation (late 1912) and the time when he finally adopted the generally covariant equations for the gravitational field that we now call the Einstein equations (late 1915). The main obstacle that held him up was his failure to realize—to use modern terminology— that, in a generally covariant theory, although the points of a manifold are *mathematically* individuated by a coordinate system, they are not *physically* individuated before one has a metric tensor field on that manifold. Consider a solution to any set of generally covariant field equations with a given distribution of sources. A second solution to these equations can always be generated from the first by the following prescription: it is identical to the first solution everywhere on the manifold except for a finite region containing none of the sources (a "hole"); in the hole, the second solution is generated from the first by dragging the latter along an arbitrary vector field that vanishes on the boundary of the hole. Originally, Einstein had considered the second solution to be physically distinct from the first, and hence rejected generally covariant field equations because they did not provide a physically unique solution for the gravitational field corresponding to a given source distribution (for more detailed discussion of the "hole" argument see Stachel 1986b, 1987, 1989). Finally, after over two years he realized that two such solutions could be considered as two mathematically distinct exemplifications of the *same* physical solution if one stipulated that the physical properties of the points of the manifold inside the hole in any solution to the field equation depend *entirely* on the nature of the solution; hence, when one creates the second solution by dragging, the physical significance of the points is dragged along with the field. It is this stipulation that constitutes the physical meaning of the concept of general covariance, which, properly speaking, has nothing to do with coordinate transformation. In the case of general relativity, one can express this idea by saying: Until one introduces a metric, a manifold does not consist of physical events but just of mathematical points with *no* physical properties; it is the metric that imparts the physical character of events to the points of a manifold.

One moral of this story is that general relativity is better formulated mathematically using a fiber bundle approach in which the metric is represented by a cross section of the bundle. If you do this, you realize immediately that, in this bundle approach, the points of the base manifold do *not* represent physical events, which instead are represented by mappings from a point on a cross section of the bundle into a point of the base manifold. If you have no cross section (no metric), you have no such mapping and hence no events (see Stachel 1986b for more details of this approach). It follows that *any* point of the base manifold can "represent" any event in an appropriate mapping.

So the problem of a correct mathematical representation of events in a generally covariant theory can be easily solved. But the moral is that if you are trying to formulate a quantum theory of general relativity (or of any generally covariant theory), you must always bear in mind that you do *not* have a manifold of events to start with, before finding a solution to the quantized equations for the metric field. This raises the following question: suppose one *had* a consistent quantum formalism for general relativity (we should only be so lucky!), and had found a solution to the quantized field equations; what could one do with it? If you have a wave function of the universe, for example, how do you interpret it physically? Usually, one interprets the quantum formalism in a background space-time, so that such a question may be answered as follows: You put an apparatus *here* to create a particle *now*, and then you put an apparatus *there* to detect it *then*; one can then use quantum probability amplitudes (such as wave functions) to compute transition probabilities between two such events. But in general relativity you have no a priori *here* and *now* or *there* and *then*. What "here and now" and "there and then" signify are among the things that one must solve the equations in order to determine. I am reminded of Gertrude Stein's last words. She roused herself from a coma to ask worriedly: "What is the answer?" Then she smiled and asked: "What is the question?" and lapsed into her final coma. In a sense, in matters of quantum gravity one must know the answer before one knows the question.

Perhaps one could find a way out of this dilemma if one could specify how to use a part of the system under consideration to measure another part, as Murray Gell-Mann suggested in his talk. But this would have to be done *without* introducing any space-time structure a priori. Gell-Mann clearly did not attempt this; he introduced an underlying space-time structure. If you introduce such a background space-time, there is of course, no problem in doing quantum mechanics on it. But then you haven't solved the problem of what to do *without* such a structure.

This question of the meaning and significance of general covariance is not a trivial problem (as I mentioned above, its solution held Einstein up for over two-and-a-half years in the development of general relativity). I feel that many people who study general relativity today still don't really understand this problem. Some recent Russian papers, for example, have unwittingly reproduced Einstein's hole argument as a criticism of general relativity.

This lack of understanding shows up particularly clearly in perturbative approaches to the theory. When one perturbs a field, one is used to automatically comparing the perturbed value of the field at a point with the value of the unperturbed field at the *same* point in the background space-time. It often requires some effort to realize that, since no a priori physical significance is attached to the points of a manifold in general relativity, no such a priori correspondence exists between points of the perturbed and unperturbed metric tensor fields; i.e., there is no a priori definition of corresponding events. This difficulty is usually masked by the use of a coordinate system: equal coordinate values are tacitly assumed to define corresponding points. But this amounts to the unacknowledged introduction of an additional "rigid" geometrical structure (see Geroch 1968), which raises the problem of whether any physical significance can be attached to the introduction of such a structure.

One might think that this problem is avoided by starting from a flat space-time solution to the gravitational field equations. Surely, it seems, one knows what one means by a definite point in Minkowski space-time! Indeed one does (leaving aside the problem of possible global topological complications), if, a priori, one is dealing with the space-time structure of the special theory of relativity. Here, points of Minkowski space-time may be physically identified by means of some additional physical elements introduced into the theory, such as rigid rods and clocks, trajectories of free particles, light rays, and so forth. These additional elements may be introduced without any problem in special relativity precisely *because* the space-time structure is given a priori, and so by definition is unaffected by the presence of additional particles or fields. Flat space as a solution to the field equations of general relativity, however, cannot be assimilated without further ado to special-relativistic Minkowski space-time: it is just as subject to the problem of physical individuation of its points as any other solution to the gravitational field equations. (In fact is is *more* subject than most; since the Riemann tensor vanishes, no invariants of the Riemann tensor can be used for this purpose). In particular, how the points of a perturbed metric are to be identified with the points of an unperturbed flat metric is an open question—unless the whole theory is treated as if it "really" were a special-relativistic theory and not a generally covariant one. In that case, however, it is not clear why one should have singled out the Einstein theory in the first place from the myriad of possible nonlinear Poincaré-invariant theories.[4]

Finally, there is a third issue which I call the "no metric, no nothing" point of view. To quote again from Einstein:

> On the basis of general relativity ... space as opposed to 'what fills space' ... has no separate existence ... If we imagine the gravitational field, i.e., the functions g_{ik} to be removed, there does not remain a space of the type [of Minkowski space in special relativity—JS] but absolutely *nothing,* not even a 'topological space' [i.e., a manifold—JS]. For the functions g_{ik} describe not only the field, but at the same time also the topological and metrical structural properties of the manifold ... There is no such thing as empty space, i.e., a space without a field. Space-time does not claim existence on its own, but only as a structural quality of the field (Einstein 1952).

Einstein was well aware that his views on this question were far from universally shared:

> It required a severe struggle to arrive at the concept of independent and absolute space, indispensable for the development of theory. It has required no less stenuous exertions subsequently to overcome this concept—a process which is by no means as yet completed (Einstein 1954b).

Even relativists have not yet fully adopted the point of view, "no metric, no nothing." If you look at the way the general theory of relativity is formulated mathematically in even the most careful treatises, for example, you see this clearly. They start out by introducing a global manifold (the points of which are usually identified forthwith with events—I have already discussed that problem), and then put such structures on this manifold as the metric tensor field. Is that the way that any one of us actually goes about solving the field equations of general relativity? Of course not. One first solves them on a generic patch, and *then* one tries to maximally extend the local solution (using some criteria for acceptable extensions) from that patch to a global manifold which is not known ahead of time. Before solving the field equations, one generally doesn't know the global manifold on which the solution will turn out to be maximally extended. So we are pulling a swindle when we tell students, as our definitions imply, that you first pick the manifold and then solve the field equations on it.

The fibre bundle approach mentioned above doesn't solve this problem because it also assumes a base manifold with given global topology to start with. So I believe that it is an important problem to formulate general relativity mathematically in such a way that it is clear from the outset that finding the maximally extended global manifold (whatever criteria are used to define such a maximal extension) is part of the problem of finding a solution.

I am *not* saying, of course, that each topology leads to a unique metric, but rather, that no metric implies no topology. You don't start out with a topology and then look for a metric; you look for a local solution to the field equations on a patch, and then investigate how far that solution patch can be extended. One will not necessarily get a unique answer for the global topology, as the example of flat space shows. But one does not start out with a global topology and then look for *all* the metrics solving the field equations that are compatible with it. The important point is that if one wants to consider *all* solutions to the field equations—or even a subclass wide enough to include solutions on topologically inequivalent manifolds—one must have a mathematical structure that allows this to be done.

Although the two ideas are not exactly the same, the moral I want to draw from this is final point may be related to something that Chris Isham talked about. Suppose you take really seriously the point of view that there is something fundamentally local about the way general relativity approaches a problem. Then there is another fundamental tension between the basic approaches of general relativity and of quantum mechanics since quantum mechanics, in a deep sense, is fundamentally global in its approach to problems. It doesn't make much sense to talk about the wave function on one patch of space-time, or the sum over all paths on one patch of space-time. In solving a quantum-mechanical problem, you have to consider the *whole* manifold from the beginning The conventional mathematical approach to general relativity, which starts with a manifold, masks this tension. If we develop a mathematical formulation of general relativity that emphasizes the element of locality from the beginning, it would emphasize this contrast more sharply. Such an emphasis on the tension may be a necessary stage in finding its ultimate resolution.

Indeed, this way of thinking also suggests new possibilities. Perhaps the existence of a global manifold is just a special case. Maybe in the ultimate theory all the patches won't always fit together to form a manifold, except in the classical limit. Here is where my point of view may connect with some of Isham's ideas. If you don't build in a global manifold at the beginning, perhaps you are better off, because you may have to get rid of it in the end anyway,

NOTES

[1] See Shelley 1901, p. 821.

[2] The excerpts from this diary, as printed in Seelig 1960, pp. 258–259, differ markedly from the text of the diary itself.

[3] On 21 February 1939, Einstein wrote to Rosen: "If you had come to me with such a proposal in the days when we were still engaged in happy daily collaboration, I would have stretched out my splendid tongue at you. Why? ...—this is the worst thing—one would no longer have any reasonable possibility of assimilating inertia to gravitation as the same thing. ... So throw it peacefully in the waste basket" ["Wenn Sir mir mit einem solchen Vorschlag gekommen wären, als wir noch täglich zu froher Arbeit zusammenkamen, hätte ich Ihnen meine prächtige Zunge lang herausgestreckt. Warum? ...—und dies ist das Schlimmste—hat man keine vernünftige Möglichkeit mehr, Trägheit und Schwere als das Gleiche aufzupassen. ... Also werfen Sie es ruhig in den Papierkorb"] (Einstein Archive, Control Index No. 20–233).

[4] Einstein made this point in a follow-up letter to Nathan Rosen on the latter's bimetric theory: "If one assumes that the second (Euclidean) metric does not express any physical reality, then the entire formalism is empty and misleading. If it does express such a reality, then one is led back to the study of special relativity with its entire manifold of theoretical possibilities. In that case, it is entirely arbitrary to describe gravitation by a *tensor,* and even more so to give any sort of preference to the *covariant* derivatives formed with this tensor over other special-relativistic modes of tensor-formation. There are then innumerable logically equally justified possibilities, from which you have selected one quite arbitrarily." ["Nimmt man an, dass die zweite (euklidische) Metrik keine physikalische Realität ausdrückt, so ist der ganze Formalismus leer und irreführend. Drückt sie aber eine Realität aus, so wird man auf das Studium der speziellen Relativität zurückgeführt mit seiner ganzen Mannigfaltigkeit theoretischer Bildungen. Dann ist es ganz willkürlich, die Gravitation durch einen *Tensor* zu beschreiben und erst recht, den mit diesem Tensor gebildeten *kovarianten* Ableitungen irgend welchen Vorzug vor anderen spezial-relativistichen Tensorbildungen den Vorzug zu geben. Es gibt dann unzählige logisch gleichberechtigte Möglichkeiten, von denen Sie ganz willkürlich eine herausgegriffen haben"] (Einstein to Nathan Rosen, 6 March 1939; Einstein Archive, Control Index No. 20–235).

REFERENCES

Bell, John (1987). *Speakable and Unspeakable in Quantum Mechanics — Collected papers on quantum philosophy.* Cambridge: Cambridge University Press.
Born, Max (1949). "Einstein's Statistical Theories." In Schilpp 1949, 163–177.
———— (1971). *The Born–Einstein Letters—Correspondence between Albert Einstein and Max and Hedwig Born from 1916 to 1955 with commentaries by Max Born.* New York: Walker and Company.
Eckert, Michel and Pricha, Willibald (1984). "Die erste Briefe Albert Einsteins an Arnold Sommerfeld." *Physikalische Blätter* 40: 29–34.
Einstein, Albert (1916). "Strahlungs–Emission und -Absorption nach der Quantentheorie." *Deutsche Physikalische Gesellschaft, Verhandlungen* 18: 318–323.
————(1918). "Motiv des Forschens." Cited from translation in Einstein 1954a, pp. 224–227.
————(1919). "What is the Theory of Relativity?" Cited from Einstein 1954a, pp. 227–232.
————(1945). "A Testimonial from Professor Einstein." In Jacques Hadamard. *An Essay on the Psychology of Invention in the Mathematical Field.* Princeton: Princeton University Press, Appendix II, pp. 142–143. Reprinted as "A Mathematician's Mind" in Einstein 1954a, pp. 25–26.
————(1949). "Autobiographical Notes." In Schilpp 1949, pp. 2–94. Cited from the corrected reprint, Einstein 1979.
————(1952). "Relativity and the Problem of Space." In ibid., *Relativity: the Special and the General Theory.* New York: Crown, pp. 135–157.
————(1953). "Introductory Remarks on Basic Concepts." In *Louis de Broglie — Physicien et Penseur.* André George, ed. Paris: Editions Albin Michel, pp. 4–15.
————(1954a). *Ideas and Opinions.* Carl Seelig, ed. New York: Crown.
————(1954b). "Foreword." In Jammer, M., *Concepts of Space.* Cambridge, Mass.: Harvard University Press, pp. xiii–xvi.
————(1979). *Autobiographical Notes—A Centennial Edition.* Paul Arthur Schilpp, transl. and ed. LaSalle and Chicago: Open Court Pub. Co.
————(1987). *Collected Papers of Albert Einstein.* vol. 1, *The Early Years (1879–1902).* John Stachel et al, eds. Princeton: Princeton University Press.
————(1989). *Collected Papers of Albert Einstein.* vol. 2, *Writings (1901–1909).* John Stachel et al, eds. Princeton: Princeton University Press.
Einstein, Albert and Besso, Michele (1972). *Correspondance 1903–1955.* Pierre Speziali, trans. and ed. Paris: Hermann.
Fine, Arthur (1986). *The Shaky Game—Einstein Realism and the Quantum Theory.* Chicago and London: The University of Chicago Press.
Fraenkel, Abraham (1954). *Bulletin of the Research Council of Israel* 3: 283–289.

Frank, Philipp (1979). *Albert Einstein—Sein Leben und Seine Zeit*. Braunschweig-Wiesbaden: Friedr. Vieweg & Sohn.

Geroch, Robert (1969). "Limits of Space-times." *Communications in Mathematical Physics* 13: 180–193.

Jammer, Max (1979). "Albert Einstein und das Quantenproblem." In *Einstein Symposion Berlin*. H. Nelkowski et al, eds. Berlin–Heidelberg–New York: Springer-Verlag, pp. 146–167.

———(1982). "Einstein and Quantum Physics." In *Albert Einstein — Historical and Cultural Perspectives*. Gerald Holton and Yehuda Elkana, eds. Princeton: Princeton University Press, pp. 59–76.

Schilpp, Paul Arthur, ed. (1949). *Albert Einstein: Philosopher-Scientist*. Evanston, Illinois: The Library of Living Philosophers.

Sciama, Dennis W.; Candelas, P.; and Deutsch, D. (1981). "Quantum Field Theory, Horizons and Thermodynamics." *Advances in Physics* 30: 327–366.

Seelig, Carl (1960). *Albert Einstein — Leben und Werk eines Genies unsere Zeit*. Zürich: Europa Verlag.

Shelley, Percy Bysse (1901). *The Complete Poetical Works of Shelley — Cambridge Edition*. Boston: Houghton Mifflin.

Stachel, John (1986a). "Einstein and the Quantum: Fifty Years of Struggle." In *From Quarks to Quasars—Philosophical Problems of Modern Physics*. Robert G. Colodny, ed. Pittsburgh: University of Pittsburgh Press, pp. 349–385.

———(1986b). "What a Physicist Can Learn from the History of Einstein's Discovery of General Relativity." In *Proceedings of the Fourth Marcel Grossmann Meeting on General Relativity*. Remo Ruffini, ed. Amsterdam: Elsevier, 1857–1862.

———(1987). "How Einstein discovered general relativity: a historical tale with some contemporary morals." In *General Relativity and Gravitation — Proceedings of the 11th International Conference on General Relativity and Gravitation*. M.A.H. MacCallum, ed. Cambridge: Cambridge University Press, pp. 200–208.

———(1989). "Einstein's Search for General Covariance, 1912–1915. In *Einstein Studies, vol. 1: Einstein and the History of General Relativity*. Don Howard and John Stachel, eds. Boston–Basel–Berlin: Birkhäuser, pp. 63–100.

von Laue, Max (1956). *Die Relativitätstheorie*. vol. 2. *Die allgemeine Relativitätstheorie*. 4th edn. Braunschweig: Vieweg.

Whitman, Walt (1949). *The Inner Sanctum Edition of The Poetry and Prose of Walt Whitman*. Louis Untermeyer, ed. New York: Simon and Schuster.

Quantum Measurements and the Environment-Induced Transition from Quantum to Classical

Wojciech H. Zurek

1. Introduction

The past decade has seen a significant resurgence of interest in the problem of quantum measurement. This can be attributed to several related developments. First and foremost, experiments that appear to cross the traditional barrier between quantum and classical, delineated by the transition between "microscopic" and "macroscopic," became possible. Thus, the usual intuitions about the behavior of physical systems became difficult to maintain, and the "leave it to the philosophers" attitude slowly began to disappear. Secondly, interpretational issues usually associated with the microscopic appeared in the most macroscopic of all possible contexts: cosmology. It is now thought that macroscopic and obviously classical structures, such as galaxies, Abell clusters, and the very large-scale "foam" in the distribution of matter, are the imprints of quantum fluctuations. Moreover, the Copenhagen interpretation, at least in its traditional garb, is obviously inapplicable to the universe as a whole. The third reason has to do with a perception that progress in the understanding of the reduction of the wave-packet—however slow and partial—is taking place. Concrete calculations rather than philosophical arguments are again worth making.

In particular, significant progress has occurred in the investigation of the role of the environment in bringing about the transition between quantum and classical. Interaction between a quantum system and its environment can often be regarded as a measurement in the course of which one of the system observables influences the evolution of the external degrees of freedom and is thus "monitored" by the environment. The resulting loss of quantum coherence erases part of the density matrix responsible for the correlations between the eigenstates of the monitored observable. This mechanism is very efficient even in the limit of weak coupling. Indeed, the classical limit of quantum theory—that is, both classically reversible equations of motion and classical states (trajectories in phase space rather than their superpositions) can be naturally attained when both the coupling strength and the Planck constant simultaneously tend to zero. The purpose of this paper is to provide a brief review of this idea of environment-induced superselection. Given the quantum-gravitational focus of this meeting, my discussion will aim at establishing whether environment-induced superselection can be of help in interpreting situations arising in the context of quantum cosmology.

I shall begin in the next section with a review of basic assumptions and key goals arising from the discussion of the quantum-classical dichotomy in the cosmological context. Section 3 summarizes the key idea of environment-induced superselection. Section 4 discusses a simple example — quantum Brownian motion—that dramatically illustrates the effectiveness of the environment in forcing quantum systems to exhibit essentially classical features. Section 5 considers the relation between environment-induced superselection and the second law of thermodynamics. Section 6 returns to cosmological applications of the idea of environment-induced superselection. Finally, Section 7 describes my conclusions.

2. Classical Reality and Quantum Cosmology

The aim of this section is to state as succinctly as possible the key assumptions, essential facts, and principal "desiderata" that force one to face the problem of quantum measurement on cosmological scales and suggest the environment as a possible solution. Not all of these "commandments" are universally accepted although some of them may appear more self-evident than others. Each of these assumptions and goals could become, under appropriate circumstances, the subject of a heated discussion. I shall attempt to justify several of the points in the following sections. The purpose of the list below is then to provide a concise description of the framework, a summary of the key themes, rather than an "introduction."

1. Quantum theory applies to the universe as a whole.

2. The Copenhagen interpretation (see Section I of Wheeler and Zurek

1983 for a collection of papers by Bohr and others on the subject)—classical, macroscopic observers measuring quantum, microscopic systems—is obviously inadequate, as the relative sizes of the observer and of the observed system are reversed when the system is the whole universe.

3. There are degrees of freedom that behave in a conspicuously classical fashion. That is, certain superpositions, allowed in principle, are never observed in practice.

4. A weak coupling between a system and its (quantum) environment can induce apparently "classical" behavior in an open quantum system.

5. Interaction with the environment has essentially the character of a measurement. The environment becomes correlated (in the terminology of von Neumann 1955, it performs the first stage of a measurement) with an observable whose eigenstates commute with the interaction Hamiltonian.

6. The interaction Hamiltonian can induce a "classical" behavior in an observable through its continuous monitoring. As a result, effective superselection rules emerge.

7. Position is often such an effectively classical observable.

8. The universe has no environment.

9. Some of the individual degrees of freedom inside the universe do have environments. Therefore, some of the observables within the universe can be forced to appear classical by the environment-induced superselection.

10. Information and information transfer are essential in these considerations. Entropy can be regarded as information that is *inaccessible,* either because it is "lost" in the correlations between systems, or because it is present only "implicitly" in a form that makes the relevant information impossible to decipher without a substantial thermodynamic cost.

The first three points, and in particular, the key assumption 1 could (and perhaps should) be discussed at length. While obvious on the surface, they could be wrong for some subtle reason related to the nature of quantum theory, gravitational theory, or their combination. Points 4–7 will be clarified in the following few sections. Point 8 was, until recently, perhaps the only clear, uncontroversial one in the whole list, not requiring further explanations. However, in view of the recent work on "baby universes," even this statement might have to be qualified. The last two points are at this stage as much subjects of research as statements of fact. I shall return to them in the last few sections.

3. Environment-Induced Superselection

The key idea of environment-induced superselection is straightforward: a quantum system S correlated with another quantum system $\mathcal{E}$ will appear "mixed" to anyone making measurements on S alone. Moreover, systems with appropriately mixed density matrices can be regarded as describing ensembles with classical probability distributions. In other words there is no operational procedure that can distinguish between a "true" mixture and an apparent mixture caused by a correlation with an inaccessible environment $\mathcal{E}$ (d'Espagnat 1971). Therefore, it is tempting to regard all mixtures as equivalent. Indeed, one could argue that there is no evidence that "true mixtures" exist. Moreover, since the mixture allows one to interpret the state of the system in terms of a classical probability distribution, it is useful to regard "mixed" systems as effectively classical.

The interaction of a quantum system S with the environment $\mathcal{E}$ results in correlations. In a sense, the environment "measures" one of the system's observables. As a result, the preferred, measured observable of the system acquires "classical reality." In a transparent notation, the evolution of the combined system-environment wave vector proceeds from the initial state $|\Phi_{S\mathcal{E}}(0)>$ to the final state $|\Phi_{S\mathcal{E}}(t)>$ as follows:

$$
\begin{aligned}
|\Phi_{S\mathcal{E}}(0)> \; &= \; |\psi_S> \times |\phi_\mathcal{E}> = (\sum_i \alpha_i |s_i>) \times |\phi_\mathcal{E}> \\
&\rightarrow \sum_i \alpha_i |s_i> \times |\phi_{\mathcal{E}_i}> = |\Phi_{S\mathcal{E}}(t)> \; .
\end{aligned}
\tag{3.1}
$$

Here, $|\phi_\mathcal{E}>$ is the initial state of the environment and the $|\phi_{\mathcal{E}_i}>$ are distinct (orthogonal) environment states correlated with the distinct states $|s_i>$ of the system. The density matrix of the system alone is now represented by

$$
\rho_S(t) = Tr_\mathcal{E} |\Phi_{S\mathcal{E}}(t)> < \Phi_{S\mathcal{E}}(t)| \cong \sum_i |\alpha_i|^2 |s_i> < s_i|.
\tag{3.2}
$$

The second equality in the above formula holds exactly when the states of the environment correlated with distinct $|s_i>$ are exactly orthogonal.

The observable Λ defined by:

$$\Lambda = \sum_i \lambda_i |s_i ><s_i|, \tag{3.3}$$

where λ_i are distinct, has the property of being effectively classical when the system is described by ρ_S, equation (3.2): Distinct eigenstates of Λ can no longer interfere with one another. Therefore, one can interpret the system described by $\rho_S(t)$ in terms of a classical "mixture," with the exact state of the system unknown to the observer.

Interaction Hamiltonians that bring about the transition into the "effective mixture" of equation (3.2) are not difficult to write down: Any Hamiltonian that commutes with Λ and mediates interaction between S and a much larger environment $\mathcal{E}$ with a coupling that is a function of Λ will have the desired effect. To demonstrate this assertion, we note that the above requirement leads to interaction Hamiltonians that depend on Λ. Moreover,

$$[H_{int}(\Lambda), \Lambda] = 0 \tag{3.4}$$

is the condition for the interaction to allow for nondemolition measurements of the observable Λ (Caves et al. 1980).

There is, of course, no guarantee that the specific interaction Hamiltonian of a randomly chosen system with its surroundings will have the form prescribed by equation (3.4). Indeed, there are interaction Hamiltonians for which the preferred "pointer observable" Λ does not exist. It seems, however, plausible, and has been confirmed by a variety of specific examples, that all observables that are classical do interact with other degrees of freedom via approximately "nondemolition" Hamiltonians.

A specific case of such a situation—a harmonic oscillator in a heat bath, modeled by a scalar field ϕ, with the interaction $H_{int} = \epsilon x \dot{\phi}$, has been solved and used to discuss environment-induced superselection by W. G. Unruh and this author (see the next paper in this volume). While the calculations are complicated by the presence of self-Hamiltonians, the mechanism of decoherence appears to be quite robust. It is also interesting to note that, in an underdamped harmonic oscillator, the resulting "preferred basis" is composed of coherent states—minimum-uncertainty wave packets—that do have the desired classical limit.

In more complicated situations, in which the interaction Hamiltonians between microscopic observables do not have the form given by equation (3.4) even approximately, new "macroscopic" observables defined by averages over large numbers of microsystems can be expected to emerge.

The advantage of such observables is their resistance to decoherence. As Gell-Mann and Hartle suggest, such hard-to-decohere properties provide a basis for macroscopic physics (e.g., hydrodynamics). Thus, in a sense, environment-induced superselection enforces a certain "coarse graining" of our description of the universe.

4. Reduction of the Wave-Packet: How Long Does it Take?

It is sometimes argued that observables of macroscopic objects, which, to a good approximation, obey reversible classical dynamics could not have been affected by their interaction with the environment and therefore could not have lost their quantum features as a result of such an interaction. I shall demonstrate here that this argument is fallacious by showing that the decorrelation time scale θ is typically much shorter than the relaxation time scale τ. In particular, in high-temperature environments, these two time scales are connected by a formula (Zurek 1986):

$$\theta = \tau(\lambda_{dB}/\Delta x)^2 \tag{4.1}$$

where λ_{dB} is the thermal deBroglie wavelength defined by:

$$\lambda_{dB}^2 = \hbar/(4mk_BT), \tag{4.2}$$

where $\hbar$ is the Planck constant, m the mass of the free particle, k_B the Boltzmann constant, and T the temperature.

To focus the discussion, we consider a free particle interacting with a heat bath at temperature T. Under appropriate conditions (see e.g., Unruh and Zurek 1989 and Caldeira and Leggett 1983 for more detailed discussion), the density matrix of the free particle in the position representation, $\rho(x, y)$, approximately obeys the master equation:

$$\dot{\rho} = \left\{ \frac{-i\hbar}{2m} \left(\frac{\partial^2}{\partial x^2} - \frac{\partial^2}{\partial y^2} \right) - \gamma(x-y) \left(\frac{\partial}{\partial x} - \frac{\partial}{\partial y} \right) - \frac{2m\gamma k_BT}{\hbar^2}(x-y)^2 \right\}. \tag{4.3}$$

To compare the relaxation and decorrelation time scales, $\tau = \gamma^{-1}$ and θ respectively, we consider an initial density matrix corresponding to a pure state consisting of two Gaussian wave-packets, each with half-width δ, separated by distance Δx:

$$|\psi> = (|\alpha> + |\beta>)/\sqrt{2}, \tag{4.4}$$

$$< x|\alpha > = (2\pi\delta^2)^{\frac{-1}{4}} \exp[-(x - \Delta x/2)^2/4\delta^2], \qquad (4.5a)$$

$$< x|\beta > = (2\pi\delta^2)^{\frac{-1}{4}} \exp[-(x + \Delta x/2)^2/4\delta^2]. \qquad (4.5b)$$

The initial density matrix

$$\rho = |\psi >< \psi| \qquad (4.6)$$

is pure. Plotted in the position representation, on the (x, x') plot, this ρ reveals four extremes. Two of them occur on the diagonal at $\pm\Delta x/2$. The remaining two are off-diagonal and lie at $x = -y = \pm\Delta x/2$. These off-diagonal extremes provide a measure of the quantum coherence between the spatially separated components of the wave-packet. Only when their size equals the size of the on-diagonal peaks, is the complete density matrix pure, satisfying:

$$\rho^2 = \rho. \qquad (4.7)$$

The master equation (4.3) was derived on the basis of an assumption that the free particle interacts with the environment. This assumption resulted in addition of two extra terms to H_0, the free particle Hamiltonian. The term proportional to the relaxation rate γ results in the slowing down of a wave-packet moving with velocity v:

$$< \dot{v} >= -\gamma < v > . \qquad (4.8)$$

Its effect on a motionless wave-packet $|\alpha >$ can be also estimated. It is characterized by the relaxation time scale τ:

$$\tau^{-1} =< \alpha_t|\dot{\rho}|\alpha_t >\simeq -\frac{\gamma}{2} < \alpha_t|(x - y)^2|\alpha_t > \left(\frac{1}{\delta^2} + \frac{1}{\lambda_{dB}^2}\right). \qquad (4.9)$$

Here, $|\alpha_t >= \exp(-iH_0t/\hbar)|\alpha >$ is used to separate evolution due to interaction with the heat bath from evolution induced by a free-particle Hamiltonian H_0, the first term in the master equation (4.3).

A similar formula describes the evolution of the off-diagonal terms:

$$\theta^{-1} =< \alpha_t|\dot{\rho}|\beta_t >\simeq -\frac{\gamma}{2} < \alpha_t|(x - y)^2|\beta_t > \left(\frac{1}{\delta^2} + \frac{1}{\lambda_{dB}^2}\right). \qquad (4.10)$$

The key difference between the relaxation and decorrelation time scales τ and θ comes from the difference in the size of the expectation values of $(x - y)^2$:

$$< \alpha_t|(x - y)^2|\alpha_t >= \delta^2, \qquad (4.11)$$

while

$$< \alpha_t | (x - y)^2 | \beta_t > = (\Delta x)^2. \qquad (4.12)$$

Therefore, the ratio of the two rates is:

$$\tau/\theta = (\Delta x/\delta)^2 \simeq (\Delta x/\lambda_{dB})^2, \qquad (4.13)$$

in accord with equation (4.1). Above, we have assumed that the typical spread δ of the wave-packet describing a free particle in contact with the heat bath of temperature T is given by λ_{dB}.

For "macroscopic" values of Δx, m and T, the ratio multiplying the relaxation scale τ in equation (4.1) is enormous and it enforces environment-induced superselection. For example, for $m = 1$g, $\Delta x = 1$ cm and $T = 300° K$, the ratio $\tau/\theta = 10^{40}$! That is, for a system with these parameters and with a relaxation time scale of the order of the age of the universe, $\tau \sim 10^{17}$s, the decoherence time scale is still only 10^{-23}s.

The usual argument for focusing the discussion on isolated systems asserts that, if a certain object cannot be considered isolated, then one can always incorporate all systems that are coupled to it and discuss the same problem in the enlarged system, including both the object of interest and its "environment." By carrying out this procedure—so the argument goes—one can always eventually find a system sufficiently weakly coupled to the environment to regard it as effectively isolated.

Equation (4.1) shows that such an "incorporation" strategy does not guarantee success. Indeed, it will most likely fail: as the size of the composite system increases, τ will presumably also increase (this is the aim of the isolation). However, at the same time, other parameters of the enlarged object—in particular, Δx and λ_{dB}—will also change. Indeed, it is easy to imagine that typically such enlargement, while increasing the relaxation time scale τ, will decrease the decoherence time θ. Consequently, the only system which is undisputedly insulated from its environment is the whole universe. This is obviously an uncomfortable place for the "incorporation" argument to succeed.

The most far-reaching corollary of the above discussion is the possibility that the classical limit can be truly attained only with the help of environment-induced decoherence. For decoherence can be instantaneous ($\theta \to 0$), even when the system is classically reversible ($\tau \to \infty$): In the limit $\hbar \to 0$, this combination is certainly consistent with equation (4.1). It is tempting to regard such a limit as a truly classical limit: not only are classical, reversible equations of motion obtained, as is always the case in the correspondence limit; classical trajectories also emerge as the only ones compatible with the limit (Zurek 1986). This is due to the fact that the decoherence term—the last term in the master equation (4.3)—is still present although it has no effect on the individual classical trajectories: Its only consequence is to "rule out" spatial superpositions by almost instantaneously damping out quantum coherence between distinct locations.

5. Reduction of the Wave-Packet, Entropy, and Information

The irreversibility of a measurement is intimately related to the second law of thermodynamics. This point was clear from the very first discussions of the "reduction of the wave-packet," (see Wheeler and Zurek 1983, Sections I and V), but its clearest early exposition is perhaps found in von Neumann 1955. More recent discussions of the relationship between irreversibility and indelibility of the measurement record are found in Zeh 1970 and Peres 1980. Our exposition will be closely patterned on an earlier paper by the present author, devoted to the discussion of information transfer in the course of measurements (Zurek 1983).

Consider the initial wave function $|\Phi_{S\mathcal{E}}(0) >$, equation (3.1). It is given by the direct product of the system and environment state vectors, $|\psi_S >$ and $|\phi_{\mathcal{E}} >$. These wave functions describe systems that are separately in definite states. It is easy to see that the entropy of each of them considered separately is initially equal to zero.

$$\mathcal{H}_S = Tr_S \; \rho_S(0) \, lg \; \rho_S(0) = 0, \tag{5.1a}$$

$$\mathcal{H}_{\mathcal{E}} = Tr_{\mathcal{E}} \; \rho_{\mathcal{E}}(0) \, lg \; \rho_{\mathcal{E}}(0) = 0. \tag{5.1b}$$

By contrast, the entropy of the system and the environment taken separately, after they have evolved into the correlated state $|\Phi_{S\mathcal{E}}(t) >$, is no longer zero: They are both in effective mixtures: $\rho_S(t)$ given by equation (3.2), with an analogous equation describing $\rho_{\mathcal{E}}(t)$. Therefore, now

$$\mathcal{H}_S(t) = Tr_S \; \rho_S(t) \, lg \; \rho_S(t) > 0, \tag{5.2a}$$

$$\mathcal{H}_{\mathcal{E}}(t) = Tr_{\mathcal{E}} \; \rho_{\mathcal{E}}(t) \, lg \; \rho_{\mathcal{E}}(t) > 0. \tag{5.2b}$$

The entropy of the two components clearly increased as a result of the correlation. However, the total entropy of the whole $\{\mathcal{S}, \mathcal{E}\}$ system could not have changed: the evolution was brought about by a unitary operator, derived from the appropriate Schrödinger equation. Moreover, the complete system is still in the pure state $|\Phi_{S\mathcal{E}}(t)>$. Therefore,

$$\mathcal{H}_{S\mathcal{E}} = Tr\ \rho_{S\mathcal{E}}(t)\ lg\ \rho_{S\mathcal{E}}(t) = 0. \tag{5.2c}$$

Where is then the missing information I_μ?

$$I_\mu = \mathcal{H}_S + \mathcal{H}_\mathcal{E} - \mathcal{H}_{S\mathcal{E}}. \tag{5.3}$$

Equation (5.3) suggests the answer to the question posed above: I_μ is known as mutual information. It describes the extent to which the states of S and $\mathcal{E}$ are correlated as a result of the interaction. The transfer of information in the course of reversible interactions accomplishes only this one goal: it increases the amount of mutual information, and as a result decreases the information—increases the entropy—associated with the subsystems. It is tempting to identify the "loss of information" needed to understand the second law with the transfer of definite information into more elusive correlations (Zurek 1982, 1983).

This last comment may be further extended to elaborate the role of amplification in the measurement process. Suppose that the environment contains many more degrees of freedom than the system, which has only two states, $|\uparrow>$ and $|\downarrow>$, and that each of these degrees of freedom becomes correlated with the $|\uparrow>$ or $|\downarrow>$ states of the system. In a transparent notation, the wave function of the correlated system-environment combination is then:

$$\begin{aligned}|\Phi_{S\mathcal{E}}(t)> = \ &\alpha\,|\uparrow> \times [|\phi_\uparrow^1> \times |\phi_\uparrow^2> \times \cdots \times |\phi_\uparrow^N>] \\ &+ \beta\,|\downarrow> \times [|\phi_\downarrow^1> \times |\phi_\downarrow^2> \times \cdots \times |\phi_\downarrow^N>].\end{aligned} \tag{5.4}$$

The *redundancy* present in the state vector given by equation (5.4) singles out the basis $\{|\uparrow>, |\downarrow>\}$ as the preferred basis of the system. The preferred pointer observable has the form:

$$\Lambda = \lambda_1|\uparrow><\uparrow| + \lambda_2|\downarrow><\downarrow|, \tag{5.5}$$

where λ_1 and λ_2 are unequal but otherwise arbitrary eigenvalues. It is very easy to find out what was the state of the system in terms of the eigenstates of Λ: Consulting only a few of the N subsystems of the environment will suffice. By contrast, finding out the state of the system in some other basis, given by some superposition of $|\uparrow>$ and $|\downarrow>$, is almost impossible: in order to accomplish that task one would have to measure prohibitively complicated superpositions of the states of the environment.

Such a redundant recording of one of the observables can be brought about by an interaction Hamiltonian given by a direct sum of Hamiltonians of the form

$$H_1 = \epsilon_1 (|\uparrow><\uparrow| \, |\phi_\uparrow^1><\phi_0^1| + |\downarrow><\downarrow| \, |\phi_\downarrow^1><\phi_0^1|) + h.c. \, . \quad (5.6)$$

Above, "h.c." stands for Hermitian conjugate. Such Hamiltonians are easy to find. (See e.g., Mott 1929 for a discussion of this issue in the context of the Wilson cloud chamber particle detector).

An additional advantage of a redundant record is the resistance of such a record to small perturbations due to the fluctuating environment: even if individual subsystems are disturbed by the perturbations, the "average" maintains the information about the outcome of the measurement for a very long time.

6. From Quantum Fluctuations to Classical Density Perturbations

Several distinct models of the inflationary epoch have been suggested. In most of them, space-time itself is regarded as classical, but its properties—e.g., expansion rate in the course of the de Sitter phase—are determined by the behavior of a quantum "inflaton" field. More ambitious models include space-time itself in the wave function. While this step satisfies the inevitable demand of consistency, it also leads to substantial difficulties. In particular, the wave function of the universe $|\mathcal{U}>$ is a solution of the Wheeler–De Witt equation,

$$H|\mathcal{U}>= 0, \quad (6.1)$$

where H is the gravitational-plus-matter Hamiltonian. Equation (6.1) is a functional differential equation in the superspace. There is no explicit time dependence in $|\mathcal{U}>$. This makes the discussion of the evolution of $|\mathcal{U}>$—the subject of discourse of cosmology—difficult to carry out. Normalization as well as other problems usually encountered in the domain of quantum gravity further complicate the interpretation of $|\mathcal{U}>$. In spite of that, several promising discussions of the role of the environment in the interpretation of simple "minisuperspace" models have been put forward recently (see Halliwell 1991 and references therein). I shall focus on a simpler and less ambitious issue, understanding the nature of the quantum-classical transition, in the course of this section.

An attractive and tractable model of the dynamical evolution of the inflaton field is afforded by the harmonic oscillator. Chaotic inflation can be modeled by the "normal" harmonic oscillator. Computation of the behavior of harmonic oscillator density matrices in a class of models, in which a single oscillator system is coupled to a "heat bath" provided by a scalar field, is discussed in more detail in Unruh and Zurek 1989. The aim of this section is to: (1) describe the behavior of an "upside-down" harmonic oscillator coupled to such a heat bath, and (2) use the analogy with the new inflationary models to draw conclusions about the quantum-classical phase transition in the early universe.

Guth and Pi 1985 consider an isolated "upside-down" oscillator for a similar purpose. Their analysis is based on the observation that the pure Gaussian wave-packet $|\psi>$ evolving in an unstable harmonic potential is stretched out by the $F = kx$ force while retaining its Gaussian shape. Moreover, the momentum and position of the particle (value of the inflaton field and its momentum) become strongly correlated:

$$(xp + px)|\psi > \cong 2\Omega_0 x^2 |\psi >, \qquad (6.2)$$

while the quantum uncertainty remains small:

$$(xp - px)|\psi > = i\hbar. \qquad (6.3)$$

Consequently, Guth and Pi argue, as soon as the correlation between p and x, equation (6.2), is substantially larger than quantum uncertainty, equation (6.3), individual pieces of the wave-packet can be regarded as effectively independent and therefore classical.

This analysis focuses on the strength of the correlations induced by the dynamics in a pure wave-packet. While its key observation is correct, it addresses only one of the aspects of the quantum-classical transition: it demonstrates that the motion of individual "pieces" of the wave-packet can be regarded in the new inflationary scenario as independent. This is clearly necessary, but far from sufficient, as it fails to address several of the issues that are crucial for a proper understanding of the dynamics of the de Sitter era and its consequences, such as density perturbations. In particular, it is not clear: (1) What is the value of the cosmological constant corresponding to the broad Gaussian wave-packet, as its fragments could be formally associated with a definite and distinct value of Λ?

(2) Will the quantum coherence still present in the wave-packet become important in the course of reheating? After all, different parts of the wave-packet stretched out by the upside-down potential eventually become compressed in the vicinity of the low-temperature broken-symmetry vacuum. (3) The inequality between the strength of the (p, x) correlation and the quantum indeterminacy given by $[p, x]$ obtains only for an upside down oscillator in which the wave-packet is stretched. Therefore, the mechanism of the transition from the quantum to the classical suggested by Guth and Pi cannot be invoked in the case of chaotic inflation, as there the potential is better approximated by a "regular" (or, sometimes, anharmonic) oscillator potential.

Environment-induced superselection avoids the difficulties pointed out above by decohering pieces of the wave-packet through a coupling between the value of the inflaton field χ and other fields, which serve as a "heat bath." It is not difficult to write down the interaction Hamiltonian that commutes with the value of the inflaton field value. Indeed, since the expansion rate is determined by the value of the inflaton (which sets the value of the cosmological constant), indirect coupling between χ and other fields (stretched by inflation at χ-dependent rates) is inevitable. Moreover, space-time itself could be regarded as an environment. In short, it is possible to propose models in which an approximate "pointer basis" for the inflaton is associated with its value χ. The number of degrees of freedom in the environment can be easily sufficiently large to model it as a "heat bath."

Under such circumstances, the analogy between the fate of the position x in the "upside-down oscillator" of Unruh and Zurek 1989 and the inflaton χ in the new inflationary scenario can be exploited. Its key features are illustrated in Figures 1–3, where the Wigner function, density matrices in position and momentum representations, as well as the entropy (which provides a measure of the decoherence), are plotted.

The oscillator was initiated in a minimum uncertainty wave-packet near the top of the potential. Its initial momentum is opposite to the force and sufficiently large to get it "over the hill." As plots of the Wigner function in Figure 1 illustrate, at large time the distribution extends exponentially in x, and, because of the (p, x) correlation, equation (6.2), also in p. For an isolated upside-down oscillator, this correlation would result in a Wigner distribution function that would be exponentially compressed in the direction perpendicular to the $p \sim x$ "major axis" of the $1 - \sigma$ contour in Figure 1. By contrast, the Wigner distribution for a damped upside-down oscillator maintains a finite spread in this direction at later times.

FIGURE 1A

FIGURE 1B

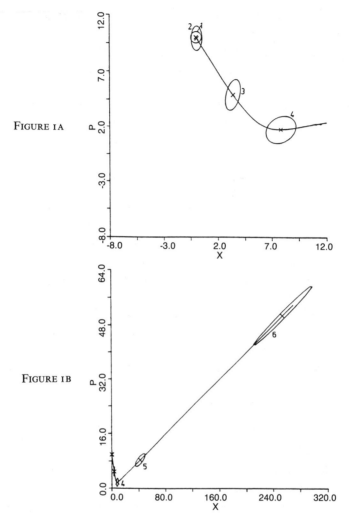

FIGURE 1. Evolution of the $1 - \sigma$ contour of the Wigner distribution function for a damped "upside-down" oscillator ($\Omega = i$, $m = 1$, $\hbar = 1$, $\gamma = 0.8$). Damping is caused by a coupling to the scalar field environment in its vacuum ($T = 0$) state. The system is initiated in a minimum uncertainty wave-packet at $x < 0$ but with a finite positive ("uphill") momentum. After a period of initial slowdown due to the combined effects of the systematic force and damping (Fig. 1a), the wave-packet accelerates and becomes stretched out along the $p \sim x$ direction (Fig. 1b). Note that at late times, while the extent of the wave-packet increases exponentially, its width in the $p \sim -x$ direction remains approximately constant. The total area covered by the $1 - \sigma$ contour— proportional to the number $\mathcal{N}$ of its uncorrelated fragments with $\Delta x \Delta p \approx \hbar$—is also increasing exponentially.

The density matrix in the position representation, illustrated for the damped unstable harmonic oscillator in Figure 2a, would look dramatically different for the isolated harmonic oscillator. Its spread would still increase exponentially, but the shape of the $1 - \sigma$ "isodensity contour" would remain similar to the one corresponding to the initial minimum uncertainty wave-packet (marked "1" in Figure 2a). In other words, for an isolated unstable oscillator, the density matrix would expand at the same rate in both the on-diagonal and off-diagonal directions. Such off-diagonal terms are constantly destroyed by the monitoring of the system by the environment. The change in the shape of the contour between the instants 1 and 2 in Figure 2a attests to the efficiency of the environment-induced superselection. At later times, the width of the contour remains approximately constant: the pointer basis consists of somewhat "squeezed" Gaussians.

The evolution of the density matrix in the momentum representation is a direct consequence of the combination of the dynamical evolution and of the position monitoring by the environment. In particular, the on-diagonal extent of the $1 - \sigma$ contour of the density matrix increases dramatically in the short time interval between the instants 1 and 2. This can be regarded as a consequence of the uncertainty principle, which mandates that better knowledge of the actual position (by the environment) will inevitably lead to a larger uncertainty of momentum. Hence, the on-diagonal extent of $\rho(p, p')$ between instants 1 and 2 in Figure 2b quickly increases. Subsequent evolution of the contours is caused by the further interplay of the dynamics and the monitoring action of the environment.

A plot of the entropy of the unstable oscillator is shown in Figure 3. A rapid initial "jump" occurs on time scales corresponding to the highest frequencies in the environment that couple with the system. It is followed by an increase that is linear in time. This linearity has a simple explanation: as the system evolves, the spread of the density matrix along the diagonal in Figure 2 increases exponentially, while the extent of spatial quantum coherence ("thickness" of the $1 - \sigma$ contour of $\rho(x, x')$) is almost exactly constant beyond the instant "2". This results in an exponential increase of the number of distinct pure microstates $\mathcal{N}$. Now, since the entropy is approximately equal to $lg \, \mathcal{N}$, it will increase proportionally to t. This accounts for the behavior observed at late times in Figure 3.

As each of the microstates has $\Delta x \Delta p \sim \hbar$, one can regard the density matrix as a description of an ensemble of systems, each of which

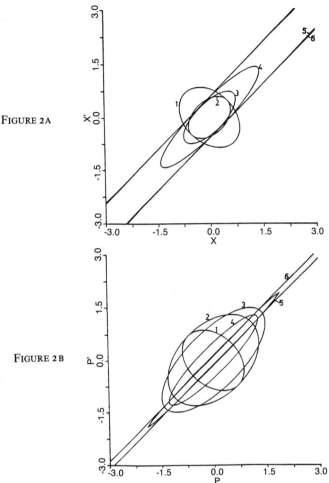

FIGURE 2A

FIGURE 2B

FIGURE 2. (a) Density matrix in the position representation for an upside down oscillator at instants corresponding to crosses 1,2, ..., 6 in Fig. 1. The $1 - \sigma$ contours were recentered to facilitate the comparison. Note the rapid decay of the off-diagonal terms between instants 1 and 2. Characteristic time scale for this process is associated with the "collision time scale" Γ^{-1}, where $\Gamma = 1000$ in the same units in which $\Omega = i$ and corresponds to the highest frequencies of the environment that couple to the system. After the initial onset of rapid decoherence, the off-diagonal spread of $\rho(x,x')$ remains approximately constant. (b) Density matrix in the momentum representation. $1 - \sigma$ contours are again recentered. Note the rapid initial spread of the contour along the diagonal. It is caused by the "measurement" of the position of the system by its environment. The principle of indeterminacy implies that a better-determined (by the scalar field environment) position must be compensated for through the increased spread of momenta.

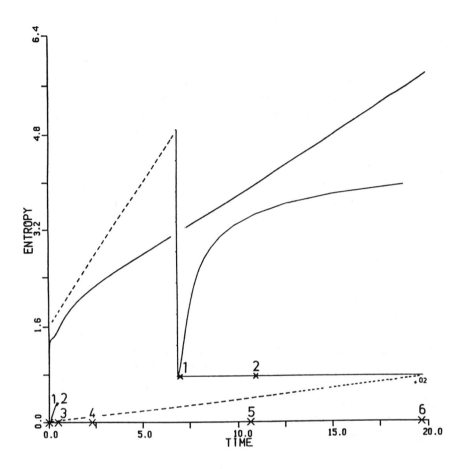

FIGURE 3. Entropy of the unstable oscillator as a function of time. Note the initial increase (insert) on the Γ^{-1} collision time scale. At later times the number of the wave-packet fragments $\mathcal{N}$ increases as $\exp(t)$. Consequently, entropy also increases as $lg\ \mathcal{N} \sim t$.

has a well-defined position and momentum. Most importantly, in contrast to the isolated case discussed in Guth and Pi 1985, the individual wave-packets can no longer interfere as the quantum coherence between them is now lost (see Cornwall and Bruinsma 1988). Hence: (1) Each of them corresponds to a definite value of the cosmological constant. (2) Even when they "meet" in the broken-symmetry vacuum, they will not interfere as they belong to distinct "branches" of the wave function of the universe. (3) This mechanism is based on decoherence rather than on the unstable nature of the evolution. Hence, it will also work for chaotic inflation.

Cornwall and Bruinsma 1988 recently considered a damped unstable oscillator model for the new inflation. They derive the high-temperature master equation using the influence-functional approach (Caldeira and Leggett 1983). One might argue that this limit is unlikely to apply throughout most of the de Sitter era as the initial temperature is exponentially damped, and therefore is likely to be small compared to the relevant cutoffs, the analogs of Γ in Unruh and Zurek 1989 and Caldeira and Leggett 1983. Nevertheless, qualitative conclusions reached on the basis of the high temperature approximation are likely to be essentially consistent with the $T = 0$ discussion presented here.

7. Summary and Conclusions

I have described how environment-induced superselection allows classical observables to emerge as a result of "monitoring" by the environment. This model can be expected to work whenever the system under consideration is not sufficiently isolated, that is, for the vast majority of macroscopic quantum systems. In particular, in the inflationary scenario, the transition between quantum and classical will be, in effect, accomplished in this model. The "environment" consisting of the other fields present will continually monitor the value of the inflaton field, thus preventing quantum superpositions.

Acknowledgments. This research was supported in part by the National Science Foundation (Grant No. PHY82-17853) supplemented by NASA funds.

REFERENCES

Caldeira, A. O., and Leggett, A. J. (1983). "Path Integral Approach to Quantum Brownian Motion." *Physica A* 121: 587–616.
———(1985). "Influence of Damping on Quantum Interference: An Exactly Soluble Model." *Physical Review A* 31: 1059–1066.
Caves, Carlton M.; Thorne, Kip S.; Drever, Ronald W.; Sandberg, Vernon D.; and Zimmerman, Mark (1980). "On the Measurement of a Weak Classical Force Coupled to a Quantum Mechanical Oscillator." *Reviews of Modern Physics* 52: 341–392.

Collet, M. J. (1988). "Exact Density-Matrix Calculations for Simple Open Systems." *Physical Review A* 38: 2233–2247.

Cornwall, J. M., and Bruinsma, R. (1988). "Quantum Evolution of an Unstable Field in a de Sitter-Space Thermal Bath." *Physical Review D* 38: 3146–3157.

d'Espagnat, Bernard (1971). *Conceptual Foundations of Quantum Mechanics.* Menlo Park, California: W. A. Benjamin.

Guth, Allen H., and Pi, So-Young (1985). "Quantum Mechanics of the Scalar Field in the New Inflationary Universe." *Physical Review D* 32: 1899–1920.

Haake, F., and Walls, D. F. (1986). "Overdamped and Quasi Quantum Nondemolition Measurements." In *Quantum Optics IV.* Proceedings of the Fourth International Symposium, Hamilton, New Zealand, February 10–15, 1986. J. D. Harvey and D. F. Walls, eds. Berlin and New York: Springer-Verlag, 181–187.

Hakim, Vincent, and Ambegaokar, Vinay (1985). "Quantum Theory of a Free Particle Interacting with a Linearly Dissipative Environment." *Physical Review A* 32: 423–434.

Halliwell, Jonathan J. (1991). "The Wheeler–DeWitt Equation and the Path Integral in Minisuperspace Quantum Cosmology." See this volume, pp. 75–115.

Joos, E. (1986a). "Quantum Theory and the Appearance of a Classical World." In *New Techniques and Ideas in Quantum Measurement Theory.* Daniel M. Greenberger, ed. New York: New York Academy of Sciences, 6–13.

———(1986b). "Why Do We Observe a Classical Spacetime?" *Physics Letters A* 116: 6–8.

Joos, E., and Zeh, H. D. (1985). "The Emergence of Classical Properties through Interaction with the Environment." *Zeitschrift für Physik B* 59: 223–243.

Milburn, G. J., and Holmes, C. A. (1986). "Dissipative Quantum and Classical Liouville Mechanics of the Anharmonic Oscillator." *Physical Review Letters* 56: 2237–2240.

Mott, N. F. (1929). "The Wave Mechanics of α-Ray Tracks." *Royal Society of London. Proceedings A* 126: 79–84.

Neumann, John von (1955). *Mathematical Foundations of Quantum Mechanics.* Robert T. Beyer, trans. Princeton: Princeton University Press.

Peres, Asher (1980). "Can We Undo Quantum Measurements?" *Physical Review D* 22: 879–883.

Savage, C. H., and Walls, D. F. (1985). "Damping of Quantum Coherence: The Master-Equation Approach." *Physical Review A* 32: 2316–2323.

Unruh, W. G. (1991). "Loss of Quantum Coherence for a Damped Oscillator." See this volume, pp. 67–74.

Unruh, W. G., and Zurek, Wojciech H. (1989). "Reduction of the Wavepacket in Quantum Brownian Motion." *Physical Review D* 40: 1071–1094.

Wheeler, John A., and Zurek, Wojciech H., eds. (1983). *Quantum Theory and Measurement.* Princeton: Princeton University Press.

Wigner, Eugene P. (1983). "Review of the Quantum Measurement Problem." In *Quantum Optics, Experimental Gravity, and Measurement Theory*. Pierre Meystre and Marlan O. Scully, eds. New York: Plenum, 43–63.

Zeh, H. D. (1970). "On the Irreversibility of Time and Observation in Quantum Theory." In *Foundations of Quantum Mechanics*. Bernard d'Espagnat, ed. New York: Academic, 263–273.

———(1986). "Emergence of Classical Time From a Universal Wavefunction." *Physics Letters A* 116: 9–12.

Zurek, Wojciech H. (1981). "Pointer Basis of Quantum Apparatus: Into What Mixture Does the Wave Packet Collapse?" *Physical Review D* 24: 1516–1525.

———(1982). "Environment-Induced Superselection Rules." *Physical Review D* 26: 1862–1880.

———(1983). "Information Transfer in Quantum Measurements: Irreversibility and Amplification." In *Quantum Optics, Experimental Gravity, and Measurement Theory*. Pierre Meystre and Marlan O. Scully, eds. New York: Plenum, 87–116.

———(1986). "Reduction of the wave-packet: How Long Does it Take?" In *Frontiers of Nonequilibrium Statistical Physics*. Gerald T. Moore and Marlan O. Scully, eds. New York: Plenum, 145–149.

Discussion

GELL-MANN: In your discussion of entropy, you referred to "spurious information." What is spurious about it?

ZUREK: Spurious in the sense that it is "uncomfortable." Too much information would make it impossible to regard apparatus as classical.

SMOLIN: Is the idea basically that the phase information runs off into the universe far away and, if I am doing a local measurement, I can't measure the correlations having to do with that information? So, as far as local measurement is concerned, it's as if the reduction had taken place, although in fact, it didn't.

ZUREK: You put it slightly differently, but I think you've got it almost exactly on the money. As soon as the measurement is done, the information is lost in the environment, although not necessarily far away, and one can't get it back.

STACHEL: Why do you lose just that much and no more? Is it put in by hand or is there a mechanism? Why doesn't it all leak out?

ZUREK: Just due to the form of the interaction Hamiltonian. Only the information about the pointer observable is communicated and, therefore, the information about the complementary observables is lost in the apparatus-environment correlations.

STACHEL: But you put the form of the Hamiltonian in by hand.

ZUREK: Yes, it's put in by hand, but I'd better build my apparatus so that I have an appropriate interaction Hamiltonian. Otherwise, it will mess up everything.

KUCHAŘ: When you considered harmonic oscillators in Sections 4 and 6, how do you know what the harmonic oscillators are, and what the system is? You have a totally quadratic Hamiltonian and you can decompose it either into individual harmonic oscillators or into harmonic modes. We can form generalized coordinates and think about the system as something described by a different generalized coordinate.

ZUREK: The brutally honest answer is, "I don't know." The identity of the system is assumed to be established and maintained independently of the interaction Hamiltonian. I appreciate your question.

KUCHAŘ: So there's some a priori input, which we are putting in by hand. The a priori input is: "what are the systems?"

ZUREK: Yes, exactly.

UNRUH: Do you see any problems left with the interpretation of quantum mechanics?

ZUREK: I wouldn't, if I were more complacent about adopting the "many-worlds interpretation." But I think there is quite a bit more that's going to emerge from a better understanding of the role of information in quantum theory. I tried to formulate things in terms of the information but I don't think this is the final word. I think one can be happy if one accepts many-worlds interpretations and doesn't worry about anything more. But I think there is a bigger message. The way it's going to emerge and modify our views will come from the information side and from the entropy side, i.e., from the relation to the second law.

HARTLE: You hinted that several times, but what is it that bothers you about the "many-worlds" view?

ZUREK: I don't think it's incomplete, but it's unsatisfying, and "unsatisfying" is a personal statement.

HARTLE: So, it was a personal question. (laughter)

ZUREK: My feeling is that if we finally understand what quantum gravity is really about, it's going to modify both gravity and quantum mechanics. Both are going to have to move a little bit. O.K.? I have — please don't ask me for justification — but I do have a feeling that perhaps one can gain an insight into how it can happen by trying to understand the role of information in this business. It pops up in so many ways.

HARTLE: I still don't understand. You haven't assigned us a problem.

ASHTEKAR: But isn't what Karel said during the talk a problem? After all, when we look at chairs, wouldn't Wojciech like to be able to say "why" it is that we see the positions of the chairs?

GELL-MANN: Why do we see them?

ASHTEKAR: No, no, I mean why is the position observable such a good observable? That's what Wojciech would like to...

GELL-MANN: He explained it.

ASHTEKAR: He explained it, provided we already singled out the position, among all possibilities, as a preferred observable. Then everything is O.K., but why did one single it out in the first place?

ZUREK: It's not that I single out position. Interaction Hamiltonians encountered in the real world happen to depend on — and, therefore, commute with — a certain observable that we end up regarding as position. Let me come back to Jim's question. I'm somehow uncomfortable with the assertion that we are always going to perceive one branch. Why are we always on a single branch of the universe? Why does the universe look the way it does when we look at it, in spite of the fact that there are all these other branches all over the place? Why do we generally perceive only one branch?

GELL-MANN: Why do you say one, when you are trying to discriminate among several alternatives of something? You're narrowing the class. No one, unfortunately, has ever succeeded in narrowing it to one. In fact, it's still infinity.

PAGE: That doesn't change the problem very much. So you have a projection onto a subspace rather than onto a ray. I don't think that changes the essential problem.

ZUREK: Let me put it this way: Can I toss a quantum coin and, depending on the outcome, can I go with a lot of money and someone to Las Vegas in one branch, and stay and do my paper on something-or-other in the other branch, and then combine the two? Why can't I be in both places? It would be so much fun.

GELL-MANN: I thought you'd just finished explaining that.

ZUREK: Not quite to my own satisfaction.

HARTLE: So what are you trying to say? Does it have to do with free will?

ZUREK: No, it's not so much free will. It's why does our consciousness choose to perceive only the single, separate branches?

KUCHAŘ: May I open yet another problem? You have a system that you couple to the apparatus and then the apparatus is coupled to the environment; that defines the preferred basis, and so the apparatus cannot be in a superposition of two distinct position eigenfunctions. Now, I understand how the apparatus functions, but I do not understand how the systems can remain quantum mechanical. The system *can* be in a superposition of two different, distinct position eigenstates. However, it's coupled to the environment, not only via the apparatus. It may be coupled to the environment directly. Why is it that these superpositions are not destroyed as fast as those for the apparatus? I'm asking now, why do we observe quantum behavior at all?

ZUREK: Thank you for bringing it up. The point that is important is a comparison between two time scales. One of them is what I would call the quantum Poincaré time scale. In an H-atom, it's the time it takes the electron to go around once. In a double-slit experiment, it's the time elapsed between the particle leaving the source and landing on the screen. That's one time scale. The other time scale that is relevant is the time in which the system leaks out a substantial amount of information to the environment. That's the time scale that's determined by the coupling constants. The point is, the quantum systems are such that the information-leakage time scales are very long compared to the dynamical time scale you see. The Schrödinger equation is therefore valid and interference can take place. In classical systems, it's the opposite. Phase information is destroyed too fast, before interference can happen.

DeWITT: I just wanted to make one historical comment. This is extremely good stuff. These problems were here sixteen years ago at the previous Osgood Hill meeting, and there has been real progress since then. But there's one important difference. I'm thinking of the discussion on the many-worlds interpretation at that conference, which was minimal, and Abner Shimony was certainly one of the opponents at that time. People had an open mind. But basically, they were against it. Why is there now so much readiness to accept this? Isn't it really because we've been concerned in the intervening years with such things as quantum cosmology and the interface between particle physics and cosmology? The other startling change that's taken place in these sixteen years is, well, I read an article the other day which said flatly that the reason that unified field theories have to be abandoned is because they do not incorporate gravity. Sixteen years ago, it was quite the other way around. Then, the prevalent view was: why bother with quantum gravity at all?

Loss of Quantum Coherence for a Damped Oscillator

W. G. Unruh

We have heard from both W. Zurek and from M. Gell-Mann at this conference on the importance of the external environment in destroying the off-diagonal quantum coherence of the density matrix expressed in what Zurek has called the pointer basis. This basis is one for which the appropriate projection operators at least approximately commute with the Hamiltonian of interaction between the system and the outside world.

Zurek and I have recently been examining a concrete example of such a process, which illustrates this loss of coherence most vividly. The example we have studied is that of a simple harmonic oscillator coupled to a heat bath. The heat bath is taken to be a one-dimensional scalar field, and the coupling between the oscillator and the heat bath is chosen so that the equations of motion of the oscillator contain the usual velocity-damping term.

We take the Lagrangian to be

$$L = \frac{1}{2} \int \left\{ [\dot{q}^2 + \Omega^2 q^2] + \int [\dot{\varphi}^2 - \varphi^2 + 2\varepsilon\,\dot{\varphi} q\,\delta(x)]\,dx \right\} dt.$$

Solving the equations of motion, we find

$$\varphi(t, x) = \varphi_0(t, x) + \frac{\varepsilon}{2} [q(t - x)\Theta(x) + q(t + x)\Theta(-x)].$$

Note that we have taken the retarded solution in keeping with our intention of specifying the *initial* rather than the final conditions for the field.

Substituting the equations for φ into the equation for the coordinate q of the oscillator, we find

$$\ddot{q} + \frac{\varepsilon^2}{4} \dot{q} + \Omega^2 q = \varepsilon\dot{\varphi}_0.$$

Note that these equations are the quantum Heisenberg equations as well as the classical equations for the oscillator.

Now, we assume that initially the system is in a product state:

$$\rho_T = \rho_H \times \rho_\varphi.$$

We calculate the reduced density matrix for the oscillator at any time by calculating

$$\rho_r(k, \Delta) = \text{Tr}\left[e^{i(kQ+\Delta P)}\rho_T\right],$$

where the Q and P are the Heisenberg operators for the position q and momentum p of the oscillator. The reduced density matrix in any other representation can be obtained from this one by appropriate manipulations. For example, the density matrix in the q, q' representation (where q is the eigenvalue of the operator Q) is given by

$$\rho(q, q') = \int e^{-ik(q+q')/2}\rho_r(k, (q-q')/2)dk;$$

in the p, p' representation, it is given by

$$\rho(p, p') = \int e^{-i\Delta(p+p')/2}\rho_r((p-p')/2, \Delta)d\Delta;$$

while the Wigner function $\rho(q, p)$ is the double-Fourier transform of $\rho_r(k, \Delta)$.

Let us examine this reduced density matrix in two cases in particular. In both, I take the initial state of the heat bath to be its vacuum state. The initial state of the oscillator is taken to be a "minimum uncertainty" pure state with the state chosen so that

$$\Delta q_0 = 8\,\Delta p_0/\Omega$$

in one case, and so that

$$\Delta q_0 = \Delta p_0/8\Omega$$

in the other. These both represent squeezed states. In the first, the squeezing stretches the state in the q direction; while in the second, it stretches the wave function in the p direction in comparison with the ground state of the oscillator.

Figures 1 and 2 show what happens to the density matrix in these two cases. The reduced density matrix for the oscillator remains Gaussian throughout. The ellipses in all of these diagrams represent the 1σ contours of the modulus, $|\rho|$, of the density matrix at a particular time and for each one of the representations of the density matrix. The first graph (a) in each set represents the Wigner function, with the track representing the motion of $\langle Q \rangle$ and of $\langle P \rangle$ over time, and the ellipses the contour of ρ at select instants of time.

The second plot, (b) in each set, gives the contours of the density matrix in the usual q, q' position representation at the same selected instants of time as in Figure (a), with the net motion of the "center of mass" taken out. Note that the off-diagonal terms in the density matrix represent off-diagonal quantum correlations.

The third plot (c) is similar to (b), except that it represents the density matrix in the p, p' representation.

Finally, the fourth plot (d) gives the entropy

$$S = \mathrm{Tr}\,(\rho \ln \rho)$$

of the reduced density matrix at various instants of time.

The parameters of the oscillator and the coupling are chosen so that the damping coefficient $\gamma = \varepsilon^2/2 = 0.1$. The oscillator is thus weakly coupled to the heat bath. Furthermore, we chose the frequency $\Omega = 1$.

It turns out that the system as written contains an ultraviolet divergence. We have thus taken a cutoff on the coupling of the oscillator to the heat bath of $\Gamma = 1000$. That is, frequencies in the heat bath over 1000 are taken not to couple to the oscillator.

The interesting features of these diagrams are the differences between the sets of Figure 1, corresponding to squeezing in the x direction, and those of Figure 2, corresponding to squeezing in the p direction. Looking at the Figures 1(b) and 2(b), we see that, in the former case, by the time represented by number 2 in the plots, the off-diagonal terms in the q, q' density matrix have been drastically reduced. The time corresponding to label 2 is less than .01 units after the initial condition corresponding to label 1. This can be interpreted as the heat bath's having made a partial measurement of Q (become correlated with the values of Q), destroying the off-diagonal coherence in the q, q' representation extremely rapidly.

The equivalent diagram in the case of squeezing in the p direction is that of Figure 2(c). Here, we see no reduction in the off-diagonal terms in the density matrix until times corresponding to labels 3 and 4. As can

Figure 1(a)

Figure 1(b)

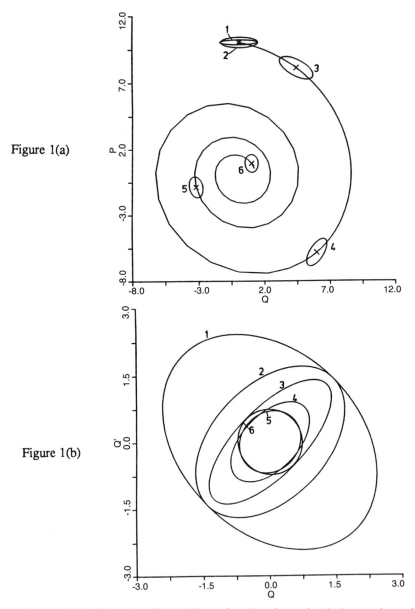

FIGURE 1 (a). The plot of the Wigner function for a simple harmonic oscillator with $\Gamma = .1$, $\Omega = 1.0$, $\langle Q \rangle_0 = 10$, $\langle P \rangle_0 = 0$, $\Delta q = 8 \Delta p / \Omega$. The ellipses are 1σ contours of modulus of the Wigner function at selected times. Figure 1(b). The plot of the 1σ contours of the modulus of the density matrix in the position representation at the labeled times of Figure 1(a). The mean motion has been removed. The labels refer to the times labeled in Figure 1(a).

Figure 1(c)

Figure 1(d)

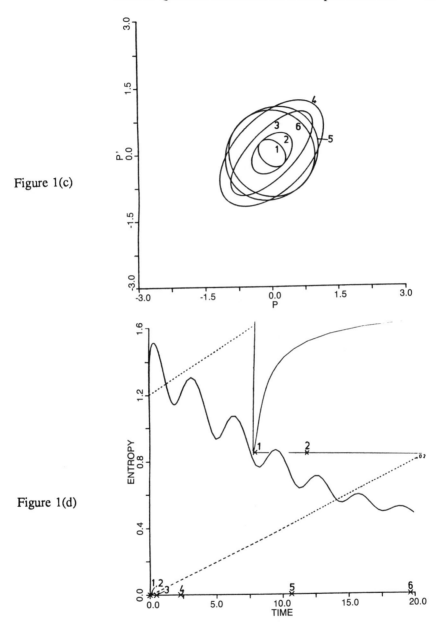

FIGURE 1 (c). Same as (b), except for the momentum representation of the density matrix. Figure 1 (d). Plot of the entropy of the reduced density matrix. The insert gives the entropy over the time interval 0 to .02, while the main plot gives the entropy over the interval 0 to 20. The X's on the axis with their labels refer to the times at which the density matrix was plotted in (a)–(c).

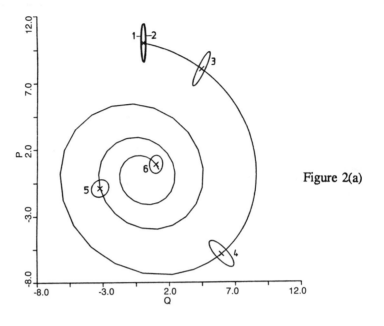

Figure 2(a)

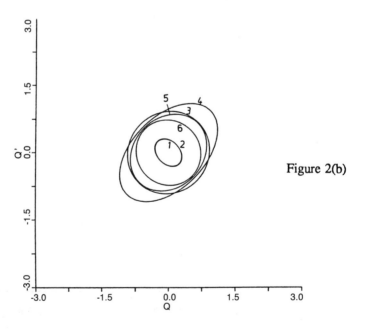

Figure 2(b)

FIGURES 2(a)–(d). The same as for Figure 1 except the initial density matrix had $\Delta q = \Delta p/8\Omega$.

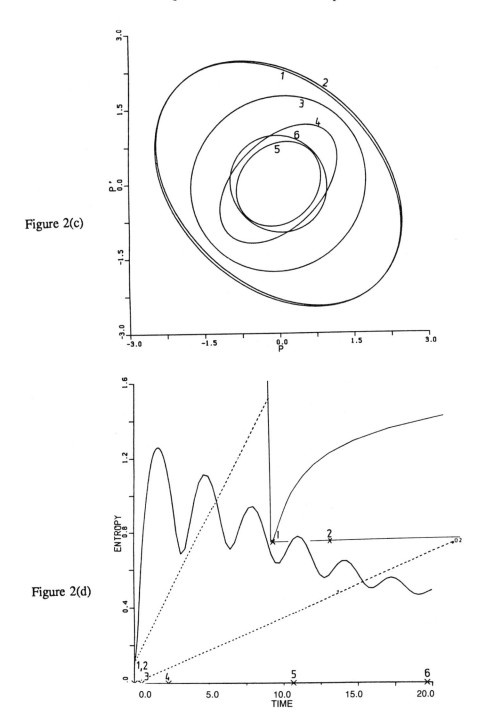

Figure 2(c)

Figure 2(d)

be seen in Figure 2(a), these correspond to times on the order of the oscillation time scale of the oscillator, the time during which the initial large momentum uncertainty has rotated into a corresponding position uncertainty, and thus has become "measurable" by the heat bath.

The coupling between the oscillator and the heat bath is via the position of the oscillator Q, and thus on short time scales Q approximately commutes with the Hamiltonian. In Zurek's words, Q is the "pointer basis" for this system. The coupling of the system to the heat bath leaves the on-diagonal terms in Q alone, while destroying the quantum coherence of the off-diagonal terms. On the other hand, as we can see in Figure 1(c), the on-diagonal terms in the momentum representation are altered, while the off-diagonal quantum coherences are left alone in this representation.

The final graph, Figures 1(d) and 2(d), show the change in the entropy of the reduced system. Because of the loss of the large off-diagonal coherences in the position representation of case 1, the entropy suffers a sudden (in less than .02 time units) increase, one which is matched in the second case only after about one quarter-oscillation time. The entropy increases to a value much higher than the final equilibrium value, and settles down to the equilibrium value only after a few damping time scales. Note also that the entropy does not settle down to 0, even though the heat bath is at zero temperature. This is an illustration of the "theorem" that a system coupled to a heat bath comes to equilibrium with a temperature equal to that of the heat bath only in the limit of infinitesimal coupling. Even our relatively small coupling results in an equilibrium entropy for the oscillator significantly different from the value of zero that one would naively expect for a zero temperature system.

For a more detailed discussion of this model and its lessons, I would draw your attention to the paper by W. Zurek and myself.

REFERENCE

Unruh, William G., and Zurek, Wojciech (1989). *Physical Review D* 40: 1071–1094.

The Wheeler–DeWitt Equation and the Path Integral in Minisuperspace Quantum Cosmology

Jonathan J. Halliwell

1. Introduction

Feynman's path-integral method provides a third formulation of quantum mechanics, very different from, but complementary to, the Schrödinger and Heisenberg canonical quantization methods (Feynman 1948). The relationship between this method and the canonical quantization methods has been well-explored for systems, such as nonrelativistic quantum mechanics, that are described by a time-dependent Schrödinger equation; that is, there exists a derivation of the Schrödinger equation from the path integral. Most field theories of current interest, however, involve constraints reflecting gauge invariance, or reparametrization invariance. The canonical quantization method is then that of Dirac, in which the constraints become operators that annihilate physical states (Dirac 1964). For these theories, the relationship between the Dirac quantization method and the path integral method does not seem to have been very well explored.

One of the most interesting constrained systems is general relativity. In its Hamiltonian formulation, the theory is portrayed as the dynamics of three-surfaces (see, for example, Hanson et al. 1976). There are four constraints, reflecting the invariance of the theory under four-dimensional diffeomorphisms. Three of the constraints, the momentum constraints, are linear in the momenta and generate diffeomorphisms within the three-surfaces. This symmetry of the theory is very similar to that of an ordinary gauge theory. The fourth constraint, however, the Hamiltonian constraint, is quadratic in the momenta, and it is this that distinguishes general relativity from an ordinary gauge theory. The Hamiltonian constraint expresses the invariance of the theory under time reparametrizations, but it also generates the dynamics; thus, the symmetry and dynamics of the theory are inextricably entangled. Because the quantization of ordinary gauge theories has been well studied, and because reparametrization invariance is a distinguishing feature of general relativity, it is of interest to those concerned with quantum gravity to study the quantization of simple models whose only symmetry is reparametrization invariance.

In this paper we will study in some detail the relationship between the Dirac and path integral quantization methods for a class of reparametrization-invariant theories described by an action of the form

$$S = \int_{t'}^{t''} dt[p_\alpha \dot{q}^\alpha - NH(p_\alpha, q^\alpha)]. \tag{1.1}$$

Here N is a Lagrange multiplier, which enforces the constraint $H = 0$. The simplest examples of systems described by such an action are the nonrelativistic point particle in parametrized form and the relativistic point particle, although the main examples we will be concerned with are the minisuperspace models of quantum cosmology. These are models for quantum gravity, in which one freezes all but a finite number of degrees of freedom of the metric. This is often done, for example, by writing the four-metric in the form

$$ds^2 = -N^2(t)dt^2 + h_{ij}dx^i dx^j, \tag{1.2}$$

where N is the lapse function, and an Ansatz is made for the three-metric h_{ij} such that it is homogeneous and is described by a finite number of functions $q^\alpha(t)$. The q^α would typically be scale factors, anisotropy parameters, etc. With this Ansatz, the momentum constraints of general relativity,

$$\mathcal{H}^i = -2\pi_{|j}^{ij} = 0, \tag{1.3}$$

are vacuously satisfied, and the Hamiltonian constraint

$$\mathcal{H} = G_{ijkl}\pi^{ij}\pi^{kl} - h^{1/2}(^3R - 2\Lambda) = 0 \tag{1.4}$$

reduces (after integration over the three-surface) to an expression of the form

$$H = \frac{1}{2}f^{\alpha\beta}(q)p_\alpha p_\beta + V(q) = 0. \tag{1.5}$$

Here, $f^{\alpha\beta}$ is the inverse of the metric on minisuperspace, $f_{\alpha\beta}$, and is of hyperbolic signature $(-,+,+,+\ldots)$. The system is therefore described by an action of the form (1.1). Some examples of minisuperspace models may be found in Amsterdamski 1985, Carow and Watamura 1985, González-Diaz 1985, Halliwell 1986, 1987a, 1987b, Hawking 1984, Hawking and Luttrell 1984a, 1984b, Horowitz 1985, Hu and Wu 1984, 1985, Louko 1987, Moss and Wright 1984, Vilenkin 1986 and Wu 1984.

In the Dirac quantization of this system, one introduces a wave function $\Psi(q)$, a function on minisuperspace. This function is annihilated by the operator version of the constraint (1.5):

$$\hat{H}\Psi(q) = 0, \tag{1.6}$$

where in (1.6) the momenta have been replaced by the corresponding operators in the usual manner. There is an amibiguity in the canonical quantization procedure because $f^{\alpha\beta}$ depends on q and one does not know how to distribute it between the momentum operators. This is the notorious operator-ordering problem. For gravitational systems, (1.6) is known as the Wheeler–DeWitt equation.

In the path-integral quantization method, the wave function is represented by an expression of the form

$$\Psi = \int \mathcal{D}p\,\mathcal{D}q\,\mathcal{D}N \exp\left(i \int dt[p_\alpha \dot{q}^\alpha - NH]\right). \tag{1.7}$$

Since the action (1.1) is reparametrization invariant, it is necessary to include ghost and gauge-fixing terms in (1.7) to ensure that equivalent histories are counted only once. It is then normally asserted that the expression (1.7) satisfies the Wheeler–DeWitt equation (1.6), although this has never been demonstrated in detail.

The purpose of this paper is to construct the path integral (1.7), including all the ghost and gauge-fixing terms, and to define it explicitly using a time-slicing procedure. We will then show, subject to certain assumptions, that it satisfies the Wheeler–DeWitt equation, and thus determine the relationship between the measure in (1.7) and the operator ordering in (1.6).

We begin in Section 2 by discussing the action, its reparametrization invariance, and gauge-fixing conditions that break it. The path integral for the propagation amplitude $G(q''|q')$ is constructed in Section 3, using the method of Batalin, Fradkin and Vilkovisky (BFV) (Fradkin and

Vilkovisky 1975, 1977; Batalin and Vilkovisky 1977; see also Henneaux 1985). In Section 4, a particular choice of gauge is made, and the integration over the ghost fields and Lagrange multipliers is performed. This yields a simple expression for the propagator, similar to the proper-time representation of Green functions of the Klein–Gordon equation (Parker 1978). In Section 5, the formalism is applied to the relativistic point particle and the nonrelativistic point particle in parametrized form. These examples clarify the range of integration of the Lagrange multiplier N (this is not fixed by the BFV procedure). In Section 6, it is shown that the issue of deriving the Wheeler–DeWitt equation reduces to a study of the derivation of the time-dependent Schrödinger equation in curved backgrounds. Previous work on this issue is described, and the relation between the path integral measure and the operator ordering in the Wheeler–DeWitt equation thus determined. Invariance under field redefinitions of the three-metric restricts the Wheeler–DeWitt equation to be of the form

$$\left(-\frac{1}{2}\nabla^2 + \xi R + V\right)\Psi = 0, \qquad (1.8)$$

where ∇^2 and R are the Laplacian and curvature in the minisuperspace metric, and ξ is an arbitrary constant. In Section 7, we evaluate the path integral exactly for a simple model, and show that it satisfies the Wheeler–DeWitt equation. In Section 8, we discuss the invariance of the quantization procedure under field redefinitions of the lapse function. We show that, for it to be invariant, one has to choose ξ in (1.8) in such a way that the operator $-\frac{1}{2}\nabla^2 + \xi R$ is conformally covariant. The operator ordering is thus completely determined. In Section 9, we discuss the problem of incorporating the condition that $\det h_{ij} > 0$ into the path-integral quantization procedure, a problem noted in the minisuperspace example of Section 7. In Section 10, we discuss the properties of the minisuperspace propagator. We show that the most naive composition law for it fails. We also show that the most naive inner product for the solutions to the Wheeler–DeWitt equation also fails. It is suggested that these features are due to the nature of time in quantum cosmology. In Section 11, we discuss the probability measure in quantum cosmology. It is argued that the measure derived from the Klein–Gordon current is the most reasonable one to use, and a possible way around the difficulties with negative probabilities, suggested by Vilenkin, is discussed. We summarize and conclude in Section 12.

This paper differs from a previous publication (Halliwell 1988) by the addition of Sections 10 and 11.

2. The Action

We consider an action of the form

$$S = \int_{t'}^{t''} dt[p_\alpha \dot{q}^\alpha - NH], \tag{2.1}$$

where α runs over D values and H is given by

$$H = \frac{1}{2} f^{\alpha\beta}(q) p_\alpha p_\beta + V(q). \tag{2.2}$$

Much of what follows will depend only on the fact that H is quadratic in the momenta. The p_α and N are free at the end points, and the q^α are fixed and satisfy the boundary conditions

$$q^\alpha(t') = q^{\alpha'} \qquad q^\alpha(t'') = q^{\alpha''}. \tag{2.3}$$

Variation with respect to p_α and q^α yields the field equations

$$\dot{q}^\alpha = N\{q^\alpha, H\} \qquad \dot{p}_\alpha = N\{p_\alpha, H\}, \tag{2.4}$$

and variation with respect to N yields the constraint

$$H(p_\alpha, q^\alpha) = 0. \tag{2.5}$$

The constraint reflects the most central feature of the action (2.1), which is that it is invariant under reparametrizations. More precisely, under the transformations

$$\delta q^\alpha = \varepsilon(t)\{q^\alpha, H\} \qquad \delta p_\alpha = \varepsilon(t)\{p_\alpha, H\} \qquad \delta N = \dot{\varepsilon}(t), \tag{2.6}$$

the action changes by an amount

$$\delta S = \left[\varepsilon(t)\left(p^\alpha \frac{\partial H}{\partial p^\alpha} - H\right)\right]_{t'}^{t''}, \tag{2.7}$$

where $\varepsilon(t)$ is arbitrary. Since, unlike the situation in gauge theories, the constraint H is quadratic in the momenta, (2.7) vanishes if and only if

$$\varepsilon(t') = 0 = \varepsilon(t''). \tag{2.8}$$

The action is therefore invariant under reparametrizations (2.6) subject to the boundary conditions (2.8).

Before constructing a quantum theory based on the action (2.1), it is necessary to impose gauge-fixing conditions that break the reparametrization invariance (2.6). As discussed by Teitelboim (Teitelboim 1982, 1983a, 1983b), the gauge-fixing condition must satisfy the following requirements:

(i) it must fix the gauge completely, i.e., there must be no residual gauge freedom;

(ii) using the transformations (2.6), it must be possible to bring any configuration, specified by p, q, N, into one satisfying the gauge condition.

These requirements are satisfied by a gauge-fixing condition of the form

$$\dot{N} = \chi(p, q, N), \tag{2.9}$$

where χ is an arbitrary function of p, q and N. This type of condition is analogous to the relativistic gauge-fixing condition used in ordinary gauge theories, $\partial^\mu A_\mu = 0$ (A_0 plays the role of a Lagrange multiplier in the Hamiltonian formulation of gauge theories). An alternative possibility is to use the analogue of a canonical gauge (e.g., $\partial^i A_i = 0$ or $A_3 = 0$), and this has been discussed by Hartle and Kuchař (Hartle and Kuchař 1984, 1986). For parametrized theories, this involves singling out one of the dynamical variables, q^0 say, as a physical time coordinate. One then imposes a condition of the form $q^0 = f(t)$, where f is an arbitrary monotonically increasing function of the time parameter t. The relationship between relativistic and canonical gauge-fixing conditions appears to be rather subtle for parametrized theories, and we hope to return to this issue in a future publication. The main difference is that the canonical condition restricts the paths to move forward in the direction of increasing q^0, whereas the condition (2.9) implies no such restrictions. In this paper we will work only with the condition (2.9).

The gauge condition (2.9) may be imposed at the level of the action using a Lagrange multiplier. One writes

$$S + S_{gf} = \int_{t'}^{t''} dt[p_\alpha \dot{q}^\alpha - NH + \Pi(\dot{N} - \chi)], \tag{2.10}$$

where $\Pi(t)$ is a Lagrange multiplier that, as we shall soon see, must vanish at the end points:

$$\Pi(t') = 0 = \Pi(t''). \tag{2.11}$$

Let us check that (2.10) yields the correct field equations. Variation with respect to p_α and q^α yields

$$\dot{q}^\alpha = N\{q^\alpha, H\} + \Pi\{q^\alpha, \chi\} \tag{2.12}$$

$$\dot{p}_\alpha = N\{p_\alpha, H\} + \Pi\{p_\alpha, \chi\}. \tag{2.13}$$

Variation with respect to Π yields the gauge condition (2.9), as intended, and variation with respect to N, subject to (2.11), yields the equation

$$\dot{\Pi} + H = 0. \tag{2.14}$$

Differentiating (2.14) with respect to t, and then using (2.12) and (2.13) one obtains the following equation for Π:

$$0 = \ddot{\Pi} + \dot{H} = \ddot{\Pi} + \{\chi, H\}\Pi. \tag{2.15}$$

Since, however, Π is required to vanish at both end points, the unique solution to (2.15) is $\Pi(t) = 0$ identically. The action (2.10) with the boundary conditions (2.11) thus yields the original field equations.

Finally, we note that the following exception to this conclusion exists. Equations (2.12) and (2.13) imply that $\{H, \chi\}$ is a constant, K say. If K just happened to take the value $\frac{n^2\pi^2}{(t''-t')^2}$, where n is an integer, then there would be nontrivial solutions to (2.15) satisfying the boundary conditions (2.11), namely the eigenfunctions of the operator $-\frac{d^2}{dt^2}$, $\sin\left(n\pi \frac{(t-t')}{(t''-t')}\right)$. However, we will eventually be working in the gauge $\chi = 0$, for which this exceptional case cannot arise.

3. The Path Integral

Given the action (2.10), one might be tempted to procede directly to a path-integral expression, since the action is now that of an unconstrained system. There is a good reason, however, why one cannot yet do this. This is that there is no guarantee the resulting path-integral expression will be independent of our choice of gauge-fixing function χ; or, in other words, we do not know what the measure is.

This problem is overcome by going to an extended phase space, which includes ghost fields. Batalin, Fradkin and Vilkovisky (BFV) have developed a very general method for doing this, based on BRS invariance (Fradkin and Vilkovisky 1975, 1977; Batalin and Vilkovisky 1977; see also Henneaux 1985). The basic idea is that one adds anticommuting ghost terms to the gauge-fixed action (2.10) in such a way that the resulting action is invariant under the global BRS symmetry. One then writes down a path integral in which the measure is taken to be the Liouville measure on the extended phase space (P, Q) consisting of the original bosonic variables plus the ghost fields. It is then possible to show, using a judiciously chosen BRS transformation on the variables (P, Q), that the path integral so constructed is independent of the choice of gauge-fixing function χ. This result is known as the Fradkin–Vilkovisky theorem.

Rather than write down the ghost action straight away, we shall try and proceed in a more systematic fashion, although it is in the nature of the subject that there are logical jumps that may only be totally justified retrospectively. BRS symmetry involves replacing the parameter $\varepsilon(t)$ in (2.6) with $\Lambda c(t)$ where Λ is a constant anticommuting parameter and $c(t)$ is an anticommuting ghost field. Eventually, we wish to generate the BRS transformations using a Poisson bracket, so it is necessary to eliminate the time derivatives from the transformations. One therefore writes $\dot{c} = \rho$, and this is imposed in the action by adding a term $\bar{\rho}(\dot{c} - \rho)$, where ρ and $\bar{\rho}$ are (anticommuting) ghost field momenta. To make ρ dynamical, one then adds a term $\bar{c}\dot{\rho}$. The (provisional) ghost action is therefore given by

$$S_{gh} = \int_{t'}^{t''} dt[\bar{\rho}\dot{c} + \bar{c}\dot{\rho} - \bar{\rho}\rho]. \tag{3.1}$$

With $\varepsilon = \Lambda c$ and $\dot{c} = \rho$, (2.6) gives the BRS transformations

$$\delta p_\alpha = -\Lambda c \frac{\partial H}{\partial q^\alpha}, \qquad \delta q^\alpha = \Lambda c \frac{\partial H}{\partial p_\alpha}, \qquad \delta N = \Lambda \rho. \tag{3.2}$$

Under these transformations, the action (2.1) changes by

$$\delta S = \int_{t'}^{t''} dt \Lambda(\dot{c} - \rho)H + \left[\Lambda c(t) \left(p_\alpha \frac{\partial H}{\partial p_\alpha} - H \right) \right]_{t'}^{t''}. \tag{3.3}$$

Since ε satisfies (2.8), we will impose the boundary conditions

$$c(t') = 0 = c(t''), \tag{3.4}$$

and thus the second term in (3.3) vanishes. The first term is not zero, however, since $\dot{c} = \rho$ only on shell.

The idea now is to find transformations of the remaining variables such that the total action is unchanged. The gauge-fixing term changes by

$$\delta S_{gf} = \int_{t'}^{t''} dt \left[\delta \Pi (\dot{N} - \chi) + \Pi \left(\Lambda \dot{\rho} - \Lambda c\{\chi, H\} - \Lambda \rho \frac{\partial \chi}{\partial N} \right) \right], \quad (3.5)$$

and the change in the ghost action is given by

$$\delta S_{gh} = \int dt [\delta \bar{\rho}(\dot{c} - \rho) + \delta \bar{c} \dot{\rho} + \bar{\rho}(\delta \dot{c} - \delta \rho) + \bar{c} \delta \dot{\rho}]. \quad (3.6)$$

If $\chi = 0$, it is easily seen that $\delta S + \delta S_{gf} + \delta S_{gh} = 0$ if one takes

$$\delta \Pi = 0, \quad \delta c = 0, \quad \delta \rho = 0, \quad \delta \bar{c} = -\Lambda \Pi, \quad \delta \bar{\rho} = -\Lambda H. \quad (3.7)$$

Moreover, for $\chi \neq 0$, it is not difficult to show that this is still the case if one adds the terms $c\{\chi, H\}\bar{c} + \rho \frac{\partial \chi}{\partial N} \bar{c}$ to the ghost action. The final form of the ghost action is, therefore,

$$S_{gh} = \int_{t'}^{t''} dt \left[\bar{\rho} \dot{c} + \bar{c} \dot{\rho} - \bar{\rho} \rho + c\{\chi, H\}\bar{c} + \rho \frac{\partial \chi}{\partial N} \bar{c} \right]. \quad (3.8)$$

The boundary conditions (3.4), supplemented with the conditions

$$\bar{c}(t') = 0 = \bar{c}(t''), \quad (3.9)$$

will ensure that the ghost fields vanish classically (ρ and $\bar{\rho}$ are free at the end points). To summarize: the total action is

$$S_T = S + S_{gf} + S_{gh}, \quad (3.10)$$

and is invariant under the BRS transformations (3.2), (3.7), subject to the boundary conditions (3.4). The action yields the field equations (2.4), (2.5) if one imposes the boundary conditions (2.3), (2.11), (3.4), and (3.9).

The BRS transformations may be concisely expressed by introducing the BRS charge $\Omega = cH + \rho \Pi$, and any BRS transformation is then of the form $\delta F = \{F, \Lambda \Omega\}$. The total action may now be written

$$S_T = \int_{t'}^{t''} dt \left[p_\alpha \dot{q}^\alpha + \Pi \dot{N} + \bar{\rho} \dot{c} + \bar{c} \dot{\rho} - \{\bar{\rho} N + \bar{c} \chi, \Omega\} \right]. \quad (3.11)$$

The BRS invariance is now manifest. The "Hamiltonian" term is invariant by virtue of the Jacobi identity (and $\{\Omega, \Omega\} = 0$) and, since the BRS transformation is a canonical transformation on the extended phase space, the "$P\dot{Q}$" term changes by at most a boundary term, which vanishes as a consequence of the boundary conditions.

Given the gauge-fixed BRS-invariant action, we may now proceed to the path integral. Let

$$G_\chi(q^{\alpha\prime\prime} \mid q^{\alpha\prime}) = \int \mathcal{D}\mu \exp(iS_T), \tag{3.12}$$

where

$$\mathcal{D}\mu = \mathcal{D}p_\alpha \, \mathcal{D}q^\alpha \, \mathcal{D}\Pi \, \mathcal{D}N \, \mathcal{D}\rho \, \mathcal{D}\bar{c} \, \mathcal{D}\bar{\rho} \, \mathcal{D}c. \tag{3.13}$$

The integral may be defined by a time-slicing procedure, and the measure (3.13) is then taken to be the Liouville measure $dP \wedge dQ$ on each time slice. The boundary conditions on the path integral are those discussed already: N, p_α, ρ and $\bar{\rho}$ are integrated over on every slice, including the end-point slices; c, $\bar{c}$ and π are fixed and equal to zero at the end-point slices; and q^α is fixed and satisfies (2.3) at the end-points. More will be said later about the details of the skeletonization and the ranges of integration.

The key point of the BFV approach is that it is now possible to show that (3.12) is independent of the choice of gauge-fixing function χ. This is achieved by changing variables in the path integral (3.12) from the variables (P, Q) to a new set of canonical variables $(\tilde{P}, \tilde{Q})$ that are related to the old ones by a BRS transformation (3.2), (3.7) with $\Lambda = -i \int dt \, \bar{c}(\tilde{\chi} - \chi)$. The action S_T is clearly invariant under this transformation; but, since Λ is a functional of the fields, the measure acquires a Jacobian factor:

$$\mathcal{D}\mu = \mathcal{D}\tilde{\mu} \exp\left(i \int dt \{\bar{c}(\chi - \tilde{\chi}), \Omega\}\right). \tag{3.14}$$

This factor has the effect of replacing χ by $\tilde{\chi}$ in (3.12). It follows that $G_\chi = G_{\tilde{\chi}}$, and thus the path integral is independent of χ, at least formally. This is the Fradkin–Vilkovisky theorem.

Certain qualifying remarks need to be made in relation to this result. Firstly, the result is only formal in that it involves some kind of generalized canonical transformation on the variables of integration, but without reference to a particular definition of the path integral — by skeletonization for example. It is very difficult to implement even genuine canonical transformations (i.e., those with $\Lambda = constant$) in phase-space path integrals, let alone the more general class of transformations involved above (see, for example, Fukutaka and Kashiwa 1988; Sheik 1987). It would be interesting to see if the above result can still be obtained using a skeletonized definition of (3.12).

The second point is that one has to be careful about the boundary conditions and the domain of integration in (3.12). It is important that the domain of integration is preserved by the above change of variables. For the bosonic variables, this will be true if they are integrated over a fully infinite range. As we shall see, however, one may wish to allow N to take a half-infinite range, in which case the theorem may not work. We hope to return to this point in a future publication.

4. The Ghost and Lagrange–Multiplier Integrations

Since the path integral is independent of χ, we may now make the gauge choice $\chi = 0$. As a result of this choice, the ghosts decouple from the other variables and the ghost integration may be performed. We define this integration by splitting the time interval into $n + 1$ equal intervals, $t'' - t' = \varepsilon(n + 1)$. With a particular choice of skeletonization, the functional integral over the ghost fields is

$$
\int \mathcal{D}\bar{\rho}\,\mathcal{D}c\,\mathcal{D}\rho\,\mathcal{D}\bar{c} \exp\left(i \int dt[\bar{\rho}\dot{c} + \bar{c}\dot{\rho} - \bar{\rho}\rho] \right)
$$

$$
= \int d\rho_{\frac{1}{2}} \ldots d\rho_{n+\frac{1}{2}} \int d\bar{\rho}_{\frac{1}{2}} \ldots d\bar{\rho}_{n+\frac{1}{2}} \int dc_1 \ldots dc_n \int d\bar{c}_1 \ldots d\bar{c}_n
$$

$$
\times \exp\left(i \sum_{k=0}^{n} \left[\bar{\rho}_{k+\frac{1}{2}}(c_{k+1} - c_k) + \rho_{k+\frac{1}{2}}(\bar{c}_{k+1} - \bar{c}_k) - \varepsilon \bar{\rho}_{k+\frac{1}{2}}\rho_{k+\frac{1}{2}} \right] \right),
$$

$$(4.1)$$

where $c_0 = 0 = c_{n+1}$ and $\bar{c}_0 = 0 = \bar{c}_{n+1}$. In this and the following expressions, it is implicit that the limit $\varepsilon \to 0$ is taken, although some of the path integrals are so simple that this is not in fact necessary. Also, the particular choice of skeletonization is not important for the simple path integrals of this and the next section. The path integrals for which it is important are discussed in Section 6.

The integrations in (4.1) are carried out according to the usual rules of Berezin integration (Berezin 1966). By shifting the variables of integration, the integration over the momenta may be performed, with the result

$$(i\varepsilon)^{n+1} \int dc_1 \ldots dc_n \int d\bar{c}_1 \ldots d\bar{c}_n$$

$$\cdot \exp\left(-\frac{i}{\varepsilon} \sum_{k=0}^{n} (\bar{c}_{k+1} - \bar{c}_k)(c_{k+1} - c_k)\right). \tag{4.2}$$

This in turn may be evaluated, to yield $(t'' - t')$. This is what one would expect, for the following reason. The path integral (4.1) is very similar to that for the free nonrelativistic point particle, for which we know that the propagator is proportional to $(t'' - t')^{-1/2}$. It differs, however, in that there are twice as many fields and they are Grassmannian. One would therefore expect (4.1) to be proportional to the squared inverse of the propagator of the point particle, which is indeed the case. The ghost integration produces precisely the factor required to ensure that the final result is independent of t'' and t', as we shall see.

We may also carry out the integrations over Π and N. Since Π is fixed at the end points, whilst N is integrated over, it seems appropriate to skeletonize Π as a coordinate and N as a momentum. That is, we define the path integral over the gauge-fixing term to be

$$\int \mathcal{D}N \, \mathcal{D}\Pi \exp\left(i \int dt \Pi \dot{N}\right)$$

$$= \int dN_{\frac{1}{2}} \ldots dN_{n+\frac{1}{2}} \frac{1}{(2\pi)^n} \int d\Pi_1 \ldots d\Pi_n \tag{4.3}$$

$$\cdot \exp\left(i \sum_{k=1}^{n} \Pi_k \left(N_{k+\frac{1}{2}} - N_{k-\frac{1}{2}}\right)\right),$$

where $\Pi_0 = 0 = \Pi_{n+1}$. This is equal to

$$\int dN_{\frac{1}{2}} \ldots dN_{n+\frac{1}{2}} \prod_{k=1}^{n} \delta\left(N_{k+\frac{1}{2}} - N_{k-\frac{1}{2}}\right). \tag{4.4}$$

We have n delta-functions and $n+1$ integrations, and thus the functional integration over $N(t)$ collapses to a single ordinary integration over N ($= N_{\frac{1}{2}}$ say). With these simplifications, the path integral (3.12) reduces to

$$G(q^{\alpha''} \mid q^{\alpha'}) = \int dN(t'' - t') \int \mathcal{D}p_\alpha \mathcal{D}q^\alpha \exp\left(i \int_{t'}^{t''} dt[p_\alpha \dot{q}^\alpha - NH]\right). \tag{4.5}$$

This is the main expression we will be working with in the following sections. It is essentially the integral over all time separations of an ordinary quantum-mechanical propagator, and thus bears a close resemblance to the proper-time representation for the Green functions of the Klein–Gordon equation (Parker 1978).

5. The Nonrelativistic and Relativistic Point Particles

We now show that the path integral expression (4.5) reproduces the familiar and expected results in two simple examples. These examples will also help establish the range of N since this is not fixed by the BFV procedure.

A. THE PARAMETRIZED NONRELATIVISTIC POINT PARTICLE

The first example is the nonrelativistic point particle in parametrized form. The usual action for the point particle is

$$S = \int_{t'}^{t''} dt \left[p_i \frac{dq^i}{dt} - h(p_i, q^i) \right], \tag{5.1}$$

where t is the preferred Newtonian time parameter and $i = 1, 2, 3$. The theory is put into parametrized form by introducing an arbitrary time parameter τ and then raising t to the status of a dynamical variable $t = q^0(\tau)$ with conjugate momentum $p_0(\tau)$, constrained to be equal to $-h(p_i, q^i)$. One thus adopts the action

$$S = \int_{\tau'}^{\tau''} d\tau \left[p_\alpha \frac{dq^\alpha}{d\tau} - N(p_0 + h(p_i, q^i)) \right], \tag{5.2}$$

where N is a Lagrange multiplier. This action is clearly of the form (2.1) with $\alpha = 0, 1, 2, 3$ and

$$H(p_\alpha, q^\alpha) = p_0 + h(p_i, q^i). \tag{5.3}$$

The path integral (4.5) for the system may be written

$$\begin{aligned}
G(q^{\alpha''} \mid q^{\alpha'}) = \int dN(\tau'' - \tau') \int \mathcal{D}p_i \, \mathcal{D}q^i \\
\cdot \exp\left(i \int d\tau \left[p_i \frac{dq^i}{d\tau} - N h(p_i, q^i) \right] \right) \\
\times \int \mathcal{D}p_0 \, \mathcal{D}q^0 \exp\left(i \int d\tau \left[p_0 \frac{dq^0}{d\tau} - N p_0 \right] \right).
\end{aligned} \tag{5.4}$$

The functional integral over p_i, q^i has the form of an ordinary quantum-mechanical propagator with time parameter $N\tau$:

$$\int \mathcal{D}p_i \mathcal{D}q^i \exp\left(i \int d\tau \left[p_i \frac{dq^i}{d\tau} - Nh(p_i, q^i)\right]\right) = \langle q^{i''}, N\tau'' \mid q^{i'}, N\tau'\rangle.$$

(5.5)

A natural choice of skeletonization for the functional integral over p_0, q^0 is

$$\int \mathcal{D}p_0 \, \mathcal{D}q^0 \exp\left(i \int d\tau \left[p_0 \frac{dq^0}{d\tau} - Np_0\right]\right)$$

$$= \int \frac{d(p_0)_{\frac{1}{2}}}{2\pi} \cdots \frac{d(p_0)_{n+(\frac{1}{2})}}{2\pi} \int dq_1^0 \ldots dq_n^0 \qquad (5.6)$$

$$\times \exp\left(i \sum_{k=0}^{n} (p_0)_{k+\frac{1}{2}}(q_{k+1}^0 - q_k^0 - \varepsilon N)\right),$$

where $\varepsilon = \frac{(\tau'' - \tau')}{(n+1)}$. The p_0 integrations yield

$$\int dq_1^0 \ldots dq_n^0 \prod_{k=0}^{n} \delta(q_{k+1}^0 - q_k^0 - \varepsilon N) = \delta(t'' - t' - N(\tau'' - \tau')),\ (5.7)$$

and our final expression for (5.4) is

$$G(q^{\alpha''} \mid q^{\alpha'}) = \int dN(\tau'' - \tau')\delta(t'' - t' - N(\tau'' - \tau'))$$

$$\times \langle q^{i''}, N(\tau'' - \tau') \mid q^{i'}, 0\rangle. \qquad (5.8)$$

Without loss of generality, let $(\tau'' - \tau') > 0$. Now consider the range of integration for N. Let us first suppose it is from $-\infty$ to $+\infty$. Then (5.8) yields

$$G(q^{\alpha''} \mid q^{\alpha'}) = \langle q^{i''}, t'' \mid q^{i'}, t'\rangle, \qquad (5.9)$$

so G is a solution to the Schrödinger equation:

$$\left(i\frac{\partial}{\partial t''} - \hat{h}''\right) G(q^{\alpha''} \mid q^{\alpha'}) = 0, \qquad (5.10)$$

where $\hat{h}''$ is the operator corresponding to h at $q^{i''}$. But now let the range of integration be 0 to $+\infty$. Then the delta function in (5.8) contributes only if $t'' - t' > 0$. It follows that

$$G(q^{\alpha''} \mid q^{\alpha'}) = \theta(t'' - t')\langle q^{i''}, t'' \mid q^{i'}, t'\rangle, \qquad (5.11)$$

so G is a Green function of the Schrödinger operator. That is,

$$\left(i\frac{\partial}{\partial t''} - \hat{h}''\right)G(q^{\alpha''} \mid q^{\alpha'}) = i\delta(t'' - t')\delta^{(3)}(q^{i''} - q^{i'}). \qquad (5.12)$$

B. THE RELATIVISTIC POINT PARTICLE

Our next example is the relativistic point particle, which is described by the action (2.1) with

$$H = \eta^{\mu\nu}p_\mu p_\nu + m^2, \qquad (5.13)$$

where $\mu = 0, 1, 2, 3$ and $\eta^{\mu\nu}$ is the usual Minkowski metric with signature $(- + ++)$. With a particular choice of skeletonization, (4.5) is

$$
\begin{aligned}
G(q^{\mu''} \mid q^{\mu'}) = &\int dN(\tau'' - \tau') \int d^4q_1 \ldots d^4q_n \int \frac{d^4p_{\frac{1}{2}}}{(2\pi)^4} \\
&\ldots \int \frac{d^4p_{n+(\frac{1}{2})}}{(2\pi)^4} \\
&\times \exp\left(i\sum_{k=0}^{n}\left[p_{k+\frac{1}{2}} \cdot (q_{k+1} - q_k)\right.\right. \\
&\qquad\qquad \left.\left. - \varepsilon N(p_{k+\frac{1}{2}} \cdot p_{k+\frac{1}{2}} + m^2)\right]\right),
\end{aligned}
\qquad (5.14)
$$

where $q_0 = q'$ and $q_{n+1} = q''$. The q integrations may be performed to yield n delta-functions, each of the form $\delta^{(4)}(p_{k+\frac{1}{2}} - p_{k-\frac{1}{2}})$; and then all but one of the p integrations may be performed, with the result

$$G(q^{\mu''} \mid q^{\mu'}) = \int \frac{d^4p}{(2\pi)^4} \int dT \exp\left(i\left[p \cdot (q'' - q') - T(p \cdot p + m^2)\right]\right), \qquad (5.15)$$

where $T = (\tau'' - \tau')N$.

If the range of N is $-\infty$ to $+\infty$, then the T integration may be performed, to yield

$$G(q^{\mu''} \mid q^{\mu'}) = \int \frac{d^4p}{(2\pi)^3}\delta(p \cdot p + m^2)e^{ip\cdot(q''-q')}. \qquad (5.16)$$

Clearly,

$$(-\Box'' + m^2)G(q^{\mu''} \mid q^{\mu'}) = 0, \qquad (5.17)$$

so G is a solution to the Klein–Gordon equation. If, on the other hand, the range of N is 0 to $+\infty$, then

$$G(q^{\mu\prime\prime} \mid q^{\mu\prime}) = -i \int \frac{d^4 p}{(2\pi)^4} \frac{e^{ip \cdot (q^{\prime\prime} - q^{\prime})}}{(p \cdot p + m^2 - i\varepsilon)}, \tag{5.18}$$

where the $i\varepsilon$ factor has been added to ensure convergence at the upper end of the T integration. This factor obliges one to integrate (5.18) along the contour in the p_0 plane corresponding to the Feynman Green function. G therefore satisfies

$$(-\Box'' + m^2)G(q^{\mu\prime\prime} \mid q^{\mu\prime}) = -i\delta^{(4)}(q^{\mu\prime\prime} - q^{\mu\prime}). \tag{5.19}$$

Both of these examples have been discussed from a somewhat different perspective, using a canonical gauge, by Hartle and Kuchař (Hartle and Kuchař 1984, 1986).

6. Quantum Mechanics in Curved Backgrounds

The results of the previous section concerning the range of N are quite general, as we now show. From there we can complete the derivation of the Wheeler–DeWitt equation.

The functional integral part of (4.5) has the form of an ordinary quantum mechanical propagator with time Nt:

$$\int \mathcal{D}p_\alpha \, \mathcal{D}q^\alpha \exp\left(i \int_{t'}^{t''} dt[p_\alpha \dot{q}^\alpha - NH]\right) = \langle q^{\alpha\prime\prime}, Nt'' \mid q^{\alpha\prime}, Nt' \rangle \tag{6.1}$$

Thus, introducing $T = N(t'' - t')$, (4.5) becomes

$$G(q^{\alpha\prime\prime} \mid q^{\alpha\prime}) = \int dT \langle q^{\alpha\prime\prime}, T \mid q^{\alpha\prime}, 0 \rangle. \tag{6.2}$$

Let us first take the range of T to be 0 to ∞. The propagator (6.1) will satisfy a time-dependent Schrödinger equation; so, operating on (6.2) with $\hat{H}''$, the Hamiltonian operator at q'', one obtains

$$\begin{aligned}
\hat{H}'' G(q^{\alpha\prime\prime} \mid q^{\alpha\prime}) &= \int_0^\infty dT \, i\frac{\partial}{\partial T} \langle q^{\alpha\prime\prime}, T \mid q^{\alpha\prime}, 0 \rangle \\
&= i\left[\langle q^{\alpha\prime\prime}, T \mid q^{\alpha\prime}, 0 \rangle\right]_0^\infty \\
&= -i\langle q^{\alpha\prime\prime}, 0 \mid q^{\alpha\prime}, 0 \rangle = -i\delta(q^{\alpha\prime\prime} \mid q^{\alpha\prime}).
\end{aligned} \tag{6.3}$$

Equation (6.2) is therefore a Green function of the Wheeler–DeWitt operator. Similarly, one may see that if T is integrated from $-\infty$ to $+\infty$, one obtains a solution to the Wheeler–DeWitt equation.

Clearly, certain assumptions are being made here concerning the behavior of the propagator (6.1) as $T \to \pm\infty$. If (6.1) is to be regarded as an ordinary function, then clearly it must go to zero as $T \to \pm\infty$ for (6.2) to converge. Typical examples of quantum-mechanical propagators behave like inverse powers of T multiplied by an oscillatory function, so do indeed have the desirable behavior. It could be, however, that (6.2) converges in a distributional sense, in which case the validity of the steps in (6.3), at least for the case $-\infty < T < \infty$, is more subtle. For the case $0 < T < \infty$, a Wick rotation to Euclidean time is possible, and the propagator (6.1) goes to zero exponentially fast as the Euclidean time approaches ∞. For the gravity case, the Wick rotation must also be accompanied by a conformal rotation in order to make the Hamiltonian positive (Gibbons et al 1978). Note that it does not seem to be possible to construct a Euclidean path integral if N has a fully infinite range.

We have now almost completed the derivation of the Wheeler–DeWitt equation. We have not, however, given a definition of the path integral in (6.1); nor have we given an explicit expression for the Hamiltonian operator $\hat{H}''$. The whole issue of deriving the Wheeler–DeWitt equation reduces, therefore, to that of constructing a skeletonized version of the path integral expression (6.1), and then determining the associated operator $\hat{H}''$ in the time-dependent Schrödinger equation

$$\left(i\frac{\partial}{\partial T} - \hat{H}'' \right) \langle q^{\alpha''}, T \mid q^{\alpha'}, 0 \rangle = 0. \tag{6.4}$$

We are thus led to study quantum mechanics in curved backgrounds, a subject that has been well studied over many years by numerous authors, including DeWitt (1957), Pauli (1973), DeWitt–Morette et al. (1980), Hartle and Hawking (1976), Cheng (1972), Parker (1979), and most recently by Kuchař (1983). Much of what follows constitutes a description of the work of the above authors, so it will be presented only in outline. The most comprehensive treatment is that of Kuchař (1983), and it is his that we will follow most closely.

In the usual canonical quantization procedure, there arises an operator-ordering ambiguity when one replaces the momenta p_α by the corresponding operators in (2.2). This is partially alleviated by demanding that the resulting Schödinger equation exhibit covariance under coordinate transformations of the q^α. One is then obliged to replace $f^{\alpha\beta}p_\alpha p_\beta$

with $-\hbar^2 \nabla^2$ in (2.2), where ∇^2 is the Laplacian in the metric $f^{\alpha\beta}$. However, the correct classical limit is still obtained, and covariance is still
respected, if one adds a curvature term $\hbar^2 R$, the scalar curvature of the
metric $f^{\alpha\beta}$. The most general form for $\hat{H}$ is, therefore,

$$\hat{H} = -\frac{\hbar^2}{2}\nabla^2 + \xi\hbar^2 R + V(q), \qquad (6.5)$$

where ξ is an arbitrary constant *whose value is not determined by the
canonical quantization procedure alone.* Particular values of ξ may be
preferred if there exist additional symmetries, such as conformal invariance or supersymmetry (Davis et al. 1984); but in general this arbitrariness remains, and the quantum theory is not unique. With this in mind,
let us turn to the path-integral description.

Historically, the path-integral formulation of quantum mechanics in
curved backgrounds began with configuration space path integrals, in
which one considers an expression of the form

$$\int \mathcal{D}q \exp(iS[q(t)]), \qquad (6.6)$$

where $S[q(t)]$ is the configuration space form of (2.1) (with $N = 1$):

$$S[q(t)] = \int_{t'}^{t''} dt \left[\frac{1}{2}f_{\alpha\beta}\dot{q}^\alpha \dot{q}^\beta - V(q) \right]. \qquad (6.7)$$

The expression (6.6) is normally defined by a time-slicing procedure.
The variables q, t are represented by discrete variables q_k, t_k, where
$k = 0, 1, \ldots, (n+1)$ and $t_0 = t'$, $t_{n+1} = t''$, $q_0 = q'$, $q_{n+1} = q''$.
Proper meaning is then given to (6.1) only after one has specified, in
terms of these discrete variables,

(i) a covariant skeletonization of the action $S[q(t)]$;

(ii) a covariant measure.

There is a natural choice for the first requirement: one uses the
Hamilton–Jacobi function $S(q'', t'' \mid q', t')$. This is equal to the action
of the classical path connecting (q', t') to (q'', t''). The Hamilton–Jacobi
function is a scalar in both of its arguments, and thus covariance is easy
to maintain using the following skeletonized approximation to the action:

$$S[q(t)] = \sum_{k=0}^{n} S(q_{k+1}, t_{k+1} \mid q_k, t_k). \qquad (6.8)$$

Adjacent vertebral points of the skeletonized path are therefore connected by classical paths.

For the measure, on the other hand, there is no unique natural choice. In fact, there is a whole family of acceptable measures. Parker, for example, used the following definition of the measure (Parker 1979):

$$\mathcal{D}q = \frac{1}{(2\pi i\varepsilon)^{\frac{D(n+1)}{2}}} \prod_{j=1}^{n} d^D q_j [f(q_j)]^{1/2} \prod_{k=0}^{n} [\Delta(q_{k+1}, q_k)]^p, \qquad (6.9)$$

where

$$\Delta(q_{k+1}, q_k) = [f(q_{k+1})]^{-1/2} \det\left[-\frac{\partial^2 S}{\partial q_{k+1} \partial q_k}\right] [f(q_k)]^{-1/2} \qquad (6.10)$$

is the Morette–Van Vleck determinant and $f = \det f_{\alpha\beta}$. This measure is covariant for all values of the arbitrary parameter p. It may be shown that the corresponding Schrödinger equation involves a Hamiltonian operator of the form (6.5) with $\xi = \frac{1}{3}(1 - p)$. We therefore see that the operator-ordering problem in the canonical quantization procedure appears in a configuration space path integral as an ambiguity in the choice of covariant measure.

It is sometimes claimed that such ambiguities in the measure can be resolved by going to a path integral in phase space, for which there is a privileged measure, namely the Liouville measure. This misconception was finally laid to rest by Kuchař (1983), who studied the covariant skeletonization of phase space path integrals for systems on curved backgrounds. Let us consider, therefore, an expression of the form

$$\int \mathcal{D}p_\alpha \, \mathcal{D}q^\alpha \exp\left(i \int dt[p_\alpha \dot{q}^\alpha - H]\right). \qquad (6.11)$$

Once again, the path integral is defined by a time-slicing procedure. The discrete representation of the q's is as above. For the p's, one introduces $n + 2$ discrete variables p_k, $k = 0, 1, \ldots, n + 1$. Only $n + 1$ of them are integrated over, because it always turns out that, as a result of the skeletonization, the integrand is independent of either p_0 or p_{n+1}. (Alternatively, one can introduce $n + 1$ variables $p_{k+\frac{1}{2}}$, $k = 0, 1, \ldots, n$ and integrate over all of them, but we shall use the former method in order to follow Kuchař.) Once again, one needs to specify both (i) and (ii). But now the difficulties are reversed. In phase space, there is indeed a privileged measure, the Liouville measure,

$$\mathcal{D}p_\alpha \, \mathcal{D}q^\alpha = d(p_\alpha)_0 \prod_{k=1}^{n} d(p_\alpha)_k dq_k^\alpha, \qquad (6.12)$$

and this ensures covariance under point transformations (the extra p integration means that the propagator is a scalar density at the initial endpoint). Kuchař observed, however, that there is no natural analogue of the Hamilton–Jacobi function. There is no function $S(q'',p'',t'' \mid q',p',t')$ equal to the action of a classical path with given q and p at both end points, since clearly such a path does not in general exist. Consequently, there is no unique natural skeletonization of the action.

Nevertheless, Kuchař showed how to skeletonize the action in a covariant manner by introducing a function $S(q'',t'' \mid q',p',t')$, defined as follows. The idea is that one first calculates the classical path $q(t)$ from (q',t') to (q'',t''). An arbitrary initial momentum, p' is then chosen (independently of $\dot{q}$), and this is extended to a function $p(t)$ by transporting it along the path $q(t)$ using a certain differential equation (e.g., the parallel transport equation, the geodesic deviation equation, etc.). The function $S(q'',t'' \mid q',p',t')$ is then defined to be

$$S(q'',t'' \mid q',p',t') = \int_{t'}^{t''} dt[p_\alpha(t)\dot{q}^\alpha(t) - H(p_\alpha(t),q^\alpha(t))]. \quad (6.13)$$

This function turns out to have all the properties one needs; in particular, it transforms covariantly under point canonical transformations, and we can now write down the covariant phase-space path integral:

$$\int d(p_\alpha)_0 \prod_{k=1}^{n} d(p_\alpha)_k dq_k^\alpha \exp\left(i \sum_{k=0}^{n} S(q_{k+1},t_{k+1} \mid q_k,p_k,t_k) \right). \quad (6.14)$$

The point is, however, that there is a whole class of functions $S(q'',t'' \mid q',p',t')$ that do the job. In fact, loosely speaking, there is a 1-parameter family. This ambiguity comes partly from the freedom to choose the differential equation with which to transport the momentum, but also from the freedom to make certain modifications to (6.13) whilst preserving covariance. On performing the momentum integration in (6.14), one reproduces the 1-parameter family of covariant measures

(6.9), to which correspond the 1-parameter family of Hamiltonian operators (6.5). The arbitrariness in the quantum theory expressed through the parameter ξ thus permeates the configuration-space and phase-space path-integral formulations, as well as the canonical quantization procedure. It is not fixed solely by covariance.

So we have described the relation between the skeletonized version of the path integral in (6.1) and the Hamiltonian operator (6.5), and our derivation of the Wheeler–DeWitt equation is complete. To summarize: with the skeletonizations described in Sections 4 and 6, the path integral expression (3.12) satisfies the equation

$$\hat{H}''G(q^{\alpha''} \mid q^{\alpha'}) = \begin{cases} -i\delta(q^{\alpha''} \mid q^{\alpha'}) & \text{if } 0 < N < \infty \\ 0 & \text{if } -\infty < N < \infty \end{cases}, \qquad (6.15)$$

where $\hat{H}''$ is of the form (6.5).

Finally, although the coefficient ξ does not appear to be fixed by covariance in the time-dependent Schrödinger equation, as we have gone to some effort to emphasize, we shall argue in Section 8 that it *is* fixed in the Wheeler–DeWitt equation, by demanding invariance under field redefinitions involving the lapse functions.

7. A Minisuperspace Example

We now consider a simple minisuperspace example. Consider a Robertson–Walker model described by the metric

$$ds^2 = \frac{-N^2(t)}{q(t)} dt^2 + q(t)d\Omega_3^2, \qquad (7.1)$$

where $d\Omega_3^2$ is the metric on the unit three-sphere. We will take the action to be the Einstein–Hilbert action with cosmological term. It is easily shown that the Hamiltonian then is (Louko 1987)

$$H = \frac{1}{2}(-4p^2 + \lambda q - 1). \qquad (7.2)$$

The path integral expression (4.5) for the propagation amplitude may therefore be written

$$G(q'' \mid q') = \int dT e^{-iT/2}$$
$$\times \int \mathcal{D}p\,\mathcal{D}q \exp\left(i \int_0^T dt \left[p\dot{q} - \frac{1}{2}(-4p^2 + \lambda q)\right]\right). \qquad (7.3)$$

One could proceed by first evaluating the functional integration over p and q, since this is a standard result— it is just the ordinary quantum-mechanical propagator for a system with a linear potential. However, the resulting T integration is rather difficult, and it turns out to be easier to perform the integrations in a different order.

Since the system is so simple, it is not necessary to resort to the sophistication of Section 6 to skeletonize the path integral. We will define the functional integral over p and q in (7.3) to be the limit as $\varepsilon \to 0$ of the expression

$$(2\pi)^{-n} \int dp_{\frac{1}{2}} \dots dp_{n+\frac{1}{2}} \int dq_1 \dots dq_n$$
$$\times \exp\left(i \sum_{k=0}^{n} \left[p_{k+\frac{1}{2}}(q_{k+1} - q_k) - \frac{\varepsilon}{2}\left(-4p_{k+\frac{1}{2}}^2 + \lambda q_k\right)\right]\right), \tag{7.4}$$

where as usual, $\varepsilon = T/(n+1)$ and $q_0 = q'$, $q_{n+1} = q''$. The q integrations are easily performed, with the result

$$(2\pi)^{-n} \int dp_{\frac{1}{2}} \dots dp_{n+\frac{1}{2}} \prod_{k=1}^{n} \delta\left(p_{k-\frac{1}{2}} - p_{k+\frac{1}{2}} - \frac{\varepsilon\lambda}{2}\right)$$
$$\times \exp\left(i\left[2\varepsilon \sum_{k=0}^{n} p_{k+\frac{1}{2}}^2 + p_{n+\frac{1}{2}}q'' - p_{\frac{1}{2}}q' + 0(\varepsilon)\right]\right). \tag{7.5}$$

The n delta functions imply that $p_{k+\frac{1}{2}} = p_{\frac{1}{2}} - k\varepsilon\lambda/2$. All but one of the p integrations may then be performed, to yield

$$\int dp \exp\left(i\left[2\varepsilon \sum_{k=0}^{n}\left(p - \frac{\varepsilon\lambda}{2}k\right)^2 + \left(p - \frac{\varepsilon\lambda}{2}n\right)q'' - pq' + 0(\varepsilon)\right]\right). \tag{7.6}$$

Carrying out the sums over k, inserting the result in (7.3) and letting $\varepsilon \to 0$, one obtains

$$G(q'' \mid q') = \int dT \int dp \times \exp\left(i\left[\left(2p^2 + \frac{(1 - \lambda q'')}{2}\right)T\right.\right.$$
$$\left.\left. - \lambda pT^2 + \frac{\lambda^2 T^3}{6} + p(q'' - q')\right]\right). \tag{7.7}$$

If desired, one could now carry out the p integration to yield the standard result for the propagator with a linear potential. However, as already stated, it turns out to be easier to integrate over T first. Let us take the range of T to be $-\infty$ to $+\infty$. Then one may write $T = \tilde{T} + 2p/\lambda$ and, after some algebra, one obtains

$$
\begin{aligned}
G(q'' \mid q') = \int_{-\infty}^{\infty} d\tilde{T} \int_{-\infty}^{\infty} dp \exp\left[\frac{i\lambda^2}{6}\tilde{T}^3 + \frac{i(1 - \lambda q'')}{2}\tilde{T}\right] \\
\times \exp\left[\frac{i4p^3}{3\lambda} + \frac{ip(1 - \lambda q')}{\lambda}\right].
\end{aligned}
\tag{7.8}
$$

G is therefore a product of two Airy functions (Abramowitz and Stegun 1965):

$$
G(q'' \mid q') = \frac{2\pi^2}{(4\lambda)^{1/3}} Ai\left[\frac{(1 - \lambda q'')}{(2\lambda)^{2/3}}\right] Ai\left[\frac{(1 - \lambda q')}{(2\lambda)^{2/3}}\right].
\tag{7.9}
$$

Since the curvature vanishes in one dimension, there is no operator-ordering ambiguity, and the Wheeler–DeWitt equation is

$$
\hat{H} G = \frac{1}{2}\left(4\frac{d^2}{dq^2} + \lambda q - 1\right) G = 0.
\tag{7.10}
$$

It is easy to verify that (7.9) is a solution to (7.10) as expected.

But we know that there is a second solution to (7.10). How do we generate it from the path integral? One way is to observe that the Wheeler–DeWitt equation (7.10) is unchanged by the transformation $(1 - \lambda q) \rightarrow e^{2\pi i/3}(1 - \lambda q)$. It follows that applying this transformation to the above solution will yield a second solution. It is

$$
Ai\left[e^{2\pi i/3}z\right] = \frac{1}{2}e^{\pi i/3}(Ai(z) - iBi(z)),
\tag{7.11}
$$

where $z = (1 - \lambda q)(2\lambda)^{-2/3}$. This second solution is clearly linearly independent of the first. Another way of generating a second solution is to integrate along a different contour in (7.8). For example, the contour running from $-\infty$ to 0 and then from 0 to $-i\infty$ yields (7.11). We have thus seen that we can use the path integral to generate a complete set of solutions to (7.10).

In performing the above calculation, however, there is an important restriction that we have failed to recognize. This is that q, being a scale factor in (7.1), is positive. Strictly speaking therefore, the problem should be treated using the formulation of quantum mechanics appropriate to restricted intervals, which is actually rather difficult. Part of the problem is that, in writing down the skeletonization (7.4), it is implicitly assumed that both p and q are integrated from $-\infty$ to $+\infty$. We will defer further discussion of this point until later.

For the moment, however, we note that the condition that $q > 0$ is not a problem if, as we have been doing here, one is simply using the path integral to generate solutions to the Wheeler–DeWitt equation. The point is that there is no mathematical inconsistency in allowing q to take a fully infinite range — one can evaluate the path integral, and solve the Wheeler–DeWitt equation. The fact that the physically relevant range of q is $q > 0$ means only that one is required to find solutions satisfying prescribed boundary conditions at the end points of the physically relevant range, i.e., at $q = 0$ and $q = \infty$. But since we have generated a complete set of solutions, this can clearly be achieved.

If one is just using the path integral to generate solutions, therefore, the fact that the physically relevant range of q is $q > 0$ presents no problem. The difficulties arise, however, when one tries to incorporate the positivity of q into the path integral from the very beginning. This will be discussed in Section 9. Before that, however, we reconsider the result (6.15) in the light of the example of this section.

8. Invariance Under Field Redefinitions

A reasonable property to demand of any quantum theory is that it be insensitive to the way we choose to define the fields involved; that is, it should be invariant under field redefinitions. The construction of Section 6 guarantees that this is the case for redefinitions of the three-metric, represented by the q's. More precisely, if one chose to perform the calculation with a different set of variables $\bar{q}(q)$, then the answer would be the same as that obtained by performing the calculation in terms of q and then substituting for $\bar{q}$ at the end. At the level of the Wheeler–DeWitt equation, this has been achieved by demanding that all quantities constructed from the metric $f_{\alpha\beta}$ be covariant.

As is stands, however, the formalism is not invariant under field redefinitions involving the lapse function N. Since any reference to N is absent in (6.15), this is not so obvious, but it may be seen by considering a simple example. Consider the minisuperspace example of the previous section for the case $\lambda = 0$, with N defined by the metric (7.1). It is described by the action

$$S = \int dt \left[p\dot{q} - \frac{N}{2} (-4p^2 - 1) \right].$$

(8.1)

The Wheeler–DeWitt equation is

$$\left[4\frac{d^2}{dq^2} - 1 \right] \Psi(q) = 0,$$

(8.2)

with solutions $\Psi(q) = e^{\pm q/2}$. Suppose, however, one uses a new lapse function $\tilde{N}$, defined by $\tilde{N} = q^{-1}N$. The four-metric is now given by

$$ds^2 = q(t)\left[-\tilde{N}^2(t)dt^2 + d\Omega_3^2 \right],$$

(8.3)

and the action is

$$S = \int dt \left[p\dot{q} - \frac{\tilde{N}}{2} (-4qp^2 - q) \right].$$

(8.4)

Classically, this action is just as good as (8.1). In each, N and $\tilde{N}$ are regarded as independent of q. The Wheeler–DeWitt equation corresponding to (8.4), however, is

$$\left(4q^{1/2}\frac{d}{dq}q^{1/2}\frac{d}{dq} - q \right) \tilde{\Psi}(q) = 0,$$

(8.5)

where, in accordance with the formalism developed so far, we replaced $-qp^2$ with the Laplacian. The operator in (8.5) is clearly not equivalent to that in (8.2) since it differs by first-derivative terms. Indeed, the solutions to (8.5) are given in terms of modified Bessel functions (Abramowitz and Stegun 1965): $\tilde{\Psi}(q) = q^{1/4}I_{1/4}\left(\frac{q}{2}\right)$ and $\tilde{\Psi}(q) = q^{1/4}K_{1/4}\left(\frac{q}{2}\right)$. There is no obvious exact relationship between these solutions and the solutions to (8.2) although the dominant terms in $\tilde{\Psi}(q)$ are of the form $e^{\pm q/2}$ for large q, so they do at least agree semi classically, as one would expect. But the point is that the Wheeler–DeWitt equations describe different quantum theories, so the procedure is not invariant under field redefinitions involving the lapse function.

More generally, the problem may be presented as follows. We showed that the action

$$S = \int dt \left[p_\alpha \dot{q}^\alpha - N\left(\frac{1}{2}f^{\alpha\beta}p_\alpha p_\beta + V(q) \right) \right]$$

(8.6)

corresponds to the Wheeler–DeWitt equation

$$H\Psi = \left[-\frac{1}{2}\nabla^2 + \xi R + V(q)\right]\Psi(q) = 0. \tag{8.7}$$

Suppose, however, one defines a new lapse function $\tilde{N}$, given by $N = \Omega^{-2}\tilde{N}$, where Ω is an arbitrary function of q. Ω may be absorbed into the Hamiltonian by defining $\tilde{f}^{\alpha\beta} = \Omega^{-2}f^{\alpha\beta}$ and $\tilde{V}(q) = \Omega^{-2}V(q)$. The action is then given by

$$S = \int dt\left[p_\alpha \dot{q}^\alpha - \tilde{N}\left(\frac{1}{2}\tilde{f}^{\alpha\beta}p_\alpha p_\beta + \tilde{V}(q)\right)\right]. \tag{8.8}$$

This is completely equivalent to (8.6) at the classical level, and there is no obvious reason why it should not be used as a starting point for quantization. The corresponding Wheeler–DeWitt equation is

$$\tilde{H}\tilde{\Psi} = \left[-\frac{1}{2}\tilde{\nabla}^2 + \xi\tilde{R} + \tilde{V}(q)\right]\tilde{\Psi}(q) = 0, \tag{8.9}$$

where $\tilde{\nabla}^2$ and $\tilde{R}$ are, respectively, the Laplacian and scalar curvature for the metric $\tilde{f}_{\alpha\beta}$.

For (8.7) and (8.9) to describe the same quantum theory, one would need $\tilde{H}\tilde{\Psi}$ to be proportional to $H\Psi$ (so that (8.9) implies (8.7), and vice-versa), where $\tilde{\Psi}$ and Ψ are related in a simple way. This will not in general be true. Nevertheless, let us look for a relationship between $\tilde{\Psi}$ and Ψ of the form $\tilde{\Psi}(q) = \Omega^\gamma\Psi(q)$, for some constant γ. Then a standard calculation shows that

$$\tilde{H}\tilde{\Psi} = \Omega^{\gamma-2}H\Psi - \frac{1}{2}(2\gamma + D - 2)\Omega^{\gamma-3}\nabla\Omega \cdot \nabla\Psi$$
$$+ \Omega^{\gamma-2}\left[A\Omega^{-1}\nabla^2\Omega + B\Omega^{-2}(\nabla\Omega)^2\right]\Psi, \tag{8.10}$$

where the dot product is with respect to the metric $f^{\alpha\beta}$ and

$$A = -\frac{\gamma}{2} + 2(D-1)\xi$$
$$B = -\frac{1}{2}(\gamma(\gamma-1) + \gamma(D-2)) + \xi(D-1)(D-4). \tag{8.11}$$

Recall that D is the dimension of the minisuperspace. Equation (8.10) shows that (8.7) and (8.9) are not in general equivalent for arbitrary γ and

ξ. However, if we choose $\gamma = (2 - D)/2$, then the coefficient of $\nabla\Omega\cdot\nabla\Psi$ vanishes, and (8.7) and (8.9) differ only in their potentials. Moreover, with this choice the coefficients A and B are then both proportional to $(D - 2) + 8(D - 1)\xi$. The main point now is that ξ is totally arbitrary — it is not fixed by demanding covariance in the q's, as we went to some length to emphasize in Section 6. We are therefore free to make the choice

$$\xi = -\frac{(D - 2)}{8(D - 1)}, \tag{8.12}$$

which implies that $A = B = 0$, and thus $\tilde{H}\tilde{\Psi} = \Omega^{\gamma - 2}H\Psi$. This means that (8.7) and (8.9) are equivalent, so the quantization procedure *can* be made invariant under field redefinitions of N, providing ξ is chosen to take the value (8.12). (A similar conclusion concerning the value of ξ, but from a different perspective, was reached in Misner 1970. See also Moss 1988.)

Equation (8.12) of course gives the value for which the operator $-\frac{1}{2}\nabla^2 + \xi R$ is conformally covariant, and the above calculation showing that this is the case is very standard. Equation (8.10) is written out explicitly so that one can see that the case $D = 1$ is exceptional. Although we motivated the discussion by a one-dimensional example, the above procedure does not work for $D = 1$. In one dimension, there is no curvature, so the terms proportional to ξ are absent from (8.10) and (8.11). One can then either choose $\gamma = 0$, so that $A = B = 0$, or one can choose $\gamma = 1/2$, so that the coefficient of $\nabla\Omega \cdot \nabla\Psi$ vanishes. But one way or another, extra terms still remain, and it does not seem to be possible to achieve invariance under field redefinitions of N for $D = 1$, at least by this approach.

In conclusion, the operator-ordering ambiguity in the Wheeler–DeWitt equation for $D > 1$ is completely fixed by demanding invariance under field redefinitions of both the three-metric and the lapse function. Note, however, that we have pursued the issue of invariance under rescalings of N only at the level of the canonical quantization procedure, not at the path integral level, although this is presumably not too difficult.

9. Quantum Mechanics on a Half-Infinite Range

In Section 7 we encountered the problem of doing quantum mechanics in terms of the variable q, whose physical range was the positive real line. This problem is not in any way an artifact of the particular model under consideration, but is a manifestation of the fact that the three-metric h_{ij} satisfies the condition $\det h_{ij} > 0$. It is therefore important to face up to this issue from the very beginning.

In the canonical quantization of gravity, this sort of difficulty was recognized a long time ago by Klauder (1970) and Klauder and Aslaksen (1970). To see the problems that arise, consider a simple one-dimensional system with coordinate q restricted to lie on the positive real line, R^+. The first problem one discovers is that the momentum operator is not Hermitian. To see that this is the case, consider the operator obtained by exponentiating the momentum operator (times i). If the momentum operator were Hermitian, one would obtain the unitary translation operator, with which one could translate into the region $q < 0$. The momentum operator cannot therefore, be Hermitian. Hermiticity of the Hamiltonian can be preserved, however, by imposing the boundary conditions $\psi'(0) + \alpha\psi(0) = 0$ at $q = 0$, where α is an arbitrary parameter. Normally, one envisages a physical situation in which the conditions at $q = 0$ are known, and thus the value of α is given.

Quantum-mechanical propagators may be constructed using an eigenfunction expansion. One finds the eigenfunctions $u_n(q)$ of the Hamiltonian subject to the above boundary conditions, and subject to the usual fall-off conditions at infinity. The propagator is then given in terms of these functions by an expression of the form

$$\langle q'', t'' \mid q', t' \rangle = \sum_n e^{-iE_n(t''-t')} u_n^*(q'') u_n(q'). \qquad (9.1)$$

The above boundary conditions will then be incorporated into the propagator.

In the path-integral approach, the problems with the restriction to $q > 0$ appear as difficulties with the skeletonization procedure. An essential property of a skeletonized path integral is that it yields the correct WKB approximation to the propagator at short time separations. The configuration-space path integral described in Section 6 with the measure (6.9) will generally have this property. Recall, however, that this construction involves the action of the classical path connecting adjacent vertebral points of the lattice. It is implicitly assumed that this classical path is unique. The problem is that this will not be true in the presence of a boundary at $q = 0$. For then, there will be two classical paths: the direct path, and a second path that is reflected off $q = 0$. This means that the WKB approximation to the propagator is not a single expression of the form e^{iS} but a sum of two such terms. The path-integral construction of Section 6 will not, therefore, reduce to the WKB approximation to the propagator at short time separations, because it involves only one factor of the form e^{iS} on each time slice. And, in general, it does not seem to be possible to construct a path-integral representation of the propagator using the usual skeletonization procedure but with q integrated from 0 to ∞ on each slice.

These problems do not mean that a skeletonized definition of the path integral consistent with the restriction $q > 0$ is not possible. Such skeletonizations are possible although they are rather complicated and will not be pursued here (Klauder 1970, Klauder and Aslaksen 1970). We merely point out that the standard skeletonization used for the case $-\infty < q < \infty$ cannot be naively applied to the case $q > 0$.

At this point, one might think that the above difficulties could be alleviated by a change of variables. One could write $q = e^x$ for example, for then x takes a fully infinite range. However, this does not work. The point is that the difficulty is not so much that $q > 0$, but rather, the fact that there exists more than one classical path connecting any two points. If, in terms of q, there are two classical paths, a direct and a reflected path, then the same will be true of x. In terms of x there will be, in addition to the direct path, a second classical path which goes to $x = -\infty$ and back in a finite period of time; thus, once again there will be two terms of the form e^{iS} in the WKB approximation to the propagator.

This example illustrates that, even if the variables one is working with have a fully infinite range, one has to do a careful analysis of the classical solutions before using the path integral to construct the propagator. Or to put it another way, one cannot escape global problems by a change of coordinates.

In the case of gravity, it has been suggested that one should work not with the three-metric h_{ij} but with the Dreibein e_i^a, where $h_{ij} = e_i^a e_j^b \delta_{ab}$, $a, b = 1, 2, 3$, for then the inequality constraint $\det h_{ij} > 0$ is automatically satisfied without any restrictions on the range of the Dreibein (Isham 1976). In the example of the previous section, this corresponds to writing $q = a^2$ and letting a take an infinite range. Once again, however, the problems still arise, because there still exist reflected paths, so there is no easy way round the difficulty in the path-integral approach. Nevertheless, these changes of variables are still useful in the canonical quantization procedure since the momentum conjugate to a variable taking an infinite range may be represented by a Hermitian operator.

These issues simply mean that it is difficult to incorporate the fact that

$q > 0$ into the path integral from the beginning. They do not prevent one from making good use of the path integral, as indeed we did in Section 7. For example, one can calculate the quantum-mechanical propagator when $q > 0$ using the method of images. First, one extends the potential into the region $q < 0$ in such a way that $V(q) = V(-q)$. One then uses the path integral to calculate the propagator $\langle q'', t \mid q', 0 \rangle$ letting the q's take a fully infinite range. This propagator will be a solution to the Schrödinger equation, and will be a delta function at $t = 0$. By the symmetry of the potential, a second such solution is given by $\langle q'', t \mid -q', 0 \rangle$, and these two solutions may be superposed to satisfy prescribed boundary conditions at $q = 0$. The propagator in the region $q > 0$ is thus obtained.

Finally, it is appropriate to remark on the validity of the skeletonization of the lapse-function integral in Section 4, because there we allowed N to be integrated over a half-infinite range. As mentioned above, the skeletonization procedure involves considering the classical path between two points of the lattice, and this was problematic for q because of the existence of reflected paths. For N, however, the "classical field equation" is just the gauge condition, $\dot{N} = 0$; thus, there can be only one path connecting N_k to N_{k+1}. There are no reflected paths, so the above problems do not arise, at least for the gauge choice we have been using.

10. Properties of the Propagator

In this section, now that we have an explicit expression for the minisuperspace propagator, we investigate some of its properties. The minisuperspace propagator (6.2) is given in terms of the quantum-mechanical propagator $\langle q'', T'' \mid q', 0 \rangle$. The quantum-mechanical propagator obeys the composition law

$$\int dq \langle q'', T'' \mid q, 0 \rangle \langle q, T' \mid q', 0 \rangle = \langle q'', T'' + T' \mid q', 0 \rangle. \qquad (10.1)$$

This composition law is a direct consequence of a special property of the paths over which one sums in the path integral representation (6.1): they move forwards in time. It is from this property that the familiar Hilbert space structure of non-relativistic quantum mechanics follows. An extensive discussion of this point is given by Hartle (Hartle 1988a, 1988b, 1990).

One might hope that a similar composition law holds for $G(q'' \mid q')$, and indeed, numerous authors in the past have assumed that it does. It is simple to show, however, that this is not the case. It follows from (6.2) and (10.1) that

$$\int dqG(q'' \mid q)G(q \mid q') = \int dqdT''dT'\langle q'',T'' \mid q,0\rangle\langle q,T' \mid q',0\rangle$$

$$= \int dT''dT'\langle q'',T'' + T' \mid q',0\rangle.$$

$$(10.2)$$

Writing $u = T'' + T'$ and $v = T'' - T'$, (10.2) becomes

$$-\frac{1}{2}\int dudv\langle q'', u \mid q',0\rangle. \qquad (10.3)$$

If T is taken to have an infinite range, then u and v have an infinite range, and (10.3) is equal to $G(q'' \mid q')$, but multiplied by an infinite factor. If T is taken to have a half-infinite range, then v ranges from $-u$ to u, and (10.3) becomes

$$-\int_0^\infty duu\langle q'', u \mid q',0\rangle. \qquad (10.4)$$

This may converge, but it will not converge to $G(q'' \mid q')$. One way or another, we conclude that

$$\int dqG(q'' \mid q)G(q \mid q') \neq G(q'' \mid q'), \qquad (10.5)$$

so a naive composition law does not hold.

Similarly, one can show that a naive inner product on the set of solutions to the Wheeler–DeWitt equation also fails. If T is taken to have an infinite range, then

$$\Psi_1(q) = \int dq'G(q \mid q')\Phi_1(q') \qquad (10.6)$$

is a solution to the Wheeler–DeWitt equation for arbitrary Φ_1. One might attempt to use the inner product

$$\int dq\,\Psi_1^*(q)\Psi_2(q) = \int dq''dq'dqG^*(q \mid q')G(q \mid q'')\Phi_1^*(q')\Phi_2(q'') \qquad (10.7)$$

However, this diverges because, as we have just shown, the integral over q diverges.

So we have seen that a naive attempt to force a Hilbert-space structure on the formalism does not work. This is perhaps due to the nature of time in quantum gravity, about which we have heard much at this meeting. Conventional quantum theory involves observations at a moment of time. The usual composition law involves integrating out the spatial coordinates on a particular time slice, as does the usual inner product. Thus, the naive composition law and inner product we have considered here are analogous to integrating over not only the spatial coordinates in non-relativistic quantum mechanics, but also over time.

It could be that time is effectively contained among the D coordinates q^α. One might then expect that the correct composition law and inner product will involve an integration over just $D-1$ of the q^α's. Indeed, the quantum mechanics of a relativistic point particle provides an example of a theory in which this is the case, and a Hilbert-space structure does exist. However, that this is so is due to the fact that time can be explicitly identified amongst the variables one uses to formulate the theory (Hartle and Kuchař 1984, 1986). In quantum cosmology, on the other hand, no one has yet succeeded in identifying one of the variables as a satisfactory time coordinate. The existence of a Hilbert-space structure for canonical quantum gravity is still very much an open question, and will not be further pursued here.

11. The Probability Measure

In this section we discuss the issue of the probability measure in quantum cosmology. This follows on from the discussion in Section 8, but it is also related to the discussion in the previous section about time in quantum cosmology.

It is probably fair to say that there is as yet no totally satisfactory scheme for interpreting the wave function of the universe. We do not offer such a scheme here. However, it seems reasonable that any such scheme must involve some kind of probability measure constructed from the wave function. It is this that we now discuss.

The situation one is most frequently interested in is that in which the wave function is oscillatory, of the form $\Psi = Ce^{iS}$. It can be shown that such a wave function is peaked around a set of classical solutions satisfying the first integral $\dot{q}_\alpha = p_\alpha = \partial S/\partial q^\alpha$ (see, for example, Halliwell 1988). One would like to have some kind of measure on this set of solutions.

A measure for this purpose was proposed by Hawking and Page (1986). It is

$$dP = |\Psi(q)|^2 (-f)^{1/2} d^D q. \tag{11.1}$$

There are, however, a number of reasons why one might object to this choice. One is that, as discussed in the previous section, time is implicitly contained amongst the metric components q^α. The measure (11.1) is therefore analogous to using the measure

$$dP = |\Psi(\mathbf{x}, t)|^2 d^3 x \, dt \tag{11.2}$$

in nonrelativistic quantum mechanics and taking it to be the probability of finding the particle between $\mathbf{x}$ and $\mathbf{x} + d\mathbf{x}$ and between t and $t + dt$. This is contrary to the usual procedure in quantum mechanics, which concerns probabilities at a given moment of time.

We also note that the measure (11.1) is not normalizable on the full superspace as follows from (10.7). This is not necessarily a problem, however, since one is usually interested in conditional probabilities, which involves looking at probability distributions within a limited region of superspace.

A further reason for rejecting the Hawking–Page measure is that it is not invariant under the rescaling transformations described in Section 8. Explicitly,

$$|\tilde{\Psi}(q)|^2 (-\tilde{f})^{1/2} = \Omega^2 |\Psi(q)|^2 (-f)^{1/2}. \tag{11.3}$$

This implies that the predictions one makes using this measure depend on the essentially arbitrary way in which one defines the fields, which is clearly unacceptable.

Finally, Vilenkin has observed that, when one considers inhomogeneous fluctuations about minisuperspace, the measure (11.1) does not imply the correct probability measure for the wave functions of the fluctuations. These fluctuation wave functions obey time-dependent Schrödinger equations, and the Hawking–Page measure implies a measure of the form (11.2) for them, which is obviously unsatisfactory.

Given all these arguments against the Hawking–Page measure, we are faced with the question of an alternative choice. A natural alternative to try is the Klein–Gordon current associated with the Wheeler–DeWitt equation (Vilenkin 1986):

$$J^\alpha = i f^{\alpha\beta} (\Psi \nabla_\beta \Psi^* - \Psi^* \nabla_\beta \Psi). \tag{11.4}$$

It is conserved, $\nabla_\alpha J^\alpha = 0$, by virtue of the Wheeler–DeWitt equation. The probability measure is usually constructed using a particular component of J^α, say J^0. (q^0 is usually something like $(\det h_{ij})^{1/2}$, and is in some ways like a time coordinate in the Wheeler–DeWitt equation; see below, however.) The probability measure is then taken to be

$$dP = J^0(-f)^{1/2}d^{D-1}q. \tag{11.5}$$

The volume element in (11.5) is that of a hypersurface of codimension 1, so this measure is more closely analogous to what one does in ordinary quantum mechanics. Moreover, it *is* invariant under the lapse rescalings described in Section 8.

There is still a problem, however. In the oscillatory region, one has

$$J^\alpha = 2|C|^2\nabla^\alpha S = 2|C|^2\dot{q}^\alpha. \tag{11.6}$$

The problem is that J^0, and indeed any other component of J^α, will not always be positive because the classical trajectories will not in general be monotonically increasing in q^0, or indeed in any of the other q^α's. This is closely related to the problem of choosing one of the q^α's as a time coordinate. It follows from this that the probability measure will not always be positive (DeWitt 1957).

This difficulty can be avoided by imposing very special boundary conditions, as has been suggested by Vilenkin (1986). Another way around it, also proposed by Vilenkin (1989), concerns the definition of (11.5). The idea is that one replaces (11.5) by the more covariant-looking expression

$$dP = J^\alpha d\Sigma_\alpha \tag{11.7}$$

(by covariance we mean covariance under coordinate transformations of the q's, of course). Here, $d\Sigma_\alpha$ is the volume element of a hypersurface of codimension 1 that is non-parallel to the classical trajectories. One could take it to be the constant S surfaces, for example, and (11.7) would then be proportional to the projection of J along the classical trajectories. This proposal avoids having to choose a particular q^α as a time coordinate. Rather, it is essentially taking as time the proper time parameter t along the classical trajectories, defined by

$$\frac{d}{dt} = \nabla S \cdot \nabla \tag{11.8}$$

(in the case where the hypersurfaces are constant S surfaces). It is claimed that this gives a positive probability measure although this was still under investigation at the time of writing (Vilenkin 1989).

Although this seems more promising, some unanswered questions remain. Firstly, it seems to make sense only in the oscillatory region, in which the universe is approximately classical. Whilst this may ultimately be the best we can do, it would be desirable to have a way of interpreting the exponential, nonclassical region also. In particular, one might hope to have a universally valid probability measure that could be used to *predict* the existence and location of the approximately classical region.

Secondly, the wave function in the oscillatory region is not usually of the precise form Ce^{iS}, but a sum of such terms. The Hartle–Hawking wave function, for example, is of the form $C \cos S$. In fact, the current J vanishes for this wave function because it is real. It seems reasonable that the different WKB wave functions should not interfere, allowing one to consider each term of the form Ce^{iS} separately in the measure (11.7). However, to the author's knowledge, no convincing interpretive scheme has been proposed that explains why one should be allowed to do this.

12. Summary and Discussion

We have discussed the relationship between the Dirac and path-integral quantization schemes for a simple class of reparametrization-invariant systems, with a particular emphasis on the minisuperspace models of quantum cosmology. We showed how to construct the gauge-fixed path integral with the correct measure using the method of Batalin, Fradkin and Vilkovisky. Our path-integral expressions were all defined explicitly using a time slicing procedure. We showed that the path-integral expression was either a Green function of the Wheeler–DeWitt operator or a solution to the Wheeler–DeWitt equation, depending on the range of the lapse function N. Our main result is that the Wheeler–DeWitt operator (and thus the path-integral measure) is uniquely fixed by demanding that the quantization procedure be invariant under field redefinitions of both the three-metric h_{ij} and the lapse function N. It is

$$\hat{H} = -\frac{1}{2}\nabla^2 - \frac{(D-2)}{8(D-1)}R + V \qquad (12.1)$$

for $D > 1$. We also discussed the problems involved in respecting the constraint $\det h_{ij} > 0$, the properties of the minisuperspace propagator, and the issue of the probability measure in quantum cosmology.

Throughout this paper, a number of remarks have been made concerning the range of the lapse function N. It is perhaps useful to draw these together. The BFV procedure described in Section 3 does not explicitly fix the domains of integration. In Sections 5 and 6, therefore, we investigated the consequences of choosing either infinite or half-infinite ranges for N. We found that the path integral then generated, respectively, either a solution to the Wheeler–DeWitt equation or a Green function of the Wheeler–DeWitt operator. It was also pointed out that one appears to be obliged to take the range $0 < N < \infty$ if one wishes to construct a Euclidean functional integral. We argued in Section 9 that the skeletonization of N was not problematic in the gauge $\dot{N} = 0$, although in general half-infinite ranges do suffer from difficulties.

We suggested at the end of Section 3 that the Fradkin–Vilkovisky theorem may not work if the range of N is $N > 0$. In fact, an example may now be given which appears to show this explicitly. Consider the example of Section 7 (the problems with the half-infinite range for q are not relevant here). Originally, we worked in the gauge $\dot{N} = 0$. Suppose, however, we work in the gauge $\dot{N} = \alpha p$, where α is an arbitrary constant, and for simplicity we let $\lambda = 0$. The Fradkin–Vilkovisky theorem implies that the path integral ought to be independent of α. Once again, the ghosts decouple and the path integral may be evaluated exactly. The result is of the form (7.7), with $\lambda = 0$, but with T replaced by $T + \alpha(t'' - t')p/2$. Clearly if N (and hence T) has an infinite range, then the α dependence may be absorbed by a shift of T and the final result is independent of α, as intended. If, however, T has a half-infinite range then the result will depend on α explicitly. This could be related to the fact that the argument given in Section 9 for the validity of the skeletonization of N no longer applies.

There are many more issues that are still to be addressed, such as the construction of a Euclidean path integral and the implementation of the boundary-conditions proposal of Hartle and Hawking (Hartle and Hawking 1983). It is also of interest to consider the much more difficult question of the derivation of the Wheeler–DeWitt equation for the full theory, with non-trivial momentum constraints and an infinite number of degrees of freedom. A formal Euclidean path integral for the full theory, using the BFV method, has been given by Schleich (Schleich 1987), although a derivation of the Wheeler–DeWitt equation was not given. These and other issues will be discussed in future publications.

Supplementary Note: Since submission of this paper, a number of relevant advances have taken place. Firstly, a completely general (i.e., not restricted to minisuperspace) derivation of the Wheeler–DeWitt equation and momentum constraints from a properly constructed path integral has been given (Halliwell and Hartle 1990b). Secondly, some progress has been made in defining the Euclidean path integrals for the wave function of the universe using complex contours (Halliwell and Hartle 1990a; Halliwell and Louko 1989a, 1989b, 1990; Halliwell and Myers 1989; Hartle 1989).

Acknowledgments. I am very grateful to Arley Anderson, Jim Hartle, Jorma Louko, Kristin Schleich and Bernard Whiting for useful conversations. I would also like to thank the conference organizers for making this a most stimulating meeting.

This research was supported in part by the National Science Foundation under Grant No. PHY82-17853, supplemented by funds from the National Aeronautics and Space Administration, at the University of California at Santa Barbara.

REFERENCES

Abramowitz, Milton, and Stegun, Irene A. (1965). *Handbook of Mathematical Functions.* New York: Dover.

Amsterdamski, Piotr (1985). "Wave Function of an Anisotropic Universe." *Physical Review D* 31: 3073–3078.

Batalin, I. A., and Vilkovisky, G. A. (1977). "Relativistic S-Matrix of Dynamical Systems with Boson and Fermion Constraints." *Physics Letters B* 69: 309–312.

Berezin, F. A. (1966). *The Method of Second Quantization.* Nobumichi Mugibayashi and Alan Jeffrey, trans. New York: Academic.

Carow, Ursula, and Watamura, Satoshi (1985). "Quantum Cosmological Model of the Inflationary Universe." *Physical Review D* 32: 1290–1301.

Cheng, K. S. (1972). "Quantization of a General Dynamical System by Feynman's Path Integration Formulation." *Journal of Mathematical Physics* 13: 1723–1726.

Davis, A. C.; Macfarlane, A. J.; Popat, P.; andd van Holten, J. W. (1984). "The Quantum Mechanics of the Supersymmetric Nonlinear σ-Model." *Journal of Physics A* 17: 2945–2954.

DeWitt, Bryce S. (1957). "Dynamical Theory in Curved Spaces. I. A Review of the Classical and Quantum Action Principles." *Reviews of Modern Physics* 29: 377–397.

DeWitt-Morette, Cécile; Elworthy, K. D.; Nelson, B. L.; and Sammelman, G. (1980). "A Stochastic Scheme for Constructing Solutions of the Schrödinger Equation." *Institut Henri Poincaré. Annales A* 32: 327–341.

Dirac, Paul A. M. (1964). *Lectures on Quantum Mechanics.* Belfer Graduate School of Science Monographs, no. 2. New York: Belfer Graduate School of Science, Yeshiva University.

Feynman, Richard P. (1948). "Space-Time Approach to Non-Relativistic Quantum Mechanics." *Reviews of Modern Physics* 20: 367–387.

Fradkin, E. S., and Vilkovisky, G. A. (1975). "Quantization of Relativistic Systems with Constraints." *Physics Letters B* 55: 224–226.

———(1977). "Quantization of Systems with Constraints: Equivalence of Canonical and Covariant Formalisms in the Quantum Theory of the Gravitational Field." CERN Report, no. TH2332.

Fukutaka, H., and Kashiwa, T. (1988). "Canonical Transformation in the Path Integral and the Operator Formalism." *Ann. Phys.* 185:301

Gibbons, G. W., Hawking, Stephen W., and Perry, M. J. (1978). "Path Integrals and the Indefiniteness of the Gravitational Action." *Nuclear Physics B* 138: 141–150.

González-Diaz, P. F. (1985). "On the Wave Function of the Universe." *Physics Letters B* 159: 19–21.

Halliwell, Jonathan J. (1986). "The Quantum Cosmology of Einstein–Maxwell Theory in Six Dimensions." *Nuclear Physics B* 266: 228–244.

———(1987a). "Classical and Quantum Cosmology of the Salam–Sezgin Model." *Nuclear Physics B* 286: 729–750.

———(1987b). "Correlations in the Wave Function of the Universe." *Physical Review D* 36: 3626–3640.

———(1988). "Derivation of the Wheeler–DeWitt Equation from a Path Integral for Minisuperspace Models." *Physical Review D* 38: 2468.

Halliwell, Jonathan, and Hartle, James B. (1990a). "Integration Contours for the No-Boundary Wave Functions of the Universe." *Physical Review D* 41: 1815–1834.

———(1990b). "Wave Functions Constructed from an Invariant Sum-Over-Histories Satisfy Constraints." ITP preprint NSF-ITP-90–97, to appear in *Physical Review D*.

Halliwell, Jonathan, and Louko, Jorma (1989a). "Steepest-Descent Contours in the Path-Integral Approach to Quantum Cosmology. I. The de Sitter Minisuperspace Model." *Physical Review D* 39: 2206–2215.

———(1989b). "Steepest-Descent Contours in the Path-Integral Approach to Quantum Cosmology. II. Microsuperspace." *Physical Review D* 1868–1875.

———(1990). "Steepest-Descent Contours in the Path-Integral Approach to

Quantum Cosmology. III. A General Method with Applications to Anisotropic Minisuperspace Models." CTP preprint CTP #1846, to appear in *Physical Review D*.

Halliwell, Jonathan, and Myers, Robert (1989). "Multiple-Sphere Configurations in the Path-Integral Representation of the Wave Function of the Universe." *Physical Review D* 40: 4011–4022.

Hanson, Andrew; Regge, Tullio; and Teitelboim, Claudio (1976). *Constrained Hamiltonian Systems: Ciclo di Lezioni Tenute dal 29 Aprile al 7 Maggio, 1974*. Contributi del Centro Linceo Inderdisciplinare di Scienze Matematiche e Loro Applicazioni, no. 22. Rome: Academia Nazionale dei Lincei.

Hartle, James B. (1988a). "Quantum Kinematics of Spacetime. I. Nonrelativistic Theory." *Physcial Review D* 37: 2818–2832.

———(1988b). "Quantum Kinematics of Spacetime II: A Model Quantum Cosmology with Real Clocks." *Physical Review D* 38: 2985.

———(1990). "Quantum Kinematics of Spacetime III: General Relativity." *Physical Review D*. Forthcoming.

Hartle, James B., and Hawking, Stephen W. (1976). "Path-Integral Derivation of Black-Hole Radiance." *Physical Review D* 13: 2188–2203.

———(1983). "Wave Function of the Universe." *Physical Review D* 28: 2960–2975.

Hartle, James B. (1989). "Simplicial Minisuperspace. III. Integration Contours in a Five-Simplex Model." *Journal of Mathematical Physics* 30: 452.

Hartle, James B., and Kuchař, Karel V. (1984). "Path Integrals in Parametrized Theories: Newtonian Systems." *Journal of Mathematical Physics* 25: 57–75.

———(1986). "Path Integrals in Parametrized Theories: The Free Relativistic Particle." *Physical Review D* 34: 2323–2331.

Hawking, Stephen W. (1982). "The Boundary Conditions of the Universe." In *Astrophysical Cosmology: Proceedings of the Study Week on Cosmology and Fundamental Physics*. H. A. Brück, G. V. Coyne and M. S. Longair, eds. Vatican City: Pontificiae Academiae Scientiarum Scripta Varia, 563–574.

———(1984). "The Quantum State of the Universe." *Nuclear Physics B* 239: 257–276.

Hawking, Stephen W., and Luttrell, Julian C. (1984a). "Higher Derivatives in Quantum Cosmology, I. The Isotropic Case." *Nuclear Physics* 247: 250–260.

———(1984b). "The Isotropy of the Universe." *Physics Letters B* 143: 83–86.

Hawking, Stephen W., and Page Don N. (1986). "Operator Ordering and the Flatness of the Universe." *Nuclear Physics B* 264: 185–196.

Henneaux, Marc (1985). "Hamiltonian Form of the Path Integral for Theories with a Gauge Freedom." *Physics Reports* 126: 1–66.

Horowitz, Gary T. (1985). "Quantum Cosmology with a Positive-Definite Action." *Physical Review D* 31: 1169–1177.

Hu, Xiao Ming, and Wu, Zhong Chao (1984). "Quantum Kaluza–Klein Cosmologies II." *Physics Letters B* 149: 87–89.

————(1985). "Quantum Kaluza–Klein Cosmologies III." *Physics Letters B* 155: 237–240.

Isham, Christopher J. (1976). "Some Quantum Field Theory Aspects of the Superspace Quantization of General Relativity." *Royal Society of London. Proceedings A* 351: 209–232.

Klauder, John R. (1970). "Soluble Models of Quantum Gravitation." In *Relativity: Proceedings of the Relativity Conference of the Midwest, Cincinnati, Ohio, June 2–6, 1969.* Moshe Carmeli, Stuart I. Fickler, and Louis Witten, eds. New York: Plenum, 1–17.

Klauder, John R., and Aslaksen, Erik W. (1970). "Elementary Model for Quantum Gravity." *Physical Review D* 2: 272–276.

Kuchař, Karel V. (1983). "Measure for Measure: Covariant Skeletonizations of Phase Space Path Integrals for Systems Moving on Riemannian Manifolds." *Journal of Mathematical Physics* 24: 2122–2141.

Louko, Jorma (1987). "Propagation Amplitude in Homogeneous Quantum Cosmology." *Classical and Quantum Gravity* 4: 581–593.

Misner, Charles W. (1970). "Classical and Quantum Dynamics of a Closed Universe." In *Relativity: Proceedings of the Relativity Conference of the Midwest, Cincinnati, Ohio, June 2–6, 1969.* Moshe Carmeli, Stuart I. Fickler, and Louis Witten, eds. New York: Plenum, 55–79.

Moss, I. G. (1988). "Quantum Cosmology and the Self-Observing Universe." *Institut Henri Poincaré. Annales* 49: 341.

Moss, I. G., and Wright, W. A. (1984). "Wave Function of the Inflationary Universe." *Physical Review D* 29: 1067–1075.

Parker, Leonard (1978). "Aspects of Quantum Field Theory in Curved Space-Time: Effective Action and Energy-Momentum Tensor." In *Recent Developments in Gravitation.* M. Levy and S. Deser, eds. New York/London: Plenum, pp. 219–273.

————(1979). "Path Integrals for a Particle in Curved Space." *Physical Review D* 19: 438–441.

Pauli, Wolfgang (1973). *Pauli Lectures on Physics.* Vol. 6, *Selected Topics in Field Quantization.* Charles P. Enz, ed. S. Margulies and H. R. Lewis, trans. Cambridge and London: MIT Press.

Schleich, Kristin (1987). "Conformal Rotation in Perturbative Gravity." *Physical Review D* 36: 2342–2363.

Sheik, A. (1987). "Canonical Transformation in Quantum Mechanics." Imperial College preprint.

Teitelboim, Claudio (1982). "Quantum Mechanics of the Gravitational Field." *Physical Review D* 25: 3159–3179.

————(1983a). "Proper-Time Gauge in the Quantum Theory of Gravitation." *Physical Review D* 28: 297–309.

————(1983b). "Quantum Mechanics of the Gravitational Field in Asymptotically Flat Space." *Physical Review D* 28: 310–316.

Vilenkin, Alexander (1986). "Boundary Conditions in Quantum Cosmology." *Physical Review D* 33: 3560–3569.

Vilenkin, Alexander (1989). "The Interpretation of the Wave Function of the Universe." *Physical Review D* 39: 1116.

Wu, Zhong Chao (1984). "Quantum Kaluza–Klein Cosmologies I." *Physics Letters B* 146: 307–310.

Interpreting the Density Matrix of the Universe

Don N. Page

In quantum gravity (at least in the canonical formulation), a central object to be assumed or computed is the density matrix of the universe. Its arguments (the labels of the rows and columns of the matrix) are pairs of three-geometries and matter field configurations on them. The density matrix cannot be arbitrary, but should obey the Wheeler–DeWitt equation in each of the two sets of arguments (except for special sets, such as degenerate three-geometries at the boundary of the domain of interest, and identical sets so that one has the diagonal elements of the density matrix). The Wheeler–DeWitt or Einstein–Schrödinger equation is the quantum version of the classical constraint equation of general relativity. Schematically, we can write it as

$$H\rho = 0 = \rho H, \tag{1}$$

where H is the Wheeler–DeWitt operator and ρ is the density matrix.

Although the Wheeler–DeWitt equation gives an uncountable number (continuum) of constraints on the density matrix (one for each point of the spatial manifold), it still apparently allows an uncountable number of solutions, at least locally. Therefore, some further proposal is necessary to specify the actual density matrix of the universe. The most highly developed proposal is the no-boundary or Hartle–Hawking proposal, first given for a wave function (Hawking 1982; Hartle and Hawking 1983; Hawking 1984) and later extended to give a density matrix (Page 1986; Hawking 1987). I will concentrate on this proposal here, and hence assume that there is a unique density matrix.

The question now arises as to how to interpret such a density matrix and how to extract testable predictions from it. I argue that the density matrix gives conditional probabilities and expectation values (Page 1986).

Normally, one interprets a density matrix ρ as giving the expectation value of any observable or Hermitian operator O as

$$E(O) = \text{tr}(O\rho). \tag{2}$$

For example, if O is a projection operator P_A onto some result A, then its expectation value is the probability of A:

$$P(A) = \text{tr}(P_A\rho). \tag{3}$$

However, if observers within the universe do not know their own probability of existing, they cannot test the absolute probability of anything they observe. All they can test are conditional expectation values and probabilities, the conditions being their own existence and the fact that they make the observations under consideration. Lumping all the conditions together into a single condition B with projection operator P_B, the conditional expectation value of O given B is

$$E(O \mid B) = \text{tr}(O P_B\rho P_B)/\text{tr}(P_B\rho), \tag{4}$$

which, for example, gives the conditional probability of A given B as

$$P(A \mid B) = \text{tr}(P_A P_B\rho P_B)/\text{tr}(P_B\rho). \tag{5}$$

Because the conditional expectation values (4) and (5) are the same as what (2) and (3) would give if ρ were replaced by the conditional density matrix

$$\rho_B = P_B\rho P_B/\text{tr}(P_B\rho), \tag{6}$$

many people would prefer to say that, once the condition has occurred, the density matrix of the universe has collapsed from ρ to ρ_B. This assumption, which is essentially von Neumann's Process 1, is certainly calculationally convenient, but I would argue that one gets a conceptually simpler ontology by assuming that the true ρ never collapses in this way. Why do I claim this? First, it is simpler to assume that there is a single ρ, which does not change (in the Heisenberg picture we are implicitly using), rather than to assume that physical reality consists of ρ_B's that change with the conditions B. Second, ρ obeys a simpler equation (the Wheeler–DeWitt equation) than ρ_B, which, in general, will not satisfy the constraint (1) if P_B does not commute with H (Page 1988). For example, the condition for the existence of observers that occur only during part of the history of the universe would give $H\rho_B \neq 0$ (so that in an amusing technical sense one could say that the existence of observers — or anything else that does not exist throughout the full history of the universe — is "unphysical," using the nomenclature that physical states are those that obey the constraint). Third, the specification of the full ρ (say by the Hartle–Hawking no-boundary proposal) may be simpler than the specification of any ρ_B. If one writes

$$\rho = \Sigma_B \text{tr}(P_B \rho)\rho_B + \Sigma_{A \neq B} P_A \rho P_B, \tag{7}$$

$$\Sigma_B P_B = I, \tag{8}$$

then it may indeed be that the whole (ρ) is much simpler than any of its parts (e.g., ρ_B).

It is for these reasons of ontological simplicity that I take essentially the Everett many-worlds interpretation: The full density matrix ρ is a sum (7) over many component worlds ρ_B (eigenstates of whatever projection operators P_B one may choose to calculate conditional probabilities). I would not, however, argue that there is any unique decomposition of unity into projection operators as in (8), so I would say the decomposition (7) is not unique. In other words, I would not say there is a preferred set of worlds into which the full universe splits. The choice of splitting is merely a convenience dictated by the choice of which conditional probability question one is attempting to answer from the formalism. (It is rather like the choice of coordinates in general relativity.) The description of the complete state of the universe should be designed to answer all possible questions of conditional expectation values or probabilities, rather than to pick out a preferred subset.

One might object to the necessity of using conditional probabilities by saying that the observer *knows* that the probability of his own existence is unity, so shouldn't that be the case? This objection is related to the question of how a complete model for the universe (such as the density matrix would purport to be) can give merely probabilistic predictions.

I prefer to answer this question by proposing the view that the probabilities given by quantum mechanics are not really probabilities but measures. That is, if the quantum-mechanical prediction gives more than one outcome A for which $P(A)$ or $P(A \mid B)$ is nonzero, then all of these outcomes actually occur (in different worlds ρ_A), and so, strictly speaking, the probability of each is unity, but the physical measure associated with each outcome is $P(A)$ or $P(A \mid B)$, which typically varies with A. But since these measures obey the axioms of probability, they can be interpreted as probabilities with no mathematical inconsistency. Further- more, since each memory eigenstate of a good observer can record only one of the possible outcomes, a prediction of the measure of each of these is essentially what the observer would consider to be a prediction of the probability that he or she would record each of the various results. Thus, quantum-mechanical probabilities are actually more like frequencies on an infinite ensemble of worlds.

Another objection to a theory that gives conditional probabilities (or, more strictly, measures, as I have just argued) is the difficulty of measuring them. Since any good observer can be aware of only one outcome A in each eigenstate of his memory, he cannot directly measure $P(A \mid B)$. All that he can deduce from getting the outcome A is that $P(A \mid B)$ is not zero. In other words, how does one measure probabilities when one can be aware of only a single outcome?

I would argue that, although the predicted $P(A \mid B)$'s cannot be mea- sured, they can be tested. That is, if the theory predicts some $P(A \mid B)$ that is extremely low (say much less than the reciprocal of the number of A's being considered possible), but if A is in fact observed, then this observation refutes the theory to a high confidence level. One may need to combine many elemental outcomes into a single composite outcome A in order to get a $P(A \mid B)$ that is very small, and this is indeed the point of repeated experiments, but if $P(A \mid B)$ is very small, one can test it by a single experiment. For example, under the condition B of a single coin toss, the conditional probability of the result A being a head is, say, one-half, which would not be testable. But if one has a repeated experiment so B consists of a million coin tosses, and each A consists of a different number of heads ($10^6 + 1$ possibilities), then the As with less than 400,000 heads each give $P(A \mid B) \ll 10^{-6}$ under the theory of a fair coin, so the observation of such an A would refute the theory to a high confidence level.

One phenomenon that may readily be discussed in terms of density matrices and conditional probabilities is the concept of time and tempo- ral evolution, approximate concepts that became highly accurate under certain semiclassical conditions. One can arbitrarily define time to be the readings of some physical system that one takes as a clock. Then, if one takes a sequence of clock readings to be a sequence of conditions B, the dependence of ρ_B, $E(O \mid B)$, and $P(A \mid B)$ upon B can be interpreted as the time evolution of the system. Even a stationary state (with respect to Schrödinger parameter time) in ordinary quantum mechanics without gravity can exhibit "evolution without evolution" in this way (Page and

Wootters 1983). In the idealized case of no interaction between the clock and the rest of the system, one can use the clock Hamiltonian to define a sequence of clock states B labeled by a clock time. Then a stationary ρ gives a ρ_B obeying the von Neumann equation of motion with respect to this clock time. In the more realistic case of an interacting clock, the conditional density matrix will not have such a simple dependence upon the clock time. However, when the universe is very large and complex, as it is today, there are a large number of clocks that interact sufficiently weakly that the dependence of the conditional density matrix upon these various clock times is nearly that given by the von Neumann equation.

Once one has time evolution conditional upon some clock readings, one can attempt to define a coarse-grained entropy for the conditional density matrix and see whether it increases with time. A complete formulation of this problem has not been given, but preliminary calculations with simple models suggest that the entropy does start out very low at the "big bang" end of the semiclassical evolution, and then increases monotonically to a large value at the "big crunch" end (Hawking 1985; Page 1985).

Thus, it does seem that one can interpret and test the density matrix of the universe in terms of conditional probabilities, and one can discuss time evolution in these terms.

Murray Gell-Mann has informed me that he had earlier expressed many similar ideas (Gell-Mann 1987).

Acknowledgment. This work was supported in part by National Science Foundation Grant No. PHY-8701860.

REFERENCES

Gell-Mann, Murray (1987). "Superstring Theory." *Physica Scripta T* 15: 202–209.
Hartle, James B., and Hawking, Stephen W. (1983). "Wave Function of the Universe." *Physical Review D* 28: 2960–2975.
Hawking, Stephen W. (1982). "The Boundary Conditions of the Universe." In *Astrophysical Cosmology: Proceedings of the Study Week on Cosmology and Fundamental Physics.* H. A. Brück, G. V. Coyne and M. S. Longair, eds. Vatican City: Pontificiae Academiae Scientiarum Scripta Varia, pp. 563–574.
———(1984). "The Quantum State of the Universe." *Nuclear Physics B* 239: 257–276.

————(1985). "Arrow of Time in Cosmology." *Physical Review D* 32: 2489–2495.

————(1987). "The Density Matrix of the Universe." *Physica Scripta T* 15: 151–153.

Page, Don N. (1985). "Will Entropy Decrease if the Universe Recollapses?" *Physical Review D* 32: 2496–2499.

————(1986). "Density Matrix of the Universe." *Physical Review D* 34: 2267–2271.

————(1988). "The Quantum State of the Cosmos." In *Proceedings of the Fourth Seminar on Quantum Gravity, May 25–29, 1987, Moscow, USSR.* M. A. Markov, V. A. Berezin, and V. P. Frolov, eds. Singapore: World Scientific, 691–699.

Page, Don N., and Wootters, William K. (1983). "Evolution without Evolution: Dynamics Described by Stationary Observables." *Physical Review D* 27: 2885–2892.

PART II

The Issue of Time in Quantum Gravity

Meditation on Time

Abner Shimony

Great clock,
Which does not turn,
And does not beat,
And does not toll,
And does not lag,
And does not cease,
From you I neither
Can nor would escape,
Because my essence
(To all else entrained)
Is your remote
Escapement.
And still I harbor
One dark wish:
To gaze upon you
From without,
As heretofore,
And now,
And always,
From within.

Is There Incompatibility Between the Ways Time is Treated in General Relativity and in Standard Quantum Mechanics?

Carlo Rovelli

1. A Lesson From Dirac

In his opening remarks, John Stachel suggested that this conference be dedicated to Dirac. Let me therefore start from a talk that Dirac gave in Italy, in Erice (Sicily), during one of the last years of his life. He gave his talk using transparencies or, more precisely, using one transparency.

The content of this single transparency was just a formula and a statement:

$$i\hbar \frac{d\,A(t)}{dt} = [A(t), H]$$

HEISENBERG MECHANICS IS THE GOOD MECHANICS.

In this talk I shall discuss the question that I pose in the title. I shall argue for a negative answer: "There is no incompatibility." The main argument that I will use is Dirac's argument: "Use Heisenberg Mechanics."

The reason for studying the problem of the compatibility of general relativity and quantum mechanics with respect to the issue of time is not at all to criticize the very interesting attempts to study modifications of quantum mechanics, such as those suggested by general relativity (Penrose 1979, 1986; Smolin 1986; Hawking 1977). It is just to point out that the program of constructing a quantum theory of gravity in the framework of standard quantum mechanics is perfectly viable, at least as far as the problem of time is concerned.

2. The Generalized Heisenberg Picture

A possible point of view about mechanics (any mechanics: classical, quantum, statistical, ...) is the following one. We have a system. This system can be in one of a set S of states s. Then we have time t, and the evolution in t, namely a map from $S \times R$ into S. Mechanics is the study of this map. Then we have observables, and a rule to assign a number, or a set of numbers, to any observable and any state. Kinematics is the study of these rules. From this point of view, a concept of time is necessary for the very definition of mechanics.

I remember that during my first year at the university I was taught, in the typical, a bit heavy, European style that, "Mechanics is the study of the evolution of physical systems in time."

However this point of view is *not* the only possible one. I will call this point of view the "generalized Schrödinger picture." The alternative point of view will then be the "generalized Heisenberg picture."

In the generalized Heisenberg picture the state of the system does not evolve. The system is in a Heisenberg state S and will stay in it "forever," by definition. In a sense, Heisenberg states have to be thought of as representative of the history of the system.

The time evolution then affects the observables. What does this mean? In the generalized Heisenberg picture, the observable O becomes a time-dependent function, $O(t)$. This means that we have now one observable for every t. One Schrödinger observable becomes a one-parameter family of observables. I want to stress that $O(t_1)$ and $O(t_2)$ are *different* Heisenberg observables.

The set of observables in the generalized Heisenberg picture is formed by all observables at all times. It is therefore naturally foliated. The leaves of the foliation are the one-dimensional sets given by all the observables that express the same "Schrödinger observable" at different times. Let me call this particular structure on the set of the observables a *time structure*.

From this point of view, mechanics is in a sense reduced to kinematics, namely to the definition of the observables. If we have the explicit definition of all the observables $O(t)$ for any t and for any Heisenberg state s, then we don't have to describe the system in terms of evolution. Clearly, the kinematics becomes nontrivial, since its specification amounts to solving the evolution equations.

The Heisenberg picture is not peculiar to quantum mechanics. There is a version of it also in classical mechanics. This is presymplectic dynamics, which is a formulation of classical mechanics that is more general than the usual canonical or symplectic dynamics. Examples of presymplectic systems are parametrized systems (Kuchar 1981) when they are thought of in a geometrical way, independently of the coordinates.

3. What is the Physical Meaning of the Heisenberg Picture?

In the Schrödinger picture we assume we are able to perform two diffent kinds of operations. We can read a clock and we can measure the value of an observable. Now let's include the reading of the clock in the list of the operations that define an observable. Then we have the Heisenberg definition of observable, in which position at time 1 is a different observable from position at time 2. For instance, a Schrödinger observable is operationally defined by the reading of a certain instrument. A Heisenberg observable is operationally defined by the reading of a certain instrument at a certain time.

I propose for clarity the following terminology:

A *partial observable* is an operation on the system that produces a number. But this number may be totally unpredictable even if the state is perfectly known. Equivalently, this number by itself may give no information on the state of the system. For instance, the reading of a clock, or the value of a field, not knowing where and when it has been measured, are partial observables.

A *true observable* or simply, an *observable,* is an operation on the system that produces a number that can be predicted (or whose probability distribution may be predicted) if the (Heisenberg) state is known. Equivalently, it is an observable that gives information about the state of the system.

A true observable may be composed of one or more partial observables. For instance, the true observable "number of spins that are 'up' at noon" (in a spin system), is composed of two partial observables, a clock reading and a counting of spins up. The true observable "value of the field at a given space-time point" is composed of the determination of the position (a partial observable), the determination of the time (a partial observable), and the measurement of the field (a partial observable).

Let me consider, for later convenience, the application of this terminology to parametrized systems. A simple system expressed in parametrized form is the relativistic free particle. We can describe it in a four-dimensional notation by means of the variables $x^\mu(\tau)$. τ is a parameter, it is not physical, and is not connected with any physical procedure. $\mathbf{x}$ and x^0 are on the same footing and are partial observables.

The true observables are $\mathbf{x}(t) = \mathbf{x}\big|_{x^0=t}$, namely the positions *at a given time* $x^0 = t$. Their explicit expressions on the phase space are obtained by solving the equations of motion. We get

$$\mathbf{x}(t) = \mathbf{x} - \frac{\mathbf{p}}{p_0}(t - x^0).$$

This is a true observable formed by two partial observables. Note that it is gauge invariant. Indeed, it has vanishing Poisson brackets with the (Hamiltonian) constraint of the system $\mathbf{p}^2 - m^2$. This is a general rule: the true Heisenberg observables in parametrized systems are the gauge-invariant observables.

4. Mechanics Without Time

Until now I've just given some definitions and noted that the Heisenberg version of quantum mechanics suggests a slight shift of point of view regarding the general definition of mechanics.

This point of view, however, leads to the following question. Is the fact that true observables are in general composed of several partial observables logically required by the very possibility of constructing a mechanics, or is it just an experimental fact of nature? In other words: is it a priori necessary that a reading of an instrument cannot be meaningful unless it is accompanied by the reading of a clock?

The main thesis of this talk is that the answer to this question is negative. Time is an experimental fact of nature, a very basic and general experimental fact, but just an experimental fact. The formal development of mechanics, and in particular Heisenberg quantum mechanics and the presymplectic formulation of classical mechanics, suggests that it is possible to give a coherent description of the world that is independent of the presence of time.

Thus, I can get closer to my first conclusion. Let me try to suggest a tentative definition for time:

From the mathematical point of view, *time* is a structure on the set of observables (the foliation that I called time structure).

From a physical point of view, time is the *experimental fact* that, in nature as we see it, meaningful observables are always constructed out of two partial observables. That is, it is the experimental fact (not a priori required), that knowing the position of a particle is meaningless unless we also know "at what time" a particle was at that position.

In the formulation of the theory, this experimental fact is coded in the time structure of the set of the observables. If true observables are composed of correspondences of partial observables, one of which is the reading of a clock, then the set of true observables can be foliated into one-parameter families that are given by the same partial observables at different clock readings.

From an operational point of view, mechanics is perfectly well defined in the absence of this time structure. It will describe a world (maybe one slightly unfamiliar to us) in which observables are not arranged along one-parameter lines, in which they have no such time structure (a kind of fixed-time world), or have more complicated structures. We must not confuse the psychological difficulties in visualizing such worlds with their logical impossibility.

Let's focus now on quantum mechanics. There isn't anything in the Heisenberg version of quantum mechanics that requires the existence of time. That is the central point of this talk. Heisenberg states, observables, measurement theory—none of these require time.

The definition of probability does not require any notion of time. The normalization condition is an a priori requirement that follows from the very definition of probability. Similarly, total probability doesn't have to be conserved in time if there is no evolution in time. Rather, conditional probability has to satisfy some requirements that follow from its physical interpretation, and it can be shown that these requirements reduce to unitary evolution if one is considering the "evolution" of the observables with respect to partial observables (clocks) that have suitable properties. But clocks, that is partial observables with suitable properties, may not exist in a system, exist only in some states, or exist but have these properties only within some approximation (see Peres 1980, Page and Wooters 1983).

A word of caution about the measurement projection is in order. As very clearly discussed, for instance, by Jim Hartle (Hartle 1986), there are two independent notions of time in ordinary quantum mechanics: the time in which the system evolves, and the "time" that orders the measurements of the observer. These two are not related, and may be non-coincident. I am talking here about the time in which the system evolves, not about the time of the observer. The standard interpretation of quantum mechanics requires a "classical" observer who reduces the wave function and measures one thing "after" the other. I am not discussing here the standard interpretation of quantum mechanics and its difficulties, but just considering the properties that the system may or may not have with respect to time. What I am proposing is that there may exist a coherent description of a system in the framework of standard quantum mechanics even if it does not have a standard "time evolution."

5. Gravity

Why should we be interested in mechanics with no time structure? Because general relativity *is* a system (a classical system) with no time structure. At least, it has no clearly defined time structure.

General relativity is a presymplectic system (a parametrized system; see Kuchař 1981). The space-time arguments $\mathbf{x}$ and t are *not* observables, they have *no operational meaning at all*, exactly like the τ of the relativistic description of a single particle.

Note, by the way, that $\mathbf{x}$ and t in general relativity have a completely different physical meaning than in a nongeneral-covariant field theory, say a scalar field theory for $\phi(\mathbf{x}, t)$. In the latter, $\mathbf{x}$ and t are partial observables that represent physical, operationally defined procedures that correspond to the existence of a physical external reference system. In general relativity, they are just unphysical parameters. The operational meaning of this remark is that general relativity is defined in such a way that it is incompatible with the presence of an external physical reference system.

Now, the components of the metric $g(\mathbf{x}, t)$ are partial observables, like $\mathbf{x}(\tau)$ or $x^0(\tau)$. The true observables are given by *gauge-invariant* functions of the metric. To explicitly find these true observables is nontrivial, since this would amount to solving Einstein's equations. It may be difficult to find them, but this is a problem of classical general relativity, unrelated to the problem of its quantization (see also Stachel 1986, Fredehagen and Haag 1986, Earman and Norton 1987).

Most other physical theories contain a precise definition of which of the partial observables of the theory is time. Even when the formalism does not manifestly indicate the time (as in the parametrized description of a relativistic particle), time is nevertheless singled out by some particular property. In certain situations, there could be more than one possible time (for instance, in any relativistic system, all the Lorentz times are good times); what matters is that some time partial observable is singled out.

In general relativity (and in any generally covariant field theory), on the contrary, there isn't any variable that plays the role of time. In particular, the independent variable x^0 is not physical at all, and one can give a complete formulation of the theory, using the Hamilton–Jacobi framework in which x^0 disappears. Further, there is no natural way to choose one of the components of the metric as "the time of the theory."

The problem of the compatibility between general relativity and quantum mechanics (on the time issue) is the following: is this very peculiar way in which time appears (more precisely: doesn't appear) in general relativity an obstacle for a quantum-mechanical description of the system?

The previous discussion shows that this situation, namely the absence of a clear definition of time, is not at all outside the framework of quantum mechanics. On the contrary, it is perfectly well-defined in this framework, provided that one uses the Heisenberg picture.

In classical general relativity, it is usually assumed (in general implicitly) that a clock is coupled to the system in terms of which local proper time can be measured. In the quantum theory, a coupled physical clock would be a nontrivial quantum object and would not give, in general, any simple notion of time. In macroscopic situations, one can have the usual time defined to within a good approximation, but at the Planck scale this is probably impossible. My main point is that, in order to deal with these very unfamiliar physical situations, there is no a priori need to give up standard quantum mechanics.

Let me add here a personal opinion. I think that exactly because of this unfamiliarity, one should not try to follow intuitions on "how it could be" (intuition reflects everyday experience), but one has to remain solidly grounded on what we know about the world, namely, the general principles of quantum mechanics.

Thus, the position that I am defending here is, at the same time, conservative and radical. Conservative, because I am suggesting that quantum gravity has still to be looked for strictly in the framework of standard quantum mechanics. Radical, because I am claiming that inside this framework, the notion of time is not necessary, and that, for quantum gravity, "time" has simply to be forgotten.

6. Scalar Product

In the actual process of quantizing the theory, the difficulty that usually leads to the statement that without time the quantum theory cannot be defined is related to the choice of the scalar product.

Let me recall the basic facts. In the Dirac quantization of constrained systems, one defines the space of unconstrained states, and one has to impose the constraint equations on these states. In a parametrized system, in which all the dynamics is contained in the constraints, one has a "Wheeler–DeWitt" equation (DeWitt 1967). The physical states that solve the Wheeler–DeWitt equation are in general infinite-norm states in the natural Hilbert structure of unconstrained states space (these issues may be studied, for instance, in the framework of minisuperspace models (see DeWitt 1967).

Therefore, one has to define a *new* physical scalar product on the space of the solutions in order to have the possibility of calculating expectation values.

It is well known that if one can select a preferred time variable t among the arguments of the unconstrained wave functions, then one can find a scalar product.

In fact one can define the Schrödinger picture (in which everything evolves in t), and the fixed t formulation suggests the form of the physical scalar product, which will then involve an integration over all the variables but t.

It would be nice to have this way of choosing the scalar product in the full theory of gravity, but we simply don't have it. This fact doesn't mean that the physical scalar product cannot be defined. It just means that we cannot use this hint for its definition. If you don't know how to reach Boston, this does not mean that Boston doesn't exist.

To find the physical scalar product is always a general problem for *any* constrained system in Dirac quantization. For instance, the same problem appears in any gauge theory.

To find the physical scalar product in gravity, where one does not know how to fix the gauge, one has either to carefully consider the conditions that the scalar product has to satisfy (the scalar product must be chosen in such a way that the physical operators become Hermitian), or look for a quantization method, other than the Dirac one, that directly gives the scalar product.

One theoretical possibility to obtain the scalar product for systems with constraints is to use group quantization by choosing a set of true observables as elementary observables (Isham and Kakas 1984, Rovelli 1987a, 1987b). In this case, one can demonstrate that group-theoretical quantization gives the physical scalar product *directly*.

In his talk on time at this conference, Lee Smolin described a model invented by Barbour and Bertotti that is a good one for a system with no time. In the discussion following that talk, I presented the group quantization of this model, which shows how this method can yield a physical scalar product for a system with no time.

7. Quantum Gravity is Not Quantum Cosmology

Before concluding, I would like to add a few words about the relation between the time issue in quantum gravity and the time issue in quantum cosmology.

I think that there is some confusion between the issue of time in quantum gravity and the issue of time in quantum cosmology (see, for instance, Hartle 1986). Now, quantum cosmology has its specific set of problems. Among these are: how to make a quantum theory for a single object, and how to have a theory that doesn't distinguish between observer and system. These are very serious and interesting problems, but they are totally unrelated to quantum gravity. These problems would be present in quantum cosmology even if the gravitational field didn't exist. For example, if the universe were flat and only inhabited by matter fields and Yang–Mills forces, one would still have the problems that come from the description of the universe as a whole, like the impossibility of having many copies of the system, and of a distinguished observer and observed system.

Conversely, one can consider gravity as a system seen by an "external" observer. One must not confuse the concept of being "external" to the dynamical system with the very different concept of being "external" to the space-time system. The observer can be dynamically separated from the system even if it is immersed in the space-time region where the system is.

The quantization of the gravitational field may be a much more circumscribed program than the program of constructing a quantum theory of the entire universe.

8. Conclusion

Let me summarize my argument. General relativity is a physical system in which there isn't any preferred notion of physical time. In the

Heisenberg picture, one can deal with a quantum system in a way that is independent of the presence of time. Therefore, there isn't any a priori incompatibility between general relativity and quantum mechanics as far as the issue of time is concerned. The difficulties with the definition of the scalar product are technical difficulties, which are always present when there are gauge invariances, and are not to be interpreted as indications that a quantum system without time is ill defined.

In more detail, we have the following conclusions:

(1) Classical mechanics and quantum mechanics both make perfect sense in the absence of time.

(2) In the framework of the Heisenberg picture, time may be interpreted as a particular structure on the set of the observables. This structure codes the experimental fact that meaningful observables are defined in terms of correspondences of partial observables.

(3) In Dirac quantization, the choice of the physical scalar product may be nontrivial. A possible alternative to Dirac quantization that gives the physical scalar product directly is group quantization.

4) There is no contradiction between the ways time is treated in general relativity and quantum mechanics. There is no a priori need to abandon the probabilistic interpretation, the notions of Hilbert structure, finite-norm states and self-adjoint operators. However one is forced to use the Heisenberg picture, as Dirac had long preached.

In conclusion, I tried to demonstrate that there is no incompatibility between general relativity and quantum gravity on the issue of time. What we have to do is simple: "forget time." The difficulties in dealing with a system in which the usual, clear notion of time is absent are, in my understanding, just psychological difficulties.

This talk is a short extract from the work contained in Rovelli 1988 and Rovelli 1990. I refer to this work for a more detailed and complete discussion of the issues discussed in this talk. (See also Rovelli 1991).

REFERENCES

DeWitt, Bryce S. (1967). "Quantum Theory of Gravity. I. The Canonical Theory." *Physical Review* 160: 1113–1148.
Earman, John, and Norton, John (1987). "What Price Spacetime Substantialism? The Hole Story." *British Journal for the Philosophy of Science* 38: 515–525.
Fredenhagen, K., and Haag, R. (1986). DESY preprint, 86–066.

Hartle, James B. (1986). "Predictions in Quantum Cosmology." In *Gravitation in Astrophysics*. NATO Advanced Study Institute, 1986, Cargèse, France. B. Carter and James B. Hartle, eds. New York: Plenum, 329–360.

Hawking, Stephen W. (1977). "Zeta Function Regularization of Path Integrals in Curved Spacetime." *Communications in Mathematical Physics* 55: 133–148.

Isham, Christopher, J., and Kakas, A. C. (1984). "A Group Theoretical Approach to the Canonical Quantization of Gravity: I. Construction of the Canonical Group." *Classical and Quantum Gravity* 1: 621–632.

Kuchar, Karel V. (1981). "Canonical Methods of Quantization." In *Quantum Gravity 2: A Second Oxford Symposium*. Christopher J. Isham, Roger Penrose, and Dennis W. Sciama, eds. Oxford: Clarendon, 329–376.

Page, Don N., and Wootters, William K. (1983). "Evolution without Evolution: Dynamics Described by Stationary Observables." *Physical Review D* 27: 2885–2892.

Penrose, Roger (1979). "Singularities and Time-Asymmetry." In *General Relativity: An Einstein Centenary Survey*. Stephen W. Hawking and W. Israel, eds. Cambridge and New York: Cambridge University Press, 581–638.

————(1986). Unpublished talk at Conference on "Approaches to Quantum Gravity," Santa Barbara, 1986.

Peres, Asher (1980). "Measurement of Time by Quantum Clocks." *American Journal of Physics* 48: 552–557.

Rovelli, Carlo (1987a). "Quantization of the 'Single-Point-Gravity' Hamiltonian System." *Physical Review D* 35: 2987–2992.

————(1987b). "Group Quantization of Constrained Systems." *Il Nuovo Cimento B* 100: 343–359.

————(1988). "Time in Quantum Gravity." University of Rome preprint.

————(1990). "Quantum Mechanics without Time: a model." *Physical Review D* 42: 2638–2646.

————(1991). "Time in Quantum Gravity: An Hypothesis." *Physical Review D* January 15.

Smolin, Lee (1986). "On the Nature of Quantum Fluctuations and their Relation to Gravitation and the Principle of Inertia." *Classical and Quantum Gravity* 3: 347–359.

Stachel, John (1986). "How Einstein discovered general relativity: a historical tale with some contemporary morals." In *General Relativity and Gravitation – Proceedings of the 11th International Conference on General Relativity and Gravitation*. M.A.H. MacCallum, ed. Cambridge: Cambridge University Press, 200–208.

Discussion

KUCHAŘ: I'm glad that in spite of your conclusion to "forget time," you didn't and finished on time.

UNRUH: In the Heisenberg picture, what do you do with the constraints? The constraint says operator A is equal to operator B, which is a different animal. How do you have an equality relation between two different animals, i.e., two different operators? I can have equality between functions because I mean numerically one equals the other. But between operators, I don't know. For example, if the constraint is $\hat{p} = \hat{x}$, I don't know what that means. So how can one implement the program in the Heisenberg picture?

ROVELLI: Well, it depends what kind of quantization scheme you are using. If you are using the Dirac quantization scheme, you implement the constraint as an operator condition on the physical states, and $\hat{A}\psi = \hat{B}\psi$ is a well-defined condition on physical states. Then you have to look for physical observables that commute weakly with the constraints. If you're using other quantization schemes, you have other prescriptions. In the group-theoretical quantization of constrained systems I referred to, you may forget completely about the constraint once you have found the group.

UNRUH: In what way is what you described different from the Schrödinger representation?

ROVELLI: Well, what I'm claiming is that if you don't have time in your theory, or if you don't have time in the general system that you're working with, then there is no Schrödinger equation. It doesn't make sense for the very good reason that there is no "t".

DEWITT: What you're trying to emphasize here is to get away from saying, "Well, we measured the position." We never measure position. We measure position at a given time.

ROVELLI: Right.

DEWITT: How is time determined? You're saying, "it can be determined internally," that is, by means of a clock. You ask, what is the position when the clock is reading such and such? If you do that, that would be a true physical observable. That would be diffeomorphism invariant in the gravity case. It would commute with all the constraints.

ROVELLI: Absolutely, yes.

TORRE: First of all, I completely agree with you in principle. However, it should be emphasized that this program is very difficult to implement. You want to find a set of observables. O.K. Finding a set of observables can be done in the following way: find a general solution to the Einstein equations as a function of initial data, then invert that relationship to get the initial data as a function of the solutions. Those will be your observables. We all know how difficult that procedure would be—more difficult than finding an observable at a certain time if there is one.

DEWITT: But why do you only consider Einstein's equations? Do you want to make everything quantum gravity?

TORRE: No, no, no. I just want a complete set of observables.

DEWITT: Then you want not only Einstein's equations, but all the other dynamical equations for all the other fields.

TORRE: I just want to find a big enough set of observables for this theory by itself, and I think that's a very fundamental point.

ANDERSON: One more remark thrown in somewhere in between Bryce and Charles. It seemed to me that the essential feature of what you said was that you're specifying solutions to constraints in advance, and that specifies well-defined observables. And I want to say that it commutes with the constraints, which of course you do want. What I would like to emphasize is that when solving a constraint equation on a manifold, if I'm doing quantum mechanics and not worrying about quantum cosmology, I get some wave function over the full manifold. Then to go back and try to interpret it, I want to look at events. I don't have to worry about time as defined from some evolving equation. I can see the history if I'm in this region on this manifold. I can start talking about what's going on there. So I can define clocks relative to behavior in that region; I don't have to have global existence, I just have to have local behavior. So I don't need to have your very physical notion of a clock, Bryce. I just have to have this notion of a space-time event and an observable that's associated with it because it has location on the manifold relative to other locations on the manifold.

KUCHAŘ: I would like to return to the question of what is an observable in a parametrized theory, either classical or quantum. I would like to object to the viewpoint that an observable is a quantity that commutes with all the constraints. There is a basic difference between constraints that arise in gauge theories and those that arise in parametrized theories. If I have a constraint that is a gauge constraint—and these are typically linear in momenta—then the orbit of the constraint vector field contains points that are merely different descriptions of the same physical state. I cannot physically distinguish between two points on such an orbit. Then, the standpoint that observables are those quantities that commute with the constraints is totally justified. On the other hand, if I have a parametrized theory, then the commutation of a quantity with the constraints implies that the quantity is a constant of motion. The trajectory that is generated by the constraint is now the dynamical trajectory. And I would say that the state of the people in this room now, and their state five minutes from now, should not be identified. These are not merely two different descriptions of the same state. They are physically distinguishable situations. This brings me to the scheme that Carlo Rovelli proposes. To every quantity at a given instant of time, one can associate an "observable." The observable is a constant of motion whose value coincides with the given quantity at that particular instant of time. I agree that this is possible. However, the problem is that, as Jim Hartle said, we know how to measure certain operators but we do not know how to measure highly complicated operators constructed out of the p's and the q's. Except at specific instants of time, Carlo's observables are highly complicated functions of the coordinates and momenta. For simple systems like harmonic oscillators with linear dynamics, one can measure such operators; this happens, for example, in the quantum nondemolition model. However, when dynamics is nonlinear, this proposal is highly unrealistic. In the nonlinear case, there is the problem of factor ordering such operators. The proposal of eliminating dynamics at the classical level and then quantizing stumbles upon the problem of factor ordering.

ROVELLI: Let me begin by recalling how one makes experiments in general relativity. How does one test, for example, that the metric around the sun is a Schwarzschild metric? The answer is that one launches a rocket. The trajectory of this rocket is not specified externally. It is dictated by the dynamical coupling of the rocket to the Schwarzschild metric. On this rocket, there would be a clock, which is also coupled to the system. We can now ask: What is the scalar curvature in the rocket when the clock shows a certain time? And this quantity is an observable; it commutes with the four constraints of general relativity.

KUCHAŘ: I didn't ask the classical question. I asked the quantum-mechanical question. I completely agree with you in the classical one.

ROVELLI: Well then, if we can only measure those things in the classical theory that commute with the constraints, I cannot see why we should assume that in quantum theory we can measure things that do not commute with constraints. Your second question was about factor ordering. Even in simple parametrized theories the observable that "measures x_1 at a given value of x_0" has a complicated expression on the phase space. In general, I agree that I don't know how to factor order such observables. However, it is not necessary to first quantize x_0 and x_1 and then construct various observables from those operators. There are alternative ways. For example, I spoke about the use of group-theoretic methods (see my paper, "Group Quantization of the Bertotti–Barbour Model"). Then, the factor ordering problem simply doesn't arise. In general relativity, I don't know how to carry out this program. The choice of the correct ordering of the true observables may be an open problem. But it is a problem related to the choice between alternative theories. For the moment, the issue is to find at least one consistent quantum theory of gravity. My understanding is that, in order to find it, searching "the" time variable is not the right direction to go.

The Problem of Time
in Canonical Quantization
of Relativistic Systems

Karel V. Kuchař

By an old sundial motto, the time thou killest will in time kill thee. I offer this as an ultimate consolation to those who may become impatient with my discourse on the role of time in canonical quantization.

Classical relativity has reduced time (as well as space) to an observer's view of a single external reality, the space-time $\mathcal{M}$. Time $\mathcal{T}$ and space $\mathcal{S}$ are not intrinsic space-time structures, but they are put into $\mathcal{M}$ by hand, time as a foliation of hypersurfaces, space as a congruence of world lines. A hypersurface is an instant of time, a world line a point of space. One can label the elements of $\mathcal{M}$, $\mathcal{T}$, and $\mathcal{S}$ by local coordinates; the events $X \in \mathcal{M}$ by X^A, the instants $t \in \mathcal{T}$ by t, and the points $x \in \mathcal{S}$ by x^a. The time hypersurface that passes through X has the label $t(X)$ and the space world line through X carries the labels $x^a(X)$. Inversely,

$$X^A = X^A(t, x^a). \tag{1}$$

The choice of $\mathcal{T}$ and $\mathcal{S}$ is limited by the causal structure of $\mathcal{M}$; the time hypersurfaces must be space-like and the space world lines time-like. Still, through a given event X one can draw infinitely many hypersurfaces and world lines compatible with these limitations; time and space are relative. In a Newtonian space-time, light cones degenerate into planes and the time hypersurfaces become unique, though the space world lines transverse to them are not; Newtonian time is absolute while Newtonian space remains relative. This underlies the differences between Newtonian and relativistic dynamics.

Dynamics studies the motion of particles or evolution of fields from given initial data. The data on $t(X) = t_{IN}$ are evolved into those on $t(X) = t_{FIN}$. In Newtonian physics, the foliation between t_{IN} and

t_{FIN} is unique. In relativistic physics, there are infinitely many foliations connecting t_{IN} with t_{FIN}, but the dynamics of particles or evolution of fields is foliation-independent. This reflects the basic philosophical tenet that space-time is real and dynamics does not depend on the way an observer may choose to view it.

The parametrized canonical formalism enables one to describe particle dynamics without splitting space-time into time and space. The particle trajectories are parametrized by an arbitrary label time τ and characterized by the Hamilton equations of motion

$$\dot{X}^A(\tau) = \{X^A(\tau), M(\tau)H\},$$
$$\dot{P}_A(\tau) = \{P_A(\tau), M(\tau)H\}. \tag{2}$$

The Hamiltonian MH is the product of a multiplier $M(\tau)$ and the super-Hamiltonian $H(X^A, P_A)$. By changing the parametrization τ of the trajectory one can arbitrarily change the multiplier $M(\tau)$. The reparametrization invariance forces on the theory the Hamiltonian constraint

$$H(X^A, P_A) = 0. \tag{3}$$

In the parametrized formalism, time and energy are included in the phase space X^A, P_A; however, what is time and what is energy depends on the split of $\mathcal{M}$ into space and time. For each such split, one can express the energy $-p_t$ in terms of the time t and the true dynamical variables x^a, p_a, and replace the constraint (3) by an equivalent one,

$$H[t] := p_t + h(t, x^a, p_a) = 0. \tag{4}$$

The classical evolution does not depend on which constraint one takes.

Classical mechanics can be reformulated in statistical terms that link it directly to quantum mechanics. An ensemble of classical systems is described by the distribution function $\rho(X^A, P_A)$, which, on the constraint surface $H = 0$, has a vanishing Poisson bracket with the super-Hamiltonian:

$$\{\rho, H\} \approx 0. \tag{5}$$

Equation (5) makes sense without any choice of space and time. After such a choice is made and the super-Hamiltonian is taken in the form (4), equation (5) acquires the meaning of a Liouville equation (Kuchař 1989b). The evolution (5) of the distribution function, like the evolution (2) of an individual classical system, does not depend on the choice of time and the resulting foliation connecting t_{IN} with t_{FIN}.

One can generalize the explained scheme from particles to fields, but I shall avoid this task and introduce instead the main topic of my talk, which is the problem of time in quantum theory. In the Dirac constraint quantization of parametrized systems, one turns the super-Hamiltonian H into an operator $\hat{H} = H(\hat{X}^A, \hat{P}_A)$ and replaces the constraint (3) by a restriction

$$\hat{H}\Psi = 0 \tag{6}$$

on the physical states Ψ of the system. Unfortunately, due to the factor-ordering problem, the specification of the physical states depends on the choice of time in equation (4), and the quantum evolution of the state function Ψ becomes thereby dependent on the choice of foliation. Alternatively, when one relies on the (quadratic) form of the super-Hamiltonian which does not require the space-time split, the problem reappears in a different form: can one endow the space of solutions of equation (6) with a Hilbert space structure that makes the inner product $\langle\Psi_1|\Psi_2\rangle_t$ independent of the time hypersurface on which it is evaluated? These problems persist and increase in complexity in quantum field theory and quantum geometrodynamics. I first want to discuss what form they take in Newtonian and relativistic particle mechanics. The basic question remains the same throughout my investigation and I better repeat it before I plunge into details: can one formulate the rules of the Dirac constraint quantization of parametrized systems so that the evolution of the state function Ψ does not depend on the choice of foliation and the inner product $\langle\Psi_1|\Psi_2\rangle_t$ remains the same on any time hypersurface?

A Newtonian particle gives an example with which all other systems can be compared and set in contrast. Such a particle freely falling in the Newtonian gravitational field is described by the super-Hamiltonian

$$\bar{H} = U^A(X)P_A + \frac{1}{2}G^{AB}(X)P_AP_B + W(X), \tag{7}$$

which is a quadratic function of the momenta (Kuchař 1980; Hartle and Kuchař 1984). The scalar term $W(X)$ is the Newtonian gravitational potential. The coefficient $U^A(X)$ of the linear term is a vector field that defines a "Gaussian" frame of reference. The coefficient $G^{AB}(X)$ of the quadratic term is the Newtonian metric; this metric is degenerate, with signature $(0, +, +, +)$, and satisfies certain integrability conditions, which I shall not write down in detail.

While equation (7) makes an explicit reference to space (through $U^A(X)$), the time function $t(X)$ does not appear in it. One can, however, reconstruct the privileged absolute time foliation from the Newtonian metric G^{AB}. The covectors t_A in whose direction G^{AB} is degenerate, $G^{AB}t_B = 0$, fill a ray, and the integrability conditions guarantee that one covector in this ray is a gradient:

$$\exists\, t(X): t_A = t_{,A}. \tag{8}$$

The scalar function $t(X)$ is the absolute time. One can change t into $\bar{t} = \bar{t}(t)$, but the foliation $t(X) = $ const. is an intrinsic space-time structure determined by the Newtonian metric.

The reference vector field U^A is supposed to be transverse to the foliation and future-pointing:

$$\Lambda := U^A t_{,A} > 0. \tag{9}$$

For simplicity, I scale the constraint (7) down by the factor Λ^{-1}, $H := \Lambda^{-1}\bar{H}$, denote the scaled coefficients by lower case letters, and work from now on directly with the scaled super-Hamiltonian:

$$H = u^A(X)P_A + \frac{1}{2}g^{AB}(X)P_A P_B + w(X). \tag{10}$$

The super-Hamiltonian (10) assumes a familiar form in the comoving coordinates $x^a(X)$ of the Gaussian frame $u^A(X)$:

$$H[t, x^a] = p_t + h, \quad h := \frac{1}{2}g^{ab}p_a p_b + w. \tag{11}$$

Here

$$g^{ab} := g^{AB}x_A^a x_B^b, \quad \text{with } x_A^a := x^a_{,A} \tag{12}$$

and

$$p_t := u^A P_A, \quad p_a := X_a^A P_A, \quad \text{with } X_a^A := X^A_{,a}. \tag{13}$$

The canonical coordinates t, x^a and momenta p_t, p_a satisfy the required Poisson bracket relations. In general, g^{ab} and w depend on t as well as on x^a.

The relation of the Dirac constraint quantization to ordinary nonrelativistic quantum mechanics is most transparent when one starts from the projected form (11) of the super-Hamiltonian. One treats $\hat{t}, \hat{x}^a$ as multiplication operators, the momenta $\hat{p}_t, \hat{p}_a$ as differential operators,

$$\hat{p}_t = -i\mu^{-1/2}\partial_t\mu^{1/2}, \quad \hat{p}_a = -i\mu^{-1/2}\partial_a\mu^{1/2},$$
$$\mu := g^{1/2}, \quad g := \det(g_{ab}), \tag{14}$$

and orders the kinetic energy term h as the Laplace–Beltrami operator:

$$\hat{h} = \frac{1}{2}\mu^{-1/2}\hat{p}_a\mu g^{ab}\,\hat{p}_b\mu^{-1/2} + w. \tag{15}$$

The constraint (6) on the states of the system has the meaning of an ordinary Schrödinger equation. The inner product of two physical states, Ψ_1 and Ψ_2,

$$\langle\Psi_1|\Psi_2\rangle_t := \int_t d^3x\,\mu(t,x)\Psi_1^*(t,x)\Psi_2(t,x), \tag{16}$$

is conserved in time. The density $\mu = g^{1/2}$ in equation (16) sets the measure under which the operators $\hat{x}^a$, $\hat{p}_a$, and $\hat{h}$ are self-adjoint. The ordering (14) of $\hat{p}_t$ is necessary for the conservation of the inner product. All of this done, the Dirac constraint quantization yields the standard quantum theory.

Because the absolute time foliation is uniquely determined by the geometry of the Newtonian space-time, there is no problem of time in nonrelativistic quantum mechanics. There is, however, a potential problem of space. Newtonian space, unlike Newtonian time, is relative, and the reference frame u^A in the super-Hamiltonian (10) can easily be changed by shifting u^A along the absolute time hypersurfaces,

$$u^A \mapsto u'^A = u^A + N^A, \quad N^A t_A = 0. \tag{17}$$

Let the coordinates $x^{a'}$ label the world lines of the new frame, $x^{a'}_{,A}u'^A = 0$. In the coordinates $t' = t$, $x^{a'}$ the super-Hamiltonian (10) assumes the form

$$H[t',x^{a'}] = p_{t'} - N^{a'}p_{a'} + \frac{1}{2}g^{a'b'}p_{a'}p_{b'} + w. \tag{18}$$

Equations (12) and (13) now hold in the new frame, and the shift vector is given by the projection

$$N^{a'} := N^A x^{a'}_A \text{ with } x^{a'}_A = x^{a'}_{,A}. \tag{19}$$

The new reference frame is in general no longer "Gaussian." This is manifested by the presence of the linear term $N^{a'}p_{a'}$ in the super-Hamiltonian (18). One can quantize the motion of the particle in the new frame by turning the constraint (18) into an operator and imposing it as a restriction on the physical states $\Psi'(t',x')$. The momenta $p_{t'}$, $p_{a'}$

are replaced by differential operators (14) and the kinetic energy term is ordered as in equation (15), but now in the primed variables. The standard anticommutator prescription

$$\widehat{N^{a'}p_{a'}} := \frac{1}{2}\left[\hat{N}^{a'},\hat{p}_{a'}\right]_{+} \tag{20}$$

takes care of the extra linear term. The problem of space can now be stated as the question of whether or not the functions $\Psi(t,x)$ and $\Psi'(t',x')$ represent the same physical state $\Psi(X)$. The answer is yes: the transformation $t' = t$, $x^{a'} = x^{a'}(t,x^{b})$ that relates the frames takes the old state function into the new one,

$$\Psi'\left(t' = t,\ x^{a'} = x^{a'}(t,x^{b})\right) = \Psi(t,x^{b}). \tag{21}$$

There is no space problem in nonrelativistic quantum mechanics.

The state functions $\Psi(t,x)$ and $\Psi'(t',x')$ describe how the two observers, u^{A} and u'^{A}, view the same state Ψ. This general statement has a couple of amusing applications (Kuchař 1980). First, take a nonrelativistic particle (Newton's apple) in inertial motion and observe it from two different reference frames, an inertial frame and an Einstein elevator. In the inertial frame, the state function $\Psi(t,x)$ of the particle is a superposition of plane waves. In the elevator, the particle is subject to a homogeneous (fictitious) gravitational field and its state function $\Psi'(t',x')$ should thus be a superposition of Airy functions. This is exactly what the transformation between the states $\Psi(t,x)$ and $\Psi'(t',x')$ yields: the Airy functions are, in a well-defined sense, the Newtonian-relativity transforms of plane waves. The transformation equation is a natural generalization of the well-known Galilei transformation between solutions of the Schrödinger equation to the case of uniformly accelerated motion. This shows how to quantize Newton's apple in Einstein's elevator.

The second application is perhaps even more amusing. Consider Newton's apple falling in a tunnel dug through the earth from Woolsthorpe to its antipode, i.e., the gravitational harmonic oscillator. One can watch the apple either from the rigid frame of reference connected with the earth, or from a synchronous "Gaussian" frame of freely falling reference points released from rest within the tunnel and oscillating to and fro through it. With respect to the rigid frame, the state function of the apple is a superposition of Hermitian polynomials. In the Gaussian frame of reference, the apple is in a free motion and its state function is a superposition of plane waves. Once again, the Hermitian polynomials can be shown to be no more than certain Newtonian-relativity transforms of plane waves.

The independence of the Dirac constraint quantization of the choice of the reference frame suggests that one can state its basic rules without explicitly mentioning any frame whatsoever. One can proceed as follows: take the alternating symbol $\delta^{A_0 A_1 A_2 A_3}$ and project it to the absolute time hypersurface,

$$\delta^{A_1 A_2 A_3} := t_A \delta^{A A_1 A_2 A_3}. \tag{22}$$

Define a covariant form $\delta_{A_1 A_2 A_3}$ of $\delta^{B_1 B_2 B_3}$ such that

$$\delta_{A_1 A_2 A_3} g^{A_1 B_1} g^{A_2 B_2} g^{A_3 B_3} = \delta^{B_1 B_2 B_3} \tag{23}$$

($\delta_{A_1 A_2 A_3}$ is not unique), and define its norm

$$\|\delta\| := \left(\frac{1}{3!} g^{A_1 B_1} g^{A_2 B_2} g^{A_3 B_3} \delta_{A_1 A_2 A_3} \delta_{B_1 B_2 B_3} \right)^{1/2} \tag{24}$$

(which is unique). From (22) and (24), construct the spatial Levi–Civita pseudotensor

$$\varepsilon^{A_1 A_2 A_3} := \|\delta\| \delta^{A_1 A_2 A_3} \tag{25}$$

and its covariant counterpart $\varepsilon_{A_1 A_2 A_3}$, as in equation (23). These objects allow one to factor order the super-Hamiltonian operator in the unprojected form (10),

$$
\begin{aligned}
\hat{H} = {}& \|\delta\|^{1/2} \frac{1}{2} [u^A, -i\partial_A]_+ \|\delta\|^{-1/2} \\
& + \frac{1}{2} \varepsilon^{A_1 A_2 A_3} (-i\partial_{A_1}) \varepsilon^A_{A_2 A_3} (-i\partial_A) + w,
\end{aligned}
\tag{26}
$$

and to impose the constraint equation (6) on the states $\Psi(X)$. They also determine the volume element $d\Sigma$ of the coordinate cell spanned by the edges $d_{(a)} X^A$ lying on the absolute time hypersurface:

$$d\Sigma := \varepsilon_{A_1 A_2 A_3} d_{(1)} X^{A_1} d_{(2)} X^{A_2} d_{(3)} X^{A_3}. \tag{27}$$

Indeed, even for $d_{(a)} X^A$ off the hypersurface, $\varepsilon_{A_1 A_2 A_3}$ takes care of the necessary projections. One can prove that the inner product

$$\langle \Psi_1 \mid \Psi_2 \rangle_t := \int_t d\Sigma \, \Psi_1^* \Psi_2 \tag{28}$$

of two solutions, Ψ_1 and Ψ_2, of the constraint equation (6) does not depend on t. This spells out the quantization rules without singling out any reference frame. When one introduces a frame at this moment, the super-Hamiltonian (26) assumes the form appropriate to it and the inner product (28) reduces back to the expression (16).

It is time to summarize what we have learned about a nonrelativistic particle. The super-Hamiltonian of the particle can be written in a geometric form (10) with help of the Newtonian metric and a Gaussian reference vector field. From the metric, one can extract the privileged absolute time. The Dirac constraint quantization can be carried out in geometric terms. Because the absolute time foliation is fixed, there is no time problem in nonrelativistic quantum mechanics. The geometric factor ordering of the super-Hamiltonian ensures that it is left unchanged by the transformation of the reference frame. This makes nonrelativistic quantum mechanics independent of the choice of space, and provides the transformation theory among the viewpoints of different observers in arbitrary accelerated motion.

Let me now try to approach a relativistic particle in the same spirit (Kuchař 1981a). The super-Hamiltonian is again a quadratic function of the momenta

$$H := \frac{1}{2m}(G^{AB} P_A P_B + m^2),\qquad(29)$$

but the space-time metric G^{AB} is now nondegenerate and equation (29) makes no reference to either time or space. Unlike the Newtonian metric G^{AB}, the relativistic metric G^{AB} does not uniquely select a privileged time foliation. If I want to bring the constraint to the form (10), I must put in the time function $t(X)$ by hand. The leaves of the t-foliation are separated by the lapse function N and their direction is given by the future-pointing unit normal n^A:

$$N := (-t_A t^A)^{-1/2}, \ n^A = -N t^A, \text{ with } t_A := t_{,A}.\qquad(30)$$

The Gaussian frame of reference perpendicular to the foliation is described by the vector field

$$u^A := N n^A, \ t_A u^A = 1.\qquad(31)$$

With the help of these variables, I cast the constraint (29) into an equivalent "square-rooted" form

$$H[t] := u^A P_A + (N^2 g^{AB} P_A P_B + m^2 N^2)^{1/2}\qquad(32)$$

adopted to the foliation. Here, the metric G^{AB} is projected to the leaves of the foliation,

$$g^{AB} := G^{AB} + n^A n^B. \tag{33}$$

Both g^{AB} and m are then scaled by the square of the lapse function. The similarity between the relativistic constraint (32) and the Newtonian constraint (10) is obvious.

Relativistic particles move along future-oriented time-like world lines,

$$u^A P_A < 0. \tag{34}$$

In the sector (34) of the phase space, the square-rooted super-Hamiltonian (32) is related to the quadratic super-Hamiltonian (29) by a positive factor:

$$H = \frac{1}{2m} N^{-2} \left(-u^A P_A + (N^2 g^{AB} P_A P_B + m^2 N^2)^{1/2} \right) H[t]. \tag{35}$$

These two forms of the constraint are thus equivalent, $H \approx H[t]$. So are any two square-rooted constraints adopted to two different foliations, $H[t_1] \approx H[t_2]$. The dynamical evolutions generated by all such constraints are mutually consistent. This is ensured by the closure of the Poisson brackets

$$\begin{aligned}
\{H[t_1], H[t_2]\} &= \text{``}C\text{''}\, H[t_1] H[t_2] + \text{``}C\text{''}\, H, \\
\{H[t], H\} &= \text{``}C\text{''}\, H[t] + \text{``}C\text{''}\, H.
\end{aligned} \tag{36}$$

The structure functions "C" are fairly hideous; they are ratios of expressions containing linear and quadratic polynomials and their square roots. However, they are all regular on the constraint surface in the sector (34). In particular, if the two time functions t_1 and t_2 yield the same initial and the same final hypersurface, the classical evolution from t_{IN} to t_{FIN} does not depend on whether one connects t_{IN} with t_{FIN} by the foliation t_1 or t_2. There is no problem of time in classical relativistic mechanics.

In the Gaussian coordinates x^a labeling the world lines of the frame u^A, the super-Hamiltonian (32) takes the form

$$\begin{aligned}
H[t, x^a] &= p_t + h, \\
h = k_g &:= (N^2 g^{ab} p_a p_b + m^2 N^2)^{1/2}.
\end{aligned} \tag{37}$$

The variables p_t, p_a and g^{ab} are defined by the same equations (12)–(13) as their Newtonian counterparts. The dynamical variable h is the true Hamiltonian with respect to the foliation t and the frame x^a.

As in Newtonian mechanics, one can study the motion in an arbitrary frame of reference (17). In the new coordinates $t' = t$, $x^{a'}$, the Hamiltonian acquires a term linear in the momenta:

$$H[t', x^{a'}] = p_{t'} + h', \quad h' = -N^{a'} p_{a'} + k'_g,$$
$$k'_g := (N^2 g^{a'b'} p_{a'} p_{b'} + m^2 N^2)^{1/2}. \tag{38}$$

The relativistic super-Hamiltonians (37) and (38) are analogous to the nonrelativistic expressions (11) and (18).

I shall now quantize the motion of the relativistic particle by the algorithm developed for the Newtonian particle. Once again, the coordinates t, x^a and their conjugate momenta p_t, p_a are turned into operators (14) that act on a Hilbert space with the inner product (16). The density μ in equation (16) is now given by the determinant of the metric $N^{-2} g_{ab}$: $\mu = N^{-3} g^{1/2}$. (Other choices of μ are also possible; later on I shall point out that $\mu = N^{-1} g^{1/2}$ is a better choice in static space-times.)

The operators $\widehat{k_g^2}$ and $\widehat{k_g'^2}$ under the square root in equations (37)–(38) are factor ordered as in equation (15),

$$\widehat{k_g^2} := \mu^{-1/2} \hat{p}_a \mu N^2 g^{ab} \hat{p}_b \mu^{-1/2} + m^2 N^2, \tag{39}$$

and the square root $\hat{k}_g$ is defined by spectral analysis. The super-Hamiltonian (37) is thereby turned into an operator, and the constraint (6) is imposed as a restriction on physical states. The inner product (16) between two such states is then conserved in t.

The same procedure can be used in an arbitrary frame of reference (38). With the linear term $N^{a'} p_{a'}$ ordered as the anticommutator (20), the state functions $\Psi'(t', x')$ and $\Psi(t, x)$ given in different frames (but in the same foliation) are again related by equation (21). Relativity of space is no problem, even for a quantized relativistic particle.

Unfortunately, relativity of time is another thing. First of all, the square-rooted constraint $\hat{H}[t]\Psi = 0$ is not equivalent to the quadratic constraint $\hat{H}\Psi = 0$ interpreted as the Klein–Gordon equation. Indeed, the action of $-\widehat{u^A P_A} + \hat{k}_g$ on $\hat{H}[t]$ as in equation (35) does not lead to the Klein–Gordon super-Hamiltonian

$$\hat{H} := \frac{1}{2m}(-\Box_G + m^2) \tag{40}$$

because $\widehat{u^A P_A} = \hat{p}_t$ does not in general commute with $\hat{k}_g$. Further, the constraints

$$\hat{H}[t_1]\Psi = 0 \quad \text{and} \quad \hat{H}[t_2]\Psi = 0 \tag{41}$$

belonging to two different foliations, t_1 and t_2, are no longer compatible. Technically, the described factor ordering does not take the classical closure relation (36) into a commutator relation in which the operators $\hat{H}[t_1]$, $\hat{H}[t_2]$ and $\hat{H}$ on its right-hand side would act on Ψ before the "C'"s and annihilate it. The quantum evolution of a state then depends on the foliation $t(X)$ chosen to connect t_{IN} with t_{FIN}. The quantization procedure which I have described yields different quantum mechanics for each choice of $t(X)$. No factor ordering prescription is known that, based purely on the geometric structures used in writing down the super-Hamiltonians (29) and (32), would make the redundant constraint system (41) consistent. There is every reason to believe that such a factor ordering does not exist.

A solution to this problem is known only in stationary space-times. A stationary metric G^{AB}, through its Killing vector field U^A, provides a geometrically privileged reference frame, just as the Newtonian metric G^{AB}, through its degeneracy direction t_A, provided a privileged foliation. The way in which a privileged *space* brings a solution to the *time* problem is indirect and rather interesting. It relies on the results that were developed in the context of complex structures and their relation to quantum field theory on curved backgrounds (Ashtekar and Magnon 1975, 1978; Magnon–Ashtekar 1976; Kay 1978). I shall present them here within the framework of the Dirac constraint quantization.

The starting point is the privileged (stationary) reference frame given by the Killing vector field **U**,

$$(L_{\mathbf{U}}\mathbf{G})_{AB} = -\nabla_{(A}U_{B)} = 0. \tag{42}$$

I label its world lines by the coordinates x^a,

$$x^a_A U^A = 0, \quad x^a_A := x^a_{,A}. \tag{43}$$

I choose a space-like hypersurface transverse to U^A, and by its Lie propagation along U^A I generate a stationary foliation $t(X)$,

$$t_A U^A = 1, \quad t_A := t_{,A}. \tag{44}$$

A different choice of the initial hypersurface leads to a different stationary foliation. The new foliation $\bar{t}(X)$ is related to the old one by a function $\sigma(x)$ that depends only on the spatial coordinates:

$$\bar{t}(X) = t(X) + \sigma(x(X)). \tag{45}$$

I thus have a fixed frame, but a large family of foliations; the transition (45) from one foliation to another can be considered as a gauge transformation.

I can decompose the Killing vector field into the lapse and the shift components,

$$U^A = Nn^A + N^a X_a^A,$$ (46)

and arrive thus at the square-rooted super-Hamiltonian (38). (The Gaussian super-Hamiltonian (37) is of no use in the present context and I shall therefore omit the primes in equation (38).) The Killing equation (42) ensures that the coefficients N, N^a and g^{ab} depend only on x^a but not on t. This does not, however, solve the factor ordering problem.

The energy

$$E := -U^A P_A$$ (47)

of the particle in the stationary reference frame is a constant of motion,

$$\{E, H\} = (L_{\mathbf{U}}\mathbf{G})^{AB} P_A P_B = 0.$$ (48)

This energy is positive ($E > 0$) in the physical sector of the phase space. The resolution of the time problem is based on turning the *quadratic* super-Hamiltonian H and the energy E into operators such that equation (48) is preserved in the quantum domain, and on using the positivity of E for the construction of the Hilbert space inner product.

In quantization, the super-Hamiltonian (29) is replaced by the Klein–Gordon operator (40), and the energy (47) by the operator

$$\hat{E} := i(U^A \partial_A + \frac{1}{2} \mathrm{div}_{\mathbf{G}} U) = iU^A \partial_A.$$ (49)

Let both operators act on the function space $\mathcal{F} = C^\infty(\mathcal{M}, \mathcal{C})$. The Killing condition (42) implies that $\mathrm{div}_{\mathbf{G}} U = 0$; as a result, equation (48) remains valid for the commutator:

$$\frac{1}{i}[\hat{E}, \hat{H}] = 0.$$ (50)

This means that the eigenvalue equation for E,

$$\hat{E}\Psi = E\Psi,$$ (51)

and the constraint equation (6) have common solutions. Span $\mathcal{F}^+ \subset \mathcal{F}$ by the positive-energy solutions of equation (51), and let $\mathcal{F}_0 \subset \mathcal{F}$ be the space of solutions of the quantum constraint (6). The intersection $\mathcal{F}_0^+ = \mathcal{F}^+ \cap \mathcal{F}_0$ of these two spaces is the physical space. It is this space that must be endowed with an inner product and completed into a Hilbert space.

The construction of the inner product starts with the current

$$J_{12}^A := \frac{1}{2i} G^{AB} (\Psi_1^* \overset{\leftrightarrow}{\partial}_B \Psi_2). \tag{52}$$

If Ψ_1 and Ψ_2 lie in $\mathcal{F}_0$, the current (52) is conserved,

$$\nabla_A J_{12}^A = |G|^{-1/2} \partial_A (|G|^{1/2} J^A) = 0, \tag{53}$$

and the integral

$$\langle \Psi_1 \mid \Psi_2 \rangle := - \int_\Sigma d\Sigma_A J_{12}^A,$$
$$d\Sigma_A := \varepsilon_{A A_1 A_2 A_3} d_{(1)} X^{A_1} d_{(2)} X^{A_2} d_{(3)} X^{A_3}, \tag{54}$$

does not depend on the choice of the hypersurface Σ. However, on $\mathcal{F}_0$ the product (54) is not positive definite.

The general solution of equation (51) is given by the expression

$$\Psi_E(X) = \phi_E(x(X)) \, e^{-iEt(X)}, \tag{55}$$

where $t(X)$ is a stationary time function (44) and $\phi_E(x)$ an arbitrary function of the space coordinates (43). Under the change (45) of $t(X)$, $\Psi_E(X)$ is kept fixed when $\phi_E(x)$ is changed into

$$\bar{\phi}_E(x) = \phi_E(x) \, e^{iE\sigma(x)}. \tag{56}$$

Stationary states are those solutions of equation (51) that simultaneously satisfy the constraint (6), (40). This requirement yields an eigenvalue problem for $\phi_E(x)$.

In the t, x^a coordinates, the Klein–Gordon super-Hamiltonian takes the form

$$\hat{H} = \frac{1}{2m} N^{-2} \left(-\hat{E}^2 - 2\widehat{N^a p_a} \hat{E} + \widehat{k_h^2} \right), \tag{57}$$

where $\widehat{N^a p_a}$ is given by equation (20) and $\widehat{k_h^2}$ by the expression

$$\widehat{k_h^2} := \mu^{-1/2} \hat{p}_a \, \mu N^2 h^{ab} \, \hat{p}_b \mu^{-1/2} + m^2 N^2. \tag{58}$$

The density μ in equation (58) is $\mu = N^{-1}g^{1/2}$. The metric

$$h^{ab} := G^{AB}x_A^a x_B^b \tag{59}$$

depends only on the frame, not on the foliation. (The line element $ds = (h_{ab}dx^a dx^b)^{1/2}$ measures the perpendicular distance between the world lines x^a and $x^a + dx^a$.) It differs from the metric g^{ab} induced on the leaves of the foliation by a shift term,

$$h^{ab} = g^{ab} - N^{-2}N^a N^b. \tag{60}$$

When I impose the Hamiltonian constraint on Ψ_E, I get the eigenvalue problem

$$(E^2 + 2E\,\widehat{N^a p_a})\phi_E = \widehat{k_h^2}\,\phi_E. \tag{61}$$

Note that if ϕ_E is a solution with a positive energy E, ϕ_E^* is a solution with the negative energy $-E$.

The operators $\widehat{k_h^2}$, $\widehat{N^a p_a}$ that specify the eigenvalue problem (61) are Hermitian under an auxiliary inner product

$$(\phi_1 \mid \phi_2) := \int d^3x\, \mu(x)\phi_1^*(x)\phi_2(x). \tag{62}$$

The physical inner product (54) can be related to the product (62) through the lapse-shift decomposition (46). By incorporating Σ into a stationary foliation, one gets

$$\langle \Psi_1 \mid \Psi_2 \rangle_\Sigma = \frac{1}{2}(\Psi_1 \mid (\hat{E} + \widehat{N^a p_a}) \mid \Psi_2)_t + \frac{1}{2}((\hat{E} + \widehat{N^a p_a})\Psi_1 \mid \Psi_2)_t. \tag{63}$$

For simplicity, I shall assume that the space is compact and the energy has a discrete spectrum. I then show that: (I) the stationary states Ψ_m, Ψ_n belonging to different energy levels $E_m \neq E_n$ are orthogonal under the product (54); (II) the norm (54) of a stationary state with positive energy is positive; and (III) the product (54) is positive-definite on $\mathcal{F}_0^+$.

(I) Write equation (61) for an eigenfunction ϕ_n and take its product (62) with an eigenfunction ϕ_m. Do the same thing in the opposite order. For $E_m \neq E_n$, this yields an identity

$$\frac{1}{2}(E_m + E_n)(\phi_m \mid \phi_n) + (\phi_m \mid \widehat{N^a p_a} \mid \phi_n) = 0. \tag{64}$$

However, by equation (63), the expression (64) is just the inner product $\langle \Psi_m \mid \Psi_n \rangle$. The stationary states Ψ_m and Ψ_n belonging to different energies are thus orthogonal,

$$\langle \Psi_m \mid \Psi_n \rangle = 0 \ \text{ for } \ E_m \neq E_n. \tag{65}$$

The energy levels are, of course, in general degenerate. By orthogonalization one can choose (a countable) basis Ψ_{nd} at each level E_n such that

$$\langle \Psi_{nd} \mid \Psi_{nd'} \rangle = 0 \ \text{ for } \ d \neq d'. \tag{66}$$

(II) Write equation (61) for an eigenfunction ϕ_E and take its product (62) with ϕ_E itself. This yields

$$(\phi_E \mid \widehat{N^a p_a} \mid \phi_E) + \frac{1}{2} E(\phi_E \mid \phi_E) = \frac{1}{2} E^{-1}(\phi_E | \widehat{k_h^2} | \phi_E). \tag{67}$$

On the other hand, from equation (63),

$$\langle \Psi_E \mid \Psi_E \rangle_\Sigma = E(\phi_E \mid \phi_E) + (\phi_E | \widehat{N^a p_a} | \phi_E). \tag{68}$$

The identity (67) casts the norm (68) into a manifestly positive form:

$$\langle \Psi_E \mid \Psi_E \rangle_\Sigma = \frac{1}{2} E(\phi_E \mid \phi_E) + \frac{1}{2} E^{-1}(\phi_E | \widehat{k_h^2} | \phi_E). \tag{69}$$

(III) Normalize the eigenfunctions Ψ_{nd} to 1 and expand the general solution $\Psi(X) \in \mathcal{F}_0^+$ of the constraint equation (6), (40) in them:

$$\Psi(X) = \sum_{nd} c_{nd} \Psi_{nd}(X) = \sum_{nd} c_{nd} \phi_{nd}(x(X)) e^{-i E_n t(X)}. \tag{70}$$

Due to the orthonormality of $\Psi_{nd}(X)$ under the inner product (54), the norm of Ψ is given by the sum

$$\langle \Psi \mid \Psi \rangle = \sum_{nd} |c_{nd}|^2; \tag{71}$$

it is obviously positive definite. One can complete $\mathcal{F}_0^+$ in the norm (71) to a Hilbert space.

While the auxiliary inner product (62) depends, through the lapse function N, on the choice of foliation, the physical inner product (54) can be shown to be gauge-invariant under the change (45), (56).

How does the explained procedure resolve the time problem in stationary space-times? At first there seems to be very little connection between the two. Recall that the time problem is the question of consistency of Schrödinger equations corresponding to square-rooted super-Hamiltonians adopted to different foliations. The procedure did not introduce any Schrödinger equation, but it was based entirely on the second-order Klein–Gordon equation: the physical space $\mathcal{F}_0^+$ was spanned by positive-energy solutions $\Psi_{nd}(X)$ of this equation. In the Schrödinger approach, the positivity of the inner product and its conservation are self-evident. In the Klein–Gordon approach, the construction of a positive inner product on the space of solutions $\mathcal{F}_0^+$ requires a long argument. How is the Schrödinger approach related to the Klein–Gordon approach?

There are two clues to follow. First, the concept of a stationary *state* $\Psi_{nd}(X)$ is intimately connected with that of a stationary *foliation*, i.e., with a stationary progress in time. Second, the inner product on the space $\mathcal{F}_0^+$ of *solutions* is introduced indirectly by means of an expression based on a hypersurface Σ; this expression is then proved to be independent of Σ, i.e., *conserved*. This leads one to suspect that there is a way of looking at things that explains the procedure in terms of Schrödinger evolutions generated by consistently factor-ordered square-rooted super-Hamiltonians. I shall reveal this connection by essentially running the conventional argument in a reversed order.

The physical states $\Psi \in \mathcal{F}_0^+$ are orthogonal to all the negative-energy eigenstates $\Psi_{nd}^* \in \mathcal{F}_0^-$:

$$\langle \Psi_{nd}^* \mid \Psi \rangle = 0. \tag{72}$$

I elevate this equation to the basic postulate of the theory, but while doing that, I change its interpretation: I do not assume that Ψ lies in $\mathcal{F}_0^+$ or even in $\mathcal{F}_0$; it is now simply some function from $\mathcal{F}$. The inner product is defined as before, by equations (53)–(54), but because Ψ does not necessarily satisfy the Klein–Gordon equation, it is no longer to be taken for granted that the bracket (72) does not depend on Σ. I thus postulate that the state $\Psi \in \mathcal{F}$ satisfies equation (72) on *any* hypersurface:

$$\langle \Psi_{nd}^* \mid \Psi \rangle_\Sigma = 0 \quad \forall \, \Psi_{nd}^* \in \mathcal{F}_0^- \ \text{ and } \ \forall \Sigma. \tag{73}$$

Because the product (73) is the flux of a current $J^A(\Psi_{nd}^*, \Psi)$ through Σ and this flux is the same, namely zero, on every hypersurface, it follows by a familiar argument that the divergence of the current must vanish, equation (53). From the structure (52) of the current and the Klein–

Gordon equation for Ψ_{nd}^*, it follows that

$$\nabla_A J^A(\Psi_{nd}^*, \Psi) = \frac{1}{2}i\ \Psi_{nd}^*(X)\big(-\Box_G \Psi(X) + m^2 \Psi(X)\big) = 0. \quad (74)$$

At any given X, at least some $\Psi_{nd}^*(X)$ do not vanish, and $\Psi(X)$ must therefore satisfy the Klein–Gordon equation. Equation (73) thus forces the state Ψ first into $\mathcal{F}_0$ and then into $\mathcal{F}_0^+$.

Now weaken equation (73) by assuming that it does not hold on every Σ, but only on the leaves of a given foliation $t(X)$:

$$\langle \Psi_{nd}^* \mid \Psi \rangle_t = 0 \quad \forall \Psi_{nd}^* \in \mathcal{F}_0^- \text{ and } \forall t : t(X) = t = \text{const.} \quad (75)$$

Due to the structure of the inner product, equation (75) is equivalent to a first-order differential equation in t (I shall give an explicit form of this equation for a stationary foliation in a moment). Therefore, the value of $\Psi(X)$ on t_{IN} determines $\Psi(X)$ everywhere. The strong equation (73) is known to have many solutions, namely, all the functions (70) from $\mathcal{F}_0^+$, and it does not restrict in any way the distribution of $\Psi(X)$ on a given hypersurface $\Sigma_0 : t(X) = t_{IN}$. The function $\Psi(X)$ that I found by solving the weaker equation (75) must therefore also solve the strong equation (73), and consequently any weaker equation obtained by restricting equation (73) to a different foliation $\bar{t}(X)$:

$$\langle \Psi_{nd}^* \mid \Psi \rangle_{\bar{t}} = 0 \quad \forall \Psi_{nd}^* \in \mathcal{F}_0^- \text{ and } \forall \bar{t} : \bar{t}(X) = \bar{t} = \text{const.} \quad (76)$$

If the foliations $t(X)$ and $\bar{t}(X)$ happen to have the same initial and the same final hypersurfaces, it follows that a given initial distribution Ψ_{IN} on t_{IN} gets evolved into the same final distribution Ψ_{FIN} on t_{FIN}, irrespective of whether it is evolved by equation (75) along the foliation $t(X)$ or by equation (76) along the foliation $\bar{t}(X)$. This resolves the problem of time in stationary space-times.

Equation (75) is equivalent to a square-rooted constraint $\hat{H}[t]\Psi = 0$; it amounts to projecting that constraint to some function basis. Similarly, equation (76) is equivalent to $\hat{H}[\bar{t}]\Psi = 0$. The consistency of equations (75) and (76), which are both entailed in a single self-consistent system (73), means that the super-Hamiltonians $\hat{H}[t]$ and $\hat{H}[\bar{t}]$ must have a commutator that satisfies an appropriate closure relation.

I shall now exhibit the super-Hamiltonian $\hat{H}[t]$ explicitly for a stationary foliation. By putting $\Psi_1 = \Psi_{nd}^*$ and $\Psi_2 = \Psi$ in equation (63), I cast equation (75) into the form

$$(\phi_{nd}^* \mid (i\partial_t + 2\widehat{N^a p_a})\Psi)_t = E_n(\phi_{nd}^* \mid \Psi)_t. \quad (77)$$

My task is to calculate from this equation the derivative $i\partial_t\Psi(t,x)$. The auxiliary inner product

$$(\phi_{nd} \mid \phi_{n'd'}) =: f_{nd\ n'd'}, \quad f^*_{nd\ n'd'} = f_{n'd'\ nd} \tag{78}$$

defines a Hermitian metric $f_{nd\ n'd'}$ in the space $L^2(\mathcal{S}, \mu d^3 x)$. Its inverse, $f^{*nd\ n'd'}$, defined by the equation

$$f^{*nd\ n''d''} f_{n''d''\ n'd'} = \delta^n_{n'}\delta^d_{d'}, \tag{79}$$

is also Hermitian. The functions

$$\phi^{nd} := f^{nd\ n'd'}\phi_{n'd'} \tag{80}$$

form a basis that is dual to ϕ_{nd}:

$$(\phi^{nd} \mid \phi_{n'd'}) = \delta^n_{n'}\delta^d_{d'}. \tag{81}$$

The standard completeness relation

$$|\phi_{nd})(\phi^{nd}| = \mathbf{1} \tag{82}$$

can be checked by applying the operator equation (82) to the basis vectors. I use it for the complex conjugate basis ϕ^*_{nd} and write it in the form

$$|\phi^*_{nd})(\phi^{*nd}| = \mathbf{1} \leftrightarrow \phi^*_{nd}(x)f^{nd\ n'd'}\phi_{n'd'}(x')\mu(x') = \delta(x,x'). \tag{83}$$

It is this completeness relation which helps me to cast equation (77) into the Schrödinger form

$$i\partial_t\Psi(t,x) = (-2\,\widehat{N^a p_a} + \hat{K})\Psi(t,x). \tag{84}$$

The action of the operator $\hat{K}$ on the state function $\Psi(t,x)$ is defined by its kernel $K(x,x')$:

$$\hat{K}\Psi(t,x) = \int d^3x'\,\mu(x')K(x,x')\Psi(t,x'), \tag{85}$$

$$K(x,x') := \sum_{ndn'd'} E_{n'}\phi^*_{nd}(x)f^{nd\ n'd'}\phi_{n'd'}(x'). \tag{86}$$

The operator $\widehat{N^a p_a}$ is Hermitian under the inner product (62), but the operator $\hat{K}$ is not: $K^*(x,x') = K(x',x)$ if and only if the metric

$f^{nd\ n'd'}$ is diagonal in nn'. This happens only if the space-time is static. In this case, $N^a = 0$ and equation (63) implies that

$$\delta_{nd\ n'd'} = \langle \Psi_{nd} \mid \Psi_{n'd'} \rangle = \frac{1}{2}(E_n + E_{n'})(\phi_{nd} \mid \phi_{n'd'}). \tag{87}$$

The metric $f_{nd\ n'd'}$ and its inverse are then given in an explicit form

$$f_{nd\ n'd'} = E_n^{-1}\delta_{nn'}\delta_{dd'} \quad \text{and} \quad f^{nd\ n'd'} = E_n\delta^{nn'}\delta^{dd'}. \tag{88}$$

In a stationary space-time, $\hat{K}$ is not Hermitian and the inner product $(\Psi_1|\Psi_2)_t$ is thus not conserved in time. This does not matter; the statistical interpretation of the theory is based on the inner product $\langle \Psi_1 \mid \Psi_2 \rangle_t$, which is a conserved quantity. These two products are related by equation (63), which, after the time derivatives are eliminated by means of the Schrödinger equation (84), yields

$$\langle \Psi_1 \mid \Psi_2 \rangle_t = \frac{1}{2}(\Psi_1 \mid \hat{K}\Psi_2)_t + \frac{1}{2}(\hat{K}\Psi_1 \mid \Psi_2)_t - (\Psi_1|\widehat{N^a p_a}|\Psi_2)_t. \tag{89}$$

It is obvious that equation (89) depends only on the Hermitian part $\hat{K}_S := \frac{1}{2}(\hat{K} + \hat{K}^*)$ of $\hat{K}$:

$$\langle \Psi_1 \mid \Psi_2 \rangle_t = (\Psi_1|\hat{K}_S|\Psi_2)_t - (\Psi_1|\widehat{N^a p_a}|\Psi_2)_t. \tag{90}$$

From the Schrödinger equation (84), I can read off the super-Hamiltonian operator

$$\hat{H}[t] = -\hat{E} - 2\,\widehat{N^a p_a} + \hat{K}. \tag{91}$$

Its comparison with the classical super-Hamiltonian (38) reveals that the operator $\hat{K}$ corresponds to the dynamical variable

$$K = N^a p_a + N(g^{ab}p_a p_b + m^2)^{1/2}. \tag{92}$$

It is the factor ordering (85)–(86) that makes the square-rooted super-Hamiltonian (91) consistent with the Klein–Gordon super-Hamiltonian (40), and the square-rooted super-Hamiltonians belonging to different stationary foliations consistent with each other. This factor ordering is different from that imposed in equation (39). Indeed, if I defined the square root in equation (92) by spectral analysis under the measure $\mu = N^{-1}g^{1/2}$ and added it to the operator $\widehat{N^a p_a}$, I would not get $\hat{K}$.

(Remember that the super-Hamiltonian (91), factor-ordered in that way, it not equivalent to the Klein–Gordon super-Hamiltonian.)

The factor ordering of equation (39) treats $N^2 g^{ab} p_a p_b$ as a single quantity. This is not what the factor ordering (92) does. By completing the Klein–Gordon super-Hamiltonian into a sum of squares,

$$\hat{H} = \frac{1}{2m} N^{-2} \left(-(\hat{E} + \widehat{N^a p_a})^2 + \widehat{k_h^2} + (\widehat{N^a p_a})^2 \right), \qquad (93)$$

one sees that the metric g^{ab} was decomposed according to equation (60) and the two pieces of k_g^2, namely, k_h^2 and $N^a p_a$, were ordered separately. These two factor orderings differ because

$$\widehat{k_h^2} + (\widehat{N^a p_a})^2 = \widehat{k_g^2} + \frac{1}{2} \left(L_{\vec{N}} \operatorname{div}_\mu \vec{N} + \frac{1}{2} (\operatorname{div}_\mu \vec{N})^2 \right). \qquad (94)$$

This illustrates the general connection of the problem of time with the factor-ordering problem. The described factor orderings coincide only in static space-times on a static foliation: the Schrödinger equation obtained from the super-Hamiltonian (37) by defining the square root of the operator (39) by spectral analysis with $\mu = N^{-1} g^{1/2}$ is then the same as the Schrödinger equation (84).

Let me summarize what we have learned about the relativistic particle. First of all, my attempt to quantize the motion of such a particle in the same way as the motion of a nonrelativistic particle failed: the constraints (41) belonging to different foliations were incompatible, and they did not lead to the Klein–Gordon equation. Such a quantization makes quantum mechanics dependent on the choice of foliation. Secondly, the resolution of this problem is known only in stationary space-times. The energy of the particle in the stationary frame gives an operator that commutes with the Klein–Gordon super-Hamiltonian and thus allows one to span the Hilbert space by stationary states of positive energy. The condition that the state function be orthogonal to all negative-energy states yields a Schrödinger equation along a given foliation and determines the factor ordering of its Hamiltonian. The Schrödinger equations corresponding to different foliations are compatible. In particular, when a given Ψ_{IN} on t_{IN} is evolved to t_{FIN} along different foliations, the Ψ_{FIN} does not depend on the foliation. Unfortunately, if the space-time is dynamic, there is no conserved energy operator, and the construction of a foliation-independent quantum mechanics breaks down. The problem of time in relativistic quantum mechanics remains unresolved within the framework of a single-particle theory.

It would take too long if I tried to explain what form the problem takes in a second-quantized theory describing a many-particle system. (I have touched on some aspects of this question in Kuchař 1989a,b). I prefer to carry my cause directly to canonical geometrodynamics, which is our real subject of interest. Do the simple particle models that I have discussed so far have any bearing on the problem of time in such a complicated and ill-understood field system?

In canonical geometrodynamics, the single super-Hamiltonian constraint (3) for the space-time position X^A and the four-momentum P_A of the particle is replaced by an infinity of super-Hamiltonian constraints

$$H(x; \mathbf{g}, \mathbf{p}] := T(x; \mathbf{g}, \mathbf{p}) + V(x; \mathbf{g}] = 0, \tag{95}$$

$$T(x; \mathbf{g}, \mathbf{p}) = G_{ab\ cd}(x; \mathbf{g}) p^{ab}(x) p^{cd}(x),$$
$$V(x; \mathbf{g}] = -g^{1/2}(x; \mathbf{g}) R(x; \mathbf{g}], \tag{96}$$

$$G_{ab\ cd} := \frac{1}{2} g^{-1/2} (g_{ac} g_{bd} + g_{ad} g_{bc} - g_{ab} g_{cd}) \tag{97}$$

for the metric $g_{ab}(x)$ and its conjugate momentum $p^{ab}(x)$ on a space-like hypersurface Σ. (Here, $R(x; \mathbf{g}]$ is the curvature scalar on Σ.) The constraints (95) are quadratic in the momenta $p^{ab}(x)$ as the relativistic particle constraint (29) was quadratic in P_A. Besides the super-Hamiltonian constraints (95), $g_{ab}(x)$ and $p^{ab}(x)$ are also subject to an infinity of linear supermomentum constraints:

$$H_a(x; \mathbf{g}, \mathbf{p}] := -2p^b_{a|b}(x) = 0. \tag{98}$$

(The vertical stroke denotes the spatial covariant derivative with respect to the induced metric g_{ab}.) The Hamiltonian $H(N, \vec{N})$ is obtained by smearing the constraints (95) and (98),

$$H(N, \vec{N}) := \int_\Sigma d^3x\, N(x) H(x) + \int_\Sigma d^3x\, N^a(x) H_a(x). \tag{99}$$

It evolves the dynamical variables $g_{ab}(x)$, $p^{ab}(x)$ along a foliation of space-like hypersurfaces separated by the lapse function $N(x)$ and displaced by the shift vector $N^a(x)$. The Hamilton equations of motion are the field-theoretical counterparts of equation (2).

One can try to quantize gravity by turning the constraints (95) and (98) into operators and imposing them as restrictions on the state functional $\Psi[g_{ab}]$:

$$H(x; \hat{\mathbf{g}}, \hat{\mathbf{p}}] \Psi[g_{ab}] = 0, \quad H_a(x; \hat{\mathbf{g}}, \hat{\mathbf{p}}] \Psi[g_{ab}] = 0. \tag{100}$$

The meaning of the second set of constraints (the supermomentum constraints) is well understood (Misner 1957; Higgs 1958): the state functional $\Psi[g_{ab}]$ can depend only on the three-geometry **g**, not on its particular representation by the metric $g_{ab}(x)$. The first set of constraints is the familiar Wheeler–DeWitt equation (Wheeler 1962, 1964; DeWitt 1967). In so far as the classical constraint (95) is quadratic in the momenta $p^{ab}(x)$, this equation resembles the Klein–Gordon equation. What does this analogy suggest about the problem of time?

In parametrized particle dynamics, not all the configuration variables X^A are dynamical degrees of freedom; one particular combination of them, $t(X^A)$, has the meaning of time. Similarly, in canonical geometrodynamics not all components $g_{ab}(x)$ of the metric are dynamical; one can surmise that ∞^3 combinations of them,

$$t(x; \mathbf{g}], \tag{101}$$

identify the time of the event that the point $x \in \Sigma$ of the hypersurface Σ carrying the metric $g_{ab}(x)$ occupies in the embedding space-time. (Indeed, another $3\infty^3$ combinations of $g_{ab}(x)$ can also be expected not to be dynamical, but serve rather as the positional indicators of that event.) The ∞^3 functionals (101) of the metric can be given the suggestive name of an "intrinsic time" ("intrinsic" because it is constructed entirely from the intrinsic metric of Σ, and "time" because it specifies how Σ is embedded in $\mathcal{M}$).

Subject to the causality restrictions, the time function $t(X^A)$ was arbitrary; similarly, the intrinsic time functionals (101) are highly arbitrary. (I will evade the question of what causality restrictions should be imposed on them.) In particle dynamics, the arbitrariness of $t(X^A)$ led to the problem of time in the Dirac constraint quantization; similarly, the arbitrariness of $t(x; \mathbf{g}]$ can be expected to lead to the problem of time in quantum geometrodynamics. In particle dynamics, the only known resolution of the time problem relies on the stationarity of the space-time background. I should thus look for a comparable situation in geometrodynamics.

While particles move in space-time, the spatial geometries **g** move in superspace (Wheeler 1968). Space-time enters into the particle super-Hamiltonian (29) through the metric $G^{AB}(X)$; superspace enters into the geometrodynamical super-Hamiltonians (95) through the supermetric $G_{ab\ cd}(x; \mathbf{g})$ and the potential $V(x; \mathbf{g})$. The resolution of the time problem in particle dynamics depended on the isometry of the metric $G^{AB}(X)$. In the canonical language, the isometry is generated by the energy variable (47), which is linear in the momentum and has a vanishing Poisson bracket (48) with the super-Hamiltonians. To decide whether the superspace structures $G_{ab\ cd}(x; \mathbf{g})$ and $V(x; \mathbf{g}]$ admit an "isometry," I should look for a dynamical variable

$$E := - \int_\Sigma d^3x\, u_{ab}(x; \mathbf{g}] p^{ab}(x) \qquad (102)$$

that is linear in the momentum $p^{ab}(x)$ and has a vanishing Poisson bracket with the super-Hamiltonians (95). Indeed, E should also have a vanishing Poisson bracket with the supermomenta (97), but this goal is easily met by requiring that the coefficient $u_{ab}(x; \mathbf{g}]$ in equation (102) be a spatial tensor field.

To be entirely general, one should also allow for the possibility that the Poisson brackets of E with the super-Hamiltonians vanish only weakly, i.e., modulo the constraints (95) and (97) themselves. (A similar caveat should have been made in particle dynamics; however, there it is easy to prove that if $m \neq 0$ and the Poisson bracket $\{E, H\}$ vanishes weakly, then it must actually vanish strongly. For $m = 0$ this is no longer true: The condition $\{E, H\} \approx 0$ does not imply $\{E, H\} = 0$ but only $\{E, H\} = \Lambda(X)H$, and G^{AB} does not then admit an isometry, but only a conformal isometry. This is actually enough for resolving the time problem for a zero-mass particle.)

To summarize, one can hope to resolve the time problem in quantum geometrodynamics along the same lines as the time problem in particle dynamics on a stationary background if there exists a dynamical variable (102) whose Poisson brackets with the super-Hamiltonians (95) weakly vanish,

$$\{E, H(x)\} \approx 0. \qquad (103)$$

It is quite difficult to prove whether such a variable exists or not (Kuchař 1981a,b), the trouble being that the "Killing" vector field $u_{ab}(x; \mathbf{g}]$ in the space of Riemannian metrics may in principle be a highly nonlocal functional of $\mathbf{g}$. The proof relies on the fact that the super-Hamiltonian $H(x)$ and the supermomentum $H_a(x)$, through which the weak equality in equation (103) is enforced, have quite a definite locality and polynomial structure. Thus, $H_a(x)$ is linear in the momentum and contains at most the first derivatives of the canonical variables $g_{ab}(x)$, $p^{ab}(x)$, while $H(x)$ is a sum of the "kinetic" piece $T(x; \mathbf{g}, \mathbf{p})$, which is an ultralocal quadratic form of p^{mn}, and the "potential" piece $V(x; \mathbf{g}]$, which is a scalar concomitant of g_{mn} of the second order. This enables one to prove first that $u_{ab}(x; \mathbf{g}]$ must be local, i.e., a tensor concomitant of $\mathbf{g}$ of a finite order n; and then, by a glissade argument decreasing the order n step by step, end with the conclusion that $u_{ab}(x; \mathbf{g}]$ must actually be ultralocal. The only ultralocal tensor concomitant of $\mathbf{g}$ is, however, (a constant multiple of) $\mathbf{g}$ itself, and hence the only candidate for the dynamical variable (105) is the integrated trace of $\mathbf{p}$:

$$E = -\int_{\Sigma} d^3 x \, g_{ab}(x) p^{ab}(x). \tag{104}$$

The Poisson bracket of E with the kinetic piece $T(x)$ of the super-Hamiltonian is proportional to $T(x)$:

$$\{E, T(x)\} = -\frac{3}{2} T(x). \tag{105}$$

In other words, for each $x \in \Sigma$, $g_{ab}(x)$ is a conformal isometry (indeed, a homothetic motion) of the supermetric $G_{ab\ cd}(x; \mathbf{g})$. Moreover, $g_{ab}(x)$ is a "time-like" supervector, i.e., $G^{ab\ cd}(x)\, g_{ab}(x) g_{cd}(x) < 0$. Unfortunately, E does not scale the potential term $V(x; \mathbf{g}]$ in the super-Hamiltonian at the same rate as the kinetic term $T(x)$:

$$\{E, V(x)\} = \frac{1}{2} V(x). \tag{106}$$

This mismatch of the scaling factors in equations (105) and (106) means that E does not weakly commute with the total super-Hamiltonians $H(x) = T(x) + V(x)$, and hence our search for an isometry of the geometrodynamical constraint system fails.

This fact has serious repercussions for the probabilistic interpretation of the Wheeler–DeWitt equation. It indicates that (similar to the case of relativistic particle dynamics) one does not know how to turn the space of solutions of the Wheeler–DeWitt equation into a Hilbert space with a conserved positive-definite inner product. (Most papers on canonical quantization of gravity simply ignore, sidestep, or postpone *ad acta* this issue.) One possible escape is the suggestion that quantum geometro-dynamics of a single metric field in superspace without an appropriate isometry is equally impossible as a single-particle interpretation of the Klein–Gordon equation in a time-dependent space-time. Such an idea naturally leads to the exploration of a "third quantization" of gravity (cf. Giddings and Strominger 1989 and references therein). Unfortunately, like second quantization in particle dynamics, such a process does not automatically solve the time problem; it merely elevates it to a technically and conceptually messier level where it is (fortunately, as some of you may say) more easily forgotten.

Made wiser by one's defeats, one can start wondering whether the parallels between a relativistic particle dynamics and geometrodynamics are indeed so compelling as they appeared to be when the world was young. There are important conceptual differences between a particle moving in space-time and a three-geometry moving in superspace. Because the mass term in the super-Hamiltonian (29) is positive, the particle obeys the causal structure charted by the light-cones of the space-time metric $G^{AB}(X)$. On the other hand, the geometrodynamical potential $V(x; \mathbf{g}]$ in the super-Hamiltonians (95) is not positive-definite, and the geometry does not need to obey the "light-cone" structure set by the supermetric $G_{ab\ cd}(x; \mathbf{g}]$. In this context, one should also remember that the signature of the supermetric has nothing to do with the signature of the space-time metric that is obtained by solving the Hamilton equations of classical geometrodynamics (Kuchař 1976, 1977).

Indeed, the relation between superspace and space-time is remarkably involved and indirect. Its nature is such that one can doubt whether the construction (101) of time from superspace is such a good idea. At the very least, one should require that the time $t(x; \mathbf{g}]$ assigned to an event in a classical Einstein space-time be independent of a space-like hypersurface that passes through that event. The functional $t(x; \mathbf{g}]$ should thus remain unchanged when one "tilts" or "bends" the hypersurface at a given event. (The formal criterion for when this happens is given in Kuchař 1976, 1982.) However, it seems impossible to construct such a functional solely from the intrinsic metric $g_{ab}(x)$; to achieve invariance under tilts and bendings, one must leave superspace for an enlarged arena, namely, the geometrodynamical phase space $g_{ab}(x)$, $p^{ab}(x)$:

$$t = t(x; \mathbf{g}, \mathbf{p}]. \tag{107}$$

The simplest example of an invariant functional (107) is any scalar built from the space-time curvature tensor. By virtue of the Einstein equations in vacuo (which include both the constraint equations (95), (98) and the Hamilton equations of motion), such a scalar can be expressed as a functional of the canonical data g_{ab}, p^{ab}. Let me illustrate this process for the square of the curvature tensor. First, the tensor is decomposed into its projections parallel and perpendicular to Σ:

$$^4R_{ABCD}{}^4R^{ABCD} = 4\left({}^4R_{\perp a\perp b}{}^4R^{\perp a\perp b} + {}^4R_{\perp abc}{}^4R^{\perp abc}\right)$$
$$+ {}^4R_{abcd}{}^4R^{abcd}. \tag{108}$$

Second, two sets of such projections are expressed in terms of the intrinsic metric $g_{ab}(x)$ and the extrinsic curvature $K_{ab}(x)$ via the Gauss–Codazzi equations:

$$^4R_{\perp abc} = K_{a[b|c]},$$
$$^4R_{abcd} = R_{abcd} + K_{a[c}K_{bd]}. \tag{109}$$

Third, the same goal is reached for the remaining set of projections by the vital use of the Einstein equations:

$$^4R_{\perp a \perp b} = R_{ab} + K K_{ab} - K_{ac}K_b^c. \tag{110}$$

Finally, the extrinsic curvature is related to the momentum through the first set of Hamilton equations:

$$K_{ab} = -g^{-1/2}\left(p_{ab} - \frac{1}{2}pg_{ab}\right). \tag{111}$$

Because the extrinsic curvature is needed to build $t(x; \mathbf{g}, \mathbf{p}]$, such a time variable deserves the title of an "extrinsic time." For reasons quite unconnected with the tilt and bending invariance, different choices of an extrinsic time proved to be convenient in linearized gravity (Arnowitt et al. 1962) and in special minisuperspace models (Kuchař 1971). The use of foliations by hypersurfaces of constant mean extrinsic curvature to identify the gravitational degrees of freedom (York 1972, 1979) is related to the idea of an extrinsic time. One should realize, however, that the introduction of an extrinsic time leads us far away from the simple analogy between a relativistic particle and canonical geometrodynamics, which rests on the assumption that time is hidden in the configuration space data. The super-Hamiltonian constraint geared to a given extrinsic time is not obtained from the quadratic super-Hamiltonian by the familiar process of square-rooting. And, to stress the main theme of this talk for the last time, the spectre did not disappear: the problem of time is still with us. How are the quantization schemas based on two different choices of an extrinsic time to be related to each other?

Another sundial motto reminds me: I know my time—dost thou know thine? To suppress its somber tone I can try to turn it into a joke:

The time assigned for this lecture is over, so now let other speakers keep their watch of the clock as well as I did. But it is more appropriate to explore its deeper level by insisting that the very same question be posed over and over again to quantum theories of relativistic systems and to quantum geometrodynamics until we hear the answer (even if it be a negative one).

Acknowledgments. I would like to thank the organizers of the Osgood Hill conference for their support. This work was also supported in part by the NSF grant PHY 85-03653 to the University of Utah. The article was written during my stay at the Institut für theoretische Physik der Universität Bern, which was partly supported by Tomalla-Stiftung and Schweizerischer Nationalfonds. I am grateful to the Institute for its hospitality and to Petr Hájíček for many enlightening discussions. Michael Kinyon has kindly read the manuscript and suggested a number of improved formulations.

REFERENCES

Arnowitt, R., Deser, S., and Misner, Charles W. (1962). "The Dynamics of General Relativity." In *Gravitation: An Introduction to Current Research.* Louis Witten, ed. New York: Wiley, 227–265.

Ashtekar, Abhay, and Magnon, Anne (1975). "Quantum Fields in Curved Space-Times." *Royal Society of London. Proceedings A* 346: 375–394.

Ashtekar, Abhay, and Magnon-Ashtekar, Anne (1978). "Remark on Quantum Field Theory in Curved Space-Times." *Académie des Sciences* (Paris). *Comptes Rendus A* 286: 531–534.

DeWitt, Bryce S. (1967). "Quantum Theory of Gravity. I. The Canonical Theory." *Physical Review* 160: 1113–1148.

Giddings, Steven B., and Strominger, Andrew (1989). "Baby Universes, Third Quantization and the Cosmological Constant." *Nuclear Physics B* 321: 481–508.

Hartle, James, B., and Kuchař, Karel V. (1984). "Path Integrals in Parametrized Theories: Newtonian Systems." *Journal of Mathematical Physics* 25: 57–75.

Higgs, Peter W. (1958). "Integration of Secondary Constraints in Quantized General Relativity." *Physical Review Letters* 1: 373–375.

Kay, Bernard, S. (1978). "Linear Spin-Zero Quantum Fields in External Gravitational and Scalar Fields." *Communications in Mathematical Physics* 62: 55–70.

Kuchař, Karel V. (1971). "Canonical Quantization of Cylindrical Gravitational Waves." *Physical Review D* 4: 955–986.

——(1976). "Kinematics of Tensor Fields in Hyperspace. II." *Journal of Mathematical Physics* 17: 792–800.

——(1977). "Geometrodynamics with Tensor Sources. IV." *Journal of Mathematical Physics* 18: 1589–1597.

——(1980). "Gravitation, Geometry, and Nonrelativistic Quantum Theory." *Physical Review D* 22: 1285–1299.

——(1981a). "Canonical Methods of Quantization." In *Quantum Gravity 2: A Second Oxford Symposium*. Christopher J. Isham, Roger Penrose, and Dennis W. Sciama, eds. Oxford: Clarendon, 329–376.

——(1981b). "General Relativity: Dynamics without Symmetry." *Journal of Mathematical Physics* 22: 2640–26.

——(1982). "Conditional Symmetries in Parametrized Field Theories." *Journal of Mathematical Physics* 25: 1647–1661.

——(1989a). "Dirac Constraint Quantization of a Parametrized Field Theory by Anomaly-Free Operator Representations of Space-time Diffeomorphisms." *Physical Review D* 39: 2263–2280.

——(1989b). "Canonical Quantization of Generally Covariant Systems." In *Highlights in Gravitation and Cosmology*. B.R. Iger et al., eds. Cambridge: Cambridge University Press, 93–120.

Mag Ashtekar, Anne (1976). Thesis, Département de Mathématiques Pures. Université de Clermont.

Misner, Charles W. (1957). "Feynman Quantization of General Relativity." *Reviews of Modern Physics* 29: 497–509.

Wheeler, John A. (1962). "Neutrinos, Gravitation and Geometry." In *Topics of Modern Physics*. Vol. 1. New York: Academic Press, 1–130.

——(1964). "Geometrodynamics and the Issue of the Final State." In *Relativity, Groups and Topology*. Lectures Delivered at les Houches during the 1963 Session of the Summer School of Theoretical Physics, University of Grenoble, Cécile DeWitt, and Bryce S. DeWitt, eds. New York and London: Gordon and Breach, 316–520.

——(1968). "Superspace and the Nature of Quantum Geometrodynamics." In *Batelle Rencontres: 1967 Lectures in Mathematics and Physics*. Cécile DeWitt and John A. Wheeler, eds. New York: Benjamin, 242–307.

York, James W., Jr. (1972). "Mapping onto Solutions of the Gravitational Initial Value Problem." *Journal of Mathematical Physics* 13: 125–130.

——(1979). "Kinematics and Dynamics of General Relativity." In *Sources of Gravitational Radiation*. Proceedings of the Battelle Seattle Workshop, July 24–August 4, 1978. Larry L. Smarr, ed. Cambridge and New York: Cambridge University Press, 83–126.

Discussion

HOROWITZ: You said that there is no Killing vector on the superspace. But I thought that the natural metric that one got from the Hamiltonian is in fact a metric of constant curvature.

KUCHAŘ: Unfortunately, the metric is not everything; the super-Hamiltonian contains also a potential term. There is a conformal Killing vector in superspace respecting the metric, but it scales the potential term incorrectly.

MASON: Why doesn't the search for an internal time go against the spirit of general covariance?

KUCHAŘ: If you search for an intrinsic time that satisfies the criterion of general covariance, i.e., that is a space-time scalar, you probably fail. There is not any that is a local functional of the metric. I feel it is highly unlikely that you will find one that is nonlocal. The search for an intrinsic time thus seems to go against the spirit of general covariance. I would say that the general covariance is somehow hidden under the surface of superspace: There are geometrically privileged variables, like the curvature scalars in a given curved space-time, which can be reconstructed from the phase-space data. Such variables can be used as privileged coordinates.

ROVELLI: The reason that you want to search for a "time" seems to be related to the fact that you want to use this "time" to define a physical scalar product. Is that true?

KUCHAŘ: I am searching for a "time" in order to recover the normal scheme of quantum mechanics, in which one attempts to measure dynamical quantities at a given instant.

ROVELLI: So, the question is: If there were another way to define a scalar product, why would you need "time"? Let me put it in another way. In the classical theory, the independence of results of the choice of foliation is coded in the algebra of constraints. Now, this is a formal way of expressing the independence. In the quantum theory, we can translate it by the absence of anomalies of the corresponding quantum constraints. So, it seems to me that if we have a theory which is free of these anomalies, then there is no need to introduce "time."

KUCHAŘ: The problem with this approach is: How do you do it in practice? Namely, how do you turn the classical constraints into operator expressions that satisfy the appropriate algebra under an appropriate ordering and hence do not engender other constraints? You cannot do it by assuming that the constraint operators are self-adjoint on some Hilbert space because this leads to a symmetric factor ordering of their commutator. Such an ordering induces some structure functions to act on the state function before the constraints, and you then get more constraints.

ROVELLI: The problem is that you have to define a new scalar product on the space of solutions to the constraints because these solutions have an infinite norm with respect to the obvious inner product. But that is a general problem of *any* gauge system. You have exactly the same problem in Yang–Mills theory, for example.

ASHTEKAR: Let me rephrase this question. You have, basically, incorporated the diffeomorphism invariance in the constraint algebra. Suppose this algebra is anomaly free in quantum theory. Then, the next problem is that of finding the inner product. If you proceed naively, the norm of physical states is infinite. Carlo Rovelli asks: Suppose, by hook or crook, he finds an inner product. Would you still be obliged to isolate a variable called "time"?

KUCHAŘ: I would say yes, to identify the meaningful observables to be measured. For myself, I want to see observables changing along my world line and therefore associated with individual leaves of a foliation. In that sense, the problem of time is shifted to the problem of constructing an appropriate class of quantities one would like to call observables. Now, what I would like to call observables probably differs from what Carlo Rovelli would like to call observables. Carlo may like to restrict that term to constants of motion, while I would like to use variables that depend on a time hypersurface. Of course, both of us know that there is a technical way of translating my observables into his observables. However, it is difficult to subject such a translated observable to an actual observation. In principle, of course, it does not matter at what instant of time one measures a constant of motion. But the constants of motion that are translations of my observables are much too complicated when expressed in terms of the coordinates and the momenta at the time of measurement. You thus have a hard time to design an apparatus that would measure such a constant of motion at a time different from the moment for which it was originally designed.

DEWITT: Isn't there another way? Just abandon the canonical formalism.

KUCHAŘ: Well, then, what questions do you ask in the interior regions of space-time? Not scattering questions...

DEWITT: Correlations.

KUCHAŘ: Correlations between what?

DEWITT: Between physical clocks and anything else you want.

KUCHAŘ: Suppose you have just the gravitational field, then correlations between what?

DEWITT: I would like to know what the Riemann tensor is in a coordinate system provided by a laboratory at the present time.

KUCHAŘ: You are then introducing external clocks.

DEWITT: Sure! because the world is full of external clocks.

KUCHAŘ: But it wasn't full of such clocks back in the past, in the early stages of the evolution of the universe.

DEWITT: I would say that "time" is probably not a useful concept in that context. You want a canonical formalism because you want a "time." You want to see something evolve.

KUCHAŘ: I do not *want* to see things evolving. I *see* things evolving, and I want to *explain* why I see them evolving.

Time and Prediction in Quantum Cosmology

James B. Hartle

1. Summary

A generalized quantum mechanics for cosmological spacetimes is suggested in which no variable plays the special role of the time of familiar quantum mechanics. In this generalization, the central role of time in familiar quantum mechanics arises, not as a fundamental aspect of the formalism, but rather as an approximation appropriate to those initial conditions of the universe that lead to classical spacetime when it is large.

2. Quantum Mechanics and Cosmology

To understand the world at a fundamental level we must apply quantum mechanics to the universe as a whole. First among the reasons is that, as far as we know them, the microscopic laws of physics conform to the framework of quantum mechanics. If these laws explain phenomena on all scales, every prediction in science must, at a fundamental level, be a calculation of a quantum-mechanical probability, whatever nonquantum approximations to these calculations may be convenient. Thus, in cosmology, even our present crude observations on large scales in principle require a description of the universe in quantum-mechanical terms. The nature of this description and its observational consequences are the subject of quantum cosmology.[1]

Were the aim of cosmology only to describe present observations on large scales, the construction of a quantum cosmology might be an interesting intellectual exercise, but it would be unlikely to yield testable predictions of an intrinsically quantum nature. A more compelling reason for a quantum cosmology is that, except for some highly idealized cases, there are no predictions of *any kind* that do not involve knowing to some degree the initial conditions of the universe. To predict with confidence that the sun will come up tomorrow morning is to estimate as low the probability that it will be destroyed before morning by collision with a neutron star now racing across the galaxy with near-light speed. What are our grounds for assigning a low probability to the existence of such a neutron star? They are, I claim, a weak theory of initial conditions. Towards the big bang, a quantum description of the universe is inescapable, and a law of initial conditions is, therefore, quantum mechanical in an essential way. It is for the search for a law of initial conditions, and for establishing the connection between these initial conditions and present observations that we need a quantum cosmology.

From the perspective of quantum cosmology, laboratory physics is the effort to find and test predictions that depend only very weakly on the initial conditions of the universe. In cosmology, by contrast, we hope to find predictions for observations that *distinguish* theories of initial conditions. These may occur on all scales. On large scales they include predictions of the approximate homogeneity, isotropy, and spatial flatness of the universe, and the spectra of deviations from these exact symmetries. On much smaller scales, it has been suggested that initial conditions may play a role in determining the basic coupling constants and even the topology and dimensionality of spacetime.[2] Beyond these traditional questions of cosmology and particle physics, however, there is another class of predictions that are crucially dependent on the initial conditions of the universe. These are predictions so familiar and so close to home that we usually take them for assumptions in our theories. The simplest example is the homogeneity of the thermodynamic arrow of time. This is the fact that presently isolated systems evolve towards equilibrium together—in the same direction of time. This cannot be a consequence of the dynamics, for that is approximately time-reversal invariant. As explained by Page (1985), Hawking (1985a), and others, it can only be a consequence of particular initial conditions that mandate initial simplicity. At a deeper level, the manifest existence of Bohr's classical world (including classical spacetime) in the late universe cannot be a general feature of quantum mechanics, for the number of states that imply classical behavior in any approximation is but a poor fraction of the total variety of states for the universe. However, the classical world can be a consequence of particular initial conditions.[3] There may be many quantum states of the universe that imply thermodynamics and a classical world, but certainly their successful prediction must be regarded as a decisive test of any proposed theory of initial conditions.

In this lecture, I would like to review the position that there is another feature of our familiar experience that is an appropriate consequence of particular initial conditions rather than fundamental dynamical laws. This is the special role played by time in quantum mechanics. In particular, I want to explore the idea that the familiar formulation of quantum theory involving a preferred time is an approximation to a more general sum-over-histories framework, which is appropriate in the late universe because of specific initial conditions that imply classical spacetime there.[4]

3. The Problem of Time

The fundamental formula of standard quantum mechanics gives the joint probability for the outcomes of a time sequence of "yes – no" questions. Such questions are represented in the Heisenberg picture by projection operators $P_\alpha(t)$ such that $P_\alpha^2 = P_\alpha$. The label α shows which question is asked, and t the time at which it is asked. Questions asked at different times are connected by the Hamiltonian H through

$$P_\alpha(t) = e^{iHt} P_\alpha(0) e^{-iHt}. \tag{3.1}$$

(Throughout we use units in which $\hbar = c = 1$.) If a sequence of questions $\alpha_1 \cdots \alpha_N$ is asked at times $t_1 \leq t_2 \leq \cdots \leq t_N$, the joint probability for a series of "yes" answers is

$$p(\alpha_N t_N, \ldots, \alpha_1 t_1) = Tr\left[P_{\alpha_N}(t_N) \cdots P_{\alpha_1}(t_1) \rho P_{\alpha_1}(t_1) \cdots P_{\alpha_N}(t_N)\right], \tag{3.2}$$

where ρ is the density matrix of the system and Tr denotes a trace over all variables. All the familiar features of quantum mechanics — state vectors, unitary evolution, the reduction of the wave-packet on an ideal measurement, and so forth — are summarized in the two formulae (3.1) and (3.2). Their utility as a compact and transparent expression of standard quantum mechanics has been stressed by many authors.[5]

This formulation illustrates very clearly the special role played by time in quantum mechanics. First, the operators in (3.2) are *time ordered*. This is the expression of causality in quantum mechanics. Among all observables, time alone is singled out for this special role in organizing the predictive formalism. Second, it is assumed that *every* observation for which a prediction is made directly by (3.2) can be assigned a unique moment of time. This is a strong assumption. Unlike every other observable for which there are interfering alternatives (e.g., position and momentum), this says that there is no observation that interferes with the determination of an observation's time of occurrence. We may, through inaccurate clocks or neglect of data, be ignorant of the precise time difference between two observations, but we assume that it could have been determined *exactly*. In such cases, we deal with ignorance as in every other case in quantum mechanics. We sum the *probabilities* over an assumed distribution of error to obtain the probabilities for the observation. We sum probabilities because we "*could* have determined the time difference but didn't."

Of course, the special role played by time finds equivalent expressions in any of the other equivalent formulations of quantum mechanics. For example, in the Schrödinger picture we deal with time-independent operators but time-dependent states. The scalar product between states is defined at one instant of time. States directly specify probabilities for observations carried out at a moment of time. Time is the sole observable not represented by an operator, but rather enters the theory as a parameter describing evolution.

An equivalent special role for time can be identified in the sum-over-histories formulation of quantum mechanics. For example, as demonstrated clearly by Carl Caves and others,[6] when the questions α_i at time t_i correspond to localization of a particle to a spatial interval Δ_i at that time, then the multitime probability $p(\Delta_N t_N, \ldots, \Delta_1 t_1)$ can be constructed from the Feynman sum of $\exp[i(\text{classical action})]$ over all particle paths that satisfy the restrictions specifying the initial conditions and move forward in time through the intervals $\Delta_1 \cdots \Delta_N$. (See Figure 1.) The special role of time is again twofold. First, only observations of position in intervals at one moment of time are considered. Second, the paths are restricted to move forward in time so that they intersect a constant time surface at one and only one position.

The preferred time in familiar quantum mechanics is associated with a number of characteristic and powerful consequences. These include causality, the notion of a complete description by a state on a spacelike surface, and unitarity. In raising the issue of the status of time in quantum mechanics, we are also raising the issue of the status of these fundamental properties.

What are the grounds for singling out one observable to play such a peculiar role in the formalism of quantum prediction? Empirically, they

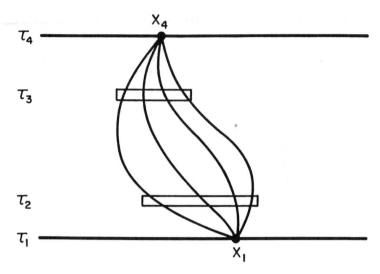

FIGURE 1. A simple experiment illustrating the construction of multitime joint probabilities in nonrelativistic sum-over-histories quantum mechanics. A particle starts localized at τ_1, registers or does not register in detectors at τ_2 and τ_3, and has its position determined at τ_4. The detectors at τ_2 and τ_3 localize the particle to intervals Δ_2 and Δ_3. The joint amplitude to start at X_1, pass through both detectors, and arrive at X_4 is the sum of $\exp(iS)$ over all paths satisfying these restrictions. The paths defining nonrelativistic quantum mechanics move forward in time in the sense that they intersect a surface of constant τ at one and only one position.

would seem to arise from the fact that, as observed on all directly accessible scales, over the whole of the directly accessible universe, spacetime has a classical geometry. In the spacetime geometry of non-relativistic physics, there is a preferred family of spacelike surfaces that unambiguously define the preferred time of nonrelativistic quantum mechanics. In the spacetime of special relativity, or indeed in any curved background, there are many families of spacelike surfaces that foliate spacetime. There is thus an issue as to which family defines the preferred time of quantum mechanics. Causality, however, implies that the quantum mechanics constructed with any one choice is unitarily equivalent to that constructed with any other. All choices give the same results. In the construction of quantum theories of spacetime, however, the choice of time variable becomes a fundamental difficulty. The classical theory distinguishes no preferred family of spacelike surfaces, and there is no evidence that the quantum theories constructed on different choices are equivalent.[7,8] There is no background spacetime to define a notion of causality. There is thus a conflict between the framework of familiar Hamiltonian quantum mechanics and the general covariance of space-time theories such as general relativity. Indeed, one suspects that any extension of general relativity, such as string theory, that does not single out a preferred background spacetime will face the same problem.

In the face of this difficulty, there are two possible routes forward. One is to keep the framework of quantum mechanics "as is" and choose from among the variables of spacetime one to play the special role of time. General covariance is thereby broken in the quantum theory. This is the traditional route of canonical quantum gravity.[7,8,9] Is it not more natural, however, to follow a second route: to seek to generalize quantum theory so that it does not require a preferred time for prediction, and to see the special role of time in familiar quantum theory, not as a fundamental property of quantum mechanics, but rather as an approximation appropriate to those special states of the universe and those special epochs when spacetime is approximately classical? Is it not more natural, in other words, to see the preferred role of time in the familiar quantum framework as an approximation appropriate to the particular initial conditions of our universe? This is the route forward that I wish to explore in this lecture.

4. What We Need

Attempts to move away from the familiar formulations of quantum mechanics by "including clocks in the system" or otherwise adjoining a dynamical time to the theory have a long history, which cannot be reviewed here.[10] A recent example is the extensive discussion of Page and Wooters.[11] The idea that our notion of time in quantum mechanics might be appropriate only to the late universe when spacetime behaves classically has been discussed by many, but emerges especially clearly from the work of DeWitt (1967), Lapchinsky and Rubakov (1979), Banks (1985), and Halliwell and Hawking (1985). In various ways these authors showed how the familiar quantum laws of evolution for the states of matter fields can emerge from the constraints of quantum gravity when applied to states that describe classical spacetime.[12] If, however, a preferred time is abandoned as a fundamental notion, what is to replace the *general* framework for quantum prediction, with which it was so closely involved? One proposal for this replacement is our subject today.

As a response to the problem of time, the idea is sometimes advanced that there is no fundamental notion of time.[13] That quantum general relativity, for example, is most properly viewed as a theory of space rather than a theory of quantum spacetime. That it is sufficient for prediction to calculate probabilities for observation from a single wave function on a single spacelike surface. That any notion of time is to be recovered from a study on that surface of the probabilities for correlations between the indicators of clocks and other variables. That our impression of the past is but an illusion, more properly viewed as correlations between records existing at the marvelous moment *now*.

However, to abandon spacetime just because it does not supply a natural candidate for the ordering parameter in quantum mechanics seems to me an overreaction. While it is no doubt true that many interesting probabilities, especially in cosmology, are for observations that are more-or-less on one spacelike surface, they do not exhaust those predicted by familiar quantum mechanics, nor those that are important. For example, a physical system can be said to behave as a good clock when the probability is high that the position of its indicator is correlated with the location of *successive* spacelike surfaces in spacetime. The predictive consequences of history are most honestly correlations between present records.[14] However, to assign a probability to a history requires the calculation of probabilities of correlations between *present* data and events in the *past*. Just to describe in an objective way the subjective experience of the passage of time requires probabilities involving different spacelike surfaces. It may be that the search for unity between gravitation and other interactions will lead us to abandon as fundamental time, space, or both. In this case, a revision of the predictive framework of quantum mechanics seems inevitable. A distinction needs to be drawn, however, between such motivations and those arising from the preferred status of time in the familiar framework. Before we invoke the conflict between that familiar framework and covariant spacetime as reason to abandon one of the most powerful organizing concepts of our experience, it may be of interest to see whether the familiar framework of quantum mechanics might be generalized a bit to apply to theories of spacetime. Here, we will assume spacetime is fundamental and seek a generalized predictive quantum framework for it.

What we need is a generalization of the fundamental formula (3.2) that supplies probabilities for observations on different spacelike surfaces, that does not prefer one set of spacelike surfaces to another, and that reduces approximately to (3.2) when spacetime is classical. We do not expect such an approximation to be valid generally, but only in the late universe as a consequence of particular initial conditions. Since all conceivable observations take place when the spacetime *is* behaving classically, it might be thought that a generalization of (3.2) is really not needed for any practical purposes. It is indeed, however, for two important purposes: first to define precisely, in terms of probabilities for correlations between spacelike surfaces, what is meant by "spacetime behaving classically", and second, to answer any question about that early period of the universe when spacetime is not classical. Without a mechanism to supply answers to such questions, our predictive framework would be incomplete.

5. A Quantum Kinematics for Spacetime

The basic framework for prediction in quantum mechanics can be stated in several different ways. There is the Schrödinger picture, the Heisenberg picture, and the sum-over-histories approach. I have argued elsewhere[4] that the sum-over-histories approach is the more natural and perhaps more general framework with which to construct a quantum kinematics of spacetime because it does not invoke a preferred time in as immediate or as central a way as the other two frameworks. A very similar position was taken by C. Teitelboim in his pioneering development[15] of a propagator calculus for quantum gravitational amplitudes, with which the sum-over-histories framework used here coincides at many points. However, the quantum kinematics—the rules for spacetime probabilities —to which one is naturally led in the sum-over-histories approach can be restated in frameworks that look formally like the Schrödinger or Heisenberg pictures by the use of a trick. I would like to try here to state the results in these terms. This will have the advantage of a certain directness and familiarity at the expense of some naturalness in motivation.

The trick is easy to state: we introduce labels for a preferred family of spacelike surfaces. We construct quantum amplitudes by the familiar rules using these surfaces as the preferred time. Since the labels are unobservable, we sum the amplitudes over them before squaring and computing joint probabilities. By introducing auxiliary labels, we break general covariance. By integrating over them we restore it.

Unobservable labels are used elsewhere in quantum mechanics. The most familiar example is with identical particles. We can begin a discussion of N identical particles by introducing N coordinates $\mathbf{X}_i$, $i =$

$1, \ldots, N$ where the label i distinguishes one particle from another. A general wave function $\psi(\mathbf{X}_1, \ldots, \mathbf{X}_N)$ characterizes a state in which one particle has different properties from another. However, when we sum that wave function over possible values of the label for each slot, we symmetrize it. The symmetry reflects the indistinguishability of the particles and the unobservability of the label.

Isham and Kuchař (1985) have given a convenient formalism for the additional labels needed to describe a preferred family of spacelike surfaces in spacetime. We use it here. A foliating family of spacelike surfaces is described by four functions $X^\alpha(\tau, x^i)$ that tell on which surface a point of spacetime lies and where it lies in the surface. We may think of these four functions as four scalar fields on the spacetime or more physically as the readings, $X^\alpha = (T, X^a)$, of ideal clocks and rods that give a system of coordinates for spacetime. If, for definiteness, we further assume that the trajectories of the clocks at fixed X^a are orthogonal to the surfaces of constant T, then the spacetime metric in these special coordinates is

$$ds^2 = -dT^2 + s_{ab}dX^a dX^b. \tag{5.1}$$

The time T of the preferred coordinates supplies the preferred time of quantum mechanics. The geometrical variables are the s_{ab}. Thus, we write for a state

$$\Psi = \Psi[T, s_{ab}(\mathbf{X}), \phi(\mathbf{X})], \tag{5.2}$$

where ϕ is a scalar matter field. This evolves by the Schrödinger equation

$$-i\frac{\partial \Psi}{\partial T} + H\Psi = 0, \tag{5.3}$$

where H is the total Hamiltonian. For Einstein gravity, we could construct H by the standard canonical procedure from the action

$$\ell^2 S_E = 2\int_{\partial M} d^3x \sqrt{h} K + \int_M d^4x \sqrt{g}(R - 2\Lambda), \tag{5.4}$$

after (5.1) has been substituted into it. (We use units in which $\hbar = c = 1$ throughout. $\ell = (16\pi G)^{1/2}$ is the Planck length.) The inner product is

$$(\Psi, \Psi') = \int \delta s_{ab} \delta\phi \Psi(T, s_{ab}, \phi)\Psi'(T, s_{ab}, \phi). \tag{5.5}$$

This, however, is relativity, and we should be able to state our results invariantly using any spacelike surface, not just the special surfaces of constant T. Indeed, it is important to do so to express the invariance of the theory under coordinate transformations. Isham and Kuchař (1985) tell us how to do it. In a general system of coordinates $x^\mu = (\tau, x^i)$ the metric (5.1) is

$$ds^2 = \left[-\nabla_\mu T \nabla_\nu T + s_{ab} \nabla_\mu X^a \nabla_\nu X^b\right] dx^\mu dx^\nu. \qquad (5.6)$$

Substitute this into the action (5.4). One obtains an action that is a functional of the $X^\alpha(x)$ and $s_{ab}(x)$. Equivalently it may be thought of as a functional of the embedding functions and the three-metric h_{ij} on a surface of constant general coordinate τ. This is because s_{ab} and h_{ij} are related by

$$h_{ij} = s_{ab} D_i X^a D_j X^b + D_i T D_j T, \qquad (5.7)$$

D_i being the derivative in the surface.

The action that results is the action for a parametrized theory in which the coordinates X^α have been elevated to the status of dynamical variables coupled to curvature. The theory is invariant under diffeomorphisms because the coordinates were arbitrary. It is not, however, generally covariant because gravitational phenomena are now described by four scalar fields X^α in addition to the metric. As a consequence of diffeomorphism invariance there are four constraints. Classically they can be written

$$n^\alpha P_\alpha + \mathcal{H}(\pi^{ij}, h_{ij}, \pi_\phi, \phi) = 0, \qquad (5.8a)$$

$$(D_i X^\alpha) P_\alpha + \mathcal{H}_i(\pi^{ij}, h_{ij}, \pi_\phi, \phi) = 0. \qquad (5.8b)$$

Here, $\mathcal{H}$ and $\mathcal{H}_i$ are the familiar Hamiltonian and momentum constraints of the classical theory—functions of the three-metric, h_{ij}, its conjugate momentum, π^{ij}, and the scalar field, ϕ, and its conjugate momentum, π_ϕ. P_α is the momentum conjugate to X^α. n^α is the unit normal to the constant T hypersurface. It can be expressed in terms of the X^α alone because $n_\alpha \propto \epsilon_{\alpha\beta\gamma\delta}(D_1 X^\beta D_2 X^\gamma D_3 X^\delta)$.

Quantum mechanically, the states are described by wave functions

$$\Psi = \Psi\left[X^\alpha(\mathbf{x}), h_{ij}(\mathbf{x}), \phi(\mathbf{x})\right], \qquad (5.9)$$

which satisfy operator forms of the constraints (5.8)

$$in^\alpha \frac{\delta \Psi}{\delta X^\alpha(\mathbf{x})} = \mathcal{H}\left(-i\frac{\delta}{\delta h_{ij}(\mathbf{x})}, h_{ij}(\mathbf{x}), -i\frac{\delta}{\delta \phi(\mathbf{x})}, \phi(\mathbf{x})\right) \Psi, \qquad (5.10a)$$

$$iD_i X^\alpha \frac{\delta \Psi}{\delta X^\alpha(\mathbf{x})} = \mathcal{H}_i \left(-i\frac{\delta}{\delta h_{ij}(\mathbf{x})}, h_{ij}(\mathbf{x}), -i\frac{\delta}{\delta \phi(\mathbf{x})}, \phi(\mathbf{x}) \right) \Psi. \quad (5.10b)$$

Equation (5.10a) is the covariant form of the Schrödinger equation (5.3). That equation follows from (5.10a) by considering only variations in Ψ that uniformly advance a surface of constant T. The additional constraints (5.10b) ensure that Ψ is independent of the choice of spatial coordinates x^i.

So far, the formalism is an exotic version of familiar quantum mechanics, but familiar in all of its basic aspects. We now turn to the calculation of joint probabilities. For simplicity, suppose that the universe is in a pure state characterized by a wave function $\Psi_C[X^\alpha, h_{ij}, \phi]$. The crucial decision in the calculation of probabilities is the status of the variables $X^\alpha(\mathbf{x})$. If, as here, they are unobservable labels, then amplitudes should be summed over them and then squared to yield probabilities for prediction. Thus, for example, the amplitude $\Phi[h_{ij}, \phi, C]$ for the existence of a spacelike surface in the geometry with metric $h_{ij}(\mathbf{x})$ and a matter field configuration $\phi(\mathbf{x})$ is

$$\Phi[h_{ij}(\mathbf{x}), \phi(\mathbf{x}), C] = \int \delta X^\alpha(\mathbf{x}) \Psi_C [X^\alpha(\mathbf{x}), h_{ij}(\mathbf{x}), \phi(\mathbf{x})]. \quad (5.11)$$

The integration is over all foliations of spacetime. In particular, the integration is over time—both positive and negative values. The probability of this three-metric and matter field is then proportional to

$$|\Phi[h_{ij}(\mathbf{x}), \phi(\mathbf{x}), C]|^2 \delta h_{ij} \delta \phi. \quad (5.12)$$

In constructing the sum (5.11), general covariance is restored, for if the X^α are integrated over a diffeomorphism-invariant range, the constraints (5.10) imply for Φ

$$\mathcal{H}\Phi = 0, \quad \mathcal{H}_i \Phi = 0. \quad (5.13)$$

The strongest argument that the $X^\alpha(\mathbf{x})$ are unobservable labels is that we know of no such special fields as the X^α, no type of matter fields coupled to curvature as in the parametrized theory under discussion, no genuinely ideal clocks existing for every time and at every place in the universe, no preferred frames. In quantum cosmology, it is not enough to ask whether such clocks *might* be constructed. We have to ask whether they exist *in fact*. One could take, with Hoyle and Hoyle (1963, p. v ff.), and Unruh and Wald (1989), the point of view that X^α exist, and are in principle observable, but that we just haven't seen them. In that case, we should sum *probabilities* over our ignorance, and the expression for the probability of a three-geometry and matter field configuration would be not (5.12) but

$$\left(\int \delta X^\alpha |\Psi_C \left[X^\alpha(\mathbf{x}), h_{ij}(\mathbf{x}), \phi(\mathbf{x}) \right]|^2 \right) \delta h_{ij} \delta \phi. \qquad (5.14)$$

General covariance would remain broken.

The prescription for the single-surface probability (5.12) can be given a simple expression in sum-over-histories quantum mechanics. While there are many questions whose outcomes can be predicted in sum-over-histories quantum cosmology, those most analogous to those of familiar quantum mechanics are framed in the arena of extended superspace— the space of three-metrics and matter fields on a spacelike surface. For example, one could ask: "Given initial conditions C, what is the probability that spacetime contains a spacelike surface whose three-geometry and field lies in a region $\mathcal{O}$ of superspace?" If $\mathcal{O}$ is a small region, that is the question we answered above. In the sum-over-histories framework, the amplitude Φ of Eq. (5.11) is given by

$$\Phi[h_{ij}(\mathbf{x}), \phi(\mathbf{x}), C] = \int_C \delta g \delta \phi \exp(iS[g, \phi]), \qquad (5.15)$$

where the integral is over all histories— four-dimensional cosmological spacetimes and matter field configurations—that satisfy the initial conditions C and are bounded by a spacelike surface with the induced metric h_{ij} and scalar field ϕ. The joint probability for C and (h_{ij}, ϕ) in a region $\mathcal{O}$ is

$$p(\mathcal{O}, C) = \int_{\mathcal{O}} \delta h \delta \phi |\Phi[h_{ij}(\mathbf{x}), \phi(\mathbf{x}), C]|^2. \qquad (5.16)$$

In fact, the amplitudes defined by (5.14) and by (5.12) coincide. The reason is that, in a familiar way, Φ_C could be constructed as a sum over $X^\alpha(\mathbf{x})$, $h_{ij}(\mathbf{x})$ and $\phi(\mathbf{x})$ that match the conditions C and the arguments of Ψ_C. Through (5.6), the sum over $X^\alpha(\mathbf{x})$ and $h_{ij}(\mathbf{x})$ is a sum over four-geometries. The remaining sum over $X^\alpha(\mathbf{x})$ in (5.11) is just that needed to sum over all four-geometries as in (5.14). In summing over all four-geometries between boundary three-surfaces, one sums over the time between them.

I argued in Section 4 that it is important for quantum cosmology to be able to define and calculate probabilities for observations on many spacelike surfaces. It would be possible to continue to exploit the trick of introducing fictitious labels to do this. The results, however, in this case are more transparent in the sum-over-histories framework. To illustrate, I shall now sketch the generalization in very schematic terms.

The construction is like the sum-over-histories construction of the multitime probabilities of ordinary quantum mechanics illustrated in Figure 1. Suppose, for simplicity, that we ask: "Given initial conditions $\mathcal{C}$, what is the joint probability that there are three spacelike surfaces in spacetime with three-geometries and fields in regions $\mathcal{O}_1$, $\mathcal{O}_2$, and $\mathcal{O}_3$?" In sum-over-histories quantum mechanics we proceed to calculate this probability $p(\mathcal{O}_1, \mathcal{O}_2, \mathcal{O}_3, \mathcal{C})$ as follows: (i) Sum $\exp(iS)$ over all four-geometries that are bounded by a surface with (h_3, ϕ_3) in $\mathcal{O}_3$ and that contain interior three-surfaces in $\mathcal{O}_1$ and $\mathcal{O}_2$ occurring in *any order*. (ii) Square this amplitude and integrate over all (h_3, ϕ_3) in $\mathcal{O}_3$. (iii) Repeat for the boundary geometry in $\mathcal{O}_1$ and $\mathcal{O}_2$, and add the results to find the joint probability $p(\mathcal{O}_1, \mathcal{O}_2, \mathcal{O}_3, \mathcal{C})$ that is the generalization of the fundamental formula (3.2).

Joint probabilities are the raw material for the calculation of conditional probabilities for prediction. Conditional probabilities are constructed according to the standard rules of probability theory by identifying exhaustive and exclusive sets of outcomes—outcomes such that one and no more than one member is certain to occur. For example, given conditions $\mathcal{C}$ the probability of an outcome $\mathcal{O}_i$ that is one member of a set $\{\mathcal{O}_i\}$ of exhaustive and exclusive possibilities (perhaps involving several surfaces) is

$$p(\mathcal{O}_i \mid \mathcal{C}) = \frac{p(\mathcal{O}_i, \mathcal{C})}{\sum_i p(\mathcal{O}_i, \mathcal{C})}. \tag{5.17}$$

The identification of sets of exhaustive and exclusive outcomes thus becomes a central question for the process of prediction in quantum cosmology.

6. Causality, States and Unitarity

The concepts of causality, state, and unitarity are features of the familiar quantum predictive framework that are closely related to each other and to the preferred time of quantum mechanics. As a consequence, their status must be re-examined in any theory in which spacetime itself is a quantum variable.

There are two distinct aspects to causality. First, we cannot send signals between two spacelike-separated events. Second, the past can influence the future but the future does not influence the past. Each of these aspects depends *for its definition* on a background spacetime. In the first case, a spacetime metric is needed to define "spacelike," and in the second case to define the distinction between forward and backward light cones. However, in quantum gravity there is no background spacetime. In sum-over-histories quantum cosmology we *sum* over all possible spacetimes. No one spacetime is distinguished over any other to give meaning to the notion of causality. We thus cannot expect a notion of causality in a covariant quantum theory of spacetime.[16]

Classical physics is built upon the concept of the state of a system at a moment of time. The state is a characterization of the system from which we can completely predict its behavior in the future and retrodict its behavior in the past using dynamical laws. In familiar quantum mechanics there is also a notion of state, but it is more restricted. At a moment of time, from the state of the system alone, we can predict the future. We cannot, however, retrodict the past. Retrodiction, the construction of history, requires a knowledge of the initial conditions in addition to a knowledge of the present state.[3] This difference is the arrow of time in quantum mechanics. In covariant quantum gravity, no such distinction between future and past is possible because there is no classical spacetime, with which to fix this distinction. We cannot, therefore, expect to recover a notion of "state of the universe" at a moment of time.

The idea of unitarity is closely connected to the ideas of state and causality. If there is a complete description of the universe, from which predictions on a future surface can be drawn, then we would expect to be able to divide the process of prediction into two parts: First, the calculation of the state on some intermediate surface, and second, the calculation from it of the state on the surface of interest. The resulting composition law is the expression of unitarity in quantum mechanics. However, if we do not expect a notion of state, we can hardly expect to find unitarity in covariant quantum gravity, at least not in its familiar form.

In the Schrödinger picture, unitary evolution implies conservation of probability. The square of the wave function gives *directly* the probability for observations on the spacelike surface on which it is defined; and if the probabilities for a complete set of observations add up to unity on that surface, they do so on all surfaces. The absence of unitarity in a covariant quantum theory of spacetime does not mean that probabilities for exhaustive and exclusive sets of possibilities fail to sum to unity. Any theory of probabilities must guarantee that. It only means that there is no direct representation of conditional probabilities for alternatives on successive spacelike surfaces in terms of a unitarily evolving wave function.[17]

The notions of a state and its unitary evolution are not primary concepts of sum-over-histories quantum mechanics. Rather, quantum amplitudes are specified directly as sums over appropriately restricted histories. Whether these amplitudes can be organized into the inner products of states on spacelike surfaces is an issue for investigation rather than a prerequisite for formulation. Let us illustrate such an investigation with nonrelativistic quantum mechanics.

The histories of nonrelativistic quantum mechanics are particle paths which move *forward in time,* in the sense that they intersect a surface of the preferred nonrelativistic time at one and only one position. The amplitude $\Phi(B, A)$ to get from a point A on an initial surface to a point B on a final one is the sum of $\exp[iS(\text{path})]$ over paths that connect these points. Because the paths move forward in time, this sum can be factored into a sum from A to a point X on some intermediate constant-t surface, a sum from X to B, and finally a sum over the intersection X. (See Figure 2.) In symbols,

$$\Phi(B, A) = \int dX \psi_B^*(Xt)\psi_A(Xt), \qquad (6.1)$$

where, for example, $\psi_A(Xt)$ is the sum of $\exp(iS)$ over all paths from A to X. This factorization defines the wave functions of states on the constant-t intermediate surface, their unitary evolution, and the inner product between them. From the perspective of the sum-over-histories formulation, the possibility of formulating nonrelativistic quantum mechanics in terms of unitarily evolving states is a consequence of a special property of its histories—that surfaces can be found that they intersect once and only once.

Quantum field theory is similar. Fields are single valued on spacetime. They thus move forward in time in the sense of having a unique configuration on any spacelike surface. The resulting composition law, analogous to (6.1), expresses unitarity in field theory.

There is no geometric variable that uniquely labels a spacelike surface in the dynamics of closed cosmologies. The extrinsic curvature

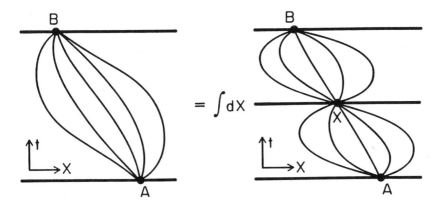

FIGURE 2. Derivation from the sum-over-histories of the composition law expressing unitarity in nonrelativistic quantum mechanics. The amplitude to get from a point A on an initial surface to a point B on a final surface is given by the sum over paths connecting them as at left. The sum is only over paths that move forward in time and intersect a surface of constant time at one and only one position. Because of this the sum can be factored, as at right, into a sum over paths from A to a point X on an intermediate time surface, a sum over paths from X to B, and a final sum over X. This decomposition defines states on the intermediate time surface, the inner product between them, and expresses their unitary evolution. If the paths did not move forward in time, it would not be possible to derive such a composition law.

scalar, K, may be such a variable for spacetimes obeying the Einstein equation, (Marsden and Tipler 1980), but the sum defining amplitudes is a sum over an infinitely larger class of histories. A general history may contain arbitrarily many spacelike surfaces with a given value of K or, indeed, of any other variable. There is thus no family of spacelike surfaces, whose members the contributing histories cross once and only once. Indeed, since the dominant histories are not differentiable, the expected number of occurrences of any particular kind of surface is infinite. Thus, in a sum-over-histories quantum mechanics of spacetime, we do not find a notion of state and a unitary composition law on *any* spacelike family of surfaces. Probabilities for all possible observations are predicted. They just cannot be taken apart to define the familiar notions of causality, state, and unitary evolution, so closely associated with the spacelike surfaces of a preferred background spacetime.

7. Classical Spacetime

Causality, unitarity, and quantum states are facts of our experience. We have argued above that they are concepts associated with classical spacetime. They are not, therefore, built-in properties of the formalism of a quantum theory of spacetime. Rather, they are approximate properties of those quantum initial conditions that imply classical spacetime in the late universe. In this section, I would like to briefly sketch how this comes about.

Most properly, the conditions for classical spacetime are defined in terms of multi-surface probabilities. Initial conditions that imply that observations on different surfaces are correlated with high probability, as would be predicted by Einstein's classical law, may be said to predict classical spacetime. In the sum-over-histories framework, multi-surface joint probabilities are constructed from multi-surface amplitudes defined as the sum of $\exp(iS)$ over all histories that satisfy the initial conditions $\mathcal{C}$ and thread the regions of superspace defining the observations on the various surfaces. For example, for a single surface we had

$$\Phi\big[h_{ij}(\mathbf{x}), \phi(\mathbf{x}), \mathcal{C}\big] = \int_{\mathcal{C}} \delta g\, \delta\phi \exp\Big\{ i(S_E[g] + S_M[g, \phi]) \Big\}, \qquad (7.1)$$

where S_E is the gravitational and S_M the matter action. Suppose that the initial conditions were such that for sufficiently large three-geometries, a single spacetime $\hat{g}$ dominated the geometric sum over histories; then

$$\Phi\big[h_{ij}(\mathbf{x}), \phi(\mathbf{x}), \mathcal{C}\big] \approx \exp(iS_E[\hat{g}]) \int \delta\phi \exp(iS_M[\hat{g}, \phi]). \qquad (7.2)$$

The remaining integral over the matter fields is a familiar sum, defining the field theory of ϕ in the background spacetime $\hat{g}$. From this, we know how to construct the states, the inner product, and the Schrödinger equation for unitary evolution. To the extent that the three-geometry h_{ij} specifies uniquely the location of a hypersurface in the background spacetime, we can write from our general rules

$$p(\phi(\mathbf{x}) \mid h_{ij}(\mathbf{x}), \mathcal{C}) = \frac{|\Phi(h_{ij}(\mathbf{x}), \phi(\mathbf{x}), \mathcal{C})|^2 \,\delta\phi(\mathbf{x})}{\int \delta\phi\, |\Phi(h_{ij}(\mathbf{x}), \phi(\mathbf{x}), \mathcal{C})|^2} \qquad (7.3)$$

for the conditional probability that, on the hypersurface labeled by h_{ij}, the scalar field has the configuration $\phi(\mathbf{x})$. Thus, for $|\Phi(h_{ij}, \phi, \mathcal{C})|^2$, we can informally read $|\Phi("t",\phi,\mathcal{C})|^2$. Similar results hold for multi-surface probabilities provided that the observations of spacetime on the intermediate surfaces do not probe the Planck scale and disturb the dominance of a single spacetime history.

Typical proposals for initial conditions do not single out a single classical history for the spacetime of the late universe. Rather, they correspond to a coherent sum of different classical spacetimes. To the extent that our failure to observe such things as very long wave length gravitons makes these alternatives decohere, these initial conditions imply a statistical mixture of classical spacetimes, for each of which the above discussion holds.[18]

In the above analysis, we have used the three-geometries of a classical spacetime to supply the condition corresponding to fixing time in the computation of the conditional probabilities that reproduce those of conventional quantum mechanics. Of course, in the real world we never measure a three-geometry, and certainly we do not carry out observations over any but extraordinarily small regions of a spacelike surface. In reality, we define time with a wide variety of mechanical systems called clocks ranging from the oscillations of a Cs maser to the Hubble expansion of the galaxies. We choose these systems because our theory tells us that their behavior is correlated with each other and with the location of spacelike surfaces through the existence of a classical spacetime on accessible scales in the late universe. It is really in this sense that we recover our notion of time.

8. Conclusions

The interpretative framework of quantum mechanics loosely subsumed under the name "Copenhagen interpretation" contains two central assumptions that seem incompatible with a quantum cosmology built on covariant theories of spacetime. The first is a distinguished class of classical systems. The second is a distinguished time variable and its associated notion of causality. The first assumption seems incompatible with the application of quantum mechanics to the whole universe. The second seems incompatible with general covariance. The particular initial conditions of our universe could imply an approximate classical reality in the late universe.[3] Is it not also possible that the distinguished time and the associated causality of familiar scales could arise, not as exact notions in the formalism of quantum theory, but as approximate notions in the late universe that are consequences of specific initial conditions in a more general sum-over-histories framework of quantum prediction?

Supplementary Note. In familiar quantum mechanics, the consistency of the prescription for assigning probabilities to a complete set of alternatives at one time is guaranteed by unitary time evolution. Such probabilities must sum to one at all times; probability must be conserved. Unitary evolution as expressed by (3.1) shows that, if $\Sigma_\alpha P_\alpha(t) = 1$ holds at one time, then it does so at all times. The resulting probabilities for α_n at t_n from (3.2) then sum to one at all t_n.

The question naturally arises as to whether there is an analogous consistency requirement for the more general sum-over-histories quantum mechanics advocated here and, if so, as to how it is satisfied. Some part of the discussion in this session was directed towards this question. It is now possible, I believe, to address this issue in greater depth and completeness. The arguments will appear in a forthcoming article, but can be briefly summarized here:

In familiar quantum mechanics, probabilities can be meaningfully assigned to histories of alternatives only when these alternatives are actually measured. In an Everett framework for quantum cosmology, there is no fundamental notion of "measured" because there is no fundamentally distinguished role for the observer. The observer is included in the predictive framework. Therefore, a generalization of the condition for a meaningful assignment of probabilities is needed in quantum cosmology. This is supplied as follows:

In a theory with a preferred time, a complete set of alternatives $\{P_\alpha(t)\}$ is said to *decohere* as a consequence of a particular initial state ρ when

$$Tr\left[P_{\alpha'_N}(t_N)\dots P_{\alpha'_1}(t_1)\rho\, P_{\alpha_1}(t_1)\dots P_{\alpha_N}(t_N)\right]$$
$$\approx \delta_{\alpha'_N \alpha_N}\dots\delta_{\alpha'_1 \alpha_1}\, p(\alpha_N t_N,\dots,\alpha_1 t_1). \quad (1)$$

The probabilities $p(\alpha_N t_N,\dots,\alpha_1 t_1)$ for decoherent alternatives, so defined, satisfy the standard sum rules, which express their consistency; that is

$$\Sigma_{\alpha_k}\, p(\alpha_N t_N,\dots,\alpha_{k+1}t_{k+1},\alpha_k t_k,\alpha_{k-1}t_{k-1},\dots,\alpha_1 t_1)$$
$$\approx p(\alpha_N t_N,\dots,\alpha_{k+1}t_{k+1},\alpha_{k-1}t_{k-1},\dots,\alpha_1 t_1). \quad (2)$$

These are the relations expressing the conservation of probabilities. A notion of decoherence may be defined for the generalized sum-over-histories quantum mechanics described here. Decoherent spacetime histories satisfy the generalization of the probability sum rules (2) even when there is no preferred time.

Additional Supplementary Note. This article was completed in the spring of 1988. The above note was added in the fall of 1988. At the time of this note, fall 1990, considerable progress has been made in developing and refining the ideas in this article, however, no attempt has been made to revise the article other than to provide more accurate references to work that was in progress or in press at the time of writing. For further and more extensive development of these ideas see the author's lectures "The Quantum Mechanics of Cosmology" in the Proceedings of the 1989 Jerusalem Winter School in Theoretical Physics: Quantum Cosmology and Baby Universes, edited by S. Coleman, J.B. Hartle, T. Piran and S. Weinberg (World Scientific, Singapore, 1991).

Acknowledgments. The author has benefited from many people through conversations on these issues. Special thanks are due to K. Kuchař and M. Gell-Mann for critical discussions over a long period of time. The work reported here was supported in part by the National Science Foundation under grant PHY 85-06686.

NOTES

[1] For a sampling of reviews consonant with the author's views on this subject see Hawking 1984, Hartle 1986a, 1986b, 1988c. For a bibliography see J. Halliwell, 1990.

[2] See, e.g., S. Coleman 1988; S. W. Hawking 1985b; S. Giddings and A. Strominger 1988.

[3] See e.g., M. Gell-Mann and J. B. Hartle 1990.

[4] This talk is an attempt at a summary of the essential points of a program that was sketched in Hartle 1986b, and that is appearing in detail in a series of papers: Hartle 1988a, 1988b. For detailed developments of the arguments adumbrated here the reader can consult these papers.

[5] Among them, e.g., Wigner 1963; Aharonov, Bergmann, and Lebowitz 1964; Unruh 1986; Gell-Mann 1987.

[6] See especially, Caves 1986, 1987; Feynman 1948; and Stachel 1986, p. 331f.

[7] Isham and Kuchař 1985.

[8] For a review and discussion see Wheeler 1979; and Kuchař 1981.

[9] For some alternatives to the traditional canonical view discussed in note 1 but still with a preferred time, see Padmanabhan 1989a, and Unruh and Wald 1989.

[10] For a sampling of such attempts known to the author see, e.g., Dirac 1926; Paul 1962; Allcock 1969a, 1969b; Peres 1980; Smolin (this volume), and Sorkin (this volume).

[11] Page and Wooters 1983; Wooters 1984, 1986.

[12] For further developments see Wada 1986; Joos 1986; Zeh 1986; Kuchař and Ryan 1986; Brout, Horwitz, and Weil 1987; D'Eath and Halliwell 1987; Halliwell 1987; LaFlamme 1987; Vilenkin 1988; Zeh 1988; Padmanabhan and Singh 1990; Vilenkin 1988.

[13] For views that in various ways, argue against any fundamental notion of time, see, e.g., Wheeler 1988; Banks 1985; the references in note 12, and the contributions by Ashtekar, Smolin, Horowitz and Rovelli in this volume.

[14] See, e.g., the reference in note 3.

[15] Teitelboim, 1983a, 1983b, 1983c.

[16] The tension between causality and covariance in quantum theories of spacetime was very clearly stated by C. Teitelboim (1983d). In his formulation he opts for causality. Here we choose covariance, recovering causality only for specific initial conditions.

[17] T. Jacobson (this volume) has stressed that causal anomalies can arise if one keeps the notion of state but permits it to evolve in a nonunitary fashion.

[18] See, e.g., Joos 1986; Zeh 1986; Kiefer 1987, 1989; Fukuyama and Morikawa 1989; Halliwell 1989; Padmanabhan 1989b.

REFERENCES

Aharonov, Yakir, Bergmann, Peter G., and Lebowitz, Joel L. (1964). "Time Symmetry in the Quantum Process of Measurement." *Physical Review B* 134: 1410–1416.

Allcock, Gordon R. (1969a). "The Time of Arrival in Quantum Mechanics I. Formal Considerations." *Annals of Physics* 53: 253–285.

———(1969b). "The Time of Arrival in Quantum Mechanics II. The Individual Measurement." *Annals of Physics* 53: 286–310.

Ashtekar, Abhay (1991). "Old Problems in the Light of New Variables." See this volume, pp. 401–426.

Banks, T. (1985). "TCP, Quantum Gravity, the Cosmological Constant and All That ....." *Nuclear Physics B* 249: 332–360.

Brout R., Horwitz, G., and Weil, D. (1987). "On the Onset of Time and Temperature in Cosmology." *Physics Letters B* 192: 318–322.

Caves, Carlton M. (1986). "Quantum Mechanics of Measurements Distributed in Time. A Path-Integral Formulation." *Physical Review D* 33: 1643–1665.

————(1987). "Quantum Mechanics of Measurements Distributed in Time. II. Connections among Formulations." *Physical Review D* 35: 1815–1830.

Coleman, S. (1988). "Why There Is Something Rather Than Nothing: A Theory of the Cosmological Constant." *Nuclear Physics B* 310: 643.

D'Eath, P. D., and Halliwell, Jonathan (1987). "Fermions in Quantum Cosmology." *Physical Review D* 35: 1100–1123.

DeWitt, Bryce S. (1967). "Quantum Theory of Gravity. I. The Canonical Theory." *Physical Review* 160: 1113–1148.

Dirac, Paul A. M. (1926). "Relativity Quantum Mechanics with an Application to Compton Scattering." *Royal Society of London. Proceedings A* 111: 405–423.

Feynman, Richard P. (1948). "Spacetime Approach to Non-Relativistic Quantum Mechanics." *Reviews of Modern Physics* 20: 367–387.

Fukuyama, T. and Morikawa, M. (1989). "Two-dimensional quantum cosmology: Direction of dynamical and thermodynamic arrows of time." *Phys. Rev. D* 39: 462–469.

Gell-Mann, Murray (1987). "Superstring Theory." *Physica Scripta T* 15: 202–209.

Gell-Mann, Murray and Hartle, James B. (1990). "Quantum Mechanics in the Light of Quantum Cosmology." In *Complexity, Entropy and the Physics of Information*. Santa Fe Institute Studies in the Sciences of Complexity, vol. VIII, Wojciech H. Zurek, ed. Reading, Mass.: Addison Wesley; also in *Proceedings of the Third International Symposium on Foundations of Quantum Mechanics in the Light of New Technology*. H. Ezawa, S. Kobayashi, Y. Murayama and S. Nomura, eds. Tokyo: Japan Physical Society.

Giddings, Steven B. and Strominger, Andrew (1988). "Loss of Incoherence and Determination of Coupling Constants in Quantum Gravity." *Nuclear Physics B* 307: 854.

Halliwell, Jonathan J. (1987). "Correlations in the Wave Function of the Universe." *Physical Review D* 36: 3626–3640.

Halliwell, Jonathan J. (1989). "Decoherence in Quantum Cosmology." *Physical Review D* 39, 2912.

————(1990). "Bibliography of Papers on Quantum Cosmology." *Journal of Modern Physics*. Forthcoming.

Halliwell, Jonathan J., and Hawking, Stephen W. (1985). "Origin of Structure in the Universe." *Physical Review D* 31: 1777–1791.

Hartle, James B. (1986a). "Quantum Cosmology." In *High Energy Physics 1985: Proceedings of Yale Theoretical Advanced Study Institute*. Mark J. Bowick and Feza Gürsey, eds. Singapore: World Scientific, 471–566.

————(1986b). "Prediction in Quantum Cosmology." In *Gravitation in Astrophysics*. NATO Advanced Study Institute, 1986, Cargèse, France. B. Carter and James B. Hartle, eds. New York: Plenum, 329–360.

————(1988a). "Quantum Kinematics of Spacetime. I. Nonrelativistic Theory." *Physical Review D* 37: 2818–2832.

————(1988b). "Quantum Kinematics of Spacetime: II. A Model Quantum Cosmology with Real Clocks." *Physical Review D* 38: 2985.

————(1988c). "Quantum Cosmology." In *Highlights in Gravitation and Astronomy*. Proceedings of the International Conference on General Relativity and Cosmology. Goa, India 1987. B.R. Iyer, A. Kembhavi, J.V. Narlikar, C.V. Vishveshwara, eds. Cambridge: Cambridge University Press.

Hawking, Stephen W. (1984). "Quantum Cosmology." In *Relativity, Groups and Topology II*. Bryce S. DeWitt and Raymond Stora, eds. Amsterdam and Oxford: North-Holland, 336–379.

————(1985a). "Arrow of Time in Cosmology." *Physical Review D* 32: 2489–2495.

————(1985b). "The Cosmological Constant is Probably Zero." *Physics Letters B* 134: 481.

Horowitz, G.T. (1991). "String Theory without Spacetime." See this volume, pp. 299–324.

Hoyle, Fred, and Hoyle, Geoffrey (1963). *The Fifth Planet*. New York: Harper and Row.

Isham, Christopher J., and Kuchar, Karel V. (1985). "Representations of Spacetime Diffeomorphisms. II. Canonical Geometrodynamics." *Annals of Physics* 164: 316–333.

Jacobson, Ted (1991). "Unitarity, Causality, and Quantum Gravity." See this volume, pp. 212–216.

Joos, E. (1986). "Why Do We Observe a Classical Spacetime?" *Physics Letters A* 116: 6–8.

Kiefer, C. (1987). "Continuous Measurement of Minisuperspace Variables by Higher Multipoles." *Class. Quant. Grav.* 4, 1369–1382.

Kiefer, C. (1989). "Quantum Gravity and Brownian Motion." *Phys. Lett. A.* 201–203

Kuchař, Karel V. (1981). "Canonical Methods of Quantization." In *Quantum Gravity 2: A Second Oxford Symposium*. Christopher J. Isham, Roger Penrose, and Dennis W. Sciama, eds. Oxford: Clarendon, 329–376.

Kuchař, Karel V., and Ryan, M. (1986). "Can Minisuperspace Quantization be Justified?" In *Gravitational Collapse and Relativity*. Proceedings of Yamada Conference XIV, Kyoto International Conference Hall, Japan, 7–11 April 1986. Humitaka Sato and Takashi Nakamura, eds. Singapore and Philadelphia: World Scientific, 451–464.

Laflamme, Raymond (1987). "The Euclidean Vacuum: Justification from Quantum Cosmology." *Physics Letters B* 198: 156–160.

Lapchinsky, V., and Rubakov, V. (1979). "Canonical Quantization of Gravity and Quantum Field Theory in Curved Spacetime." *Acta Physica Polonica B* 10: 1041–1048.

Marsden, Jerrold E., and Tipler, Frank J. (1980). "Maximal Hypersurfaces and Foliations of Constant Mean Curvature in General Relativity." *Physics Reports* 66: 109–139.

Padmanabhan, T. (1989a). "Quantum Cosmology–The Story So Far." In *Gravitation, Gauge Theories and the Early Universe.* B. R. Iyer, N. Mukunda, and C. V. Vishveshwara, eds. Dordrecht and Boston: D. Reidel, 373–403.

———(1989b). "Decoherence in the density matrix describing quantum three-geometries and the emergence of classical spacetime." *Phys. Rev. D* 39: 2924–2932.

Padmanabhan, T., and Singh, T. P. (1990). "On the Semiclassical Limit of the Wheeler–DeWitt Equation." *Class. Quant. Grav.* 7:411.

Page, Don N. (1985). "Will Entropy Decrease If the Universe Recollapses?" *Physical Review D* 32: 2496–2499.

Page, Don N., and Wootters, William K. (1983). "Evolution without Evolution: Dynamics Described by Stationary Observables." *Physical Review D* 27: 2885–2892.

Paul, Harry von (1962). "Über Quantenmechanische Zeitoperatoren." *Annalen der Physik* 9: 252–261.

Peres, Asher (1980). "Measurement of Time by Quantum Clocks." *American Journal of Physics* 48: 552–557.

Rovelli, Carlo (1991). "Is There Incompatibility between the Ways Time is Treated in General Relativity and in Standard Quantum Mechanics?" See this volume, pp. 126–140.

Smolin, Lee (1991). "Space and Time in the Quantum Universe." See this volume, pp. 228–291.

Sorkin, Rafael D. (1991). "Problems with Causality in the Sum-Over-Histories Framework for Quantum Mechanics." See this volume, pp. 217–227.

Stachel, John (1986). "Do Quanta Need a New Logic?" In *From Quarks to Quasars: Philosophical Problems of Modern Physics.* Robert G. Colodny, ed. Pittsburgh: University of Pittsburgh Press, 229–347.

Teitelboim, Claudio (1983a). "Quantum Mechanics of the Gravitational Field." *Physical Review D* 25: 3159–3179.

———(1983b). "Proper-Time Gauge in the Quantum Theory of Gravitation." *Physical Review D* 28: 297–309.

———(1983c). "Quantum Mechanics of the Gravitational Field in Asymptotically Flat Space." *Physical Review D* 28: 310–316.

———(1983d). "Causality versus Gauge Invariance in Quantum Gravity and Supergravity." *Physical Review Letters* 50: 705–708.

Unruh, W. (1986). "Quantum Measurements." In *New Techniques and Ideas in Quantum Measurement Theory.* Daniel M. Greenberger, ed. New York: New York Academy of Sciences, 242–249.

Unruh W. G. and Wald, Robert M. (1989). "Time and the Interpretation of Canonical Quantum Gravity." *Physical Review D* 40: 2598-2614.

Vilenkin, Alexander (1988). "Quantum Cosmology and the Initial State of the Universe." *Physical Review D* 37: 888–897.

———(1989). "The Interpretation of the Wave Function of the Universe." *Physical Review D* 39: 1116.

Wada, Sumio (1986). "Quantum Cosmological Perturbations in Pure Gravity." *Nuclear Physics B* 276: 729–743.

Wheeler, John Archibald (1979). "Frontiers of Time." In *Problemi dei fondamenti della fisica, Scuola internazionale di fisica "Enrico Fermi."* Corso 52. G. Toraldo di Francia, ed. Amsterdam: North-Holland, 395–497.

———(1988). "World as system self-specialized by quantum networking." *IBM Journal of Research and Development* 32: 4–15.

Wigner, Eugene P. (1963). "The Problem of Measurement." *American Journal of Physics* 31: 6–15.

Wootters, William K. (1984). "'Time Replaced by Quantum Correlations." *International Journal of Theoretical Physics* 23: 701–711.

———(1986). In *Fundamental Questions in Quantum Mechanics*. Proceedings of the Conference on Fundamental Questions in Quantum Mechanics, held at the State University of New York at Albany, April 12–14, 1984. Laura A. Roth and Akira Inomata, eds. New York and London: Gordon and Breach.

Zeh, H. D. (1986). "Emergence of Classical Time from a Universal Wavefunction." *Physics Letters A* 116: 9–12.

———(1988). "Time in Quantum Gravity." *Physics Letters A* 126: 311–317.

Discussion

KUCHAŘ: In your problem with path integrals in Newtonian physics, you integrated over x's at a given time. However, in your problem on geometrodynamics, you integrated over all of the three-geometries, which seems like integrating over spacetime regions.

HARTLE: That is a point that I didn't stress. Let me recall the fundamental formula once again:

$$p(\alpha_N t_N, \ldots, \alpha_1 t_1) = Tr \left[P_{\alpha_N}(t_N) \cdots P_{\alpha_1}(t_1) \rho \, P_{\alpha_1}(t_1) \cdots P_{\alpha_N}(t_N) \right].$$

A preferred time enters here not only in the ordering of the operators but also in the fact that every operator for which a prediction is made is characterized by a value of the preferred time. It's a remarkable property when you come to think of it. Time is just as much an observable as any variable represented by an operator. Why should every prediction that is made have that particular observable as a label? In order to generalize the formula, we not only have to generalize it with respect to the ordering, but we also have to generalize the nature of the observables; because, if we find a formula that is perfectly symmetric and doesn't involve any particular ordering, but we are only allowed to discuss observables that are confined to one special class of spacelike surfaces, then we once again have introduced a preferred time. Considering spacetime regions is one proposed generalization. I think Rafael Sorkin is going to explain some difficulties with causality in this generalization, in the nonrelativistic framework. But, it seems to me that, if we are going to avoid a preferred time, we need somehow to have both aspects of the generalization.

SMOLIN: In the canonical theory, there is a very clear notion of what is a physical observable. One can assert that there is such a notion. In this theory, is there also a clear notion of what is a physical observable?

HARTLE: It is characteristic of path-integral formulations that they single out a particular configuration space if you are going to regard them as the fundamental starting point for a theory of quantum mechanics. In ordinary quantum mechanics, we deal with paths that are in x-space so that, at the start, position space is preferred. We then have to analyze experiments to show how you can discuss other measurements, for example, of momentum. Thus, there is not an immediate transformation theory in sum-over-histories quantum mechanics, although it can be derived, as Feyman showed. Of course, it's also true, I think, of the traditional theory that, while it predicts directly probabilities for a wide class of observables like $\hat{x}^3\hat{p} + \hat{p}\hat{x}^3$, it is not very good at explaining *how* one could observe most of them. In the sum-over-histories formulation of cosmology that I have discussed, we have integrals over geometries. It is spacetime, therefore, that supplies the basic observables, and we can construct others, like superspace regions, from spacetime. How far one can go in recovering any transformation theory is, I think, at this moment unclear. It's also not so clear how necessary it is.

PAGE: I just wanted to know if you had any more explanations as to why we can't restrict ourselves to the present hypersurface. I mean, this all looks very beautiful; well, maybe it is more fundamental. But in practice, whenever we test things, it seems that we are really just comparing our records of one kind of clock, our records of one history as recorded by one means, and records of that history as recorded by other means. If we get agreement on that, it seems good enough.

HARTLE: How do you know it's a clock? What if I say, "This is a clock." [Points to a paper cup with 12:00 written on it.]

PAGE: Are there a lot of things that agree with it, that would give the same value?

HARTLE: I guess I am slightly skeptical. If I found a hundred paper cups and they all said it's 12:00 like this one does, I still wouldn't know that it's a clock until I have done a calculation and showed that it actually *behaves* like a clock.

[Unrecorded question on the role of the brain.]

HARTLE: I am skeptical of the point of view that the internal states of our brain are a complete description of the physical world, and the only things about which we need to make predictions in physics. I am skeptical that we can reduce *all* our discussions to this "marvelous moment," and never do another calculation in time, even though the marvelous moment is what's accessible to us directly. It seems strange, for example, to believe that a spacelike hypersurface is even defined by the internal state of our brain. Could we believe it is defined to an accuracy of 10^{-43} seconds? I am reluctant to take psychological notions as fundamental elements of physical theory. I just don't think that the brain is actually defining time intervals. I don't think our brain *can* define time intervals to arbitrary precision. But let me say the following: Should it turn out that we can formulate physics at the marvelous moment, nothing that I have said need be changed. It would just be slightly superfluous. People who believe in the marvelous moment would still have the obligation of supplying the measure on the configuration space, or equivalently, the inner product. The sum-over-histories framework I have described is one such proposal for this. The measure is the one induced on those particular amplitudes by the sum over histories.

STACHEL: Neither I nor any of the people I know really gather information on a global spacelike hypersurface. If anything, I gather fluxes of information coming through the surface of my body, a spacelike two-surface evolving through time. This is one of the reasons why I think a $(2 + 2)$ formalism might be advantageous (laughter). At any rate, my first observation is quite independent of that. If you can tell me how to get information on an evolving spacelike hypersurface, I would be very happy

HARTLE: The sum-over-histories framework is consistent with your point of view because these regions of superspace that I have defined probabilities for are only weakly constrained by our observations. We don't have access, certainly, to information over a whole spacelike hypersurface. That just means these superspace regions are extraordinarily big. There are a whole lot of geometries that could fit within them. Their probabilities, however, are still predicted.

MASON: I have a question that Roger Penrose likes to ask. The expressions that people like you and Steven Hawking use for the path integral to calculate, say, "A given B", and "B given A" are identical, modulo complex conjugation. The result of the calculation gives the same answer for the probabilities for "A given B" and "B given A". The problem is that, in the real world, relative probabilities are *not* symmetric like that. So, it seems to me that you should have an asymmetric expression for the path integral if you want both a consistent interpretation and correct probabilities. Quantum mechanics is time asymmetric. It works in only one direction.

HARTLE: What I was saying yesterday is that one might try to look for a quantum *framework* such that one could calculate joint probabilities that were symmetric, but in which the observed time asymmetry would emerge as a consequence of particular initial conditions that define a preferred time for the universe.

GELL-MANN: Is that what you were saying? I thought your point of view was that, initially, everything was symmetric but then shortly afterwards a few measurements, two or three, would select e^{iS} over e^{-iS}.

HARTLE: One possibility is that the initial conditions single out one classical universe with a preferred time direction. This happens, for example, in models where the universe actually has a clock. The arrow of time in quantum mechanics is then encoded in those initial conditions. Now, these are not the kind of initial conditions that "no boundary" proposal gives. Those result in something more like a superposition of various classical solutions. In that case, we have to talk about the semiclassical regime that Murray Gell-Mann and Jonathan Halliwell are talking about, and understand how such classical solutions decohere so that one doesn't have interference between them. The prediction of the "no-boundary" proposal is, as Don Page, Steven Hawking and others have shown, that a semiclassical solution corresponds to an *ensemble* of classical trajectories one end of which is simple while the other end is complex. The *ensemble* is time symmetric; there are just as many trajectories that run one way as the other. But each individual universe has a preferred time direction. We live in one.

MASON: Is that the same as saying that the relative probability interpretation works if one of the ends is simple?

HARTLE: To the extent that I follow that statement, I think I would agree. Thus, I would claim that we are, in some sense, implementing Roger Penrose's proposal. It is just that it is not exactly in the way that he would like (laughter).

GELL-MANN: Does anybody here know the answer to a question about the variations of the "no boundary" proposal? It is conjectured that the Hartle–Hawking proposal may have a mixture of e^{iS} and e^{-iS} and therefore both directions of time. One of them is then selected by observations, or something like that, and one hopes that the other one will go away quietly. But is there another boundary condition, offered by Vilenkin and anticipated perhaps by Starobinski and Linde? And does it just have e^{iS}?

ISHAM: Well, the claim is that it has only outgoing waves in super-space.

GELL-MANN: I know. But the question is whether that is the same as having just e^{iS}.

ISHAM: I think he claims that it does, doesn't he?

GELL-MANN: Well, when I asked him, he said he hadn't thought about it that way. But maybe subsequently he had. I would guess so, actually. So that would be an illustration of the difference between a gradually emerging time asymmetry version and one in which there is time asymmetry from the beginning.

HARTLE: In any case, what I was trying to do yesterday was to extend the *whole framework* of quantum mechanics in such a way that it can accommodate all such proposals generally, and the familiar arrow of time would emerge not as part of the formalism but as a consequence of the particular initial conditions, at least in one branch. The quantum-mechanical arrow of time won't be a separate arrow. The symmetric formulation would seek to *explain* the correspondence between the quantum-mechanical arrow and the familiar cosmological and thermodynamic arrows.

GELL-MANN: I would like to emphasize, however, that there is a huge difference between having only e^{iS}, and having both e^{iS} and e^{-iS} and then picking one.

PAGE: I am afraid I don't understand what e^{-iS} means. Can you give an example in quantum mechanics where it features?

GELL-MANN: It isn't quantum mechanics. That is just the point. Taken literally, it is a generalization of some ill-understood kind, where the composition law doesn't hold, etc., and yet that may be what Hawking and Hartle have suggested.

ASHTEKAR: Jim, yesterday, you started saying something in response to Ted Jacobson's comment (see Jacobson's contribution, "Unitarity, Causality, and Quantum Gravity"), but we ran out of time. Could you respond now?

HARTLE: I don't recall now exactly what I was going to say. So, let me make a general response. I think that, for a more general sum-over-histories quantum mechanics, if these ideas that Rafael Sorkin and I have discussed fail, then problems with unitarity and causal anomalies are the most likely ways in which they will fail. However, the universe is extraordinarily classical after the first 10^{-43} seconds or so. To change the "clock" given to us by spacetime to some nonclassical behavior, we have to get this whole large system to run backwards a little bit in time. Therefore, the hope is that these anomalies will be extraordinarily small. If, as Ted Jacobson suggests, large causal anomalies are produced for familiar scales, we certainly will have to give these ideas up. The causal anomalies discussed by Rafael Sorkin are of a somewhat different sort. They are not anomalies associated with the formalism, but with the definition of observables. There is a whole untouched area that bears on these questions, which in quantum mechanics, is called "measurability." In quantum mechanics we say that every Hermitian operator — like $\hat{x}\hat{p}^3 + \hat{p}\hat{x}^3$ — is an observable, but it is quite a different thing to demonstrate how you can measure it. As Murray was explaining on Monday, the class of things that you can actually observe is of course much smaller than the class of all Hermitian operators. That is a fact, and in quantum cosmology one is proposing an explanation for why that is true. What has to be investigated is whether such quantities, which occur naturally in the formalism, are limited by our ability to construct apparatus to measure them. Actually, in this respect, in quantum cosmology one is not in a situation very different from that in ordinary quantum mechanics. There, too, it is no use just declaring that something is an observable without giving a prescription as to how one might observe it.

ASHEKAR: Except that we don't get into these causality and unitarity problems in ordinary quantum mechanics.

ANDERSON: May I comment on that? One often overlooks the difference between doing quantum mechanics with an external apparatus and doing it in a way in which both the system and the apparatus evolve quantum mechanically. Now, we have a presumption that it is meaningful to calculate the expectation values of observables in the absence of a measuring apparatus, whose presence is not even discussed. The hope is that these expectation values actually will be extracted by the measuring apparatus when the two systems are actually coupled together. Now, in general, the expectation values give us the averages of the possible outcomes and not the specific coupled outcomes for the system and the apparatus. So, if you are really careful, you have to evolve the joint system, which is what you do in quantum cosmology because you are a part of the whole thing. So you have to be careful when you say that ordinary quantum mechanics doesn't have these problems.

SMOLIN: Let me respond to that. In ordinary quantum mechanics, it is exactly the presence of an inner product and a Hermitian Hamiltonian that guarantees that the probabilities defined by the expectation value are meaningful. In particular, if you have a set of mutually exclusive outcomes, the probabilities that the theory predicts for them add up to one in a way that is consistent with causality. You may try to get predictions out of quantum mechanics that are probabilistic without using an inner product along the lines proposed, for example, by Jim Hartle, Finkelstein, Graham and others, in which one extends the definition of the relative-frequency operator. However, if one does have an inner product, the definition of the relative-frequency operator *depends* on that inner product in a nontrivial way. You need to know not only the state you are considering, but also what is "not that state," i.e., what is orthogonal to it. Thus, it uses the notion of orthogonality in the state space. Therefore, the burden of showing that you can have a sensible probability interpretation in the absence of an inner product and unitary evolution is on the people proposing to do that.

HARTLE: Where there is no time, there is no causality. The interesting question is whether, when we go to the limit where these notions are approximately well-defined, large causal anomalies can arise. That is what Ted Jacobson was suggesting.

SMOLIN: I think that is a fair statement.

Time in Quantum Cosmology

Jonathan J. Halliwell

1. Introduction

We have heard in this workshop a number of points of view about the nature of time in quantum gravity, and the related issue of the existence of a Hilbert space structure. These points of view may be loosely divided up into two camps. Either (A) time is something we have at the beginning that goes into the theory, or (B) time is not something we have at the beginning, but that emerges from the theory, possibly in a restricted set of circumstances. In this short note, I would like to discuss this latter point of view in the context of the Dirac quantization approach to quantum cosmology. This is the approach that is adopted in most modern treatments. One can also consider the ADM approach to quantum cosmology, in which the constraints are solved before quantizing. This corresponds to the former point of view mentioned above, and will not be discussed here.

2. Minisuperspace Models

We will begin by considering minisuperspace models. The general background to this type of model may be found, for example, in Halliwell 1987, 1991. Minisuperspace models are described by a finite number of functions $q^\alpha(t)$, representing certain components of the three-metric. Their wave function $\Psi(q)$ satisfies the Wheeler–DeWitt equation

$$H\Psi = \left(-\frac{1}{2}\nabla^2 + U(q)\right)\Psi(q) = 0, \qquad (1)$$

where ∇^2 is the Laplacian operator on minisuperspace.

When solving the Wheeler–DeWitt equation, one is interested first of all in determining the regions in which the wave function is oscillatory, and the regions in which it is exponential. The location of these regions will depend on the nature of the potential $U(q)$, but it will also depend to some extent on the boundary conditions. The general conditions under which Ψ is exponential/oscillatory are discussed in Halliwell 1986.

The regions in which the wave function is exponential are regarded as classically forbidden—Ψ does not correspond in any way to a Lorentzian four-geometry. The regions in which the wave function is oscillatory, on the other hand, are regarded as classically allowed—Ψ is in some sense peaked about a set of classical Lorentzian four-geometries. This provisional interpretation of the wave function will be discussed further in the final section.

In the oscillatory region, on which we will now concentrate, one may use the WKB approximation, in which one writes,

$$\Psi(q) = C(q)e^{iS(q)}, \tag{2}$$

where S is a rapidly varying phase and C is a slowly varying pre-factor. Inserting (2) in (1), one obtains

$$\frac{1}{2}(\nabla S)^2 + U(q) = 0, \tag{3}$$

$$\nabla S \cdot \nabla C + \frac{1}{2}C\nabla^2 S = 0, \tag{4}$$

where, in accordance with the WKB approximation, we have dropped a term of the form $\nabla^2 C$. One can show that the wave function (2) is peaked about the set of trajectories for which

$$p = \nabla S, \tag{5}$$

where p is the momentum conjugate to q (Halliwell 1987). Using the Hamilton–Jacobi equation (3), one can show that these are classical trajectories—(5) is a first integral of the field equations.

One can introduce a parameter with which to label the points along these trajectories, which we will denote suggestively by τ. The tangent vector in configuration space for the paths about which Ψ is peaked is then

$$\frac{d}{d\tau} = \nabla S \cdot \nabla. \tag{6}$$

τ then corresponds to the *proper time* along the classical trajectories. We note the following:

(1) Time emerges merely as the parameter with which one labels the points along the trajectories about which the wave function is peaked.

(2) Reparametrization invariance (a symmetry of the underlying classical theory) emerges as the freedom to choose this parameter—one could equally introduce a parameter t such that

$$\frac{1}{N(t)}\frac{d}{dt} = \nabla S \cdot \nabla \tag{7}$$

for some arbitrary function $N(t)$, and reparametrization invariance is then the freedom to choose $N(t)$.

(3) Time, and indeed space-time, are approximate notions that emerge only in certain regions of minisuperspace, namely, those in which the wave function oscillates.

(4) The location and very existence of the oscillatory regions, and hence of classical space-time, will generally depend on the boundary conditions on the wave function.

3. Perturbations About Minisuperspace

We have considered the situation in which all of the variables q^α become classical in some region of minisuperspace. Now let us consider the situation in which just some of them become approximately classical, while the rest remain quantum mechanical. This situation arises when one considers inhomogeneous perturbations about a homogeneous minisuperspace background (Banks et al. 1985; D'Eath and Halliwell 1987; Halliwell and Hawking 1985; Vachaspati and Vilenkin 1988; Wada 1986). Let us consider the perturbations caused by an inhomogeneous scalar field $\Phi(\mathbf{x}, t)$. The Wheeler–DeWitt equation is

$$\left(-\frac{1}{2}\nabla^2 + U(q) + H_M(q, \Phi)\right)\Psi(q, \Phi) = 0, \tag{8}$$

where H_M is the scalar field Hamiltonian. We will look for solutions of the form

$$\Psi(q, \Phi) = \Psi_0(q)\Psi_M(q, \Phi), \tag{9}$$

where Ψ_0 is a (possibly approximate) solution to the background minisuperspace Wheeler–DeWitt equation (1). We are interested mainly in the region of superspace in which the gravitational field becomes approx-imately classical; thus, we take Ψ_0 to be of the form

$$\Psi_0(q) = Ce^{iS}. \tag{10}$$

Inserting (9) and (10) into (8), once again equations (3) and (4) are obtained for S and C. In addition, however, one obtains the following equation for Ψ_M:

$$i\nabla S \cdot \nabla \Psi_M = H_M \Psi_M. \tag{11}$$

The operator $i\nabla S \cdot \nabla$ we have seen before. It is the tangent vector (6) to the classical minisuperspace trajectories. Equation (11) is therefore a time-dependent functional Schrödinger equation for the scalar field wave function

$$i\frac{\partial \Psi_M}{\partial \tau} = H_M \Psi_M. \tag{12}$$

Consider next the question of the existence of an inner product and a Hilbert space. One can introduce an inner product for solutions to (12):

$$(\Psi_M, \chi_M) = \int d[\Phi] \Psi_M^*(q, \Phi) \chi_M(q, \Phi), \tag{13}$$

and this is conserved,

$$\frac{\partial}{\partial \tau}(\Psi_M, \chi_M) = 0, \tag{14}$$

by virtue of (12). Equation (13) is generally well-defined because the so-lutions in which one is interested are square-integrable in Φ. We have thus recovered from the Wheeler–DeWitt equation, in the WKB approxima-tion, the familiar formalism of quantum field theory in curved space-time for the scalar field in the functional Schrödinger picture.

Note that we have not attempted to introduce an inner product for the solutions to the full Wheeler–DeWitt equation (8), involving an inte-gration over the q's. Such an inner product would probably be ill-defined since the wave functions are generally not square-integrable in q. The existence of a Hilbert space structure for canonical quantum gravity is still very much an open question.

So we have seen that the standard quantum field theory for the scalar field, including its Hilbert space structure, can emerge from quantum cosmology in the WKB approximation. However, it emerges only in regions of superspace in which the gravitational field is approximately classical, and thus a notion of time exists. In this approach, time and Hilbert space do not go into the theory at the beginning, but come out of the theory only under certain conditions. This point of view, in the context of the path integral approach, is currently being developed by Hartle (1988a, 1988b, 1990).

4. Decoherence and Classical Space-Time

In Section 2, it was stated that an exponential wave function corresponds to a classically forbidden region, and an oscillatory wave function corresponds to a classically allowed region. We are now in a position to discuss this interpretation more thoroughly.

It is generally the case that classical behavior arises not only because the wave function is strongly peaked about a particular configuration, or because the configuration is macroscopic, but also as a consequence of a particular interaction with the environment (Joos and Zeh 1985; Gell-Mann and Hartle 1990). A system becomes classical when its reduced density matrix, obtained by tracing out the environment, decoheres (i.e., becomes diagonal).

For the entire universe, there is of course no external environment. However, some of the degrees of freedom used to describe the universe may be regarded as environment. These degrees of freedom then "monitor" the remaining degrees of freedom, which may then become classical as a consequence (Zurek 1991). One ought to demonstrate explicitly that this happens in quantum cosmology.

A natural model to use in order to study this question is the perturbative model of the previous section. We shall regard the scalar field modes as the environment, and look for decoherence in the reduced density matrix, implying that the gravitational field becomes classical. Consider, therefore, the reduced density matrix

$$\rho(q, q') = Tr_\Phi \Psi^*(q, \Phi)\Psi(q', \Phi)$$
$$= \Psi_0^*(q)\Psi_0(q') \int d[\Phi]\Psi_M^*(q, \Phi)\Psi_M(q', \Phi). \tag{15}$$

What one would hope to find is that ρ decoheres in regions of minisuperspace for which Ψ_0 oscillates, but does not in regions where Ψ_0 is exponential. This would put the usual interpretation on a sounder basis.

Kiefer (1987) has calculated the reduced density matrix for the Halliwell–Hawking model, which consists of a homogeneous, isotropic minisuperspace background with scalar field and gravitational wave perturbations (Halliwell and Hawking 1985). He found that decoherence does indeed occur in the oscillatory region, so the gravitational field does indeed become classical there, although he does not consider the exponential region. This is currently under investigation (Halliwell 1989). A similar calculation is that of Mellor and Moss (1988), who considered a Kaluza–Klein minisuperspace model. They took the environment to be the fluctuation modes on the internal space. They also observed decoherence, but again considered only the oscillatory region. There are potentially a number of things that can be explained using decoherence arguments. These will be discussed in future publications.

Acknowledgments. I am very grateful to the organizers for creating the very stimulating atmosphere in which this workshop took place.

This research was supported in part by the National Science Foundation under Grant No. PHY82-17853, supplemented by funds from the National Aeronautics and Space Administration, at the University of California at Santa Barbara.

REFERENCES

Banks, Tom, Fischler, W., and Susskind, Leonard (1985). "Quantum Cosmology in $2 + 1$ and $3 + 1$ Dimensions." *Nuclear Physics B* 262: 159–186.
D'Eath, P. D., and Halliwell, Jonathan J. (1987). "Fermions in Quantum Cosmology." *Physical Review D* 35: 1100–1123.
Gell-Mann, Murray and Hartle, James B. (1990). "Quantum Mechanics in the Light of Quantum Cosmology." In *The Physics of Complexity, Entropy and Information.* Santa Fe Institute Studies in the Sciences of Complexity, vol. IX, Wojciech H. Zurek, ed. Reading, Mass.: Addison-Wesley. Also in *Proceedings of the Third International Symposium on Foundations of Quantum Mechanics in the Light of New Technology.* S. Kobayashi, ed. Tokyo: Japan Physical Society.
Halliwell, Jonathan J. (1986). "The Quantum Cosmology of Einstein–Maxwell Theory in Six Dimensions." *Nuclear Physics B* 226: 228–244.
———(1987). "Correlations in the Wave Function of the Universe." *Physical Review D* 36: 3626–3640.
———(1989). "Decoherence in Quantum Cosmology." *Physical Review D* 39: 2912.
———(1991). "The Wheeler–DeWitt Equation and the Path Integral in Minisuperspace Quantum Cosmology." See this volume, 75–115.

Halliwell, Jonathan J., and Hawking, Stephen W. (1985). "Origin of Structure in the Universe." *Physical Review D* 31: 1777–1791.

Hartle, James B. (1988a). "Quantum Kinematics of Spacetime I. Nonrelativistic Theory." *Physical Review D* 37: 2818–2832.

———(1988b). "Quantum Kinematics of Spacetime: II. A Model Quantum Cosmology with Real Clocks." *Physical Review D* 38: 2985.

———(1990). "Quantum Kinematics of Spacetime: III. General Relativity." *Physical Review.* Forthcoming.

Joos, E., and Zeh, H. D. (1985). "The Emergence of Classical Properties through Interaction with the Environment." *Zeitschrift für Physik B* 59: 223–243.

Keifer, Claus (1987). "Continuous Measurement of Mini-Superspace Variables by Higher Multipoles." *Classical and Quantum Gravity* 4: 1369–1382.

Mellor, Felicity and Moss (1988). "Decoherence in Quantum Kaluza–Klein Theories." Newcastle preprint NCL-89-TP/19.

Vachaspati, Tanmay, and Vilenkin, Alexander (1988). "Uniqueness of the Tunneling Wave Function of the Universe." *Physical Review D* 37: 898–903.

Wada, Sumio (1986). "Quantum Cosmological Perturbations in Pure Gravity." *Nuclear Physics B* 276: 729–743.

Zurek, Wojciech H. (1990). "Quantum Measurements and the Environment Induced Transition from Quantum to Classical." See this volume, pp. 43–66.

The Role of Time in the Interpretation of the Wave Function of the Universe

Robert M. Wald

The brief talk I presented at the Osgood Hill Conference described some recent work done in collaboration with W. G. Unruh. The unsatisfactory status of the interpretation of the wave function of the universe in canonical quantum gravity was reviewed. It was argued that an appropriate notion of "time" is needed for a viable interpretation of quantum gravity, and, further, that the dynamical variables of the theory are not suitable for playing this role. This situation led us to seek a new theory, in which a non-dynamical time variable is explicitly present (as in ordinary Schrödinger quantum mechanics), but such that, when a "good clock" dynamical variable is present, the correlations between dynamical variables are described by the Wheeler–DeWitt equation. (The latter requirement is presumably necessary in order that the theory correspond, in the classical limit, to general relativity.) An attempt at formulating such a theory was described; it had the desired qualitative properties, but it failed to produce the Wheeler–DeWitt equation. A preliminary account of the work can be found in Unruh 1988. A more detailed account—which includes an analysis of a theory of the type we seek that corresponds classically to general relativity with an arbitrary cosmological constant—can be found in Unruh and Wald 1989.

REFERENCES

Unruh, W. G. (1988). "Time and Quantum Gravity." In *Proceedings of the Fourth Seminar on Quantum Gravity*. M. A. Markov, V. A. Berezin, and V. P. Frolov, eds. Singapore: World Scientific Press.
Unruh, W. G., and Wald, Robert M. (1989). "Time and the Interpretation of Canonical Quantum Gravity." *Physical Review D* 40: 2598–2614.

Unitarity, Causality and Quantum Gravity

Ted Jacobson

A quantum state is a nonlocal concept; nevertheless, quantum theory is consistent with causality. How is this possible? Since the recipe for extracting predictions from the state is nonlocal (in a sense to be made clear below), it is not enough that evolution proceeds by local partial differential equations. We shall argue here that, in fact, quantum evolution must be unitary in order for predictions to be consistent with causality. This crucial role of unitarity is important to keep in mind whenever contemplating changes in the ordinary framework of quantum theory. We first show that, with a fixed causal structure on space-time, causality implies unitarity. We then extend the argument to quantum gravity by considering superpositions of states that are semiclassical except in localized regions.

Suppose then that a fixed causal structure exists, and that quantum evolution is given by a linear operator $\mathcal{E}$.[1] Let a quantum system be prepared on a space-like hypersurface Σ in a superposition $\psi = \psi_1 + \psi_2$, where ψ_1 and ψ_2 have support in disjoint regions $\mathcal{R}_1$ and $\mathcal{R}_2$, so that $\langle \psi_1 \mid \psi_2 \rangle = 0$ (see Figure 1).
The probability of finding the system in region $\mathcal{R}_i$ ($i = 1, 2$) is then given by

$$ p_i = \frac{\|\psi_i\|^2}{\|\psi_1\|^2 + \|\psi_2\|^2}. $$

Note the nonlocal nature of this formula: p_1 depends upon the norm of ψ_2 in region $\mathcal{R}_2$, and vice versa. This nonlocality will lead to violation of causality if evolution is not unitary. If the state is evolved to a space-like hypersurface Σ' on which the future $\mathcal{R}_i'$ of the regions $\mathcal{R}_i$ remain disjoint, then in the new state $\mathcal{E}\psi = \mathcal{E}\psi_1 + \mathcal{E}\psi_2$, $\mathcal{E}\psi_1$ and $\mathcal{E}\psi_2$ still have disjoint support, and are therefore orthogonal. (We assume that $\mathcal{E}$

propagates states within the causal cone.) The probabilities p_i' for finding the system in $\mathcal{R}_i'$ are thus given by

$$p_i' = \frac{\|\mathcal{E}\psi_i\|^2}{\|\mathcal{E}\psi_1\|^2 + \|\mathcal{E}\psi_2\|^2}.$$

Now causality implies that the probabilities p_i' are the same as p_i, which will be the case if and only if

$$\frac{\|\mathcal{E}\psi_1\|}{\|\psi_1\|} = \frac{\|\mathcal{E}\psi_2\|}{\|\psi_2\|}.$$

We thus conclude that the norms of ψ_1 and ψ_2 must be scaled by the *same* factor in the evolution from Σ to Σ'.

A much stronger conclusion can be reached by considering evolution from Σ to the family of space-like surfaces that coincide with Σ everywhere in region $\mathcal{R}_1$. For all these surfaces, ψ_1 is unchanged, so the above argument shows that the norm of ψ_2 must in fact be *unchanged* if the probability p_1' is to be independent of how the surface is extended beyond region $\mathcal{R}_1$. Since the argument could be applied to *any* state ψ_2, we conclude that the linear operator $\mathcal{E}$ must preserve the norm of all states, and therefore must be unitary.

To recapitulate, if $\mathcal{E}$ fails to be unitary, predictions in the region $\mathcal{R}_1$ will become ambiguous, due to the freedom in extending the space-like surface beyond $\mathcal{R}_1$. Moreover, if a preferred surface were somehow to be selected, causality would be violated, since predictions in $\mathcal{R}_1$ would depend on what was happening at space-like separations from there.

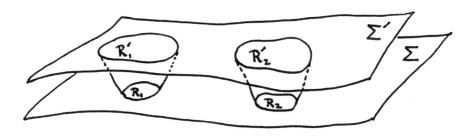

FIGURE 1.

Until now we have been considering quantum mechanics of a particle in the Schrödinger picture. Our considerations clearly apply to quantum field theory as well, since they concern only the issue of locality and state vector normalization, and the particle position operator plays no crucial role. Whether our conclusions apply to other formulations of quantum theory is another question however. They clearly do apply to the path integral formulation, since the path integral essentially plays the role of the linear operator $\mathcal{E}$, and the initial and final boundary conditions on the paths specify the corresponding states.

In the Heisenberg picture, the issues are somewhat different. In order to preserve Hermiticity, we can presume that the observables evolve via $\mathcal{O}' = \mathcal{E}^\dagger \mathcal{O} \mathcal{E}$. If $\mathcal{E}$ fails to be unitary, then the identity operator does not remain the identity under evolution, and $(\mathcal{O}_1 \mathcal{O}_2)' \neq \mathcal{O}_1' \mathcal{O}_2'$. If evolution does not respect operator products, the correspondence between operators and physical observables becomes ambiguous. It is conceivable that this ambiguity can be fixed by defining the observables on some preferred space-like surface (e.g., before the nonunitarity sets in, if such a surface exists) and then evolving them "whole" thereafter. In any case, all global conservation laws would be expected to fail. But perhaps these pathologies are physically acceptable under physical conditions that produce the hypothetical unitarity violation. If so, the Heisenberg picture does seem to allow a nonunitary evolution while remaining consistent with causality.[2]

How do the above considerations apply to quantum gravity? In the fully quantum regime, there is no well-defined causal structure, so of course we cannot argue as above. However, when such fully quantum behavior is confined to a spatially localized region Q, we can use the surrounding classical causal structure to argue that evolution must be unitary even in the region Q, whatever "evolution" might mean there. To see how, consider the following experiment.

A pair of spins is prepared in the correlated state $(|\uparrow\uparrow\rangle + |\downarrow\downarrow\rangle)/\sqrt{2}$. The spins are then taken off to space-like separated regions $\mathcal{R}_1$ and $\mathcal{R}_2$ where interactions that do not disturb the spin state may occur. The final state then has the form $(|\uparrow\uparrow\rangle|+\rangle + |\downarrow\downarrow\rangle|-\rangle)/\sqrt{2}$ where $|\pm\rangle$ is the state of all other degrees of freedom. Now the relative probability p_+/p_- of "up" and "down" for, say, the spin in region $\mathcal{R}_1$ is given by $p_+/p_- = \langle+|+\rangle/\langle-|-\rangle$. If the dynamics is not unitary somewhere in region $\mathcal{R}_2$, then p_+/p_- will depend on which "space-like" surface is used to evaluate the state in region $\mathcal{R}_2$. This effect could in principle be used to communicate over arbitrary space-like intervals if quantum gravity were not unitary and if the surface ambiguity were somehow fixed. For example, if for an ensemble of such states a messenger in region $\mathcal{R}_2$ always creates a black hole when the spin there is found to be "up," this would be communicated (statistically) to $\mathcal{R}_1$. The probability of "up" in $\mathcal{R}_1$ would be affected by the change in $\langle+|+\rangle$ induced by nonunitary evolution near the singularity, where quantum gravity effects presumably come into play.

In full quantum gravity it is not known how to identify a time variable and an inner product, in which the "evolution" described by the Wheeler–DeWitt equation or by a path integral is unitary. It is therefore tempting to try to extract a probability interpretation in the absence of unitarity, simply by picking some set of alternatives, computing probabilities from the corresponding relative amplitudes, and normalizing these probabilities "by hand." It is sometimes argued that, if this is done in a way that agrees in the semiclassical limit with the usual interpretation, it provides an adequate extension of the interpretation to the fully quantum domain. That unitarity is abandoned is regarded as acceptable since, outside the semiclassical regime, the causal structure is in any case not well-defined. From the previous paragraph we see that there is a serious problem with this proposal, however, since even the local predictions of an observer far from the region of unitarity violation would be rendered ambiguous or acausal.

In conclusion, unitarity is required in order to keep the nonlocality involved in the normalization of quantum amplitudes from manifesting itself in ambiguity or acausal behavior. Schrödinger or path integral formulations of quantum gravity that abandon unitarity are therefore probably not acceptable. Nonunitary Heisenberg formulations, although pathological in other respects, may possibly remain unambiguous and causal as far as predictions within regions of unitary evolution are concerned.

Acknowledgments. I am indebted to Doug Eardley for his suggestion that nonunitary evolution might lead to problems with locality. I thank Carlo Rovelli for raising the question about the Heisenberg picture. The text of this article was modified in March 1990 to reflect some changes in my point of view since the original talk was given. This work was supported by NSF grants PHY88-04561, PHY89-10226, and a University of Maryland GRB Semester Research Award.

NOTES

^{1} In this article only linear evolution is considered. Nonlinear modifications of quantum mechanics that preserve locality do exist, however they allow coupling between the evolution of different macroscopic states. See the paper by J. Polchinski, "Weinberg's Nonlinear Quantum Mechanics and the EPR Paradox," ITP preprint NSF-ITP-90-101 (1990), and references therein.

^{2} The contribution of Carlo Rovelli to this volume also discusses the possibility of a preferred status for the Heisenberg picture in quantum gravity. See also his paper, "Time in Quantum Gravity: Hypothesis," *Phys. Rev. D* to appear 1991.

Problems with Causality in the Sum-over-Histories Framework for Quantum Mechanics

Rafael D. Sorkin

1. Introduction

The sum-over-histories or "path-integral" framework is a natural one in which to attempt to overcome the contradiction between general covariance and the need that quantum theories seem to have for a distinguished notion of time. One aspect of this need is the difficulty of defining quantum observables in the absence of pre-specifiable space-like hypersurfaces to which to refer them. But the sum-over-histories framework does without such hypersurfaces, and thereby offers us a way to define quantum observables that meaningfully ask the type of questions we would hope a theory of quantum gravity would be able to answer.

Thus, for example, an understanding of the endpoint of black-hole evaporation presupposes a definition of horizon; but this notion obviously has a global character that would be very hard to capture in any but an explicit space-time framework. Similarly, a question like: "How many cycles of expansion, recollapse, and 'bounce' have occurred before the present cosmological era?" would be exceptionally difficult to formulate in terms of canonical variables (and would not obviously be any easier to express in terms of the noncommuting local metric operators of so-called "covariant" quantum gravity). In contrast, the sum-over-histories approach allows us to frame such questions in terms of individual classical space-times, in which within each the meaning of concepts like horizon and "bounce" is immediately clear.

Indeed, one can imagine "ideal measurements" that ask which one of some well-defined set of alternatives is realized by the actual history; and the sum-over-histories prescription in principle provides amplitudes in terms of which the relative probabilities of the various alternatives can be computed. We will see, however, that carrying out analogous measurements in nonrelativistic quantum mechanics would, in general, exert an influence on the past! To realize the full promise of the sum-over-histories framework, therefore, people will probably have to develop its measurement theory much further, trying especially to understand the place of causality (together with locality and unitarity) in this formulation that takes as its starting point the fundamentally nonlocal notion of probability amplitude.

2. The Sum-over-Histories Framework in Nonrelativistic Quantum Mechanics

In order to introduce the sum-over-histories in the form I have in mind, let us consider the simplest nonrelativistic system, a point particle in one spatial dimension. Rather than thinking in terms of "observables" such as position and momentum, we treat the particle's world line itself as real, and therefore contemplate measuring its properties more generally than just where it intersects a pre-specified $t = $ constant hypersurface. Such measurements in fact are precisely the kind that Feynman described in introducing the path integral idea (Feynman 1948). There, he took as his "first postulate" that: "if an ideal measurement is performed to determine whether a particle has a path lying in a region of space-time, then the probability that the result will be affirmative is the absolute square of a sum of complex contributions, one from each path in the region." Earlier he explains that an ideal measurement is one "that is capable only of determining that the path lies somewhere within R," without yielding any *additional* information that the experimenter would then have to "ignore."

The question of whether the path lies in a specified region is fairly general, but one can also think of other questions of potential relevance. For example, one can ask whether (subject to the condition that it be at the origin at $t = 0$) the path makes an excursion to $x > 1$ and then returns to $x = 0$ at any time before (say) $t = 20$. Such a question is analogous to the question of whether the universe expands to a large size and then recollapses in less than (say) 200 billion years. In order to cover all such possibilities, let us simply specify *any* partition of the path space into mutually exclusive and exhaustive possibilities C_1, C_2, $C_3, \ldots$, and ask for the relative probability of occurrence of each class C_i. It is clear that this procedure includes as one special case the ordinary position observable $q(t)$, which just divides the set of all world lines into classes C_q labelled (in one spatial dimension) by a real-valued index q, parametrizing the possible locations of the particle at time t.

The viewpoint just indicated can be summarized by saying that quantum mechanics is a theory of stochastic processes, but one that uses a nonclassical rule for computing probabilities. To sharpen this comparison, let us consider for a moment the classical theory of the random walk, which is the discrete-time analog of the quantum-mechanical free particle. Here, one begins with a space of possible "paths" or "histories" and introduces "dynamics" into this kinematical framework by giving a rule for assigning probabilities to certain ("measurable") subsets of that space. For classical processes, this rule is in essence very simple. We merely attach a positive real number p to each history and assign to the class C_i the composite probability

$$Prob(C_i) = \sum_{\gamma \in C_i} p(\gamma). \tag{1}$$

The only real complication is that the paths in question are unbounded in time, and therefore the individual probabilities $p(\gamma)$ do not truly exist. One can make them exist by truncating the histories at some "final" time, but unlike in the quantum situation that follows, it is not necessary to do so. The class probabilities $P(C_i)$ are rigorously well-defined even if the individual $p(\gamma)$ are not.

Now, the quantum rules for assigning probabilities differ from the classical ones mainly because they attach complex *amplitudes* to individual paths rather than positive real numbers. If the above quoted "first postulate" of Feynman's article were correct, then this would be essentially the only difference; but unfortunately a straightforward interpretation of this postulate leads to rules that fail to reproduce the ordinary quantum probabilities, and are not even internally consistent (Sorkin 1987). Instead, one can adopt the following prescription, which will be the basis of our remaining reflections. Although it may appear slightly ad hoc, it does furnish the correct probabilities for arbitrary sequences of position measurements that are, in general, the only direct point of contact between the sum-over-histories framework and the usual quantum-mechanical formalism.

Aside from the role of complex amplitudes, the rule I am about to state differs from the classical procedure (equation (1)) in two important ways. First of all, the paths it works with are truncated at some future time; and second (although there are equivalent ways to describe this), they are "return paths" that consist of precisely one forward-moving and one backward-moving portion. For simplicity let us impose the initial condition $q(t_0) = q_0$ and then pick some fixed partition $\{C_i\}$ of the set of all paths that originate at the point (t_0, q_0). I will assume that there exists a time t_C, such that the conditions defining the C_i refer only to the portions of the paths between t_0 and t_C. To each forward-directed path extending from t_0 to t_C, we attach an amplitude $A(\gamma) = \exp(iS)$, and to the time reverse of each such path, we associate the conjugate amplitude (which is in fact just what $\exp(iS)$ works out to be when time is taken to run backward and S is the uncorrected classical action). The rule is then that: *the (unnormalized) probability of class C_i (relative to the other classes) is the sum of the amplitudes of all return paths whose "upward" and "downward" portions both fulfill the conditions defining C_i.*

In symbols, this rule,

$$P(C_i) = \sum_{\gamma \in C_i} A(\gamma), \tag{2}$$

looks almost identical to the classical rule (1); but, of course, the notation hides the facts that the $A(\gamma)$ are complex, that the γ are "two-way" paths that turn around at $t = t_C$, and that the P_i are only relative probabilities that have to be divided by $\sum_i P_i$ before becoming true probabilities.

Notice that the total amplitude itself yields the probability; it does not have to be squared because the amplitudes of return paths are already products of one-way amplitudes. Notice also that Feynman's question as to whether the path lies in some region R is included here as a special case, in which there are only two classes C_i, namely, the class C_1 of paths that do lie in R, and the complementary class C_2 of paths that do not.

The least satisfactory aspect of the rule just stated is probably that it refers to an arbitrary time t_C, which I call the "collapse time," because of the close connection between its role and what is usually referred to as "collapse of the wave function." To some extent, this introduction of t_C reintroduces into our framework a kind of distinguished time, thereby threatening to bring back the contradiction with general covariance that we were trying to fend off. It also places severe constraints on our choice of the amplitudes $A(\gamma)$, which otherwise would have been completely free. Specifically, because the value of t_C has no meaning in itself (in particular we do not assume that any measurement occurs at that time), it is necessary for consistency that the expression (2) be independent of the choice of t_C as long as the latter is sufficiently far in the future. As we will see in a moment, this independence appears as the expression of unitarity within the present framework.

Despite the drawbacks just mentioned, truncation of the histories does have a beneficial effect from the standpoint of causality. Since amplitudes refer to whole histories, a priori you might expect even a localized change of conditions somewhere to have nonzero effects everywhere else. In particular, there seems to be no reason why interfering with the system tomorrow wouldn't affect the probabilities of events that will occur today. However, if all of today's probabilities can be computed from truncated histories that end before midnight, then clearly the potential acausality will not in fact be present. A formulation without any collapse time would not be able to escape such potential acausalities in this way, although one might hope that another way would nonetheless be found to escape them.

Incidentally, in terms of two-way histories, the "decoherence" of which Gell-Mann spoke in his talk signifies that only paths whose upward and downward portions nearly coincide contribute much to the probabilities $P(C_i)$.

3. Independence of t_C

Let us see exactly how unitarity, in the ordinary sense, implies that probabilities computed according to (2) are independent of our choice of collapse time. To this end, let t_1 and t_2 be two possible such choices, and suppose that initially we use the later time t_2 as collapse time. Each one of the probabilities $P(C_i)$ will then be a sum of amplitudes of the form $A(\ldots x_1, z_2, y_1, \ldots)$, by which I mean the sum of the amplitudes of all return paths that pass through $(t = t_1, q = y)$ on the way "up," reverse their direction at $(t = t_2, q = z)$, and pass through (t_1, x) on the way back down (and also belong to C_i, of course).[1]

Now, the sum to be done is over all three positions x, y, and z. Performing the sum over z first will yield the delta-function $\delta(x, y)$, thereby causing the remaining sum to reduce to what it would have been had t_1 been chosen as collapse time. In formulas, the basic equality is[1]

$$\sum_z \langle x_1 | z_2 \rangle \langle z_2 | y_1 \rangle = \sum_z \langle z_2 | x_1 \rangle^* \langle z_2 | y_1 \rangle = \delta(x, y), \qquad (3)$$

where $\langle x_1 | z_2 \rangle$, for example, is the sum of the amplitudes of all partial paths going downward from z at t_2 to x at t_1. The desired reduction then occurs as follows:

$$\sum_{xyz} A(\ldots x_1, z_2, y_1 \ldots)$$

$$= \sum_{xyz} A(\ldots x_1) A(x_1, z_2, y_1) A(y_1 \ldots)$$

$$= \sum_{xy} \sum_{z} A(\ldots x_1) \langle x_1 | z_2 \rangle \langle z_2 | y_1 \rangle A(y_1 \ldots)$$

$$= \sum_{xy} A(\ldots x_1) \delta(x, y) A(y_1 \ldots)$$

$$= \sum_{x} A(\ldots x_1 \ldots).$$

Notice that it was also crucial here that the amplitude $A(\gamma)$ of any path be the product of the amplitudes of its segments, i.e., that the "action" S be additive.

4. Causal and Acausal Measurements

In much the same way as we just established independence of collapse time, we can also prove that no sequence of position measurements can influence the probabilities of events occurring entirely to its past. However, this respect for causality does not persist for more general measurements. In fact, we will see that it fails already for putative measurements of properties that are nontrivial logical combinations of positional properties at two or more different times. First, however, let us consider the case of positional measurements at a single time.

In order to formulate the conditions for causality in the present framework, let us take for simplicity a pair of yes–no questions E and F, the first referring to events occurring before t_1 and the second referring to events occurring between t_1 and t_C. Let $P(E)$ be the probability that E has the answer yes (i.e., that the actual path satisfies the condition in question) *assuming that F remains indefinite* (i.e., that no measurement is made to determine it). Similarly let $P(E, F)$ be the probability that both E and F are yes if *both* are measured; and let $P(E, F')$, $P(E', F)$, and $P(E', F')$ be defined analogously, where E' and F' are the logical complements of E and F. The condition that measuring F not affect $P(E)$ is that

$$P(E, F) + P(E, F') = P(E). \qquad (4)$$

In the shorthand adopted above, this last condition can be written as

$$\sum_z A(E, F, z_C, F, E) + \sum_z A(E, F', z_C, F', E) = \sum_z A(E, z_C, E). \qquad (5)$$

Here as before, $A(E', F, z_C, F, E)$, for example, is the combined amplitude of all return paths which satisfy E and F on the way up, turn around at (t_C, z), and satisfy F but not E on the way back down.

Now, let us take the special case in which F is a positional observable at a single time. Specifically, let us suppose that it is the question whether, at $t = t_1$, the path lies in a specified subset F of configuration space. The first amplitude in (5) then becomes

$$\sum_{x \in F} \sum_{y \in F} A(E, x_1, z_C, y_1, E),$$

and the entire left hand side of (5) is then

$$\sum_z \left(\sum_{x \in F} \sum_{y \in F} + \sum_{x \notin F} \sum_{y \notin F} \right) A(E, x_1, z_C, y_1, E). \qquad (6)$$

If the sums over x and y were unrestricted, then the last expression would immediately reduce to $A(E, z_C, E)$, which is just $P(E)$ as required for causality. In fact, there are two contributions lacking, namely, those for which one of the points x, y lies in F and the other doesn't. But we found above that each of these contributions actually vanishes because

$$\sum_z A(\ldots x_1, z_C, y_1, \ldots) = 0$$

when $x \neq y$, and in particular when x and y belong to disjoint subsets of configuration space. Hence the sums in (6) *are* in effect unrestricted, and the condition (4) for causality is fulfilled.

In the more general situation where the question F does not concern merely the position q at a single time, the equality (4) will in general fail. For example, suppose that F is the property, "$q \in A$ at $t = t_1$ and

$q \in B$ at $t = t_2$," where $t_1 < t_2 < t_C$, and where A and B are specified subsets of configuration space (just as F itself was in the case we just considered). Our new question F is an instance of precisely the type of observable Feynman described, although his region R has here been reduced to a particularly simple form, comprising just a pair of subsets of two distinct $t = \text{constant}$ hypersurfaces.

To see that (4) is false in this situation, we need merely attempt to repeat the steps by which we established its truth when F was a single-time position observable. In a notation similar to that already used, the (unnormalized) probability $P(E, F)$ that both E and F hold is

$$P(E, F) = \sum_z \sum_{(x,u) \in F} \sum_{(y,v) \in F} A(E, x_1, u_2, z_C, v_2, y_1, E), \quad (7)$$

where, of course, "$(x, u) \in F$" is short for "$x \in A \& u \in B$," and similarly for (y, v). As before, we can simplify this expression using unitarity and multiplicativity of path amplitudes:

$$\begin{aligned}
P &= \sum_z \sum_{(x,u) \in F} \sum_{(y,v) \in F} A(E, x_1, u_2) A(u_2, z_C, v_2) A(v_2, y_1, E) \\
&= \sum_{(x,u) \in F} \sum_{(y,v) \in F} A(E, x_1, u_2) \delta(u_2, v_2) A(v_2, y_1, E) \\
&= \sum_{(x,u) \in F} \sum_{(y,u) \in F} A(E, x_1, u_2) A(u_2, y_1, E) \quad (8) \\
&= \sum_{x \in A, u \in B, y \in A} A(E, x_1, u_2) A(u_2, y_1, E) \\
&=: ABA,
\end{aligned}$$

where the final expression, ABA, is a convenient shorthand for the expression above it.

In the same way the relative probability $P(E, F')$ can be transformed into

$$\begin{aligned}
P(E, F') &= \sum_{(x,u) \notin F} \sum_{(y,u) \notin F} A(E, x_1, u_2) A(u_2, y_1, E) \\
&= A'BA' + AB'A + A'B'A + AB'A' + A'B'A'.
\end{aligned} \quad (9)$$

Now, $A'BA'$ and $A'B'A'$ combine to yield in exactly the same manner

$$\sum_{x \notin A} A(E, x_1, E);$$

and $AB'A$ combines with ABA from (8) to yield

$$\sum_{x \in A} A(E, x_1, E).$$

Finally, these last two expressions add to the unrestricted sum,

$$\sum_x A(E, x_1, E) = \sum_x A(E, x_C, E) =: P(E),$$

in which the penultimate equality used the independence of collapse time that we established earlier.

Overall then, four of the six terms represented in (8) and (9) add up to yield precisely $P(E)$, but there remain the two residual terms $A'B'A$ and $AB'A'$. In order for the causality condition (4) to hold, therefore, these two terms would have to vanish, or at least sum to zero. However, it is clear that in general they don't, the first one being explicitly

$$\sum_{x \notin A} \sum_{z \notin B} \sum_{y \in A} A(E, x_1, z_2, y_1, E),$$

and the second being its complex conjugate. Thus, causality would be violated if the observable F could actually be subjected to an "ideal measurement."

Strictly speaking, we have not proved this last statement because the probabilities we have computed are only relative probabilities and ought to be properly normalized before being required to add up correctly. This turns out not to be a real loophole, however, as it is easy to check that the ratios $P(E)/P(E')$ and $[(P(E,F) + P(E,F')]/[P(E',F) + P(E',F')]$ are also unequal in general.

In concluding, let me emphasize that what we have found does not mean that any measurement which can give a truth-value to F violates causality in some way. On the contrary, successive separate measurements of A and B will render F either true or false, and also must respect causality, as follows directly from our proof of (4) for the case when F is a single-time property. But successive separate measurements are not ideal observations for the dichotomy $F-$ not $-F$, because they determine extraneous information, namely *separate* truth-values for A and B, rather than a single truth-value for the combination $F = A\&B$. They partition the set of all histories into four classes instead of just two as a "minimally disturbing" measurement would do.

In connection with quantum gravity, it is perhaps not necessary that the relevant questions be able to appear as dichotomies. Maybe it is enough that the partitions of interest be refinable in some way to partitions that are actually compatible with causality, just as $\{F, F'\}$ is refinable to $\{(A\&B), (A\&B'), (A'\&B), (A'\&B')\}$. Even such a resolution of our problem, however, would leave open the question of why the more general, acausal measurements are not also possible (if in fact they are not). To understand this, one will have to better understand the meaning of measurement itself within the sum-over-histories framework, and specifically to understand what distinguishes the "physically possible" partitions from the "impossible" ones. This question becomes especially perplexing in the "cosmological" context, in which the partition of histories cannot be "physically implemented" by any intervening external apparatus.

Whether these reflections must lead in the end to the conclusion that the sum-over-histories framework suffers from the same contradictions as all other interpretations of quantum mechanics seems unclear at the moment. However, the originality of the way old questions appear within that framework offers hope that, even if the old contradictions do recur, they will do so in a form that will suggest, more readily than otherwise, the modifications needed to overcome them.

Acknowledgment. This research was partly supported by NSF grant PHY 8700651.

NOTE

[1] This notation of using subscripts to indicate values of t may be slightly unusual, but I hope it makes the resulting expressions easier to read. For an expanded notation, one could replace x_1 by (t_1, x) and z_C by (t_C, z), etc.

REFERENCES

Feynman, Richard P. (1948). "Space-Time Approach to Non-Relativistic Quantum Mechanics." *Reviews of Modern Physics* 20: 367–387.
Sorkin, Rafael D. (1987). "On the Role of Time in the Sum-Over-Histories Framework for Gravity." In *History of Modern Gauge Theories*. Proceedings of the Conference on the History of Modern Gauge Theories, Logan, Utah, July 1987. M. Dresden and A. Rosenblum, eds. Singapore: World Scientific. Forthcoming.

Discussion

GELL-MANN: Could you please summarize, in one sentence, what the point was? I didn't get it.

SORKIN: I'm very unhappy with this result. I'm unhappy because I believe that the sum-over-histories offers our best hope for a version of quantum mechanics that is compatible with gravity.

GELL-MANN: But it seems to have nothing to do with any special theory. Are you unhappy with sum-over-paths generally?

SORKIN: That's right. I'm unhappy with the sum-over-paths in general if you try to use it as anything more than a technical tool for propagating wave functions, i.e., computing functions like $\psi(q)$, that give us probabilities in the usual way for position measurements at a fixed time. So when you try and use it as Jim Hartle does, and as I would also like to do, to provide answers to more generally phrased questions, such as: "did the particle pass through the following space-time region?" you get this problem. If you ask two questions, one of which is totally to the past of the other, the very fact that you set up the apparatus to measure the later question has an effect on the probability of getting a "yes" or "no" to the earlier question.

Space and Time
in the Quantum Universe

Lee Smolin

"I am granted this important principle, that nothing happens without a sufficient reason why it should be thus rather than otherwise..."

"As for me, I have more than once stated that I held space to be something purely relative, like time; space being an order of coexistence as time is an order of successions."

–Leibniz (1696, 1714).

"A consideration that is of the greatest importance in all philosophy and in theology itself is this: that there are no purely extrinsic denominations because of the interconnection of things, and that it is not possible for two things to differ from one another in respect of place and time alone, but that it is always necessary that there shall be some other internal difference... then, if place by itself does not make a change, it follows that there can be no change that is merely local. In general, place, position and quantity... are mere relations... To be in a place seems, abstractly at any rate, to imply nothing but position. But, in actuality, that which has a place must express place in itself; so that distance and the degree of distance involves also a degree of expressing in the thing itself a remote thing..."

–Leibniz (1696).

"The physical space I have in mind (which already includes time) is therefore nothing but the dependence of the phenomena on one another. A completed physics that knew this dependence would have no need of separate concepts of space and time because these would already have been encompassed."

–Ernest Mach (1866).

"Imagine, then, a series of superimposed ways in which things may be ordered in nature. Whenever one order dissipates, another emerges to occupy its place. Each of these variables of natural order require a different style of explanation. Each supervening order presents explanatory opportunities that compensate for the difficulties created by the loosening of the previous orders. These discontinuities in the type of order and explanation do not coincide in any simple fashion with the boundaries between inanimate and living matter, or between the biological and the social. Nor do the supervenient forms of order and explanation ever entirely displace the preceding and weakened ones.

"In the course of this overlaying of forms of order and explanation, there is a waning of the distinction between what must be accepted as given (by way of initial conditions or the value of the variables) and what can be rationally explained. The contrast between order and disorder is softened, together with the contrast between the nature of order in theory and in the phenomena that the theory explains. Our conception of intelligible relations becomes increasingly subtle and capacious."

–Roberto Mangabeira Unger (1984)

"We are, quite literally, in a new world, a much more peculiar place than it seemed a few centuries back, harder to make sense of, riskier to speculate about, and alive with information which is becoming more accessible and bewildering at the same time. It sometimes seems that there is not just more to be learned, there is *everything* to be learned."

–Lewis Thomas (1988)

1. Summary

This paper is devoted to the problem of constructing a quantum theory that could describe a closed system—a quantum cosmology. I argue that this problem is an aspect of a much older problem—that of how to eliminate from our physical theories what I call "ideal elements," which are elements of the mathematical structure whose interpretation requires the existence of things outside of the dynamical system described by the theory. I argue that both Mach's principle and the measurement problem in quantum mechanics are aspects of this problem, so that both of these problems are linked in the problem of constructing a sensible quantum cosmology.

Part of this paper is critical, and is aimed at uncovering criteria that a theory of quantum cosmology must satisfy if it is to give physically sensible predictions. I propose three such criteria and show that conventional quantum cosmology can only satisfy them if there is an intrinsic time coordinate on the phase space of the theory. Along the way, I show that approaches based on correlations in the wave function that do not use an inner product cannot satisfy these criteria, so they cannot, in my view, lead to sensible theories. In addition, I argue that these approaches cannot be falsified by any observation that would not also provide a falsification of classical general relativity (possibly coupled to quantum matter fields).

I also show that the Everett interpretation is a red herring, in the sense that its use cannot lead to an acceptable measurement theory in any situation where there is not a conserved inner product, in which case the conventional interpretations also suffice.

To illustrate the discussion, I discuss the problem of quantizing a class of relational dynamical modes invented by Barbour and Bertotti. The dynamical structure of these theories is closely analogous to general relativity, and the problem of their measurement theory is also similar. I conclude that these models can only be sensibly quantized if they contain an intrinsic time.

In the last part of the paper I put forward some very tentative suggestions about how a quantum theory without ideal elements might be constructed, using examples based on statistical graph theory. I suggest a dynamical principle based on the complexity of the graphs and conjecture that it may explain how the dimensionality of space emerges from a purely relational theory without any background space.

2. Introduction: The Problem of Ideal Elements in Physical Theory

In this contribution I would like to argue that the problem of quantum gravity is an aspect of a much older problem, that of how to construct a physical theory which could be a theory of an entire universe, and not just a portion of one. This problem has a long history. It was, I believe, the basic issue behind the criticisms of Newtonian mechanics by Leibniz (1696, 1714), Berkeley (1721), and Mach (1893). It is also, I will argue below, the basic issue behind the problem of the interpretation of quantum mechanics. Quantum gravity is an aspect of this problem because the task of quantum cosmology is precisely to construct a quantum theory of a whole universe. It is therefore not surprising that the basic problems that are encountered in quantum cosmology, at both the conceptual and technical level, can be seen to be reflections of these two older problems:

Can a sensible dynamical theory be formulated that does not depend on an absolute or background space?; and *Can quantum mechanics be understood in a way that does not require the existence of a classical observer outside of the system?*

It is my contention that quantum cosmology remains a complete enigma precisely because we have not learned how to resolve these two questions. That is, we do not know how to construct a theory that could be interpreted as a theory of an entire universe. All of the theories that we understand presently are, of necessity, theories of only a portion of a universe. This is because, as I will explain below, they all incorporate *ideal elements*. I will use this term to denote absolute or background structures that are contingent, in the sense that they may be altered without altering the basic character of the theory, play a role in the dynamical equations of the theory, and are not themselves determined by solving any dynamical equations. I will argue below that, generally, when one has ideal elements in a theory the physical interpretation of that theory requires that the world contain things besides the dynamical system described by the theory.

That the theories we use in physics depend on ideal elements is no shame as the whole context of physical science, outside of cosmology, is the study of systems that are imbedded in a much larger universe. What is wrong is to attempt to use theories that are successful when applied to a small portion of the universe to describe the whole universe, ignoring the presence of the ideal elements and the crucial role they play for the interpretation of these theories. If we want a theory that could be a theory of an entire universe, it must refer to nothing outside of itself—it must be a theory without ideal elements.

To my knowledge, no one has ever formulated a physical theory that is completely without ideal elements. There are, however, some important examples of theories in which some, but not all, of the ideal elements one finds in Newtonian mechanics are eliminated. The most important of these examples is in fact general relativity in the case where cosmological boundary conditions have been imposed. Indeed, classical general relativity is, I believe, a great deal closer to the kind of theory we need in order to do cosmology than is quantum mechanics. Thus, the main obstacles to a quantum cosmology come from the quantum-mechanical side of the theory rather than from general relativity.

However, not all of the obstacles come from the quantum-mechanical side. This is first of all because general relativity with cosmological boundary conditions is not very well understood as a dynamical system, in ways which are important for the problem of constructing the quantum theory. Thus, for our purposes, it is important to emphasize that pure general relativity, as a theory without background structures, is almost never used as a basis of predictions in observational cosmology. What is used are certain restrictions of general relativity, in which additional assumptions have been imposed, for example, the assumption of spatial homogeneity. The resulting theory has ideal elements coming from the symmetry assumptions used in the reduction, and these ideal elements play, in practice, an important role in extracting predictions from the theory.

What is at issue here is not whether the reduced theory gives a good approximation to full general relativity. I assume that, for the purposes it is usually used for, it does. (It is, however, not a trivial problem to show precisely how the reduced theory gives an approximation to the full theory—and this is an aspect of the problems I am concerned with here.) What is at issue is why we go about giving general relativity a physical interpretation by reducing it to a theory with ideal elements. I claim we do this because, while there is no difficulty in principle, it is not easy in practice to extract physical predictions from the full theory. It is not easy because we are used to relying on background structures to extract physical predictions from a theory; we have no experience working directly with the observables of theories without ideal elements.[1]

An aspect of these problems is that, in the cosmological case, we do not know any physical observables of the theory explicitly in terms of the phase space coordinates, let alone a complete set that could be used as the basis for an intrinsic physical interpretation. This ignorance concerning the classical dynamics of general relativity turns out to be an important obstacle to quantizing the theory at the nonperturbative level, as I discuss in my other contribution to this volume.

Thus, there is much that classical general relativity has yet to teach us about doing physics without ideal elements.[2] However, because the problem of physical observables in quantum cosmology is a difficult one, it is very useful to have some examples of theories which go some way towards eliminating the dependence on ideal elements but are simpler than general relativity. A very interesting class of such theories, which will be discussed below, was invented by Barbour and Bertotti (Barbour 1974a, b, 1975; Barbour and Bertotti 1977, 1982). In addition to providing models in which to study some of the issues that arise in quantum cosmology, the Barbour–Bertotti models are good to study because they give us some insight into the kinds of things that can happen in a theory without ideal elements. In particular, they suggest that a universal consequence of eliminating ideal elements from a theory is that certain aspects of local physics are determined by global properties of the particular solution to the dynamical equations. Mach's principle, that the structure of the local inertial frames should be determined by the global distribution of matter in the universe, is one consequence of this (Mach 1893, Einstein 1916). Another consequence that one sees for the first time in these models is that the values of some of the physical constants are determined by the global solution rather than being put in by hand. In my concluding remarks, I will argue that this last feature may have profound consequences.

In this article, I discuss several topics connected with the problem of how to construct theories without ideal elements, and what we may hope to get from them. The discussion is split into two parts, a first, critical part, and a second, speculative part. In the first part, I discuss whether or not quantum cosmology, as conventionally formulated, can lead to a physically sensible quantum theory of the entire universe. (By conventional, I mean a formulation in which the states of the theory live in a linear vector space, the linear operators of which are associated with the observables.) I conclude that any physically meaningful formulation of conventional quantum cosmology must involve an inner product structure, so that the attempts to base an interpretation on correlations,[3] in the absence of an inner product structure, cannot possibly succeed. I further show that conventional quantum cosmology can only have a sensible physical interpretation if it is the case that an intrinsic time exists in the phase space of the theory, such that the Hamiltonian constraint can be read as a unitary evolution equation in that intrinsic time (for a review, see Kuchař 1982).

If an intrinsic time does not exist, then a quantum theory of the entire universe cannot be constructed without abandoning the basic linear structure of conventional quantum mechanics.

The second part of the paper contains some speculations on how I believe this might be done by constructing a theory that has, to begin with, no ideal elements of the kind that are required in the measurement theory of quantum mechanics. The ideas I discuss in this second part are speculative, and some of them are extremely speculative. All of them are ideas I have been thinking about for a long time. I describe them here because, although I have not so far been able to carry them to the point of formulating a definite theory, I believe they are the kind of ideas we will need to construct a physically sensible quantum cosmology.

More specifically, the critical argument begins in the following two sections in which I describe the ways in which classical mechanics and quantum mechanics depend on ideal elements. Framing the problem of the interpretation of quantum mechanics in terms of the larger issue of the elimination of ideal elements allows me to explain why the Everett interpretation (Everett 1957) does not provide a firm ground for a physical interpretation for quantum cosmology in the absence of an intrinsic time. The crucial point, I believe, is not whether the Everett interpretation gives a meaningful alternative to the usual measurement theory for ordinary quantum mechanics. It is rather whether it can do so in circumstances in which the usual measurement theory cannot be used, such as when there is no conserved inner product. I show in detail that the answer to this question is no, so that the Everett interpretation can only lead to a physically sensible measurement theory when a conserved inner product is present.

For clarity, the discussion of Section 4 is set mostly in the context of ordinary quantum mechanics, by which I mean the case in which there is a Schrödinger equation that describes evolution in an external time. In Section 5, I discuss the special circumstances of theories, such as quantum cosmology, in which the Hamiltonian is proportional to a constraint. I propose here three criteria that I believe any physically sensible measurement theory for quantum cosmology must satisfy. I then show that these criteria can only be satisfied if there is a conserved inner product, which in the case of quantum cosmology means that there must be an intrinsic time.

In Section 6, I discuss the Barbour–Bertotti models as examples of classical dynamical systems that are, at least partially, without ideal elements. I give the Hamiltonian formulation of these models, and show that they are very similar in structure to Hamiltonian general relativity. They then provide good models for the study of the physical interpretation of quantum cosmology. I also show how, in these models, local properties of a theory with ideal elements are determined by the global solution of a theory without ideal elements, as just mentioned.

In Section 7, I discuss the quantization of the Barbour–Bertotti models. I show that although the constraints of the theory can be formulated and solved quantum mechanically, it is not at all clear how to give a sensible physical interpretation to the resulting "physical" state space. This is because, with the elimination of the ideal background space and time, the conditions necessary for the ideal structures of quantum mechanics to be introduced and given a physical interpretation are absent. This situation is completely analogous to what occurs in quantum cosmology. As in quantum cosmology, the theory can only be saved if there is, in its phase space, an intrinsic time coordinate. Whether or not one exists is a problem left for future study.

The discussion in the sixth and seventh sections is a condensation of material that appears, in greater detail, in a paper by Julian Barbour and myself (Barbour and Smolin 1988).

In the eighth section, I begin the more speculative part of this essay. I outline what I believe a physical theory constructed without ideal elements might look like, and I indicate how quantum mechanics could be recovered from such a theory.

In order to make the discussion more concrete, I present in Section 9 an example of a theory that is, I believe, without any ideal elements. This is a very simple model, based only on the notion of relation, which mathematically is expressed in terms of graph theory. Of course, it is very difficult to see how one would recover our usual notions of quantum physics taking place in a background space of a certain dimensionality from such a model. But that is really the point; this difficulty is a measure of both the nontriviality and the potential payoff of doing physics without ideal elements. In particular, I raise the issue of what kind of dynamical principle may be formulated for a model of this kind. I will argue here that notions of complexity and variety of structure are very natural candidates for analogues of energy in a theory without ideal elements. In the context of the model system, I propose one such principle, which I call the *principle of maximal variety,* and show how it could lead to a dynamics that could determine, among other things, the dimensionality of space.

In the final section, I speculate that a cosmological theory based on this principle of maximal variety might have profound consequences for a problem that is not usually seen as part of cosmology: why are the physical conditions of the universe amenable to the self-organization of matter leading to the existence of life? I hope by these final remarks to convince the reader that this might be an appropriate question for study in the context of some future, physically sensible, quantum theory of the whole universe.

3. Ideal Elements in Classical Mechanics

The paradigmatic example of an ideal element is the use of a background, absolute space in Newtonian dynamics (Newton 1962). From a modern point of view, the role of the background space is subsumed under the notion of a set of background, global inertial frames. To specify the evolution of a dynamical system in classical mechanics, one needs to give initial coordinates and momenta with respect to an inertial frame. As discussed in detail in Barbour 1982 and 1989a, this involves the specification of more information than would be given were one to give just the relative distances and velocities of the particles in the system. Furthermore, the choice of which frames are inertial must be made before the dynamics of the system of particles is formulated; it cannot be a consequence of those dynamics. Thus, to give correctly the evolution of a Newtonian dynamical system from initial data, one must explicitly refer to something outside of that system. If one does not know beforehand which set of frames is to be considered inertial, and which are not, one cannot make predictions with Newtonian mechanics.

Thus, while Newtonian mechanics, as well as its special-relativistic generalizations, asserts the existence of a special family of preferred frames, which are then used in the specification of the dynamics, the theory does not contain any dynamical mechanism that might provide a basis for understanding why those frames, and not others, are preferred. This is the sense in which I would like to call the inertial frames ideal elements: they are required for the formulation of the dynamics of the system under study, and they are asserted to have certain properties that influence the evolution of the system under study but that, however, cannot be determined dynamically within the theory. In short, as expressed by Einstein in the introduction to his first review paper on general relativity, they are ideal because they act, but they cannot be acted upon (Einstein 1916).

Now, suppose one wanted to understand the question of why certain frames are inertial while others are not. This is, indeed, a natural question to ask; the most puzzling thing about first year mechanics for many students (and teachers) is why there is a relativity of velocity, but not of acceleration. Why is it that one can measure acceleration absolutely? The only answer Newtonian mechanics gives is that one is accelerating with respect to some pre-existing absolute structure, which Newton called absolute space and we call the background inertial frames. Thus, if we want to understand why there are preferred frames whose existence can be detected by local experiments, we must look outside the Newtonian dynamics of a local system.

Mach's name is usually associated with the hypothesis that the choice of the preferred frames is dynamically determined by some process that has something to do with the distribution of matter in the universe at large.[4] There are two reasons why one might believe that this is the case: (1) The distant matter that we can see does seem to be, at least approximately, nonrotating with respect to local inertial frames. (2) The choice of inertial frames seems to be affected to only a slight degree by the local distribution of matter (by this I am referring to the smallness of the frame-dragging effects). Thus, if the choice of inertial frames is determined by the distribution of matter, it is reasonable to think it should involve far away matter.

Thus, if we want to understand dynamically how the distinction arises between accelerating and inertial in the dynamics of a local system described by Newtonian mechanics, it is natural to look for an explanation outside of that system. This is an example of how the presence of an ideal element in the description of a particular dynamical system prompts us to look outside that system for a dynamical explanation of that ideal element.

The Barbour–Bertotti models, which I will describe below, are explicit examples of theories in which the hypothesis that the distribution of matter in a closed system determines the choice of inertial frames for its subsystems is explicitly realized.

There is another ideal element that plays a basic role in Newtonian dynamics, namely, time. The time with which the evolution of a Newtonian system is described is time measured on a clock that is outside of the system and does not interact with it. Just as in the case of the inertial frames, one must assume that the clock exists outside of the system before the initial conditions of the dynamical system are prescribed. It cannot be derived from, or deduced from, the purely relational data that can be extracted from measurements of relative distances within the system.

4. Ideal Elements in Quantum Mechanics

The problem of the interpretation of quantum mechanics has been for many years a rather discouraging area for research, at least partly because the determined attention of clever and imaginative people to this problem has not often been repaid with real insight into its nature or its solution.[5] Part of the problem may be that it has never been very clear what form the solution to this problem should take. Many discussions point in the direction of a new epistemological or ontological principle, but why should a contingent physical theory either create problems that are resolved by, or provide the basis for conversion to, a radical philosophical belief? I would like to argue here that the reason for this situation is that the problem of the interpretation of quantum mechanics, as it is usually formulated, has no solution. That is, while the formalism of quantum mechanics produces results in close agreement with experiment, I believe that no interpretation of this formalism is possible that both presumes that quantum mechanics is the final physical theory and is within the bounds of the generally accepted epistemological and ontological principles of natural science.

The reason for this is that the formal elements responsible for the puzzling aspects of quantum mechanics are ideal elements. Among these I will focus here on three, the correspondence of quantum-mechanical operators with classical observables, the inner product, and the notion of time in quantum mechanics. In the conventional interpretations of quantum mechanics, these elements are interpreted as referring to the action of classical systems outside the system under study. Thus, the conventional interpretations label these elements explicitly as ideal.

However, there are other interpretations of quantum mechanics, such as the Everett interpretations (Everett 1957; Geroch 1984), in which these elements are not explicitly labeled as ideal. Instead, the proponents of these interpretations claim that an interpretation of these purported ideal elements can be constructed from inside the formal structure of the theory in a way that does not refer to any outside elements. This, indeed, is the basis of the claim that the Everett interpretation may be the basis for an interpretation of quantum cosmology (Everett 1957, Wheeler 1957, and the references in note 3).

If this claim were true, then quantum mechanics would be a theory without ideal elements. However, as I will argue below, this claim is false.[6] The elements that the conventional interpretation labels as ideal can be seen to be ideal directly from their formal relations to the other elements of the theory. Indeed, had the original interpretation of the theory been the Everett interpretation it would have been necessary to expose the ideal elements by an internal analysis of the logic of the theory in the same way that Leibniz, Berkeley, and later Mach, exposed the ideal elements of Newtonian mechanics.

Thus, my argument in this section will consist of three parts. First, I will separate the ideal elements of the formal structure of quantum mechanics from the nonideal elements by means of an internal, formal analysis. Second, I will argue that, as in the case of Newtonian mechanics, any sensible interpretation of the theory must associate the ideal elements with the action of things outside of the state space of the dynamical system whose evolution is under consideration. Consequently, the advocates of the Everett interpretation cannot succeed in giving these elements of the formalism an interpretation completely internal to the dynamical system. Third, I will argue that those more radical advocates of the Everett interpretation who concede my second point, but argue that quantum mechanics can be given a sensible interpretation in which ideal elements such as the inner product play no role (see the references in note 3), are also wrong.

Thus, my conclusion will be: not only does the theory not provide its own interpretation, it is structually incapable of doing so.

In the first two steps, the discussion will be restricted to conventional Schrödinger quantum mechanics. In the third step, when I am showing why quantum theory cannot be given a sensible physical interpretation without an inner product, the discussion will apply to both Schrödinger quantum mechanics and quantum cosmology. In the next section, I will discuss the additional issues that emerge in quantum cosmology as a consequence of the fact that the Hamiltonian is a constraint.[7]

4.1. THE IDEAL ELEMENTS OF QUANTUM MECHANICS

Let us begin by recalling the definition of an ideal element. A formal element of a physical theory is ideal if it is (1) contingent, (2) plays a role in determining the evolution of the physical degrees of freedom of the theory, and (3) is itself nondynamical. By contingent I mean that it reflects an arbitrary choice, and that a different choice could have been made without altering the basic mathematical structure of the theory. By nondynamical I mean that the choice that is made is not arrived at by solving any dynamical equations, and cannot be influenced by the state of the dynamical degrees of freedom.

If a structure is not ideal I will call it intrinsic.

The importance of the condition that an ideal element involves an arbitrary choice that could be made differently without altering the mathematical structure of the theory is to separate the issues I am concerned with here from more general issues of why the basic structure of the theory is chosen as it is, or why we construct a theory at all. For example, the fact that in classical mechanics the dynamics is specified by second-order differential equations does not reflect an ideal structure as

I mean it here.

In quantum mechanics, once the structure of a linear state space is given, the intrinsic elements are the physical states themselves and the algebra of linear operators that acts on the physical state space.

Now, a linear state space and an associated algebra of operators are not enough to construct a physical theory. (If it were, we would have a quantum theory of gravity!) Quantum mechanics becomes a physical theory with the addition of three other elements, which I will argue are, by their formal properties, ideal.

The first of these is the correspondence between elements of the operator algebra and certain observables from the classical phase space of the theory. This correspondence is ideal in that it (1) might be set up in more than one way; (2) by the association of a certain operator with the Hamiltonian, determines the dynamics of the system; and (3) is nondynamical in the sense that nothing that happens within the quantum dynamics can alter or modify this correspondence.

The second ideal element in the formalism of quantum mechanics is time. We may first show that the notion of time in quantum mechanics satisfies the three formal requirements of being ideal. The choice of time is reflected in the formalism of quantum mechanics in what Carlo Rovelli, in his discussion of the problem of time in this volume, calls time structures (Rovelli 1991a). These are one-parameter families of either states or operators (depending on whether one is in the Schrödinger or Heisenberg representation) that are parametrized by the time coordinate. This choice is ideal in that it is nondynamical. Nothing in the dynamics of the theory can change the specification of the time structures once they are given by the specification of the Hamiltonian. Equally to the point, as many people have pointed out, there is no time operator in quantum mechanics. That is because the value of the time has nothing to do with the expectation value of any observable of the quantum system. If time refers to anything at all, it refers to the value of an observable of a system that is outside of the system being studied and is not dynamically connected to it.

The third ideal element is the inner product. It satisfies the same three formal criteria for being ideal. One has to begin with a free choice of the inner product, as the vector space of states admits an infinite number of inner-product structures. The choice influences the evolution of dynamical quantities if by that we mean what is actually measured. In ordinary quantum mechanics the choice of an inner product is not dynamical, i.e., the correct inner product is not the solution to any dynamical equation, and nothing in the evolution of the system plays a role in the choice.

The choice of Hamiltonian, the choice of a time structure, and the choice of an inner product structure are not independent of each other. They are constrained by the requirement that the chosen Hamiltonian be Hermitian with respect to the chosen inner product. In this case, the time structure is that which is generated by the action of the Hamiltonian.

In the conventional interpretation, all three of these elements are interpreted as referring explicitly to things outside of the dynamical system modeled by the linear state space and its operator algebra. Elements of the operator algebra are interpreted as generating the action, on the states of the quantum system, of classical systems that are external to it. The notion of time is explicitly taken to mean the time on a classical clock outside of the quantum system. Finally, the numbers produced by use of the inner product are interpreted as being probabilities that a classical external measuring system will produce certain numbers upon interaction with the quantum system.

4.2. WHY THE IDEAL ELEMENTS MUST REFER TO AN OUTSIDE AGENCY IF THE THEORY IS TO HAVE A SENSIBLE PHYSICAL INTERPRETATION

Now, I come to the second point of my argument. *Must* these ideal elements be interpreted as referring to something outside of the system? This is the hardest of my arguments since it concludes that a certain type of interpretation of the formal structure of the theory is impossible. Such an argument must, to the extent that it is definitive, be the kind of a priori argument that physicists, from long experience, have learned to distrust.

Thus, before introducing an a priori argument that applies to ideal elements in general, I will lower my sights and explain why each of the three ideal elements we have been discussing cannot be given a sensible physical interpretation that does not refer to something outside of the quantum system.

Let us begin with the first ideal element, the correspondence between elements of the operator algebra and physically measured quantities. We note first that this correspondence is essential if any connection is to be made between the formalism of quantum mechanics and the body of experiments that are usually adduced as giving support to quantum mechanics. Now, any such correspondence results, in the case of a particular observable, in the picking out of a preferred basis in the linear space of states that are associated with definite values of the physical quantity.

An interpretation of quantum mechanics in which no reference is made to anything outside of the system must provide a mechanism for picking out these preferred bases in a way that is completely internal to the intrinsic structure of the theory. That this is a fundamental problem for the Everett interpretation is recognized by its own advocates, as is evidenced by the fact that several of them have attempted to give a construction of such a preferred basis. Included in these attempts is the "measurement basis" of David Deutsch (1985), and the "pointer basis" of Zurek (1981, 1982); work on similar lines was also done by Simon Kochen (1985).

Zurek solves the problem by an explicit use of the coupling of the measurement system to the external environment; his pointer basis is picked out by an assumption that it consists of eigenstates of an operator that commutes with the interaction Hamiltonian, which describes the coupling between the measuring system and the external environment. Because it explicitly uses the coupling to an external environment, Zurek's work can be seen as an elucidation of the role of ideal elements in the measurement process; it is not an attempt to provide an intrinsic measurement theory.

Deutsch and Kochen, on the other hand, do attempt to solve the problem of giving a preferred basis intrinsically, without reference to an external environment. (However, they do make use of other structures such as the Hamiltonian and the inner product.) Deutsch, in particular, attempts to give intrinsic conditions to pick out a preferred basis in the Hilbert space of a system coupled to a measuring system. His criteria express the idea that in this basis the interaction Hamiltonian between the system and observer should be diagonal, so that once a measurement was made (which the Copenhagen interpretation would describe by collapse of the state into some element of this basis) no further interaction would take place. Deutsch shows that criteria can be chosen such that this proposal succeeds for simple systems involving finite-dimensional Hilbert spaces. However, he also shows that in realistic systems, for which the Hilbert spaces are always infinite dimensional, these criteria are not sufficient to distinguish a unique basis. The problem seems to be that for these systems the intrinsically given structure is not sufficient to break the original symmetry of the state space strongly enough to produce a preferred measurement basis.

As for the second ideal element, time, Jim Hartle has discussed in his contributions to this volume and elsewhere what happens when one attempts to interpret the time parameter in quantum mechanics in terms of a dynamical variable internal to the quantum system (Hartle 1988a, b; 1990, 1991). It becomes immediately clear that, if this is done, very little of the usual formal structure of quantum mechanics can be maintained, if for no other reason than the following: if a measurement of time can result in a dispersion of values that can only be specified probabilistically, then it becomes impossible to maintain the notion of conservation of probability under evolution in time. Jim Hartle's response to this is to try to construct a quantum theory that is based on the path integral that is in general not equivalent to any conventional structure in terms of a linear vector space of states and observables. An analysis of the success of this program is outside of my scope here; more relevant for my purposes is Hartle's demonstration that the usual structure of quantum mechanics cannot be maintained if time is to refer to a clock inside of the quantum system.

Thus, if the time coordinate that is used in conventional quantum mechanics is to refer to the value of any dynamical variable, it must be a variable that commutes with all of the dynamical variables of the quantum system.

This brings us to the last ideal element, the inner product. We begin with the observation that, since there are an infinite number of inner products on any linear space, the choice of the correct inner product requires physical input. In the textbook treatments of quantum mechanics, the inner product is chosen so that the physical observables of the theory whose classical counterparts are real are Hermitian. The most important of these is, of course, the Hamiltonian. Thus, the choice of an inner product is, with this condition, partially dependent on the earlier choices of a correspondence between classical and quantum observables, and of the Hamiltonian, or time structures. Since the choice of an inner product depends on these choices of ideal elements, the inner product is ideal. Furthermore, to the extent that those choices depend on the existence of things outside the quantum system, the choice of the inner product also depends on them.

Furthermore, the notion of the inner product may be seen implicitly to involve a notion of time in the following way. The inner product gets its physical meaning from the notion of probability, which in turn gets its physical meaning from the notion that one can give a complete and mutually exclusive set of occurrences for each observation that is described probabilistically. In order to be able to compare the result of an experiment with a prediction from quantum mechanics, the theory must assign probabilities in such a way that the probabilities predicted for each of the complete and mutually exclusive outcomes add up to one. This is because the experimentalists will define the observed probabilities so that they add up to one by dividing the number of each outcome by the total number of trials. Thus, unless the probabilities predicted by the theory satisfy this property, they cannot agree with the experimentally observed probabilities.

However, the specification of a complete and mutually exclusive set of outcomes implicitly involves a notion of simultaneity. This simultaneity must be a classical notion. This is because, as just mentioned, if one attempts to describe a simultaneity of mutually exclusive events purely on the basis of correlations with the state of a quantum-mechanical clock, one does not end up with probabilities that add up to one. Thus, the notion of the inner product depends on the existence of a classical time, which in turn requires the existence of a clock outside of the quantum system.

These attempts, therefore, to give the three ideal structures a physical interpretation in terms of the intrinsic structure of the quantum theory seem to fail. If conventional quantum mechanics is to be a physical theory, these three elements must have a physical interpretation, and in each case, an acceptable interpretation seems to be possible only if it refers to the existence of something outside of the quantum system.

I now turn to a more general argument for the connection of ideal elements with external agencies.

4.3. IDEAL ELEMENTS AND THE PRINCIPLE OF SUFFICIENT REASON

While I do not expect to convince an audience of physicists with a priori arguments, let me not leave this topic without indicating what sort of arguments I have in mind for the general proposition: Formally ideal elements of a physical theory necessarily require that there be things in the universe outside of the system under study if that theory is to have a sensible physical interpretation. The prototype for this argument is, I believe, an argument that Leibniz used against Newton's concepts of absolute space and time (Leibniz 1696, 1714). This argument is based on his *Principle of Sufficient Reason,* which states loosely that, in a complete theory of the universe, every question of the form: *Why is the world this way rather than that way?* must have an answer. Examples of such questions are: Why are there three rather than more or less spatial dimensions? Why are there at least four families of quarks and leptons? In contrast, the question: why is there only one time dimension?, is not of this type because it is intrinsic to the notion of time that it refers to a one-dimensional structure.

The idea behind this principle then is that there should be no arbitrary choices made in the construction of a complete physical theory. Any time such a choice must be made, one is putting in by hand something that one would like explained by a deeper, more fundamental, theory. Thus, the principle of sufficient reason can be taken to mean that, in a complete physical theory, there should be no ideal elements.

Another way to put the principle of sufficient reason is that God should not have had any choice in the construction of our universe. Leibniz attacked Newton's notions of absolute space and time by noting that, in Newton's theory, it is possible to ask why God did not create the universe ten minutes later and three feet to the left (1696, 1714). This question fails the test of the principle of sufficient reason; to eliminate it, Leibniz proposed that the notions of space and time should be modified in such a way that the questions cannot be asked. This can be done if space and time are made strictly relational notions.

Newton's defenders replied that God, or at least their God, could do whatever He pleased; it was not for human begins to inquire into His reasons (Clarke in Leibniz 1973). I will not follow the course of this argument, although I do not doubt that the arguments we are having about these issues at this meeting will seem as "theological" to people three hundred years in the future as the Clarke–Leibniz correspondence seems to us. Instead, I will note that the contemporary theoretical physicist cannot take refuge in this and the other arguments used by Newton's defenders because the history of physics since Newton's time teaches us that any theory, no matter how well-confirmed it may be, will in time be replaced by a better one. Any question we may ask in a given theory that fails the test of sufficient reason thus represents an opportunity for a future theory. As long as we do not have this future theory, we must concede that whenever such a question is left open, a theory might arise that resolves this question and yet agrees with the testable parts of our present theory. And, as long as the possibility of this happening is open, we cannot claim with confidence that the present theory is complete.

This being admitted, it remains to show how the resolution of a question that fails the test of the principle of sufficient reason requires the existence of something outside of the system described by the theory that gave rise to the question. Again, the example of absolute space and time will serve as a prototype. The point is that these notions were not there purely as a result of Newton's fancy. People before Newton had advocated a purely relational notion of space, in particular, Descartes. Newton rejected these ideas in favor of absolute space because *they did not work*. A purely relational notion of space, as used by Descartes and others, failed to account for the fact that rotation and acceleration seem to be absolute quantities in that they can be determined by an observer undergoing these motions without reference to anything outside his local frame. Absolute space and time were introduced by Newton to answer the question: when someone is rotating or accelerating and observing the consequent inertial effects, what is it that he or she is accelerating with respect to?[8]

A purely relational theory would seek to answer this question by substituting some dynamical entity for absolute space. But this entity cannot be anything in the local system, since once the frame of absolute space is given, that local dynamics is already completely determined in terms of the locally observable quantities. Further, that dynamics leads to predictions that agree with observations. Thus, if there is such an entity, it must not lie in the local system under study, but must reside in something outside of the dynamical system under study. This was essentially Mach's argument for the distant stars playing a role in the principle of inertia.

Let us abstract the formal structure of this argument. The ideal element is present because it is necessary to explain something in the internal or local dynamics of the system under study. The theory that incorporates that ideal element may be assumed to be complete in the sense that once the ideal element has been specified, the internal dynamics is completely determined and produces predictions that agree with experiment. (If it were not complete in this sense, we would not be very interested in its further development). Therefore, if Leibniz's argument impells us to replace that ideal element with a dynamical entity, that entity cannot be one of the dynamical degrees of freedom of the intrinsic system because that dynamics has already been completely specified. Hence, it must be something that is outside of the system under study.

I believe that this argument applies, point by point, to the ideal elements in quantum mechanics. However, for lack of both time and space, I will defer the details of this argument to another place (Smolin 1991b).

4.4. WHY QUANTUM MECHANICS CANNOT BE A PHYSICALLY SENSIBLE THEORY WITHOUT A CONSERVED INNER PRODUCT

Now, I turn to the third part of my argument, which is that a sensible interpretation of quantum mechanics cannot be made without the use of the ideal elements we have been discussing. In particular, it is sometimes asserted that in the absence of an external observer that gives meaning to the inner product in the usual quantum measurement theory, predictions can be recovered directly from correlations in the wave function (see note 3).

Such proposals are usually, although not always, associated with the Everett interpretation. Now, as I mentioned above, my purpose here is not to attack the Everett interpretation, as it was originally given in Everett's papers. I am willing, for the purposes of this discussion, to concede that for ordinary quantum mechanics and in situations where there are classical measuring apparatuses, the Everett interpretation provides a workable alternative to the usual measurement theory. That is to say, I concede that in the cases in which the usual measurement theory is applicable, the Everett interpretation will also be applicable.[9] However, as I have argued above, in these circumstances the Everett interpretation is relying on ideal elements no less than is the Copenhagen interpretation to establish a connection between the formalism of quantum mechanics and observations. My purpose here is rather to examine the question of whether an interpretation based on the correlations in the state functions can allow us to extract predictions from quantum mechanics in situations, such as quantum cosmology, where the conventional interpretations fail.

My conclusion is negative. I will show that the Everett interpretation can only give sensible physical predictions in those circumstances in which the conventional interpretations also apply. In situations in which the ideal elements necessary for the conventional interpretations to work are absent, the Everett interpretation will fail as well. Thus, while it may or may not be attractive for allowing an interpretation in which collapse of the wave function need not be invoked, the Everett interpretation does not offer any help to us in solving the problem of inventing a physically sensible quantum cosmology.

One finds in the literature on the Everett interpretation two closely related proposals for how physical predictions are to be extracted from quantum states without the use of the projection postulate. These involve use of the relative frequency operator, and the technique of "precluded regions."

4.5. The Relative Frequency Operator and the Inner Product

In this first proposal one shows, following a theorem discovered independently by Finkelstein (1963), Hartle (1968), and Graham (1973), that the statement of the probability interpretation, that a measurement of a state $\langle\psi|$ will produce a value associated with a state $\langle i|$ with a probability $|\langle\psi|i\rangle|^2$, can be reduced to the determination of an eigenvalue of a certain operator called the relative frequency operator. This is interesting because the rule that if a state is an eigenstate of an operator, then a measurement of that operator will produce with certainty the eigenvalue involves neither a notion of probability nor the existence of an inner product on the Hilbert space. Thus, it appears as if the use of probability in quantum mechanics is being reduced to a more primitive, nonprobabilistic notion, that eigenstates of an operator have definite values of the quantity associated with that operator.

The theorem in question can be stated as follows. In order to measure a probability we need to make a large (ideally an infinite) number of measurements on identically prepared systems. Since, in the Everett interpretation, all measurement interactions are to be described within the quantum state formalism, this must be described explicitly; we must construct a state $|\psi^N\rangle$ associated with N systems prepared identically in the state $|\psi\rangle$. The state $|\psi^N\rangle$ lives in the state space $\mathcal{H}^N$ where $\mathcal{H}$ is the state space of a single system. Now, let A be an observable in the single-particle state space that we wish to measure, and let its eigenstates be $|i\rangle$, $i = 1, \ldots, m$. A basis for $\mathcal{H}^N$ is then given by $|i_1, \ldots, i_N\rangle = |i_1\rangle \otimes \ldots \otimes |i_N\rangle$. The *relative frequency operator for an outcome* k is then defined by,

$$f_N^k |i_1, \ldots, i_N\rangle = N(k; i_1, \ldots, i_N)|i_1, \ldots, i_N\rangle, \qquad (4.1)$$

where $N(k; i_1, \ldots, i_N) = \sum_\alpha \delta_{i_\alpha k}/N$ is equal to the number of times k occurs in the list $i_1, \ldots, i_N$ divided by N.

Now, the theorem is that in the limit $N \to \infty$, $|\psi^N\rangle$ is an eigenvalue of $f^k \equiv \lim_{N\to\infty} f_N^k$ with the eigenvalue $|\langle\psi|k\rangle|^2$. In this way, probabilities for a single system are reduced to eigenvalues of an operator on the quantum states of the ensemble.

This theorem has one puzzling aspect. The condition that a state be an eigenstate of a given operator does not involve an inner product. How can the eigenvalue of f^k be equal to $|\langle\psi|k\rangle|^2$, an expression that involves an inner product? The answer is that the definition of the operator f^k refers implicitly to an inner product.

In order to see that this is the case, let us see if f^k could be defined on a space of states without an inner product. Without an inner product, we cannot say that a set of basis elements $|i\rangle$ are orthonormal; they are only linearly independent. In this case, we could replace them by another linearly independent set, given by

$$|i'\rangle = \begin{cases} |k\rangle, & \text{if } i = k \\ |i\rangle + \alpha_i|k\rangle, & \text{if } i \neq k \end{cases} \qquad (4.2)$$

We then can construct a new relative frequency operator f_N^{kl}, using the $|i'\rangle$ in (2) instead of the $|i\rangle$. It is easy to see that $f_N^{kl} \neq f_N^k$. Thus, we see that the relative frequency operator depends on the basis $|i\rangle$ having more structure than linear independence.

In order to pick out a meaningful f_N^k, we need in fact to impose the condition that the basis $|i\rangle$ is orthonormal. Equivalently, we can demand that the basis elements be eigenstates of a particular operator A. This is equivalent to assuming the existence of an inner product, as it is true that for any operator A with a complete set of eigenstates there will exist an inner product in which it is Hermitian. Then its eigenstates will be orthonormal in that inner product.[10]

Thus, we learn that in any context in which a meaningful relative frequency operator can be defined on $\mathcal{H}^N$, there will exist an inner product on $\mathcal{H}$ such that the Finkelstein–Graham–Hartle theorem holds. However, in this context we could also use expectation values computed directly in $\mathcal{H}$ along with the ordinary probability interpretation to get the same answers. Thus, in any context in which probabilities can be derived from a relative frequency operator, there also must exist a physical inner product, implicitly used in the construction of the relative frequency operator, that will allow the same probabilities to be derived in the normal way.

4.6. PRECLUDED REGIONS AND MEASURING INSTRUMENTS

A second approach to measurement theory in Everett-type interpretations involves the use of what we will call the principle of "precluded regions." This principle (which is used in the proof of the Finkelstein–Graham–Hartle theorem—hence the equivalence of the two approaches) is the following. We follow the very clear exposition of Geroch (1984). We assume that, with the observer included in the quantum state description, the Hilbert space is a direct product, $\mathcal{H}_{\text{system}} \otimes \mathcal{H}_{\text{detector}}$. In addition, we assume the existence of a preferred "measurement basis," consisting of direct products of observer states with eigenstates of the observable measured by the detector. Let us consider the expansion of a state $|\Psi >$ in terms of this basis. If any components in this expansion vanish (or are very small)[11] then we may say that the associated basis elements describe correlations that do not occur in the world described by $|\Psi >$. In particular, the detector will never be seen in states that occur only in these precluded direct products.

Now, as Geroch shows, one can couple a measuring instrument to the system through a relative frequency operator of the kind that we have been discussing. Then, if the measurement instrument is picked in an appropriate initial state before its interaction with the system, the principle of precluded regions can be used, along with the Finkelstein–Graham–Hartle theorem, to derive probabilistic predictions concerning the results of experiments.

Now, we must ask a crucial question. What if none of the components of the state $|\Psi >$ in the measurement basis vanish? Then, according to Geroch, the principle of precluded regions does not enable us to make any prediction. This, he would further say, is a virtue rather than a problem because such a state will correspond to a situation in which the detector is initially in a state such that its pointer would not have a definite, or nearly definite, value after the interaction with the system. In this case, the detector has not been prepared in a state in which it could function as a good detector, and the theory makes no prediction.

This is fine as far as it goes. But it shows that the principle of precluded regions can only be used if certain conditions are fulfilled. These are[12]: (1) there must be a clear separation of the degrees of freedom of the composite system into detector and system degrees of freedom; (2) the detector must be prepared in a state initially such that after its interaction with the system there are precluded regions. Moreover, the possible precluded regions that could result, depending on the initial state of the system, must correspond to states in which the detector has, to a very good approximation, a single classical value. Otherwise, we could not interpret the precluded regions in terms of detector readings.

Generally, the second condition means that the state after the interaction must be a semiclassical state, as far as the degree of freedom of the detector's pointer is concerned. It must be possible to say where the pointer is not. However, we know that, for the quantum mechanics of an ordinary particle, the only states that vanish (i.e., are exponentially small) on most of the configuration space of the particle are semiclassical states associated with one or more classical values of the position.[13]

Now it is possible, with a small extension of the principle of pre-cluded regions, to apply it in cases in which no components of the state in the measurement basis vanish. Given the assumptions we have already seen we must make, we can proceed by decomposing the state into a sum of direct products, each term of which involves a direct product of a semiclassical state of the detector with its corresponding relative state of the system.

One can then apply the principle of precluded regions to each term in the sum. Although this will not lead to predictions in experiments with one detector, it can be used to predict correlations between the responses of several detectors if more than one detector interacts with the system. Of course, an application of this extension of the principle depends on the preferred set of semiclassical states for the detectors. But, as any application of the principle depends on such states, there is no reason not to go ahead and make this extension.

We thus must conclude that the principle of precluded regions re-quires us either to assume that a part of the system, to be called the detector, can be prepared in an initial state such that it will be, after the measurement, in a semiclassical state; or, if that is not the case, to decompose the state according to such semiclassical states of the detec-tor before applying the interpretation. Thus, the principle of precluded regions is not really different in application from the conventional inter-pretations. It requires that there be a distinguished part of the system, called the detector, which is assumed, for the purpose of the measurement theory, to behave classically. If these assumptions cannot be made, the interpretation cannot be applied. However, if these assumptions can be made, then any of the conventional interpretations could also be applied. Thus, there is no situation in which the principle of precluded regions could wrest a prediction from quantum mechanics and the conventional interpretations could not.

The implications of our discussion of relative frequency operators and precluded regions for quantum cosmology are then identical. If the conventional interpretations, in which a portion of the universe is sim-ply excluded from the Hilbert space description and treated classically, won't lead to sensible predictions, then neither will an Everett-type in-terpretation. In addition, the Everett interpretation relies no less than the conventional interpretations on the existence of an inner product on physical states in extracting any probabilistic predictions from the theory. The Everett interpretations may be attractive, and they may be helpful heuristically. But, as a matter of principle, any problem concerning the interpretation of quantum states that they could solve could have been solved using a conventional interpretation.

4.7. The Page–Wootters Approach to Time

I would like to briefly mention one additional topic concerning interpretations based on correlations. This is the proposal of Page and Wootters (1983) concerning dynamics in stationary states. They give an alternative proposal for quantum dynamics that does not explicitly use the Schrödinger equation, and that can be applied when one has a composite wave function, including the clock as well as the system, that is stationary. Without going into the details of the argument, which parallel exactly the cases we have been discussing, I would like to note that the situation with regard to this proposal is the same as we have been discussing with regard to Everett interpretations in general.

There is nothing wrong with the proposal of Page and Wootters, as far as it is presented in their paper. In any case in which there is a physical inner product and a Hermitian Hamiltonian, predictions regarding time evolution could be extracted from quantum states using either the Page and Wootters proposal or the conventional method, which uses time-dependent states. However, in any situation in which the conventional method will fail to give a well-defined evolution with conserved probabilities, such as when there is no inner product such that the states in question are normalizable or no nonvanishing Hermitian Hamiltonian, the proposal of Page and Wootters will similarly fail. Thus, it cannot help us solve the problems of the interpretation of the solutions to the Wheeler–DeWitt equation in the absence of an inner product on physical states that is exactly conserved under evolution in some time.

5.   Is There a Measurement Theory for Quantum Cosmology?

Up till this point I have been speaking mostly about conventional quantum mechanics, i.e., about a system in which the dynamics is driven by the Schrödinger equation. In quantum cosmology there is no Schrödinger equation, there are only constraint equations. This is related to the fact that there is no possibility of studying the evolution of the system in terms of an external time; therefore, the fact that time cannot play the same role in quantum cosmology as it does in ordinary quantum mechanics is consistent with the notion that time in ordinary quantum mechanics is an ideal element.

Because there is no notion of external time and no external observers, the usual measurement theory for quantum mechanics cannot be applied to quantum cosmology. The question we must then answer is: is there some alternative measurement theory that would allow us to extract physical predictions from quantum cosmology?

To give this question force, we need to define what we would be willing to consider as an acceptable measurement theory for quantum cosmology. To this end I would like to introduce three criteria that I propose must be satisfied by any acceptable measurement theory for quantum cosmology.

I will assume to begin with that the quantum constraint equations of the theory have been solved to yield a space of physical states, $\mathcal{S}_{\text{Phys}}$ and an algebra of physical observables $\mathcal{A}_{\text{Phys}}$.[14]

(1) There should be a criterion to distinguish a linear subspace $\mathcal{S}_{\text{Sensible}} \subseteq \mathcal{S}_{\text{Phys}}$ of *physically sensible* states. We do this to take care of the possibility that not every solution of the constraints can be given a physical interpretation, just as not every solution to the time-independent Schrödinger equation corresponds to a physical state in quantum mechanics. We require that the measurement theory apply fully to every element of $\mathcal{S}_{\text{Sensible}}$. That $\mathcal{S}_{\text{Sensible}}$ be a linear space means that if one can apply the measurement theory to any two states, then one must be able to apply it to an arbitrary linear superposition of them.

(2) If the measurement theory labels any quantities as probabilities, all of these quantities should be numbers between zero and one, and the sum of the probabilities for any complete set of mutually exclusive occurrences (as defined by the measurement theory) should equal exactly one.

(3) The measurement theory must allow us to extract some predictions from quantum cosmology that are not already predictions of classical general relativity, or of quantum field theory coupled to classical general relativity.

The meaning of this last point is that a measurement theory for quantum cosmology must yield more than a choice of boundary conditions for the classical Einstein equations (possibly coupled to quantum matter). A measurement theory for quantum cosmology that can do no more than posit conditions that lead to certain "semiclassical" states, and then extract predictions that follow from the classical Einstein equations applied to the classical solution on which the semiclassical states are built, has no more predictive power than does the classical Einstein theory itself. In particular, it cannot be falsified by any observation that would not also be a falsification of classical general relativity.[15]

In the previous section, I examined claims that a measurement theory for quantum mechanics can be constructed without an inner product. I showed that such a theory cannot yield any probabilistic predictions except in the trivial case in which the predicted probabilities are always either zero or one. I would like to claim that all purported applications of quantum cosmology in which predictions are extracted without use of an inner product are exactly of this type. In terms of real predictive power, "quantum states" play a role in these formalisms only to justify the choice of certain classical solutions to Einstein's equations from which appropriate predictions (or, more properly, retrodictions) are extracted.

Any acceptable measurement theory for quantum cosmology must do more than this. It must be falsifiable by some observations that would not falsify classical general relativity. This is the reason for criterion three.

Given the conclusions of the previous section, any acceptable measurement theory for quantum cosmology must employ an inner product structure. However, here we must face the fact that the problem of choosing the inner product in constrained theories such as quantum gravity and the Barbour–Bertotti models is highly nontrivial. This is because it is usually the case that the physical states are not normalizable with any inner product that one might pick on the larger, unconstrained state space.

There is a simple mathematical reason for this, which is that by satisfying constraints, the physical wave functions must be constant on some degrees of freedom on which unconstrained wave functions can depend. This very often leads to the solutions to the constraints having infinite norm in the inner product on the unconstrained state space.[16]

Thus, if there is to be an inner product on the physical states for quantum cosmology, it must be defined directly on that space, and hence it must be chosen after the constraints are solved. This is interesting from our present point of view because it means that the choice of the inner product must reflect dynamical information, as the constraints code all of the local dynamics.

The question to be answered in constructing a measurement theory for quantum cosmology then becomes: what is the criterion by which the inner product on the physical state space is constructed? There is in the literature a program for making such a choice, which is called the program of an internal or intrinsic time (Kuchař 1981b). From the point of view I have been developing here, one might construct a justification for this program in the following terms:

A sensible physical interpretation of quantum cosmology requires an inner product. Any choice of an inner product will lead to a notion of probability. As I discussed earlier, any notion of probability is only sensible if it comes together with a notion of time, such that the probabilities can satisfy a condition of being conserved under evolution in time. Otherwise, one does not have a criterion for picking out mutually exclusive sets of occurrences whose probabilities are required to add up to one. (This is necessary, by criterion (2).) Thus, to invoke an inner product is to invoke a notion of time.

Now, as I discussed above, and as Jim Hartle argues at length (1988a, 1991), there can be no strict implementation of the principle of conservation of probability for a time that is the value of a dynamical variable of a quantum system. Therefore, a sensible measurement theory for quantum cosmology can only be constructed if there is a time variable that is not a dynamical variable of the quantum system that describes the universe.

Does this mean that quantum cosmology is impossible, since there is no possibility of a clock outside of the system?

There is, as far as I know, exactly one loophole in this argument, which is the one exploited by the program of an intrinsic time. This is that one coordinate on the phase space of general relativity might be singled out and called time in such a way that the states, represented by functions on the configuration space, could be read as time-dependent functions over a reduced configuration space from which the privileged time coordinate is excluded.

If this proposal is to lead to a sensible measurement theory, this *intrinsic* time coordinate must satisfy certain properties. In particular, it must be true that a canonical transformation can be made on the phase space such that the Hamiltonian constraint can be written in the form

$$C = p_0 + h(q_i, p_i, q_0), \qquad (5.1)$$

where p_0 is the momentum conjugate to the intrinsic time coordinate q_0 and $i \neq 0$. *If* such a canonical transformation exists, then the Hamiltonian constraint

$$C\psi(q_0, q_i) = 0 \qquad (5.2)$$

can be interpreted as a Schrödinger equation,

$$\left(-\imath\partial_{q_0} + h(q_i, q_0, -\imath\partial_{q_i})\right)\psi(q_0, q_i) = 0. \qquad (5.3)$$

An inner product can then be chosen such that h is Hermitian, in which case the principle of conservation of probability will be satisfied.

It is important to stress that, for this program to be successful, there must be a canonical transformation that brings the Hamiltonian constraint *exactly* to the form (5.1). If that form is only approximate, then the conservation of probability will be, at best, approximate. But this is absolutely unacceptable for the measurement theory of what is supposed to be the fundamental theory of nature.

If such a canonical transformation does exist, then the time coordinate, and hence the inner product, has been dynamically determined. If this were to happen, we would say that the notions of time and the inner product, which are purely ideal in ordinary quantum mechanics, have been dynamically determined in terms of completely intrinsic structures in quantum general relativity.

The conclusion is then that quantum cosmology can have an acceptable measurement theory only in the case in which a canonical transformation exists that puts the Hamiltonian constraint exactly into the form (5.1).[17] This is a dynamical problem; it could, for example, be true for some forms of the field equations and for some choices of matter couplings, and false for others.

What is the status of this problem? At present, the situation is that, despite a great deal of effort, no such canonical transformation has been found. At the same time, it has not been proven that no such canonical transformation exists. Julian Barbour has given arguments that an intrinsic time should not exist in the compact case (1986), but these, unfortunately, are not conclusive. As the whole issue of whether there is an acceptable measurement theory for quantum cosmology rests on this issue, I would like to suggest that it deserves attention, especially in the light of the recent progress in canonical gravity.[18]

In the next sections, we will examine this issue in the context of some model systems: the Barbour–Bertotti theories. As the question of the existence of an intrinsic time is a dynamical one, it probably cannot be settled by an examination of model systems. At the same time, examination of this issue in model systems may help to develop strategies or arguments that may apply to the case of general relativity.

6. The Barbour–Bertotti Theory: Dynamics Without Ideal Elements

I would like now to describe a form of classical dynamics in which the ideal elements we discussed in Section 3 have been eliminated. These are the Barbour–Bertotti models (Barbour 1974a, b, 1975; Barbour and Bertotti 1977, 1982); they are classical dynamical systems that depend only on relational space-time quantities. We are interested in these models for two reasons. First, they share certain features with general relativity that are consequences of the absence of an absolute or background space-time, and as such, provide good models for quantum cosmology. Second, they provide an example of what kinds of things can happen when ideal elements are eliminated from a physical theory.[19]

We will be interested first in a set of Barbour–Bertotti models that involve N nonrelativistic point particles. The particle models have the following properties:

(1) The dynamics depends only the relative distances $r_{ij}(\lambda) \equiv |x_i(\lambda) - x_j(\lambda)|$ between the particles, and their velocities, $dr_{ij}/d\lambda$, where λ is an arbitrary time parameter. In particular, there is no reference to a global inertial frame.

(2) The dynamics is invariant under arbitrary monotonic rescaling of the time parameter, $\lambda \to \lambda'(\lambda)$. Thus, there is no invariant notion of velocity or kinetic energy for the system of particles.

It is possible to formulate a set of dynamical models directly in terms of the relative distances between the particles (Barbour 1974a, b, 1975; Barbour and Bertotti 1977). I will discuss such models below; however I would first like to describe a set of models that are simpler to analyze (Barbour and Bertotti 1982). These are constructed on the model of gauge theories in which fictitious coordinates are introduced and then eliminated by the imposition of an appropriate gauge invariance. The theory is then formulated in terms of coordinates $x^i(\lambda)$ in a fictitious three-dimensional space. To implement the requirement that the theory only depend on the relative distances and their rates of change, we require that the theory be invariant under the following gauge transformations,

$$x^i \to x^{i\prime} \equiv x^i + a(\lambda) + (\Sigma(\lambda) \times x)^i \qquad (6.1)$$

where $a(\lambda)$ and $\Sigma(\lambda)$ are time-dependent translations and rotations of the Euclidean frame. To find the appropriate expression for the velocities, let us note that, in a short time $\Delta\lambda$, the coordinates of the particles x^i

change due both to the changes in the particle positions and in the frames. Assuming the changes in the latter are first order in $\Delta\lambda$, we have to first order

$$\delta x^i = \frac{\partial x^i}{\partial \lambda}\Delta\lambda + b(\lambda)\Delta\lambda + (\sigma(\lambda) \times x)^i \Delta\lambda, \tag{6.2}$$

where $b(\lambda)\Delta\lambda$ and $\sigma(\lambda)\Delta\lambda$ are the first order changes in the frame in the time $\Delta\lambda$. We may then define a total time rate of change as

$$\frac{Dx^i}{d\lambda} \equiv \frac{\partial x^i}{\partial \lambda} + b(\lambda) + (\sigma(\lambda) \times x)^i. \tag{6.3}$$

We now construct a gauge-invariant, or *intrinsic* time derivative of the configuration $x^i(\lambda)$ in the following way. For each λ, we may vary b and σ to maximize the expression

$$\sum_i \frac{Dx^i}{d\lambda} \cdot \frac{Dx^i}{d\lambda}. \tag{6.4}$$

If we indicate by b_0 and σ_0 the values of these parameters that minimize (6.4), we define the intrinsic derivative to be

$$\left.\frac{dx^i}{d\lambda}\right|_{\text{int}} \equiv \frac{Dx^i}{d\lambda}(b_0, \sigma_0). \tag{6.5}$$

We may now write an action principle invariant under the transformations (6.1) and $\lambda \to \lambda'(\lambda)$

$$S_{BB}(x^i(\lambda)) = \int d\lambda\sqrt{T}\sqrt{-V}, \tag{6.6}$$

where

$$\sum_i T = \frac{m_i}{2}\left.\frac{dx^i}{d\lambda}\right|_{\text{int}} \cdot \left.\frac{dx^i}{d\lambda}\right|_{\text{int}} \tag{6.7}$$

and

$$V = \sum_{i\neq j} V(x_i - x_j), \tag{6.8}$$

where $V(x_i - x_j)$ is a potential energy function that is assumed to be negative and to depend only on the relative coordinates. Alternately, the same dynamics can be obtained by varying the action

$$S(x^i(\lambda), b(\lambda), \sigma(\lambda)) = \int d\lambda\sqrt{\sum \frac{m_i}{2}\frac{Dx^i}{d\lambda} \cdot \frac{Dx^i}{d\lambda}}\sqrt{-V} \tag{6.9}$$

with respect to x^i, b and σ.

Having written down a dynamical system that eliminates the ideal notions of space and time in Newtonian dynamics, we would like to know what the solutions of the theory describe. Perhaps surprisingly, one may choose a gauge in which the solutions to this theory are exactly those of Newtonian dynamics with the same interactions, and with the total energy, momentum and angular momentum fixed to vanish. To see this, one proceeds as follows.

One first extremizes the action (6.9) with respect to b and σ at each moment in time. This has the effect of replacing the derivatives (6.3) with the intrinsic derivative (6.5). This establishes a preferred frame in which the intrinsic derivatives are the ordinary time derivatives. To see how to proceed, let us next consider the Euler-Lagrange equations generated by (6.6) once this substitution is made. They are

$$\frac{d}{d\lambda}\left(m_i\sqrt{\frac{-V}{T}}\frac{dx_i}{d\lambda}\right) - \sqrt{\frac{T}{-V}}\frac{\partial V}{\partial x_i} = 0. \tag{6.10}$$

We see that this equation is nonlocal because the expression for the acceleration of any one particle contains the ratio of the total kinetic energy to the total potential energy of the system. We still have, however, the freedom to fix the time coordinate λ, which up to this point has been arbitrary. We may use this freedom to fix the time coordinate to eliminate the nonlocal dependence of the equations of motion. We do this by choosing

$$T + V = 0. \tag{6.11}$$

In this gauge, the system evolves according to Newton's laws. However, the time coordinate with respect to which the evolution takes this simple form is now no longer a background, ideal quantity. It is determined by the total dynamical system, by solving the equation (6.11). Through this equation, the rate at which a Newtonian clock ticks is dynamically determined by the distribution of energy in the entire system.

Similarly, the process of casting the theory in Newtonian form produces a set of inertial frames with respect to which the kinetic energy takes the normal Newtonian form. Again, as in the choice of time, these frames are no longer background, ideal elements in the theory. Instead, they have been dynamically determined by the process of extremizing b and σ in the action (6.9). Thus, this theory explicitly realizes Mach's idea that the choice of inertial frames is determined by the distribution of matter in the whole universe. In a large system described by the Barbour–Bertotti dynamics, the distinction between accelerated and nonaccelerated motion for any one particle is determined dynamically by the total distribution of matter and motion in the system as a whole.

I believe that these features are characteristic of any situation in which one constructs a new theory in order to eliminate an ideal element from a previous theory. The new theory is initially more complicated than the old theory. Typically, the complications necessary to eliminate the ideal elements involve both nonlocal couplings, as in the action (6.6), and additional gauge symmetries. The nonlocal couplings can be eliminated by choosing a particular gauge; however, this choice depends on the configuration of the system as a whole. When this has been done, the equations take the form of the old theory with the additional requirement that what was ideal is now dynamically determined in terms of the state of the system as a whole.

There is another set of Barbour–Bertotti models in which the values of the physical constants are actually determined by this process (Barbour 1974a, b, 1975; Barbour and Bertotti 1977). These theories were proposed by Barbour and Bertotti several years before they proposed the class of models I have been describing. They differ from these in that they do not reduce exactly to a Newtonian dynamical system, even when the appropriate gauge choices have been made. Instead, Newtonian physics emerges as an approximation in the case that the distribution of matter in the universe is sufficiently uniform. This means that the models predict corrections to Newtonian dynamics coming from inhomogeneities in the universe. In particular, the asymmetry of the distribution of matter in the galaxy leads to a contribution to the precession of the planets that apparently falsifies the model.[20]

Whether or not these models are physically viable is not, however, relevant to the question of what they have to teach us about how a relational model may determine the parameters that govern local physics. I would thus like to indicate how these models are constructed (Barbour and Bertotti 1977). The action is taken to be of the form

$$S = \int d\lambda \sqrt{T} W. \tag{6.12}$$

The kinetic energy is taken to depend explicitly on the relative distances r_{ij} between pairs of particles.

$$T = A \sum_{i<j} \frac{m_i m_j}{r_{ij}} \left(\frac{dr_{ij}}{d\lambda} \right)^2. \tag{6.13}$$

The term W contains the interactions; it is of the form

$$W = \sum_{i<j} \frac{m_i m_j}{r_{ij}}. \tag{6.14}$$

We may note that these two terms do not separately have the conventional dimensions of energy; however, (6.12) does have the dimensions of action if we give to the constant A the dimensions length^5/mass^{4}time^2. Note, however, that by reparametrization invariance, nothing in the classical dynamics can depend on the value of A.

Barbour and Bertotti show that if we consider initial conditions corresponding to a universe of total mass $M = \sum_i m_i$ uniformly distributed inside a sphere of radius R, the local dynamics is well approximated by Newtonian physics with a gravitational attraction between two nearby bodies given by Newton's law with a gravitational constant

$$G = \frac{4R\dot{R}^2}{M}, \tag{6.15}$$

where dot specifies the time derivative in an appropriate Newtonian time gauge (Barbour and Bertotti 1977).

Finally, I would like to discuss briefly how Barbour–Bertotti (1982) field theories are constructed. These are interesting because they illustrate an important issue that arises in any attempt to derive known physics from a nonlocal or relational theory, which is how Lorentz invariance arises.

As I have been discussing, in a relational model one wants to explain certain ideal elements of local physics dynamically in terms of equations of motion that are to begin with nonlocal. If, however, we want dynamical equations of the familiar form, this nonlocality must be in space but not in time, and thus we require that the definition of the theory make use of a preferred space-like surface. The question we must immediately face then is how a Lorentz-invariant description could arise out of such a theory after one has fixed the time gauge in order to reduce the dynamics to some conventional, local form.

Let us consider then a Barbour–Bertotti formulation of a field theory. We consider, for simplicity, the theory of a single scalar field, ϕ on R^3. The kinetic energy and the potential energy will be defined to have the usual forms,

$$T = \int d^3x \frac{1}{2}(\dot{\phi}^2) \tag{6.16}$$

$$V = \int d^3x \left(\frac{1}{2}(\partial\phi)^2 + U(\phi)\right) \tag{6.17}$$

where dot denotes derivative with respect to an arbitrary time coordinate λ and $U(\phi)$ is some potential energy function. In order to get reparametrization invariance in λ, the action will be taken to be of the usual Barbour–Bertotti square root form

$$S = \int \sqrt{T} \sqrt{E - V}; \tag{6.18}$$

here, we allow the possibility of a non-zero total energy E, which may be necessary for the theory to have nontrivial solutions.

This theory does not have manifest Lorentz invariance in which the arbitrary time coordinate λ is mixed with the spatial coordinates. But it does, nevertheless, have a Lorentz-invariant dynamics. To see this, let us write the equations of motion,

$$\frac{d}{d\lambda}\left(\sqrt{\frac{E - V}{T}}\,\dot{\phi}\right) - \frac{\partial}{\partial x^a}\left(\sqrt{\frac{T}{E - V}}\frac{\partial \phi}{\partial x}\right) - \sqrt{\frac{T}{E - V}}U' = 0. \tag{6.19}$$

where prime represents differentiation by ϕ.

If we choose a time gauge such that

$$E = T + V, \tag{6.20}$$

the equation of motion becomes simply

$$\Box \phi = U'(\phi), \tag{6.21}$$

which we know has manifest Lorentz invariance. However, all we have done is to pick a convenient gauge in which to write the equation of motion. Thus, in any other gauge the theory must also be Lorentz invariant; but the transformation laws will be complicated, nonlocal, and nonlinear in the dynamical degrees of freedom.

The Lorentz invariance may be seen from the Hamiltonian point of view without picking a gauge, as the reader may verify.

7. Quantization of Relational Dynamical Systems

I would now like to discuss what happens when one attempts to quantize the Barbour–Bertotti models (Barbour and Smolin 1988). I will show that the attempt is successful in that one can give a formal quantization by following the Dirac procedure, which results in a space of physical states and their associated physical operators. Thus, the attempt succeeds in constructing all of the *intrinsic* elements of quantum dynamics. Whether the formalism can be given a physical interpretation is, however, a more difficult question because none of the ideal elements we discussed in Section 4 are given by this construction.

Nor can they be. The reason is that by eliminating the ideal elements from the classical dynamics, we have formulated a dynamics that necessarily applies to a closed system. There is no possibility of interaction with particles outside of the system because the theory has no background notion of space and time, which would be necessary for a description of an interaction involving elements of the system and any outside agency. But, without the possibility of interactions with outside agency, there is no possibility of introducing into the quantum theory the ideal elements that are needed for the theory to have a meaningful physical interpretation.

As in the case of quantum cosmology, the theory can be completed if only if the elements of the theory that are usually ideal can be reconstructed as intrinsic elements. In particular, if there is no external classical clock, the inner product must refer to probabilities that are measured in terms of an intrinsic time. As in the case of general relativity, whether or not an intrinsic time exists for the model is a dynamical question. At present, this question is unresolved for the Barbour–Bertotti models, as it is in general relativity.[21]

To quantize the Barbour–Bertotti model we need to express it as a constrained Hamiltonian system. We begin by defining canonical momenta,

$$\pi^i = \frac{\delta S}{\delta(\partial x^i/\partial \lambda)} = m_i \sqrt{\frac{-V}{T}} \frac{Dx^i}{d\lambda}. \tag{7.1}$$

Because the action contains no time derivatives of b and σ, we have six primary constraints,

$$P \equiv \sum_i \pi^i = 0, \tag{7.2}$$

and

$$L \equiv \sum_i x^i \times \pi^i = 0. \tag{7.3}$$

In addition, the momenta satisfy, by virtue of their definition, an additional primary constraints,

$$h \equiv \sum (\pi^i)^2 + V = 0. \tag{7.4}$$

It is simple to show that the Hamiltonian is, as we expect, a linear combination of these constraints,

$$H = -b \cdot P + \sigma \cdot L + \sqrt{\frac{T}{-V}} h. \qquad (7.5)$$

We can formally quantize this model in the standard way. We pick a representation consisting of the space of functions, $\psi(x^i)$. The coordinates x^i are represented in the usual way as multiplication operators, while the canonical momenta are represented as $\hat{\pi}^i \psi = i\hbar \partial \psi / \partial x^i$. The constraints may then be represented as

$$\hat{h}\psi(x^i) = \left[-\sum_i \frac{\hbar^2}{2m_i} \frac{\partial^2}{\partial x^{i2}} + V \right] \psi(x^i) = 0, \qquad (7.6)$$

$$\hat{P}\psi(x^i) = \sum_i \frac{\partial \psi(x^i)}{\partial x^i} = 0, \qquad (7.7)$$

$$\hat{L}\psi(x^i) = \sum_i x^i \times \frac{\partial \psi(x^i)}{\partial x^i} = 0. \qquad (7.8)$$

We may note here the similarities with the Hamiltonian formulation of general relativity. (1) Because of a time-reparametrization invariance, the Hamiltonian is a linear combination of constraints; (2) the constraints contain a set that are linear in the momenta, which generate the gauge symmetries other than time reparametrization; and (3) the generator of time reparametrization is quandratic in the canonical momenta.

This last point is important. As has been noted many times, any theory with a background time parameter t may be rewritten in a parametrization-invariant way by introducing the background time as an additional canonical coordinate together with an associated canonical momenta p_0 (Kuchař 1981b). The Hamiltonian evolution equation may then be replaced by a Hamiltonian constraint. However, the cost of doing this is that the Hamiltonian constraint contains a term linear in p_0, indicating that t plays a preferred role in the dynamics of the theory.

However, in theories without a background time parameter, such as general relativity and the Barbour–Bertotti models, there is no background time parameter, and as a result the Hamiltonian constraint only contains the canonical momenta quadratically. Whether or not there is a canonical transformation that "deparametrizes" the dynamics, i.e., puts the Hamiltonian constraint in the form of (5.1), is then a dynamical question. For this reason, I believe the Barbour–Bertotti models to be better models for the role of time in cosmological general relativity than are the parametrized particle models.

These models share one more important element with cosmological general relativity. The theories contain no nonvanishing additive constants of motion. All of the usual additive constants of motion, energy, momentum, angular momentum, charges, and so forth, vanish. The importance of this feature for the problem of the quantization of both general relativity and the Barbour–Bertotti models has been stressed by Barbour in 1989b.

The next step in the study of these models is to study particular model systems in which one can find explicit solutions to the constraints (7.6–7.8). The simplest of these systems involve three particles moving in one dimensions. Several models of this kind with step function potentials were studied in Barbour and Smolin 1988, where some examples of explicit solutions are given.

All of these solutions confirm the expectation that the physical states are not normalizable in the inner product of the unconstrained state space. The problem of the interpretation of these models then rests on the problem of how a physical inner product is to be constructed. For the reasons discussed in Section 5, this depends on the existence of a canonical transformation that allows the Hamiltonian constraint (7.4) to be put in the form (5.1). Some work has been put into a search for models in which this does happen, but the results are so far inconclusive.

Finally, I would like to mention that these models may be used to test various proposals that have been made concerning the interpretation of quantum states in quantum cosmology. Most of the proposals in the literature, not being dependent on the specifics of the dynamics (as is the proposal of intrinsic time), should make sense when applied to these models if they are to make sense when applied to quantum cosmology. Such a study was carried out in Barbour and Smolin 1988, where it was shown that the "semiclassical" measurement theories proposed by a number of authors (DeWitt 1967, Banks 1985) do not yield acceptable physical interpretations of these models. They fail all three of the criteria I introduced in Section 5 for an acceptable measurement theory for quantum cosmology. The proposed "semiclassical" measurement theories do not apply to any linear subspace of solutions to the constraints; rather, they can be applied only to semiclassical states built on classical solutions, and, in particular, not to arbitrary superpositions of those states. They fail the second criteria explicitly, as the numbers that these theories want to label as probabilities fail to satisfy a conservation principle; indeed, the violations of conservation of probability are explicitly computable in any given state. Thus, the theories can only yield sensible predictions in cases in which the predicted probabilities are either exactly zero, or exactly one. This is enough to indicate that these proposed measurement theories have no more predictive power than the classical dynamics on which the quantization is based, and have nothing really to do with quantum mechanics, if by that is meant anything that includes the full meaning of the superposition principle.

I would like to close this section by offering a challenge to anyone engaged in constructing measurement theories for quantum cosmology. First, apply the proposed measurement theory to a model such as one of the Barbour–Bertotti models in which all of the solutions to the quantum constraint equations can be explicitly constructed. I propose that only a prospective measurement theory that can satisfy the three criteria I gave in Section 5 when applied to these model theories should be considered a possible candidate for a measurement theory for quantum cosmology. If it fails this test in a case in which the classical dynamics is completely transparent and the full solution space to the constraints can be explicitly found, I do not see how we can reasonably expect it to make sense when applied to general relativity or any other realistic gravitational theory.

8. Quantum Physics Without Ideal Elements

The conclusion of the previous sections is that, if we want a quantum theory of a closed universe, it is not enough to apply the standard quantization procedure to a classical theory in which the ideal elements associated with a background space and time have been eliminated. The resulting theory cannot be meaningful because quantum mechanics itself contains ideal elements that require for their physical interpretation that there be things outside of the dynamical system described by the theory. But, if the ideal elements in the classical theory have been eliminated, the possibility of interaction with an outside system has been precluded.

Thus, to make a quantum theory of the entire universe we have to rid quantum mechanics itself of its ideal elements. The Everett interpretation is an attempt to do this; however, as we have seen, this attempt fails. One cannot rid a theory of its ideal elements by changing the words we use to talk about it. Being ideal is a structural property of the formal relations among the elements of the theory. The only possibility is to modify the formal structure of quantum mechanics itself.

The intrinsic time proposal is a proposal for doing exactly that, by making the choice of the inner product dependent on a particular set of coordinates in the phase space. These coordinates are, in turn, defined to be the solution to a certain dynamical equation, which is that a canonical transformation exists that puts the Hamiltonian constraint into the form (5.1). As I have stressed above, whether or not an intrinsic time exists is a dynamical question; the results that have so far been found are negative, but inconclusive. It would be, I believe, a very important step to have either an example of an intrinsic time, or a theorem that one does not exist.

If no intrinsic time exists, we must then search for other ways in which the structure of quantum mechanics should be modified so that a quantum theory of a whole universe would be possible. I would like to use the rest of this article to discuss some ideas as to how this might be achieved.

The task that I would like to consider is how we might invent a new physical theory that could be a quantum theory of the universe as a whole. Such a new theory should be based on new principles and new hypotheses about nature. I do not have such a theory. However, I do have a list of hypotheses that I believe may guide us in the discovery of such a theory. If the reader is willing to indulge a bit of speculation, I would like to describe some of these hypotheses. Of necessity, the discussion will be schematic. I will simply list the hypotheses I have in mind, following each with a few sentences of discussion.

(1) *The randomness of quantum mechanics is a consequence of its being necessarily a theory of only a small portion of the universe.* This, I believe, is the only possibility that is fully consistent with the principle of sufficient reason. It is hard to believe that the randomness of quantum mechanics is the final word; in any case, I think it is probably fruitful to choose not to believe it. If instead one insists that there be underlying causes for the outcomes of experiments that quantum mechanics can only predict statistically, one must conclude that those causes cannot be local. The strongest reason for this belief is the experimental disproof of the Bell inequalities. However, even without that, one would be led to this by examination of a variety of thought experiments that illuminate the nonseparable nature of quantum states of many-particle systems.

If then one embraces the notion that the irreducible fluctuations of quantum systems are due to some kind of nonlocal interactions, one is led immediately to the conclusion that one could only have a deterministic description of a whole universe. Bell's theorem tells us that once two particles have interacted, their quantum states are forever tied together in a nonseparable state so that only the whole system can have definite properties. This chain must then be extended to all the particles that those two interacted with, and so on, until the whole universe is included in the nonseparable state. Only at this level could one hope to have a complete description of the individual events in the system.

Suppose that such a description exists. This would be a deterministic but global theory in the sense that its deterministic evolution would involve nonlocal interactions tying the whole system together. Now, consider what would happen if one attempted to construct a local theory to describe experiments that observers could do in a small corner of the universe. Of necessity, such a theory would have to be statistical, as a completely deterministic theory of a portion of the universe is impossible because of the nonlocal interactions tying this portion to the rest of the universe. This theory would be derived by introducing a statistical ensemble for all of the degrees of freedom in the rest of the universe that could not be controlled by an experimenter working in our little corner and then averaging out those degrees of freedom. The suggestion is that the resulting statistical theory would be quantum mechanics.[22]

What would a theory of the whole universe that quantum mechanics was derived from look like?

(2) *There is a deterministic theory that describes the universe as a whole. It is completely relational and nonlocal.* This is the theory without ideal elements that we are seeking. It would be relational because any theory without ideal elements must contain only dynamical variables that involve relational quantities. It would necessarily involve nonlocal interactions; first, because that is necessary in order to get around Bell's theorem, and second, because nonlocal interactions are essential for the process of replacing universal ideal elements, such as the background spatial and temporal structure, by purely dynamical elements, as we saw in our discussion of the Barbour–Bertotti models.[23]

(3) *The causal topology of this fundamental theory reflects the history of interactions among particles in the universe rather than the present spatial relations.* This is another consequence of the requirement that the theory evade the experimental disproof of the Bell inequalities and reproduce the nonlocal quantum correlations that arise from the direct-product structure of the Hilbert spaces of many particle systems. Nonlocal quantum correlations are predicted by quantum mechanics to arise every time

two systems interact and share conserved quantities. Unless one believes that these correlations weaken in time and space (which would imply that the Schrödinger equation is inexact), one must assert that these correlations presently involve an intricate network of sets of particles stretching throughout the universe, each set containing particles that interacted at one time in the history of the universe.

Now, any deterministic theory that hopes to reproduce these correlations must involve a set of nonlocal variables or interactions for every set of correlations built into the present wave function. If we considered the topology induced by these interactions, it would follow the lines of this network; two variables would be neighbors if they described correlations among sets of particles that overlap. This topology is extremely nonlocal and intricate, but if the quantum correlations predicted by quantum mechanics are real, it reflects some real properties of the world. I am proposing that it is this topology that governs the interactions among the dynamical variables of the fundamental, nonlocal, deterministic theory I am describing.

What is the place of space in this universe? Why is the world then not just a complicated interconnected network of interactions, without any continuum structure?

These are, obviously, the hardest questions for my program to answer. I propose that

(4) *The dynamics of the fundamental theory are such that in the thermodynamic limit in which the number of fundamental particles goes to infinity, spatial relations emerge as a good approximate description of the causal relations among particles in the system.* Thus, space is a completely emergent property; there is no fundamental space in the world. Furthermore, in this limit, classical physics should emerge as the limit of the dynamics of the fundamental theory.

The spatial description that would emergy from such a theory must be relational since there is no way a background frame could emerge from the network of relations.

(5) *The relative distance between two particles is a measure of their closeness in terms of the topology of the network of relations of the fundamental theory.* More specifically, the location of a particle is determined by its relations to the other particles in the network. Each particle has a *view*, which is a description of how it is tied into the rest of the network. The relative distance between any two particles will be a measure of the differences between their views.

I will shortly describe a model system in which these concepts can be realized. Having discussed space, I must now mention time.

(6) *The fundamental nonlocal dynamical theory has a description in terms of an evolution of its state in a relational time.* Thus, I am choosing to keep time as a fundamental concept, while space becomes an emergent property. Why am I doing this? Why not make the theory nonlocal in both space and time?

The first reason for this is that I want to know how the physics I understand, such as Newton's laws and the Schrödinger equation, might emerge from the kind of theory I am proposing; and if the notion of time is left intact, then that is one less thing I have to recover. If I can achieve the goals I've outlined and construct a theory without ideal elements, without abandoning the notion of a dynamical time, I would prefer to do that. A second reason to keep time is that it allows me to keep one thing from the physics I understand while throwing much of the rest away. If I still have states, initial conditions and evolution, I am still doing some kind of physics, whether that evolution is governed by differential equations or a cellular automaton.

However, I must confess that the main reason I want to keep time is that I believe that the most fundamental observation we make about the world is that events are structured in terms of a flow in time. Our perceptual space is a construction in a way that our perceptual time is not. I believe that, in spite of relativistic invariance, time *is* different from space.

Having said this, I have to explain how I can hope to recover relativistic invariance from a theory in which time and space are treated so differently. The idea is that relativistic invariance on scales much smaller than the universe should emerge in the limit that we have been discussing, as a consequence of the fact that the physics of the limit is described by second-order field equations. The recovery of Lorentz invariance in the Barbour–Bertotti field theory (Barbour and Bertotti 1982) I discussed in Section 6 is a model for how this could happen.

(7) *The universe consists of a finite number of particles or fundamental entities.* This has as a consequence that the thermodynamic limit in which classical physics in a background space-time emerges is not exact. There are fluctuations around this limit. This means that the localizability of any one particle, relative to the rest of the universe, has limitations that come from these fluctuations.

If we apply the usual rule that the size of fluctuations are of the order of $1/\sqrt{N}$ relative to the system as a whole (where N is the number of degrees of freedom), we find that the scale of these fluctuations should be 10^{28}cm $\times 1/\sqrt{10^{80}} = 10^{-12}$cm.

This suggests the following hypothesis.

(8) *Quantum physics emerges as a description of the fluctuations around the thermodynamic limit in which classical mechanics on a spatial background emerges from the fundamental theory.* This, I would like to suggest, is the meaning of Dirac's large number coincidences (Dirac 1937, 1938). Of course, this is not to be taken literally, i.e., I am not suggesting that Planck's constant should vary with the radius of the universe. But this order-of-magnitude estimate suggests that, if the rest of the structure I have described could be realized, quantum mechanics could have something to do with its fluctuations (Smolin 1983, 1986a).

This is the rough picture. If it is to be realized, several glaring problems must be confronted. These are:

(i) How can the dynamics of the fundamental relational theory be chosen so that a low-dimensional space emerges as an approximate description of the network of relations in the thermodynamic limit. And, why should the dimension of that space be three?

(ii) Furthermore, this dynamics must be picked so that classical mechanics governs the dynamics of the apparent spatial relations among the dynamical entities. In particular, we must explain how the principle of inertia and the relativity of inertial frames emerges from a theory in which all distances to begin with are both relational and absolute.

(iii) Finally, we would like to know where in this scheme gravity comes from, and why it is weaker by many orders of magnitude than the other forces.

The second question is the easiest one, as we have in the Barbour–Bertotti theories a class of models in which exactly this happens. The first question is much harder; it is, I believe, the whole crux of the matter. Indeed, why a network of relations should produce a low-dimensional space with very large ratio between the largest and smallest relative distance *is,* I would claim, the cosmological constant problem, realized in this purely relational context.

To the third question I have at present no good answer. I will make some very speculative suggestions in the conclusion.

9. A Model Without Ideal Elements

I would like to suggest a model that realizes several of the basic features of the imagined theory I have been outlining. This model contains a list of N basic entities, which we will call monads (Leibniz 1714). The state of the system will consist of a description of relations among these N entities. For the simplest models, these relations will be of the simplest possible type: for every pair of monads, i and j, there will be a relational variable, A_{ij}, which can be on or off.

Thus, the state of the system will be given by a symmetric matrix A_{ij} of "0"s and "1"s, corresponding to the relations being on or off, respectively. (The diagonal elements are defined to be off, by convention.) Each state also has a graphical description, which is a graph with N labeled points representing the monads, in which we draw a line between two points if the corresponding relation is on. See Figure 9.1 for an example. The states of the model are then the possible graphs on n points (Harary 1969, Cvetkovic et al. 1980). One can play several kinds of games with these models. While I have achieved no startling results, I would like to indicate a little of the flavor of the subject.

First, we can assign a law of evolution, in which each link can change its state or not at each moment of time depending on some function of the connectivity of the graph. These models work like cellular automata in an indefinite number of dimensions. Second, one can specify an ensemble of graphs and then study statistical properties of the ensemble in the limit that the number of monads goes to infinity. This last game is actually a well-developed area of research, which is called statistical graph theory (Erdös and Rényi 1961, Erdös and Spencer 1974).

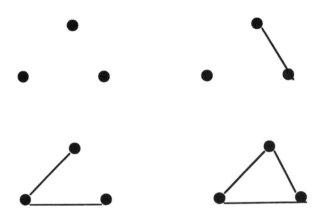

FIGURE 9.1. An example of a system with three monads, with some of its possible states.

For example, one may ask the following question. Let us consider, on the set of graphs with N vertices, the ensemble that is specified by saying that each bond has a probability of p to be on. The ensemble is then specified by N and p. We can ask what the probability is that a graph in this ensemble is connected. The answer, well known to graph theorists, is illustrated in Figure 9.2. There is a critical value of p, called p_c, which is proportional to $1/N$, such that for $p > p_c$ almost every graph is connected, while for $p < p_c$ almost every graph is disconnected.

Most of the games that I have been playing have to do with how the concepts of relative distance and dimension could emerge from these models. There are, in fact, two kinds of concepts of distance that one can work with. The first, topological distance, has been studied by graph theorists (see the previous references). The idea is that the distance between two points is a measure of how easy it is to walk from one to the other by stepping on links in the graph. There are many possible measures of this; one common definition is $D_{ij} =$ *the minimal number of steps between i and j.*

Given a concept of distance, we want to know how far away points in the graph can be. This is captured in the concept of the diameter of a graph, which is the maximal distance between any two points in the graph (see the previous references).

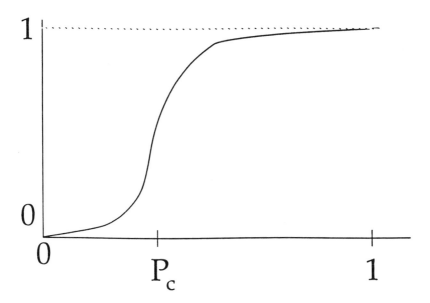

FIGURE 9.2. The probability that a graph is connected, as a function of p.

It is interesting to ask how the average value of the diameter varies with p, the probability for a link to be on. The qualitative behavior of this, drawn from several numerical experiments, is displayed in Figure 9.3. It is easy to understand this result qualitatively. For p equal to 1, every point is connected to every other one, and the diameter is 1. For p close to 1 most links are still on, and the diameter grows slowly. Even for p around a half, most pairs of links can be connected in two steps, and a diameter above 3 is rare. As we decrease p, graphs with large diameters only become common as we approach p_c.

Below p_c, the typical graph consists of a large number of disconnected pieces, and the diameter cannot grow very large. Thus, large diameters only become probable in the region just above p_c, where the typical graph is just barely connected.

Now, let us consider what concepts of dimension could be introduced into this system. Given a point set on which we have a metric, we could first ask whether we could imbed the set isometrically in some Euclidean space. There is a simple theorem that says that this can always be done in R^{N-1}, but this is not very interesting if we are contemplating $N = 10^{80}$. This definition also has an important disadvantage; if we succeed in having a graph with large N imbed isometrically in R^d with $d << N$, we can always increase the dimension of minimal embedding dramatically by adding one point.

Thus, we would like a more statistical measure of dimension that becomes very good for very large graphs, but cannot be easily perturbed by small changes in the graph. Here are two such definitions: (1) Consider $n(i, s) =$ *the number of points a distance s from the i-th vertex.* If the graph has a well-defined dimension, we would expect that, around every point, $n(i, s) \sim s^{d-1}$ for some d, which would then be called the statistical dimension of the graph around the point i. (2) Each R points in the graph define a simplex; given distances d_{ij} among these points, the volume of the simplex is defined by a well-known determinant. We will say that a graph has a simplicial dimension d_s if every R simplex with $R > d + 1$ has volume much less than one. Alternatively, we could define a statistical simplicial dimension by using here the average volume over all of the R simplices in the graph.

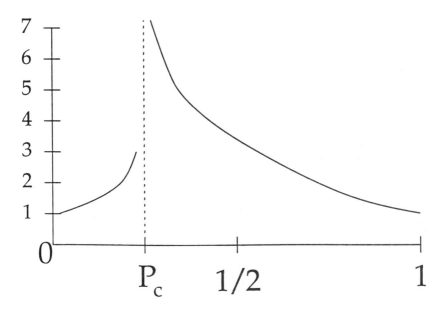

FIGURE 9.3. The dependence of the maximal distance in a graph with p.

Now, intuitively, the diameter of a graph is connected inversely to its dimension; only if the dimension is low, can the graph stretch out and have large distances between its points. Thus, we would expect to find graphs of low dimensions to be dominant in the ensemble I described for p just above p_c. Although we cannot go into details here, numerical experiments do confirm this intuition. However, there is a problem, which is that, in order to get good results, one finds very quickly that one must go to fairly large graphs. At the same time, the computer time required to check these hypothesis grows like large powers of the number of points because we are checking properties of all R simplices of the graph. Thus, one very quickly realizes that, with modest amounts of computer time available, one is not going to get great results by using simple ensembles such as the one I've described.

This suggests a search for other ensembles in which the properties we are interested in might be more common. One might despair of finding any ensembles which exhibit low dimensionality, but there are actually some encouraging results by mathematicians. In particular, certain ensembles of trees are known in which the typical graph has dimension two, given various definitions of distance and dimension, including the ones I've mentioned (Meir and Moon 1970, 1978). This is actually not surprising as trees are minimally connected graphs, which can be disconnected by turning off any of their bounds.[24]

One might then search for an ensemble of graphs that is slightly more connected than trees, in the hope that these might exhibit dimension three. So far I have not succeeded at this, but I am still trying.[25]

Another possibility might be altering the definition of distance that we are using. I would like to suggest another definition that is inspired by a loose reading of Leibniz's *Monadology* (1714). Recall the adjacency matrix A_{ij}. Consider the i-th column, $(v^i)_j \equiv A_{ij}$; it tells us how the i-th particle is connected to the rest of the graph. From the Leibnizian point of view, this contains all the information that the i-th monad could have about a universe of relations of the simple form we are describing: it is a list of all the other monads with which it is connected. Thus, following Leibniz, we will call this vector the *view* of the i-th monad. Now, following the *Monadology* we should expect that what we call the position in space of a monad is a reflection of its view, i.e., it is determined by what it sees when it looks out at the rest of the universe.

Now, consider two monads with the same view. This means that they are connected in the same way to the rest of the universe; what they see when they look out is identical. This means that if space is an emergent property of the network of connections, they should be at the same point of space. Now, consider two particles whose views are very similar. Again, if spatial relations are a reflection of the connectivity, their spatial relations to most things in the universe should be very similar. Hence, they should be close to each other. This suggests a notion of distance which is based on the differences between views. More particularly, we will define the network distance

$$D_{ij}^{\mathrm{network}} \equiv |\vec{v}^i - \vec{v}^j|, \tag{9.1}$$

where the norm here is the Euclidean norm.

It would be interesting to examine whether this notion of distance gives more satisfactory results for the problem of finding ensembles of low dimensionality.

Let us ask what features a graph with low dimensionality would have if this new notion of distance is used. Let me remark that a graph that embeds in a low-dimensional space contains a very large variety of distances in it, with the ratio of the largest to the smallest distance expected to be of the order of $N^{1/d}$, with d the dimension. Furthermore, the distribution of distances over all pairs of points is expected to be proportional to r^{d-1}, up to the point where r approaches the scale of the maximal distance.

Now, if D_{ij}^{Network} is taken to define the notion of distance, such a distribution could only be achieved in a graph that had a very large distribution of views. That is, if we compare views, we want many more pairs of views to be dissimilar than are similar. Now, since the total of all of the views is a description of the connectivity of the graph, what this means is that we want graphs in which very few points are connected to the rest of the graph in a similar way. Let me quantify this idea. Define the *variety* of a graph, $\mathcal{V}$ to be

$$\mathcal{V} \equiv \frac{1}{N^2} \sum_{i<j} \left(D_{ij}^{\text{Network}} \right)^2. \tag{9.2}$$

Now, for each N, there will be a maximal $\mathcal{V}$ possible for graphs with N points. By the rough arguments made above, it is among these graphs with maximal variety that we should find the graphs with low dimension.

Now, to put this more precisely, consider a collection of ensembles, $E(N, \mathcal{V})$, with N points and variety $\mathcal{V}$. Given any of the notions of dimension that we have defined, let us define $\langle d \rangle_{(N,\mathcal{V})}$ to be the average of the dimension over the ensemble. Then I conjecture that, for large N, this average dimension is a decreasing function of $\mathcal{V}$, and for a region around the maximal variety the dimensionality is either two or three.

Certainly the converse of this is easy to establish. In graphs with low variety, there are many sets of points that are all the same distance from each other. When this happens, the dimensionality will be large. When the variety is zero all points are connected to the rest in the same way and the graph is either totally connected or totally disconnected. In the former case the dimension (by most measures) will reach its maximal value; the second case is undesirable for other reasons. Furthermore, one can see by inspection that trees have high values of $\mathcal{V}$.

It seems to me that this notion of variety, or something like it, is what we need to understand how physics in a low-dimensional space could emerge from a purely relational theory of the type we have been discussing. I would now like to make my last hypothesis, and elevate this notion of variety to the level of a dynamical principle.

(8) *In a discrete theory of relations, the variety is the fundamental dynamical quantity. The fundamental dynamical principle is that the universe evolves in such a way as to maximize its variety.*

This principle of maximal variety is in a very real sense the opposite of the principle of maximal symmetry that has motivated much recent work in string theory and particle physics. In our case, symmetry is what we don't want; a graph that has a lot of symmetry has a lot of points that are connected to the rest of the universe in the same way, and that means that the dimension will probably be high. To get a low dimensionality we want very little symmetry; we want all points to have a distinct view of the rest of the universe. To put it very bluntly, a purely relational world that has a lot of symmetry must be boring, while only a really interesting world can have a low dimensionality. If a world has high symmetry, one view tells us a lot about it. Worlds with high variety are the most interesting; in such a world, the view of any given particle (monad, observer) tells us the least about the world seen by the other ones. I am suggesting that the principle of maximal variety can tell us how a world built in a way that is consistent with the principle of sufficient reason can be both low dimensional and interesting.

10. Final Remarks

If the reader is still with me, I have one more very speculative thing to say before closing. I would like to use my proposed principle to try to answer the last question I posed above: why is there gravity in the world?

Let me begin by pointing out that in a sense slightly different from the one I have just been using, a world without gravity would be a world without a lot of variety. This is because, if gravitation were not present, the universe would long ago have reached thermal equilibrium. It is only the peculiar properties of gravitation, its universal coupling and universal attraction, as well as the fact that its coupling is much weaker than the other forces, that makes possible a world in which metastable configurations can exist arbitrarily far from equilibrium. Without gravity we would have no hot stars radiating into cold space; we would have no planets with warm, enveloping, atmospheres; we would have, certainly, no life.

This has always puzzled me. If we ask, what features must the laws of nature have if life is to exist, part of the answer is that gravity must exist and have a strength relative to the average density of matter which is neither too strong nor too weak. This is, by the way, not the only thing that is necessary; people who have looked into this have concluded that the various physical constants and masses must be very close to what they actually are if carbon is to exist, and carbon chemistry is to be as rich as it is in structural possibilities (Morowitz 1968, 1987).

Thus, as has been observed by a number of people, the laws of nature, and the values of the various physical constants, seem to be fine tuned so that life can exist. This is a very disturbing observation; it suggests the presence of an order in the universe far beyond that to which our present science can aspire. Contemplation of this observation usually drives people to talk either of God or of the anthropic principle. This, in turn, has the further unfortunate effect of driving most other people away from contemplation of this problem.

I would like to contemplate the possibility that this observation could be explained scientifically, although most likely not within our lifetimes. Let us consider what physics might be like in 500 years, when many of the problems that engage us so keenly have the status that the problem of planetary orbits has for us now. Let us ask what kind of physics could explain the fine tuning of the laws of physics to account for the existence of life?

What is required is a physics in which what are now considered physical constants have become dynamical variables, subject to a dynamical principle. Now, one of the themes of the earlier sections of this paper was that this can happen when one replaces an old theory with a new one by replacing ideal elements with dynamical quantities. The Barbour–Bertotti models give us some examples of this, as I mentioned. If we consider that physics progresses by a series of such replacements, it seems plausible that in the future people will have a theory with many less ideal elements than we have at present, in which case what are now physical constants may be dynamically determined.

One of the themes of the later sections of this paper was that a key problem that must be solved to construct a quantum theory without ideal elements, i.e., a meaningful quantum cosmology, is to understand how a low spatial dimension might emerge from a purely relational theory. I suggested that a principle of maximal variety, or something very much like it, could solve this problem.

Now I would like to make one last suggestion.

(9) *The principle of maximal variety drives the tuning of the physical constants so that systems of arbitrary complexity can evolve and maintain themselves in the universe. In some future physics, it will account for both the existence of gravitation and the existence of life.*

This requires that the notion of variety, as I have just been loosely using it, and the technical notion of variety I introduced in the network theory, have something to do with each other. Of course, without a theory

I cannot claim that they do. However, I can perhaps make it plausible by pointing to the fact that if a future purely relational physics were some-day constructed along the lines I have indicated here, the mathematics used in this fundamental physical theory would resemble very closely the mathematics used in the study of complex biological systems such as ecologies and neural networks. My suggestion is that this might not be a coincidence.

If, at some time in the far future, these hypotheses turn out to be more than late-night fancies, then the unity of Aristotle's science, in which biology and physics were both part of a basic, cosmological science, could be recovered as a consequence of the development of physics. Recently, James Lovelock has suggested that a complete theory of the earth's ecology must incorporate the notion that life is a property, not of any particular organism, or species, but of the earth itself. Contemplation of the importance of gravitation in creating the conditions for life to exist on earth suggests that, to the extent that Lovelock is right, life is a property, not of the earth, but of the whole solar system. Is it not then possible that cosmology will someday, in the natural course of its development, require the notion that life is an emergent property of the universe as a whole?

Acknowledgments. First of all, I would like to thank Julian Barbour for many years ago awakening my incipient Leibnizian tendencies, and for conversations and collaboration since then. Anyone familiar with his writings will see the extent to which his views have influenced those expressed here. My thinking about the issues of quantum cosmology has also been strongly influenced by my interactions with Abhay Ashtekar, Louis Crane, Ted Jacobson and Carlo Rovelli during the last several years. John Dell provided invaluable advice in carrying out the numerical experiments mentioned in the last section. I am grateful also to him and to Bryce DeWitt, David Finkelstein, Jim Hartle, Chris Isham, Ted Jacobson, Karel Kuchař, Harold Morowitz, Roger Penrose and Rafael Sorkin for discussions over the years which have influenced my views on these matters.

This work was supported by an NSF grant to Syracuse University.

Since this article was written, there have been some more develop-ments concerning the ideas discussed in the last sections of this paper. Julian Barbour (1989b) has studied several new mathematical models in which a notion of variety is extremized. Very recently, a model of how spatial structure may arise as a result of extremizing a mathematical notion of variety was invented by Barbour. This new model, which was inspired in part by a suggestive comment of David Deutch, is more easily studied on the computer than the model that is discussed here, and is presently under investigation.

Recently, Ashtekar and Rovelli have argued that the reality conditions of quantum gravity, when correctly applied, should exactly determine the inner product. This has been shown to work out in $2 + 1$ gravity, in linearized general relativity and in a variety of other systems. I have been reasonably convinced by this point of view, and now feel that the problem of how the inner product on the physical state space is most likely resolved by this condition. This still leaves the problem of the interpretation of the inner product in quantum cosmology, which I still believe requires the construction of a measurement theory for quantum cosmology in which time, or some generalization of that concept that allows us to give meaning to the notion of a complete set of mutually exclusive occurrences (as used in my second criteria of section 5), plays a role.

Finally, an idea about the role of biological ideas in cosmology, which was partly inspired by the ideas discussed in section 10, is described in L. Smolin, "Did the Universe Evolve?", Syracuse Preprint, 1990.

NOTES

[1] Of course, we also impose symmetries on Einstein's equations because they make those equations easier to solve.

[2] As far as general relativity goes in reducing the dependence on ideal elements, it may not go far enough. For example, the theory has very little to say about what matter it may couple to, or about how that matter may be self-coupled. Thus coupling to matter is done, in the classical theory, simply by introducing a stress-energy tensor with certain properties. Many people, from Einstein to the present-day advocates of string theory, have felt that this was not sufficient; a theory that purports to be a theory of the whole universe, they argue, should also have something to say about what is in that universe. This issue is also, I believe, an aspect of the problem of eliminating ideal structures from our physical theories.

[3] See Hartle 1986; Geroch 1984; Tipler 1986; Halliwell 1987.

[4] Whether Mach actually made this suggestion, or was making a different

point is not germane here; for a discussion of this see Barbour (1986).

[5] In spite of this, and the difficulty of the problems in this area, it is also true that a great deal, of lasting value, that is crucial for our understanding of quantum mechanics has come out of work in this area.

[6] I would like to stress here that I am not arguing that the Everett interpretation itself is wrong if by the Everett interpretation we mean the structure described in Everett's original paper (Everett 1957), in which the notion of relative states is used to construct an interpretation of the theory in which collapse of the wavefunction need not be invoked. This is, I believe, a legitimate approach to an interpative scheme for quantum mechanics (see Smolin 1984). It is also, partly, an attractive one as collapse of the wavefunction is an ugly addition to the theory, which does better without it. What I believe is wrong are the claims made about the Everett interpretation, which suggest that it provides a measurement theory for quantum mechanics that does without explicit reference to outside agencies in its measurement theory, i.e., that it turns quantum mechanics into a theory without ideal elements. This mistaken claim is then the basis of the further mistaken claim that the Everett interpretation can serve as the basis for a sensible quantum cosmology. The Everett interpretation relies on ideal elements in the same way that any other interpretation does, through its use of the three ideal structures described here. In the Everett interpretation (as in any other interpretation), any physical interpretation given to these ideal elements that is sufficient to serve as a basis for comparison between theory and experiment necessarily involves reference to things outside of the system under study.

[7] The argument I have just outlined has many ins and outs, and cannot be done complete justice in the space I have here. What follows is just a sketch of the argument; a more complete version will appear elsewhere (Smolin 1991b).

[8] For a very good account of the history of this question, with an emphasis on just the issues we are discussing, see Barbour (1989a).

[9] My views on the applicability of the Everett interpretation to ordinary quantum mechanics are discussed at length in Smolin 1984.

[10] Another way to see this is to use directly the formula for f^k given in Hartle, which is $f_N^k = \Sigma_{i_1,\ldots,i_N} \mid i_1,\ldots, i_N > N(k; i_1,\ldots, i_N) < i_1,\ldots, i_N \mid$. This definition is equivalent to (4.1) only if it is the case that $\langle i \mid j \rangle = \delta_{ij}$.

[11] I do not discuss here the problem of the meaning of "very small." A good discussion of this is found in the paper by Stein (1984).

[12] These points are brought out clearly in the exchange between Geroch (1984) and Stein (1984).

[13] While this point is not stressed in Geroch's exposition, we find there, as well as in other discussions that refer to the principle of precluded regions, the detector is always put in a semi-classical state. Our claim is that this is not for the convenience of the exposition, but that this is a necessary assumption if the interpretation is to be successfully applied.

[14] The procedure for quantizing a constrained Hamiltonian system is dis-

cussed in many places. I use here the language and notation from my other contribution to this volume (Smolin 1991a).

[15] That such a scheme's predictive power is identical with that of the classical Einstein theory can be seen from the following argument. Given any classical solution to Einstein's equations with cosmological boundary conditions, one can always construct a semi-classical wavefunction based on this state. At the formal level, in which constraints are not really regularized or solved, but one only makes formal manipulations of functional integrals, one may show that any such state is *formally* a solution to the constraints of quantum gravity. (However, whether such a formal state is actually a solution to the constraints in a nontrivial sense, in which they are defined through the use of a regularization procedure and appropriate care is taken about the interaction of the regularization procedure with the requirements of diffeomorphism covariance, is, I believe, not at all clear.) One can then always invent boundary conditions for the quantum constraints which have the intended semi-classical state as a formal solution.

[16] Another way to see why this happens is to note that the constraints correspond to the condition that conserved quantities such as energy have a fixed value. This means that the wave function will have a dependence on t, the quantity conjugate to the fixed quantity, as e^{iEt}. Now, if t is a physical degree of freedom in the unconstrained state space, the inner product on that space will often include an integral $\int dt$. ... This integral will diverge for eigenstates of E. In ordinary quantum mechanics, problems such as this are resolved by constructing physical states as superpositions of eigenstates of E. But this is exactly what we cannot do in this case because the constraint fixes E for all physical states. Thus, in these cases, only unphysical states can be normalizable under the inner product on the unconstrained state space.

[17] It has also been proposed that, in the absence of an intrinsic time, a conserved probability can be defined for any choice of time simply by dividing the wave function at each time by a normalization factor which is constructed by integrating over all the other variables in the wave function at that time. This program almost certainly leads to unacceptable violations of causality, as is discussed by Ted Jacobson in his contribution to this volume (Jacobson 1991).

[18] In the asymptotically flat case, Ashtekar has made progress in identifying an intrinsic time in the linearized theory (Ashtekar 1991). This is related to earlier work on extrinsic time (Kuchař 1981b).

[19] These models depend on still other ideal elements such as a fixed Euclidean metric on space. It is much harder to construct theories with no ideal elements; for an attempt, see Section 7.

[20] Assuming that the distribution of visible light is the same as the distribution of gravitating matter in the galaxy! Of course, this is now widely questioned; if the distribution of gravitating matter in our galaxy is more spherically symmetric, then these models may be consistent with observations.

[21] The quantization of these models is discussed also in a contribution to

this volume by Carlo Rovelli, who argues that the group quantization method of Isham leads here to a preferred physical inner product, apparently resolving a problem I am about to raise (Rovelli 1991b). The problem with this, from the point of view I have introduced here, is that to simply eliminate an ideal element from a theory, as Carlo does with the notion of time, does not resolve the problem. What is required is a theory in which the ideal element is transformed into an intrinsic element in a way that explains dynamically how the arbitrary elements in the choice of the ideal elements arise. Our world seems to exist in time; what is needed is a consistent quantum theory with no ideal elements that also contain a notion of time. A formulation that denies the existence of time does not, at least at first sight, do this for us. (I must add here that Carlo does not agree with this—see his contribution in this volume.) Quantization of these models is also being studied by Jacobson and collaborators.

[22] A hidden variables theory for quantum mechanics of N particles in two dimensions was constructed according to these principles in Smolin 1986a. A related set of ideas is discussed in Smolin 1983.

[23] Why deterministic? Why not? If we are willing to make a non-local theory, then we have removed the constraints against making a deterministic theory, so why not try to make one?

[24] The statistical mechanics of trees is known to people in statistical mechanics as the infinite-dimension limit of diffusion-limited aggregation (Witten and Sandor 1983).

[25] In this regard, an intriguing paper (Nagels 1985) should be mentioned that claims the existence of an ensemble of graphs with a distribution of distances like that on S^3. If true, this is extremely interesting.

REFERENCES

Ashtekar, Abhay (1991). "Old Problems in the Light of New Variables." See this volume, pp. 401–426.

Banks, T. (1985). "TCP, Quantum Gravity, the Cosmological Constant and All That . . . ." *Nuclear Physics B* 249: 332–360.

Barbour, Julian B. (1974a). "Relative-Distance Machian Theories." *Nature* 249: 328–329.

———(1974b). "Erratum." *Nature* 250: 606–606.

———(1975). "Forceless Machian Dynamics." *Il Nuovo Cimento B* 26: 16–22.

———(1976). "Historical Background to the Problems of Inertia." Unpublished.

———(1982). "Relational Concepts of Space and Time." *British Journal for the Philosophy of Science* 33: 251–274.

———(1986). "Leibnizian Time, Machian Dynamics, and Quantum Gravity." In *Quantum Concepts in Space and Time*. Roger Penrose and Christopher J. Isham eds. Oxford: Clarendon, pp. 236–246.

————(1989a). *Absolute or Relative Motion? A Study from the Machian Point of View of the Discovery and the Structure of Dynamical Theories.* Cambridge and New York: Cambridge University Press.

————(1989b). "Maximal Variety as a New Fundamental Principle of Dynamics." *Foundations of Physics* 19: 1051–1073.

Barbour, Julian B., and Bertotti, Bruno (1977). "Gravity and Inertia in a Machian Framework." *Il Nuovo Cimento B* 38: 1–27.

————(1982). "Mach's Principle and the Structure of Dynamical Theories." *Royal Society of London. Proceedings A* 382: 295–306.

Barbour, Julian B., and Smolin, Lee (1988). "Can Quantum Mechanics Be Sensibly Applied to the Universe as a Whole." Yale University preprint.

Berkeley, George (1710). *A Treatise Concerning the Principles of Human Knowledge.* Cited from 1957 edition, New York: Liberal Arts Press.

————(1721). *De Motu.* Cited from 1971 edition, in *The Works of George Berkeley.* Oxford: The Clarendon Press.

Bertotti, Bruno, and Rovelli, Carlo (1990). Manuscript, in preparation.

Bondy, J.H., and Murty, U.S.R. (1976). *Graph Theory with Applications.* Amsterdam and New York: North-Holland.

Cvetkovic, Dragos M.; Doob, Michael; and Sachs, Horst (1980). *Spectra of Graphs: Theory and Application.* Pure and Applied Mathematics, vol. 87. Samual Eilenberg and Hyman Bass, eds. New York, San Francisco, and London: Academic.

Deutsch, D. (1985). "Quantum Theory as a Universal Physical Theory." *International Journal of Theoretical Physics* 24: 1–41.

DeWitt, Bryce S. (1967). "Quantum Theory of Gravity. I. The Canonical Theory." *Physical Review* 160: 1113–1148.

Dirac, Paul A.M. (1937). "The Cosmological Constants." *Nature* 139: 323–332.

————(1938). "A New Basis for Cosmology." *Royal Society of London. Proceedings A* 165: 199–208.

————(1964). *Lectures on Quantum Mechanics.* Belfer Graduate School of Science Monographs, no. 2. New York: Belfer Graduate School of Science, Yeshiva University.

Einstein, Albert (1916). "Die Grundlagen der allgemeinen Relativitätstheorie." *Annalen der Physik* 49: 769–882. English translation in *The Principle of Relativity.* New York: Dover, 1962, pp. 111–164.

Erdös, P., and Rényi, A. (1961). "On the Strength of Connectedness of a Random Graph." *Acta Mathematica* 12: 261–267.

Erdös, P., and Spencer, J. (1974). *Probabilistic Methods in Combinatorics.* Probability and Mathematical Statistics: A Series of Monographs and Textbooks, Z.W. Birnbaum, and E. Lukacs, eds. New York: Academic.

Everett, Hugh III (1957). " 'Relative State' Formulation of Quantum Mechanics." *Reviews of Modern Physics* 29: 454–462. Reprinted in *The Many Worlds Interpretation of Quantum Mechanics.* Bryce S. DeWitt and Neill Graham, eds. Princeton: Princeton University Press, pp. 141–149.

Finkelstein, David (1963). "Logic of Quantum Physics." *New York Academy of Sciences. Transactions* 25: 621–637.

Geroch, Robert (1984). "The Everett Interpretation." *Nous* 18: 617–633.

Graham, Neill (1973). "The Measurement of Relative Frequency." In *The Many Worlds Interpretations of Quantum Mechanics.* Bryce S. DeWitt and Neill Graham, eds. Princeton: Princeton University Press, pp. 229–253.

Halliwell, Jonathan J. (1987). "Correlations in the Wavefunction of the Universe." *Physical Review D* 6: 3626–3640.

Harary, Frank (1969). *Graph Theory.* Reading, Massachusetts: Addison-Wesley.

Hartle, James B. (1968). "Quantum Mechanics of Individual Systems." *American Journal of Physics* 36: 704–712.

———(1986). "Predictions in Quantum Cosmology." In *Gravitation in Astrophysics.* NATO Advanced Study Institute, 1986, Cargèse, France. B. Carter and James B. Hartle, eds. New York: Plenum, pp. 329–360.

———(1988a). "Quantum Kinematics of Spacetime. I. Nonrelativistic Theory." *Physical Review D* 37: 2818–2832.

———(1988b). "Quantum Kinematics of Spacetime. II. A model Quantum Cosmology with Real Clocks." *Physical Review D* 38: 2985–2999.

———(1990). "Quantum Kinematics of Spacetime. III. General Relativity." *Physical Review D.* Forthcoming.

———(1991). "Time and Prediction in Quantum Cosmology." See this volume, pp. 172–203.

Jacobson, Ted (1991). "Unitarity, Causality, and Quantum Gravity." See this volume, pp. 212–216.

Kochen S. (1985). "A New Interpretation of Quantum Mechanics." In *Symposium on the Foundations of Modern Physics: 50 Years of the Einstein–Podolsky–Rosen Gedankenexperiment.* P. Lahti and P. Mittelstaedt, eds. Singapore/Philadelphia: World Scientific, pp. 151–170.

Kuchař, Karel V. (1981a). "General Relativity: Dynamics without Symmetry." *Journal of Mathematical Physics* 22: 2640–2654.

———(1981b). "Canonical Methods of Quantization." In *Quantum Gravity 2: A Second Oxford Symposium.* Christopher J. Isham, Roger Penrose, and Dennis W. Sciama, eds. Oxford: Clarendon, pp. 329–376.

———(1982). "Conditional Symmetries in Parametrized Field Theories." *Journal of Mathematical Physics* 23: 1647–1661.

Leibniz, Gottfried Wilhelm (1696). "On the Principle of Indiscernibles." In Leibniz 1973, pp. 133–135.

———(1714). "Monadology." In Leibniz 1973, pp. 179–194.

———(1715–1716). "Correspondence with Clarke (Selections)." In Leibniz 1973, pp. 205–238.

———(1973). *Leibniz: Philosophical Writings.* G.H.R. Parkinson, ed. M. Morris and G.H.R. Parkinson, trans. London and Melbourne: Dent.

Lovelock, J.E. (1979). *Gaia: A New Look at Life on Earth.* Oxford and New York: Oxford University Press; 2nd ed., 1987.

———(1988). *The Ages of Gaia: A Biography of Our Living Earth.* New York: W.W. Norton.

Mach, Ernst (1866). "Bemerkungen über die Entwicklung der Raumvorstellungen." *Fichtes Zeitschrift für Philosophie* 49: 227–232.

———(1893). *The Science of Mechanics: A Critical and Historical Exposition of its Principles.* Thomas J. McCormack, trans. Chicago and London: Open Court.

Meir, A., and Moon, J.W. (1970). "The Distance between Points in Random Trees." *Journal of Combinatorial Theory* 8: 99–103.

———(1978). "On the Altitude of Nodes in Random Trees." *Canadian Journal of Mathematics* 30: 997–1015.

Morowitz, Harold J. (1968). *Energy Flow in Biology.* New York: Academic.

———(1987). *Cosmic Joy and Local Pain: Musings of a Mystic Scientist.* New York: Scribner.

Nagels, G. (1985). "Space as a bucket of dust." *General Relativity and Gravitation* 17: 545–558.

Newton, Isaac (1962). *The Mathematical Principles of Natural Philosophy and the System of the World.* Andrew Motte, trans. Florian Cajori, ed. Berkeley: University of California Press.

Page, Don, and Wooters, William K. (1983). "Evolution without Evolution: Dynamics Described by Stationary Observables." *Physical Review D* 27: 2885–2892.

Rovelli, Carlo (1991a). "Is there Incompatibility between the Ways Time is Treated in General Relativity and in Standard Quantum Mechanics?" See this volume, pp. 126–140.

———(1991b). "Group Quantization of the Barbour–Bertotti Model." See this volume, pp. 292–296.

Rovelli, Carlo, and Smolin, Lee (1990). "Loop Representations for Quantum General Relativity." *Nuclear Physics B* 331:80–152.

Smolin, Lee (1983). "Derivation of Quantum Mechanics from a Deterministic Non-Local Hidden Variable Theory I. The Two-Dimensional Theory." Institute for Advanced Study preprint.

———(1984). "On Quantum Gravity and the Many Worlds Interpretation of Quantum Mechanics." In *Quantum Theory of Gravity: Essays in Honor of the 60th Birthday of Bryce S. DeWitt.* Steven M. Christensen, ed. Bristol: Adam Hilger, pp. 431–454.

———(1986a). "Stochastic Mechanics, Hidden Variables, and Gravity." In *Quantum Concepts in Space and Time.* Roger Penrose and Christopher J. Isham, eds. Oxford: Clarendon, pp. 147–173.

———(1986b). "On the Nature of Quantum Fluctuations and Their Relation to Gravitation and the Principle of Inertia." *Classical and Quantum Gravity* 3:

347–359.

————(1986c). "Quantum Gravity and the Statistical Interpretation of Quantum Mechanics." *International Journal of Theoretical Physics* 25: 215–238.

————(1991a). "Nonperturbative Quantum Gravity via the Loop Representation." See this volume, pp. 440–489.

————(1991b). *Space and Time in the Quantum Universe.* In preparation.

Stein, Howard (1984). "The Everett Interpretation of Quantum Mechanics: Many Worlds or None?" *Nous* 18: 635–652.

Thomas, Lewis (1988). "Foreword." In Lovelock 1988.

Tipler, Frank J. (1986). "The Many-Worlds Interpretation of Quantum Mechanics in Quantum Cosmology." In *Quantum Concepts in Space and Time*. Roger Penrose and Christopher J. Isham, eds. Oxford: Clarendon, pp. 204–214.

Unger, Roberto Mangabeira (1984). *Passion: An Essay on Personality.* New York: Free Press.

Wada, S. (1987). "Interpretation and Predictability of Quantum Mechanics and Quantum Cosmology." DAMTP preprint.

Wheeler, John A. (1957). "Assessment of Everett's 'Relative State' Formulation of Quantum Theory." *Reviews of Modern Physics* 29: 463–465.

Witten, T.A., and Sandor, L.M. (1983). "Diffusion-Limited Aggregation." *Physical Review B* 27: 5686–5697.

Zurek, Wojciech Hubert (1981). "Pointer Basis of Quantum Apparatus: Into What Mixture does the Wave Packet Collapse?" *Physical Review D* 24: 1516–1525.

————(1982). "Environment-Induced Superselection Rules." *Physical Review D* 26: 1862–1880.

Discussion

DEWITT: What's the ideal element in the asymptotically flat case?

SMOLIN: The time of the observers in the laboratory at infinity.

KUCHAŘ: I think that Murray Gell-Mann should have made this remark, but because he didn't, I will make it. Because Leibniz didn't believe in the ontological existence of time, he dropped the letter "t" from his name.

SMOLIN: Is that true?

KUCHAŘ: Yes! He spelled his name with "z".

DEWITT: It's a good thing that he did believe in space because the "z" would've gone too (laughter).

HARTLE: Just one comment. In this sum-over-histories approach, which I'm going to discuss this afternoon, I looked at a model due to Banks that is equivalent, at least classically, to the model you describe. And I think I would claim I know how to interpret it. I had lots of discussions with your collaborator, Julian Barbour, and I wonder if you had any comment on the connection between the two.

SMOLIN: Well, I have two comments. One of them is that, as you've said in everything you've written, you've accepted that the canonical theory does not have a good measurement theory. You're trying to go on and solve the next problem, which is how or whether you can make sense of quantum mechanics outside of it. We're still trying to really convince ourselves whether or not the canonical theory has a good measurement theory. Secondly, whether your approach, which involves doing things with path integrals that have no canonical expression, gives a good measurement theory for these models or not is something that we need to discuss in detail. Julian and I spent a lot of time talking about what was in your paper on this topic and didn't agree. So I think I can't comment on that.

PAGE: If you do have a stationary state as in your model, I think you can define a time if you introduce a clock. If you consider the simplest example, in this case where you've got three particles in one dimension, and you fix the potential to be a negative constant, then the constraint says that the total kinetic energy is fixed. And that means (supposing the masses are equal) that the sums of the squares of these three momenta add up to a constant, which is the negative of the negative potential. And then the momentum is constant. So all you have is those two relative distances. Suppose you take the first relative distance to be a clock. Then, the distance of the third particle from the center of mass is my observable. And now we ask: given that the clock has some particular reading, what's the value of the observable? Suppose I choose states for the clock, e.g., suppose I choose some wave-packet state, I project the full state onto this wave-packet state, which will now be normalizable in the ordinary Schrödinger inner product. Then, if I define some other clock states, the Hamiltonian will connect them. The Hamiltonian will be the sum of two parts, one for the first relative separation and one for the separation of the third particle from center of mass. And then, if I want to calculate expectation values for the relative state of this other function, and I ask how that depends on the reading of the clock, then in this particular case we just get a Schrödinger equation for the motion of the other thing in terms of this clock variable. So, as long as I can divide up the system into one thing that I call a clock and something else that I call my physical observable, I can ask: what's the correlation between the clock and the other thing? I can even get an equation relating them. And in the simple cases, it's just the Schrödinger equation. In the general case, of course, it won't be such a simple equation. And I can ask whether the fact that it's not exactly the Schrödinger equation gives results that are contrary to observation. In this case, it's a very simple description. One can describe this perfectly well in quantum-mechanical terms and in terms of these conditional expectation values.

SMOLIN: We also looked very carefully at this point of view. And I agree with what you say, in the case that we have an inner product on the physical Hilbert space, i.e., on the space of solutions. But the problem I'm thinking of is: "What states within the space of the solutions to the constraints are going to be normalizable given a particular choice for the physical inner product and what states are not going to be?" This is the same thing, in some sense, as deciding what time variable you're going to use. The choice of the inner product then depends on what time variable you use. If one wants unitary evolution, one must recall that what operators are Hermitian depends on the choice of inner product. So I think it's not so simple given the fact that the solution to the constraints are not normalizable in the obvious inner product. It's not simple because any further choice of the physical inner product depends on having some idea about what you mean by time. Just as in quantum cosmology you'd like things not to be normalizable when they grow in the time dimension, but normalizable when you arrange them to grow "perpendicular" to that.

But perhaps this needs a longer discussion.

ROVELLI to PAGE: You were taking three particles and you were saying, let's make the relative distance of two of them a clock. And you were not saying, let's use the position of the first particle, because I think you know the position of one particle would not commute with the constraint, which tells you that the total momentum is zero. The relative position of the two particles does commute. However, we've another constraint, which is the Hamiltonian constraint. The difference of these two does not commute with the Hamiltonian constraint. So the question is: "Why do you say that we want an object that commutes, say, with the diffeomorphism constraint, and then introduce a variable, time, which by definition is something that doesn't commute with the Hamiltonian? I want to show that there is a way, I think, a formal way to quantize this model using Chris Isham's ideas (see Rovelli's article that follows).

DEWITT: So you say the problem is for us to find the canonical quantization group. That's a formidable task.

ROVELLI: Absolutely, yes.

Group Quantization
of the Barbour–Bertotti Model

Carlo Rovelli

In my talk on the issue of time in this conference, I claimed that systems without a preferred time, like general relativity, can in principle be quantized in the framework of standard quantum mechanics.

I also claimed that the difficulty of defining the scalar product is not a difficulty of principle. Rather, it is a *technical* problem that can be overcome by suitable technical methods; a method for defining the scalar product is given by group-theoretical quantization.

Lee Smolin has presented an extremely interesting model, the Barbour–Bertotti model, which is a finite-dimensional model that is genuinely without time, so that it reproduces the "no time" features of gravity theories.

Thus, I felt challenged to demonstrate that the group method is able to provide a scalar product in this model and therefore to define a complete "no time" quantization, at least in this relatively simple finite-dimensional example.

A quantizing group can indeed be found. Thus, I present here a complete quantization of the simplest nontrivial version of the Bertotti–Barbour (1977, 1982) relational model by using the group method. I refer to my other talk in these proceedings for the general philosophy and for the references.

In my opinion, this quantum model could be used as a model system to study how to extract information from a no-time quantum system.

In the canonical framework, the dynamics of the model is contained in the two constraints

$$H = T + V = 0, \tag{1}$$

$$P = \sum_{i=1}^{n} p^i = 0, \tag{2}$$

where

$$T = \frac{1}{2} \sum_{i=1}^{n} (p^i)^2, \tag{3}$$

and V is a "potential energy" that is a function of the canonical coordinates x_i. We consider the simplest nontrivial realization of the model in which $n = 3$ and the potential is constant and equal to minus a positive constant that we call (for later convenience) $\frac{3}{2}U$.

The crucial conceptual point is that, before proceeding to the quantization, one has to get a clear understanding of the physics contained in the model and, in particular, to understand what are the true gauge-invariant observables.

Any physical observable must commute with *all* the constraints. I stress this fact, which is too often forgotten. The constraints are all on the same footing as far as the requirement that physical operators must commute with them is concerned. If we assumed that an observable that didn't commute with all the constraints could be measured, we would get a nondeterministic theory; that is, a theory in which the initial conditions do not uniquely determine the evolution.

Thus, for instance, the difference $x_1 - x_2$ commutes with the momentum constraint but does not commute with the Hamiltonian constraint, and therefore it cannot be a physical observable of this model.

Now, according to the general prescription for the group–theoretical quantization of constrained systems, I have to find a complete algebra of physical observables, namely a set of observables big enough to parametrize the physical phase space that is closed under Poisson brackets, and commutes with the constraints.

The problem is best faced from a geometrical point of view. The Hamiltonian constraint fixes a spherical surface in momentum space. Any $O(3)$ transformation in momentum space sends the sphere into itself, and therefore its generators

$$L^i = \varepsilon^{ijk} x^j p^k \tag{4}$$

commute with H, as can be easily verified. The three L^i and the three p^i form a closed algebra of observables that commute with the Hamiltonian constraint. The momentum constraint fixes a plane in momentum space. The intersection of the sphere with the plane is a circle. A one-dimensional subalgebra of the algebra of the L^i sends the circle into itself. From the geometrical picture, it is straightforward to see that this is generated by

$$L = \frac{1}{\sqrt{3}}(L^1 + L^2 + L^3) \tag{5}$$

where the normalization has been added for later convenience. L and p^i commute with the constraints. Because of the constraints themselves, the three p^i are not independent; in fact, they are constrained to lie in a one-dimensional circle. It is worthwhile to reduce the redundancy of the elementary observables. Because of the topology, it is not possible to find a unique, continuous, single-valued coordinate for the circle, therefore we must content ourselves with two coordinates. We note that, if we define

$$s = p^1 - p^2,$$
$$c = \frac{1}{\sqrt{3}}(p^1 + p^2 - 2p^3), \tag{6}$$

then the algebra of the set s, c, L closes:

$$\{s, c\} = 0$$
$$\{s, L\} = c$$
$$\{c, L\} = -s. \tag{7}$$

Given the way we have chosen the observables s, c, L, it is clear that they commute with the constraints

$$\{s, H\} = \{c, H\} = \{L, H\} = 0$$
$$\{s, P\} = \{c, P\} = \{L, P\} = 0; \tag{8}$$

these equations can also be easily checked explicitly.

Thus, we have found a complete closed algebra of physical observables. The algebra given by equation (7) is the Lie algebra of the group E_2. The quantization of the system is therefore given by an irreducible unitary representation of the group E_2. The irreducible unitary representations of E_2 can be expressed on the Hilbert space $\mathcal{H}[S_1, d\phi]$. They are generated by the self adjoint operators

$$\hat{L} = -i\frac{d}{d\phi}$$
$$\hat{s} = u \sin \phi$$
$$\hat{c} = u \cos \phi, \tag{9}$$

where u is a parameter that labels the different representations. There are several equivalent ways to choose the correct representation. Here, the problem can be solved very directly by considering the classical limit of the theory. In fact, we note that the elementary observables satisfy

$$s^2 + c^2 = \frac{2}{3}\left(P^2 + H + \frac{3}{2}U\right) \sim U, \tag{10}$$

and therefore it is clear that, in order to get the correct classical limit, we should put

$$u = \sqrt{U} \tag{11}$$

in equation (9).

Note the advantages of the group quantization procedure for constrained systems. First of all, we did not need to face order ambiguities for the constraints. We skipped the technical difficulties of solving the constraint equations, as well as the technical difficulty of ordering the physical observables so that they commute with the constraints. More important is the fact that we have got a unique prescription for the physical scalar product. It is possible, in fact, to compare our result with the ones presented in Lee Smolin's talk. We obtain directly the same linear space of physical states obtained by *solving the constraint equations,* and our operators can be obtained as a suitable ordering of the elementary operators defined there; therefore, our results essentially agree. But there, the quantization could not be completed because a unique prescription for the physical scalar product was lacking, while in the present approach the group representation theory fixes uniquely the physical scalar product.

Thus, the quantization of the theory is concluded; we have a well-defined Hilbert space of states and a set of self-adjoint operators that corresponds to classical physical observables. At this point, one may begin to study the quantum theory and make physical predictions. For instance, as a physical prediction of the quantum theory we may note that the spectrum of the L observables is quantized.

The important point is that the system is well defined just in terms of the observables (7), which do not show any relation with a time evolution. This is not in contradiction with anything we know in physics, and in particular it is not at all in contradiction with the probabilistic interpretation of quantum mechanics.

In conclusion, I have quantized the simplest version of the Barbour–Bertotti model by using a modification of the group-theoretical quantization. This method provides the physical states directly and, what is more important, the physical scalar product. Given the fact that one can define all the basic ingredients for a complete definition of a quantum-mechanical system and for its probabilistic interpretation, the result shows that the individuation of a preferred "time" variable is not conceptually necessary for the definition of a quantum-mechanical system. Time is a useful concept only in some specific systems, or in some specific states. The difficulties in dealing with a quantum system with no time are just psychological difficulties.

REFERENCES

Barbour, Julian B., and Bertotti, Bruno (1977). "Gravity and Inertia in a Machian Framework." *Il Nuovo Cimento B* 38: 1–27.

———(1982). "Mach's Principle and the Structure of Dynamical Theories." *Royal Society of London. Proceedings A* 382: 295–306.

PART III

Strings and Gravity

String Theory Without Space-Time

Gary T. Horowitz

Many people have observed that the development of a new theory of physics not only answers existing open questions but makes some of them obsolete. This is because the new theory introduces improved concepts for describing the physical world and renders some of the old concepts meaningless. Familiar examples include the lack of absolute simultaneity in special relativity, and the lack of a local energy density in general relativity. Quantum gravity will undoubtedly produce similar changes. The most likely (in my opinion) is that our basic concepts of space and time will become obsolete. It is often remarked that, since general relativity identifies gravity with the curvature of space-time, a quantum theory of gravity must involve a quantum theory of space and time. In other words, our usual notions of space and time must be just a low-energy classical approximation to something more fundamental. This same conclusion follows (even without quantization) from attempts to unify gravity with other forces. In any unified theory involving gravity, a fundamental four-dimensional space-time does not exist. A simple illustration of this is provided by Kaluza–Klein theory. A general solution in this theory is a D-dimensional Ricci-flat space-time. There is no way to unambiguously extract a four-dimensional space-time from it. Only for special solutions, e.g., those of the form $M \times K$ (where M is a four-dimensional manifold) can a four-dimensional space-time be defined.

If space-time is to be a derived concept, the theory must be formulated in terms of something more fundamental. Since string theory has the potential for being our first consistent quantum theory of gravity, it is important to ask what is its prediction for this more fundamental structure. In many discussions of string theory, one has the impression that there isn't any. One considers strings moving in a fixed background space-time. One mode of the string corresponds to a massless spin-two particle called the graviton, and the scattering amplitudes for these gravitons are calculated. In effect, the metric has been divided into a kinematical background space-time and a dynamical gravitational field described by the string. If this was the complete description of the theory, one might conclude that string theory represented a step backward rather than forward in our understanding of space-time.

Of course, this is not the complete theory. The graviton-scattering amplitudes mentioned above are calculated perturbatively, and the background space-time represents the solution one is perturbing about. In a more fundamental, nonperturbative formulation of string theory, both the dynamical and the kinematical aspects of space-time should be derived concepts.

Unfortunately, there is still no consensus as to what is the most appropriate nonperturbative formulation of string theory. Various proposals have been made based on different fundamental structures. These include the collection of moduli spaces for all Riemann surfaces, completed in an appropriate way to include "infinite-genus" moduli spaces (Friedan and Shenker 1987), the space of two-dimensional quantum field theories (e.g., Banks and Martinec 1987), or the space of all maps from the interval (or circle) into a manifold (e.g., Witten 1986a; Bowick and Rajeev 1987a, b). I would like to describe an approach based on this last suggestion. It is closely related to a string field theory proposed by Witten (1986a), and is known as the purely cubic action. Unfortunately, it is not possible to describe in detail all aspects of this theory in the time available. Instead, I will review the main ideas involved, emphasizing those aspects of the theory that relate to space-time. For more details, the reader is referred to the literature.[1] Rather than try to motivate the form of the theory, I will simply present it as it is currently understood. The discussion will be limited to the bosonic string. It should thus be viewed as a model for a more realistic superstring theory.

A space-time, of course, consists of both a metric and a manifold. It is convenient to divide the problem of formulating a space-time-independent string field theory into two parts. I will first fix a manifold M and later argue that the theory is indepedent of the choice of M.

The basic ingredients of the theory are the following. Let $\mathcal{M}$ denote the space of maps from the interval $[0, \pi/2]$ to M. This is the space of (half) string configurations and will be called the *string space*. The *string field* Φ is defined on[2] $\mathcal{M} \times \mathcal{M}$. It is a density of weight 1/2 in each argument. Classically, Φ is complex valued and satisfies a reality condition. Quantum mechanically, it is operator-valued. The operations of multiplication and integration of string fields are defined as follows:

$$(\Phi * \Psi)(X, Y) \equiv \int \Phi(X, Z)\Psi(Z, Y)DZ$$

$$\int \Phi \equiv \int \Phi(X, X)DX,$$

where X, Y, and Z denote points of $\mathcal{M}$. These are clearly not the conventional definition of multiplication and integration of fields. They are more analogous to infinite-dimensional matrix multiplication and trace. These operations are (formally) well defined without any extra structure such as a metric or measure since the string fields are densities of the appropriate weight.

Having described the key ingredients, I can now write down the theory. The action is simply

$$S = \int \Phi * \Phi * \Phi. \tag{1}$$

Extremizing this action yields the equation of motion

$$\Phi * \Phi = 0. \tag{2}$$

In direct analogy to matrix algebra, the star product is not commutative in general, but does commute inside the integral. It follows immediately that the action S is invariant (to first order) under the transformation

$$\delta\Phi = \Phi * \varepsilon - \varepsilon * \Phi, \tag{3}$$

where ε is an arbitrary string field. This is the infinitesimal gauge invariance of the theory.

We now discuss some solutions to the field equation (2). It is convenient to reformulate the problem of finding solutions as follows. Let I denote the identity of the star product, i.e., $\Phi * I = I * \Phi = \Phi$ (roughly speaking, I is a delta function $I = \delta(X, Y)$). If one can find operators Q_L and Q_R satisfying ($Q = Q_L + Q_R$):

$$QI = 0, \tag{4a}$$

$$(Q_R\Phi) * \Psi - \Phi * (Q_L\Psi) = 0, \tag{4b}$$

$$\{Q, Q_L\} = 0, \tag{4c}$$

then one can show that $Q_L I$ is a solution to (2). Furthermore,

$$Q_L I * \Phi + \Phi * Q_L I = Q\Phi. \tag{5}$$

So, expanding about this solution, $\Phi = Q_L I + A$, the action becomes

$$\tilde{S} = \int \frac{3}{2} A * QA + A * A * A.$$

If Q is a second-order differential operator, $\tilde{S}$ essentially takes the form of a conventional field-theory action. However it has been obtained from the purely cubic action (1) by shifting about a solution to the classical field equations. It has been argued that all solutions to (2) can be expressed in the form $Q_L I$ for some Q_L (Woodard 1988). The linearized equation of motion about this solution is $QA = 0$, and the linearized gauge invariance is $\delta A = Q\varepsilon$.[3]

Now we will consider several examples. Consider a D-dimensional flat space-time, and set $Q_L = \int_0^{\pi/2} j^{BRST}(\sigma) d\sigma$ where j^{BRST} is the BRST current of the first-quantized (open) string in this background. Explicitly,

$$j^{BRST} = c^0 (\eta^{\mu\nu} P_\mu P_\nu + \eta_{\mu\nu} X'^\mu X'^\nu) + c^1 (P_\mu X'^\mu) + \text{pure ghost term},$$

where the σ-dependence is suppressed, a prime denotes derivative with respect to σ, c^0 and c^1 are the two components of the ghost field, and $P_\mu(\sigma) = -i\delta/\delta X^\mu(\sigma)$. If one defines Q_R by a similar integral over the right half of the string, then one can show that (4a,b,c) are satisfied if and only if $D = 26$. So, the critical dimension of the first-quantized string in flat space-time is associated with a particular solution to the field equation $\Phi * \Phi = 0$. Expanding about this solution ($\Phi = Q_L I + A$), one finds that the linearized fluctuations correspond to the spectrum of states of the first-quantized open string. Futhermore, the scattering amplitudes calculated in the string field theory correctly reproduce the standard results to all orders in perturbation theory (Giddings, Martinec, and Witten 1986).

This is a crucial test for any proposed nonperturbative formulation. The best understood aspect of string theory is still its perturbative expansion. General arguments have been given that, for the superstring, each term in this expansion should be finite. But no complete proof is yet available.[4] Clearly, if a nonperturbative approach fails to reproduce the usual perturbation expansion, there is little reason to expect the new theory to be well behaved.

Note that one is using a notion of time provided by the solution $Q_L I$ to construct the quantum theory perturbatively. The current j^{BRST} explicitly contains $\eta_{\mu\nu}$, which gives us a notion of asymptotic states, positive frequency, etc. Actually, this is not unique to string theory. Even in general relativity, the quantum perturbation theory (to the extent that it can be defined) uses the space-time structure provided by the solution about which one is perturbing.

As a second example, define Q_L and Q_R as before except with the flat metric $\eta_{\mu\nu}$ replaced by a curved metric $g_{\mu\nu}$. If $D = 26$, the curvature is much less than the Planck curvature, and the Einstein vacuum field equation is satisfied; then one can show $Q_L I$ is an approximate solution to $\Phi * \Phi = 0$. In this way, one recovers Einstein's theory as the low-energy limit of string theory. One can also add matter fields such as an electromagnetic field. In this case, $Q_L I$ is an approximate solution provided the above conditions are satisfied with Einstein's vacuum equation, replaced by the Einstein–Maxwell equations.

The above solutions are associated with 26-dimensional space-times. To make contact with observations, one would have to assume that most of the dimensions are compactified as in a standard Kaluza–Klein theory. However, there are other solutions associated with space-times of any dimension.[5] For example consider a D-dimensional flat space-time. Set

$$Q_L = \int_0^{\pi/2} c^0(\sigma) P^\mu(\sigma) P_\mu(\sigma) d\sigma,$$

and define Q_R similarly. These operators satisfy (4a,b,c) with no restriction on D. So $Q_L I$ is a solution. One can show that linearized fluctuations about this solution take the form of infinite-dimensional plane waves $e^{i \int_0^\pi k_\mu(\sigma) X^\mu(\sigma) d\sigma}$, with $k^2(\sigma) = 0$. The momentum associated with the center of mass, $P_0 = \int_0^\pi P(\sigma) d\sigma$, is clearly never space-like on these states; so, unlike the previous solution, there is no tachyon. Unfortunately, the scattering theory does not seem to be well behaved.

All of the solutions we have mentioned so far have a well-defined space-time metric. They are thus analogous to solutions of Kaluza–Klein theory of the form $M \times K$. However we argued earlier that there should exist solutions with no space-time. A simple class of solutions of this type are obtained from

$$Q_L = \int_0^{\pi/2} c^0(\sigma) f(\sigma) d\sigma,$$

where f is any function satisfying $f(\pi - \sigma) = f(\sigma)$ and $f(\pi/2) = 0$. The special case $f = 0$ corresponds to the solution $\Phi = 0$. This is clearly invariant under the transformation (3) for all ε. It is unusual in a field theory to have a state that is left invariant by all local gauge transformations. It can be viewed as a new type of state of unbroken symmetry. (This may be related to recent discussions of a state of unbroken diffeomorphism invariance in string theory, see Witten 1988). All the solutions with $f \neq 0$ turn out to be gauge equivalent.[6] In fact, one can show that all the linearized fluctuations about a solution with $f \neq 0$ are pure gauge. This shows that the space of solutions to $\Phi * \Phi = 0$ is disconnected. The solution with $f \neq 0$ cannot be continuously joined to any other solution.

Now we turn to the question of whether string field theory is independent of space-time topology. (For more details see Horowitz and Witt 1987). The only place that the manifold M entered the previous discussion was in the definition of the string space $\mathcal{M}$. To see whether this is essential, we must specify the class of strings more precisely. Sobolev spaces are convenient for this purpose. Recall that, for maps X^μ from the interval $[0, \pi/2]$ into R^n, the Sobolev space of weight s is defined by expanding $X^\mu(\sigma) = \sum x_n^\mu e^{4in\sigma}$ and setting

$$\Omega_s R^n \equiv \left\{ (x_n^\mu): \sum_{n=-\infty}^{\infty} |x_n^\mu|^2 (1 + n^2)^s < \infty \right\}.$$

This is a Hilbert space for each real number s. The Sobolev embedding theorem says that $\Omega_s R^n$ is contained in C^k for $s > k + 1/2$. So for $s > 1/2$, the space $\Omega_s R^n$ contains only continuous strings.[7]

We now wish to extend this definition from R^n to a general manifold M. To do so, recall that every manifold can be embedded in R^n for a sufficiently large n. We can thus define

$$\Omega_s M = \left\{ X^\mu(\sigma) \in \Omega_s R^n : X^\mu(\sigma) \in M \text{ almost everywhere} \right\}.$$

This is no longer a Hilbert space since we have lost the linear structure. For $s > 1/2$, it has been shown that $\Omega_s M$ is an infinite-dimensional manifold which is locally based on a Hilbert space. Objects of this type are known as *Hilbert manifolds*. In many respects, Hilbert manifolds are much simpler than finite-dimensional ones. For example, the following theorem has been established (Eells and Elworthy 1970).

Theorem: *If two Hilbert manifolds are homotopic, then they are diffeomorphic.*

This is a remarkable result, which has no analog in finite dimensions (even when restricted to manifolds of the same dimension). This theorem has several consequences, which include: (1) Every Hilbert manifold has a unique differentiable structure; (2) the infinite sphere, i.e., the set of all unit vectors in a Hilbert space, is diffeomorphic to the Hilbert space; and (3) the tangent bundle of a Hilbert manifold is diffeomorphic to the manifold. For our purposes, the most important consequence is (4): If M and N are homotopic then $\Omega_s M$ and $\Omega_s N$ are diffeomorphic. In particular,

$$\Omega_s M \equiv \Omega_s(M \times R^n).$$

So, in this sense, string space is independent of the dimension of space-time. But, since we have been assuming $s > 1/2$, the strings are continuous. So $\Omega_s M$ does depend on some of the topology of M.

However, there are several reasons why discontinuous strings should be included in the string space. Since string field theory must at least reproduce the perturbative scattering amplitudes about each background, the domain of the string field must include the domain of the first-quantized wave functions for each background. But the first-quantized string is itself a two-dimensional field theory. Thus, one can argue quite generally that the configuration space should include distributional strings (Glimm and Jaffe 1981). More specifically, for the case of 26-dimensional flat space, one can show that continuous strings form a set of measure zero in the first-quantized inner product. (The proof is similar to the demonstration that, for a simple harmonic oscillator, paths of finite action have zero measure). Finally, even if the domain of a first-quantized string in a particular space-time background could be shown to involve only continuous strings, one can still argue that the domain of the string field theory should be larger than the first-quantized configuration space for any particular background.

In terms of Sobolev spaces, allowing discontinuous strings means considering $\Omega_s M$ for $s < 1/2$. In this case, one can show that $\Omega_s M$ is contractible for any M.[8] If this space were a Hilbert manifold, then the above theorem would show that it is diffeomorphic to a Hilbert space, and hence completely independent of the topology of M. Unfortunately, this is not yet known. It is clear from the definition that $\Omega_s M$ is a metric space (since it is a subset of a Hilbert space), but it may not have a smooth differentiable structure. Nevertheless, the fact that it is contractible shows that it retains very little information about the topology of M. Assuming that it is a manifold, we can take the string space $\mathcal{M}$ to be the underlying manifold of a Hilbert space. This is one definition of R^∞, and is perhaps the simplest infinite-dimensional manifold. With this choice, the purely cubic action is independent of both the metric and topology of space-time.

At this stage, one may reasonably ask if too much structure has been lost. By using discontinuous paths, one has lost the distinction between open and closed strings. Yet, it is well known that these two types of strings have very different physical properties. The resolution is that the type of string should be encoded into the solution. As we have discussed, it is known how to construct a solution which reproduces the open string states and scattering amplitudes. It is not yet known if there is an analogous solution that describes the closed string,[9] or whether the closed string will involve more fundamental changes in the theory (Strominger 1987b). Even more surprisingly, one has also lost any notion of a string! The space of square integrable maps from a higher-dimensional object into a manifold is also contractible, and hence can be identified with R^∞. The notion of a string must also come from the solutions. It is conceivable that there are other solutions whose linearized fluctuations correspond to oscillations of higher-dimensional objects, but I believe this is unlikely.

Very little is known about how to quantize the purely cubic action nonperturbatively. A standard canonical quantization does not seem possible since there is no initial-value formulation, and hence no canonical formulation of the theory. (However, it may be possible to have a canonical quantization based on the space of solutions). An alternative approach is to borrow ideas from ordinary field theory and construct an effective action. Formally, this can be defined as follows. One starts with the generating function

$$e^{iW[J]} = \int D\Phi e^{iS[\Phi]+i\int J*\Phi},$$

where the integral is over all string fields Φ, and J is an appropriate source. Let $\Phi_B = \delta W/\delta J$, and invert to express J in terms of Φ_B. Then the effective action is

$$\Gamma[\Phi_B] = W[J] - \Phi_B J.$$

The full quantum equation of motion would then be

$$\delta\Gamma/\delta\Phi_B = 0.$$

Unfortunately, outside of perturbation theory, it is not clear how to interpret these equations.

However, this approach is potentially of interest for the following reason. There has recently been a great deal of discussion about the effects of "baby universes branching off from a parent universe" in quantum gravity. Indeed, Coleman has argued at this meeting that this process might provide an explanation for the vanishing of the cosmological constant. These effects are usually calculated using functional integrals over Euclidean four-geometries. But it has been argued that a complete description of this process requires a third-quantized theory of gravity (Giddings and Strominger 1989). A theory of this type is likely to be considerably more complicated than string field theory. However, since string field theory may be independent of space-time topology, third quantization may not be needed. A string quantization based on functional integrals will include an integral over all Φ, which should include an integral over all space-time metrics *and topologies*. Thus, baby universes would be automatically included. This should allow one to calculate all effects of baby universes without resorting to third quantization.

A more unconventional approach to nonperturbative quantization is the group theory approach discussed at this meeting by Isham. This may be particularly appropriate for the purely cubic action, considering the fact that this theory is fundamentally more algebraic than geometric in structure.

In addition to understanding how to quantize the cubic action nonperturbatively, there remain many open problems, which can roughly be divided into three classes. The first are technical problems associated with the proposal described here. These include showing that $\Omega_s M$ for $s < 1/2$ is a manifold (or at least understanding its structure if it is not a manifold). It is also important to understand how to rigorously define densities of weight $1/2$ on infinite-dimensional spaces. If it turns out that there are inequivalent ways to do this, it would not be unreasonable to include string fields of all possible types. The different possibilities might be interpreted as corresponding to different choices of some background structure.

A second class of open problems concerns how to extend the theory described here to closed strings and superstrings. Some progress in these directions have been made (Strominger 1987b; Witten 1986b; Qiu and Strominger 1987; Awada 1987; Martin 1988), but some problems (e.g., Wendt 1989) remain to be overcome.

Finally, and perhaps most important, is the issue of whether the topology-independent string field theory discussed here can reproduce all the known effects of space-time topology in perturbation theory about certain solutions. For example, if $\pi_1(M) \neq 0$, then there should be additional states of the (first-quantized) closed string corresponding to winding around the noncontractible loops. Before one can investigate how these additional states arise from a topology-independent formalism, one needs a better understanding of closed string field theory.

Despite these open problems, the general framework described here has broad implications for thinking about quantum gravity. Since space-time plays no role in the fundamental theory, questions such as "Does space-time undergo Planck-scale fluctuations in topology?" are meaningless, except in a semi-classical approximation about a solution with a well-defined space-time. Similarly, since there is no fundamental space-time, there is no fundamental time, so the question (Penrose 1981): "Are the laws of quantum gravity time-reversal invariant?" is also not well defined, except semi-classically. A physical theory without time can be viewed as the logical extension of earlier-established theories, which have continually degraded the importance of time. Special relativity replaced absolute time with only a family of preferred inertial frames. General relativity replaced these inertial frames with only a local time along each observer's world-line. It is perhaps natural for quantum gravity to remove time altogether.

Of course, removing time is only the first step. One must next understand how to ask and answer physical questions in this theory. Our everyday experience clearly shows that it is necessary to be able to recover space-time under certain conditions. But the most interesting predictions of quantum gravity are likely to come from the sector of the theory without space-time. Understanding what these are is perhaps one of the greatest challenges facing us.

Acknowledgments. I wish to thank the organizers of the Osgood Hill conference for the opportunity to participate in this lively meeting. This work was supported in part by the National Science Foundation under Grant No. PHY85-06686 and by the Alfred P. Sloan Foundation.

NOTES

[1] The main ideas were presented by Horowitz, Lykken, Rohm, and Strominger (1986). Further details were discussed by many people, including Strominger (1987a), Woodard (1987), Romans (1988), and Horowitz and Martin (1988). A similar approach was also investigated by Hata, Itoh, Kugo, Kuni-

tomo, and Ogawa (1986), and Kikkawa, Maeno, and Sawada (1987).

[2] It also depends on a (space-time-independent) anticommuting ghost field, which we will suppress.

[3] If one tries to derive this by expanding the general gauge transformation (3) about the solution $Q_L I$, one encounters an apparent sign discrepancy with (5). The resolution is that, due to the ghosts, the string fields Φ are Grassmann odd while the gauge parameters ε are Grassmann even. Thus, the sign in (5) is indeed reversed for the gauge parameters.

[4] However, it should be noted that there is no argument that the perturbation expansion converges. Indeed there is evidence that it does not (Gross and Periwal 1988).

[5] For further discussion of the next two solutions, see Horowitz et al. 1988.

[6] Even though one can make f arbitrarily small by a gauge transformation, it can never be set equal to zero since the solution $\Phi = 0$ is gauge invariant.

[7] Actually, these are continuous half strings. Since we have defined the string field on $\mathcal{M} \times \mathcal{M}$ with no restriction that the two half strings join, most strings are discontinuous at the midpoint. The justification for this will be given shortly when we argue that strings should in fact be discontinuous.

[8] If one does not wish to introduce a manifold structure on M, one can consider the case where M is simply a metric space, i.e., a set with a distance function $\rho(x, y)$. Since every metric space can be embedded in a Hilbert space, the argument of Horowitz and Witt 1987 can be used to show that $\Omega_s M$ for $s < 1/2$ is again contractible.

[9] This could perhaps be viewed as a solution to a purely cubic version of the Lykken–Raby (1986) theory. It is known that the original theory did not reproduce the closed-string scattering amplitudes, but a purely cubic version may be different due to associativity anomalies (Horowitz and Strominger 1987).

REFERENCES

Awada, M. (1987). "Pregeometry of Open Superstring Field Theory." *Physics Letters B* 188: 421–424.

Banks, T., and Martinec, Emil (1987). "The Renormalization Group and String Field Theory." *Nuclear Physics B* 294: 733–746.

Bowick, M. J., and Rajeev, S. G. (1987a). "String Theory as the Kähler Geometry of Loop Space." *Physical Review Letters* 58: 535–538.

———(1987b). "The Holomorphic Geometry of Closed Bosonic String Theory and Diff S'/S'." *Nuclear Physics B* 293: 348–384.

Eells, J., and Elworthy, K. D. (1970). "Open Embeddings of Certain Banach Manifolds." *Annals of Mathematics* 91: 465–485.

Friedan, Daniel, and Shenker, Stephen (1987). "The Analytic Geometry of Two-Dimensional Conformal Field Theory." *Nuclear Physics B* 281: 509–545.

Giddings, Steven B., and Strominger, Andrew (1989). "Baby Universes, Third Quantization, and the Cosmological Constant." *Nuclear Physics B* 321: 481–508.

Giddings, Steven B.; Martinec, Emil; and Witten, Edward (1986). "Modular Invariance in String Field Theory." *Physics Letters B* 176: 362–368.

Glimm, J., and Jaffe, A. (1981). *Quantum Physics: A Functional Integral Point of View*. New York: Springer-Verlag.

Gross, David J., and Periwal, Vipul (1988). "String Perturbation Theory Diverges." *Physical Review Letters* 60: 2105–2108.

Hata, Hiroyuki; Ito, Katsumi; Kugo, Taichiro; Kunitomo, Hiroshi; and Ogawa, Kaku (1986). "Pregeometrical String Field Theory: Creation of Space-Time and Motion." *Physics Letters B* 175: 138–144.

Horowitz, Gary T., and Martin, Stephen P. (1988). "Conformal Field Theory and the Symmetries of String Field Theory." *Nuclear Physics B* 296: 220–252.

Horowitz, Gary T., and Strominger, Andrew (1987). "Translations as Inner Derivations and Associativity Anomalies in Open String Field Theory." *Physics Letters B* 185: 45–51.

Horowitz, Gary T., and Witt, Donald M. (1987). "Toward a String Field Theory Independent of Spacetime Topology." *Physics Letters B* 199: 176–182.

Horowitz, Gary T.; Lykken, Joseph; Rohm, Ryan; and Strominger, Andrew (1986). "Purely Cubic Action for String Field Theory." *Physical Review Letters* 57: 283–286.

Horowitz, Gary T.; Morrow-Jones, Jonathan; Martin, Stephen P.; and Woodard, R. P. (1988). "New Exact Solutions for the Purely Cubic Bosonic String Field Theory." *Physical Review Letters* 60: 261–264.

Kikkawa, Keiji; Maeno, Masahiro; and Sawada, Shiro (1987). "String Field Theory in Curved Space." *Physics Letters B* 197: 524–530.

Lykken, Joseph, and Raby, Stuart (1986). "Non-Commutative Geometry and the Closed Bosonic String." *Nuclear Physics B* 278: 256–275.

Martin, Stephen P. (1988). "Conformal Field Theory Representation and Background Independent Action for Superstring Field Theory." *Nuclear Physics B* 310: 428–460.

Penrose, Roger (1981). "Time Asymmetry and Quantum Gravity." In *Quantum Gravity 2: A Second Oxford Symposium*. Christopher J. Isham, Roger Penrose, and Dennis W. Sciama, eds. Oxford: Clarendon, 245–272.

Qiu, Zongan, and Strominger, Andrew (1987). "Gauge Symmetries in (Super) String Field Theory." *Physical Review D* 36: 1794–1799.

Romans, L. J. (1988). "Operator Approach to Purely Cubic String Field Theory." *Nuclear Physics B* 298: 369–413.

Strominger, Andrew (1987a). "Point Splitting Regularization of Classical String Field Theory." *Physics Letters B* 187: 295–302.

———(1987b). "Closed Strings in Open-String Field Theory." *Physical Review Letters* 58: 629–632.

————(1987c). "Closed String Field Theory." *Nuclear Physics B* 294: 93–112.

Wendt, C. (1989). "Scattering Amplitudes and Contact Interactions in Witten's Superstring Field Theory." *Nuclear Physics B* 314: 209–237.

Witten, Edward (1986a). "Non-Commutative Geometry and String Field Theory." *Nuclear Physics B* 268: 253–294.

————(1986b). "Interacting Field Theory of Open Superstrings." *Nuclear Physics B* 276: 291–324.

————(1988). "Topological Sigma Models." *Commun. Math. Phys.* 118: 411–449.

Woodard, R. P. (1987). "Surface Terms in the Cubic Action." *Physical Review Letters* 59: 173–175.

Woodard, R. P. (1988). Private communication.

Discussion

ASHTEKAR: In the first half of your talk, what structure does M have? Is it more than a point set?

HOROWITZ: I am using the manifold structure.

ASHTEKAR: Differential manifold?

HOROWITZ: Yes.

ASHTEKAR: Then, I have to fix a dimension or I can't say anything.

HOROWITZ: You fix everything about M, you fix it's dimension, topology, everything.

ASHTEKAR: And the dimension can be arbitrary?

HOROWITZ: Yes.

DEWITT: In the definition of the star product and integration, are X and Y maps?

HOROWITZ: They are points in the space of maps.

DEWITT: So they are maps.

HOROWITZ: They are maps. True. But I integrated over maps, it's like a path integral.

SMOLIN: Gary, would you care to comment on the extent to which those integrals are well defined without a background space-time?

HOROWITZ: Okay; I said very glibly that I needed my Φ to be a density of weight 1/2 in order for this to be well defined. If you push me and ask what do I mean by that rigorously or mathematically, I have to confess that I don't know in this infinite-dimensional space that I want to work in. However, there is a very nice regularization of this $*$-product, which was proposed by Richard Woodard and Mark Srednicki. What they do is essentially what you do in your loop-space approach. They discretize the σ space, the space of parameter values, to reduce Φ to functions on a finite-dimensional space; then this is completely well defined, it's independent of any background space-time structure.

KUCHAŘ: Do you know if every solution is of the form $Q_L I$?

HOROWITZ: I do not know whether every solution can be written in this form. I know that a large class can.

KUCHAŘ: Is there any conjecture about it?

HOROWITZ: There is. In fact Richard Woodard has some ideas on how one might try to prove that all solutions can be written in this form. So I guess it's a possible conjecture, one that maybe Richard would want to make, that all solutions can be written in this form.

KUCHAŘ: You started with a theory that had no metric in it. You wrote down the equations, you rewrote them in a certain form, and then you found that, *if* you introduce the metric, you can construct a particular solution of those equations, which didn't have a metric. This somehow doesn't say that the metric emerges from the metric-free equations. It only shows that there is a particular solution of these equations that is constructed from a metric. So the metric is put in by hand, it's not coming necessarily from "above." Am I right or not?

HOROWITZ: Well, I wouldn't quite put it that way. I would say that the solutions with a metric are analogous to the product solutions of the form $M \times K$ in the Kaluza–Klein theory. In other words, there are particular classes of solutions in which one has a well defined space-time metric. There are other solutions for which one does not.

ASHTEKAR: Gary, where did Newton's constant come from?

HOROWITZ: There's a dimensional parameter in the formula for Q, which I did not write down explicitly, and that dimensional parameter is what sets the scale, and you essentially fix it to be the observed Newton's constant.

ASHTEKAR: You only get something like a length, so one doesn't distinguish between Planck's constant and G and so on.

HOROWITZ: There is basically one dimensional constant in the theory, so that sets the scale and everything else can be related to that.

SORKIN: I have a short question. Is there some sense in which the solutions could be pure gauge?

HOROWITZ: There is a sense in which solutions are pure gauge.

SORKIN: Gauge equivalent to another.

HOROWITZ: Right, there are such solutions. There are gauge transformations on the space of solutions; you can ask which ones are gauge equivalent to other solutions. As far as I know, none of the solutions I've discussed are gauge equivalent to each other, but there certainly are ones that you could construct that would be gauge equivalent to them.

HARTLE: Are there any solutions that are cosmological solutions?

HOROWITZ: Yes, I gave an example. If you take a radiation-dominated Robertson–Walker space-time, where the curvature is small compared to the Plank scale, you can write down an approximate solution that would be good down to the Plank scale.

DESER: Just to keep you honest, you ought to mention that this is only for open strings.

HOROWITZ: I would say that the theory that I have been describing is a theory of open and closed strings, which is why we're able to change the space-time metric. It is true that one has difficulty constructing a purely closed string field theory. I don't know if the difficulty is fundamental.

GELL-MANN: There are beautiful technical tricks here that were all invented for the open string, and most of the attempts to make a theory with purely closed strings have run up against the fact that all those technical tricks don't make any sense if there aren't any open strings. So, nobody knows how it will work.

DEWITT: Saying that the theory is independent of space is actually too strong a statement until you know a little bit more about the space of solutions. You've given us some examples, but some very important solutions were ones that had a background metric of some kind, even with the appropriate signature. And also you don't know yet what your ground state is.

HOROWITZ: I think all of those questions have to do with whether string theory describes the real world. We all agree that we're living in a universe that is very well approximated by a four-dimensional space-time of Lorentzian signature and all of that. If the theory is constructed nonperturbatively and has a ground state that has no space-time interpretation, then one should throw it out; it's not describing our universe. But I agree, what I would like to say is that fundamentally "time" is not there.

DESER: But you can hope to construct it.

HOROWITZ: Only semi-classically.

UNRUH: You *have* to be able to do that. But it looks like you may not even be able to do that generically.

HOROWITZ: But I *don't* want "time" for any solution. I only want time to come out from solutions which describe space-time. Certainly I need to know why the universe is the way it is. Why do we see four-dimensional space-time? So that would be the statement that there is a unique ground state that is...

GELL-MANN: Ground state, meaning something funny?

HOROWITZ: Well, classically meaning one of these solutions.

WOODARD: How do we know how to get the quantum ground state? I mean, the action for all solutions is zero.

HOROWITZ: Well, one should include surface terms. Look, we know so little about the nonperturbative formulation that I don't want to...

WOODARD: We know a lot. We know they're all characterized by certain representations.

HOROWITZ: There is a way of classifying the solutions. This is in response to the earlier question whether all solutions could be put into this $Q_L I$ form, and I know that you have some ideas on that, so maybe in the discussion session you could say something about that.

ASHTEKAR: It seems to me that there is a big difference between the viewpoint that you have proposed here and the viewpoint that people—some of us at least—have in the canonical-quantization program of gravity. The difference seems to be the following: We feel that, while there will be many states in quantum gravity that have no classical analog at all, there really *is* a microscopic structure of space-time. So, even though today on the macroscopic scale space-time looks continuous, and so on, if we did get to the microscopic level, it wouldn't be so simple. Whereas, in your description, space-time always arises out of a classical solution to the string field equations, and therefore the statement is that, either we have these other things, in which space-time has no meaning at all, or the space-time is as we know it today; there is no room for discussing things like: what is the "microscopic structure of space-time?"

HOROWITZ: Well, I don't think that's quite fair. Let me try this line of argument on you. Suppose you wanted to probe the structure of space-time on a very short scale today: one would imagine sending in particles of very high energy and looking at the results. Well, we certainly know that, in string theory, if you want to do a calculation of what happens when you send in a particle of very high energy, you would get something very different from what you get in ordinary field theory; because, when particles collide, all the other modes of the string will come into play and will affect the answer. One could then interpret that as resulting from some difference in space-time, or something like that;

but I don't have a clear idea of what it would mean to have this funny structure on space-time at small scales.

FRIEDMAN: It does seem, though, that in your solutions you at least have things you interpret as classical metrics; for other solutions, there is no such interpretation. Can you talk about a quantum solution in which the metric that you feed in has a bunch of nearby metrics, so that the space-time looks foam-like? It doesn't seem clear to me from the presentation. Once you feed in an arbitrary metric in your space of solutions, then I don't see why you don't talk about fluctuations of that metric and then fluctuations of topology.

HOROWITZ: Well, if I understand what you're suggesting—that one take a classical metric that has small-scale, very highly curved, Planck-size worm holes and things like that—then I guess I'm saying: I don't believe that is a reasonable picture. It's not that one has classical metrics on classical manifolds with very, very strong curvature. I think the whole idea of a metric and a manifold will only make sense on scales that are large compared to the Planck length.

FRIEDMAN: I understand what you're saying. It just looks like that from the solution.

HOROWITZ: I can't get a solution like that because of this assumption that the curvature is everywhere small compared to the Planck curvature.

SMOLIN: The structure that you end up with is very general. You have a function space on a space of maps that is to satisfy certain properties, and on it there could be a certain algebraic structure with non-commutative multiplication and a trace. So, one can imagine many other things: not only spaces of maps of loops into manifolds satisfying the structure, but spaces of maps on the spaces of loops, etc. So, there could be many other types of solutions to this kind of algebraic structure.

HOROWITZ: I agree. In some sense, strings no longer play such a fundamental role in the evolved picture: but I guess I would agree that one still has to get down to experiment, and I think quantum-gravity people have long since gotten rid of that restriction. But one has to relate the theory to something, and what we're trying to do here is to relate it to perturbative string-theory calculations, which agree with experiment in the sense that they do correspond to gravity. Perturbative scattering of gravitons in string theory produces exactly what you would get if you calculated it in Einstein's theory.

DESER: In 26 dimensions.

SMOLIN: My conjecture is that when we understand the space of solutions to the constraints of quantum gravity, there will be a structure of this form, i.e., a space of maps of some space of loops on some manifold that has the appropriate structure. They will be a solution to, perhaps, your classical equation; that is, every state of quantum gravity may be a solution to your equations. The algebraic structure will come from the knot-theoretic structure of the solutions, and a relation between string theory and the quantum-gravity theory will come from the representation theory of the braid group in the Virasoro algebras.

HOROWITZ: Well, of course, it would be very interesting if that were true. What one would have to wonder is: where are the analogous solutions in your theory that would describe scattering or have fields other than gravity?

SMOLIN: Well, it includes that, and the idea is that in some sense they come from algebraically representing nontrivial topology.

KUCHAŘ: In the first-quantized version of the theory, the gravity field was just one of the many excited modes. In your construction, you start with a theory that doesn't have any space-time structure, and then you expand the string field about that solution to the theory based on the gravity field. There is no mention in the solution of the other fields. So, is the gravity field somehow preferred, or can you construct solutions using the other fields as well?

HOROWITZ: Sure. I mentioned that you could put in the electromagnetic field, for example. You can put in background fields associated with the other modes of the string as well. There is nothing that's picked out. I used gravity because that's what people are interested in.

KUCHAŘ: Is there then some solution which doesn't need to talk about individual fields, but somehow takes an overall picture of all excitations of the first-quantized string, in which you do not distinguish between the fields?

HOROWITZ: I gave one example, which is just this function of the ghosts. I constructed Q, which had just the ghosts and no X's at all. So there is no way that one could construct space-time or any of the usual components of the string field about that solution. So I think it is true that you have solutions where you don't have anything like this usual decomposition.

KUCHAŘ: So all the fields have a semi-classical meaning only?

HOROWITZ: Yes, the string field is the fundamental object. It lives on this big manifold; you can decompose it into the usual field on space-time only about particular solutions.

STACHEL: The reason we choose to study the initial value problem is because it gives us a handle on the data needed to specify a solution. It's very closely connected to the question of the degrees of freedom of the system. Is there anything analogous to be expected in this kind of theory?

HOROWITZ: As far as I know, there's been very little progress on that. I think it's a very interesting question, and because the structure looks so algebraic, I would think that the algebraic approach would be more promising than to go back to the differential-equation type of evolution. But something like a representation theory could classify all solutions, how many there are, etc. I think that would be wonderful.

ROVELLI: One of the main messages that I got from your talk, at least the second part, is that there is this formulation of the classical theory that depends very little on the structure that you're starting from. You stressed the fact that it depends very little on the target space, and that the action is just Φ^3. I think that you could have started from anything, e.g., the position of a rabbit in this room or something like that. If I understand what you're saying, the physical content of all that is somehow coded in this little $*$. So, my question is: first, I would like you to comment about that; and second, how much does the definition of this $*$ depend on this being a functional of loops, and how would it be affected if you started from functionals of loops with a different target space with different dimension?

HOROWITZ: I was careful when I first set up the theory to define my string space to be simply the space of half strings, so that my $*$-product never required me to use any extra structure on one of these half-string spaces. The more conventional picture is to use full strings, and then you have to define a midpoint, you have to know the right and left halves of the string and it clearly would make a difference whether you're mapping strings into one dimension or another. But the way I set it up was that you have these two half-string spaces, and then I don't think there is much structure in the individual half-string spaces.

ANDERSON: I too think that the most important thing you said was how you get this independence of the target space. As you know, I have an argument that, semi-classically, you can get a notion of extended ho-motopy or discontinuous paths, and I would like to ask for your remarks on that, and about how that affects your argument?

HOROWITZ: I guess the key point is "semi-classically." As I un-derstand it, you take a path that is actually a smooth path and look at deviations from it. I think that starting with a smooth path is the extra structure that allows you to extend this homotopy notion and doesn't al-low me to do so because I don't have it. I'm just taking all paths that are square integrable.

ANDERSON: But say that square integrability—at least where it arises in my argument—is: we do this Gaussian approximation, expand-ing in deviations. You're expanding deviations too. You've got a back-ground solution.

HOROWITZ: No, no. It's a very different thing I'm expanding about. You're actually expanding the paths; I'm expanding in fields on that path space.

UNRUH: You emphasized in setting up the ∗-product that it was just like matrix multiplication. Can you do this for a finite-dimensional matrix space?

HOROWITZ: I emphasized that, but there are unfortunately, a few subtleties. For example, these fields are supposed to be Hermitian, and you can not have a Hermitian matrix squared and set it equal to zero and get anything other than zero. So there is no way that I could realize this just in terms of finite-dimensional matrices. Now the thing that saves you is the ghosts, in the sense that these fields not only are Hermitian, but they are also Grassmann-valued. So, it's the combination that allows this to work. Now, that's in some sense what one gets in this discretized version, so I think the answer would be: at least you've gone from Φ on an infinite-dimensional space to fields on a finite-dimensional space. Now, whether you could actually get it down to finite-dimensional matrices, I'm not so sure. But I agree that you can at least go down one step by using this regularization that Richard Woodard and Mark Srednicki developed. So your fundamental string fields now just depend on $\mathbf{R}^n$, the value of X on discrete points, and you can then set up the same ∗-algebra.

ASHTEKAR: It may be possible to do it, not in complex numbers but in quarternions.

HOROWITZ: Because of the noncommutativity or something?

ASHTEKAR: Yes.

MASON: I didn't notice any $E_8 \times E_8$ in there; does that disappear or is it somehow suppressed?

HOROWITZ: Neither. I was describing the bosonic string, because that theory is simpler and I have to suppress less in order to present it. It's the superstring that picks out $E_8 \times E_8$, and that's still very much around in current discussions of superstring theory. It has been extended somewhat in the following two senses: initially, say four years ago, it was believed that there were only two theories in a ten-dimensional flat space-time that were consistent: $E_8 \times E_8$ and the $SO(32)$ theory. It has since been noticed that there are other theories, e.g., $SO(16) \times SO(16)$, which make sense and are consistent in ten-dimensional flat space. It has also been discovered that there are many ways of changing the gauge group upon compactification. So you can get much larger and smaller gauge groups in lower dimension; and, as I briefly mentioned, there are these new ideas about how to construct four-dimensional string theories, in which you don't have to mention internal compactification. And those theories have lots of different possibilities for the gauge group. My reaction to all of this is that I think various claims for the uniqueness of string theory are very premature. It's still conceivable that there's one grand superstring field theory, or some other nonperturbative formulation, and all of these different theories people have constructed over the past couple of years will be viewed as just different classical solutions. That has not been shown yet, and even if it had been, one could ask, why am I working with just the scalar fields in this string space and why not with spinor, vector, or tensor fields? So there are lots and lots of ways of extending string theory. My interest in it is not so much uniqueness; I believe it will provide a quantum theory of gravity.

GELL-MANN: But Gary didn't give the answer to your question, Lionel. The reason he didn't apply the method to that theory, which has only closed strings, is because QFT has never been applied to such theories.

HOROWITZ: Well, the $SO(32)$ theory is a theory of open strings.

MASON: You said two things about curved-space solutions. The first thing you said was that it's an approximate solution, and you said that the curvature must be small. I deduced from that that, if the solution were exact, the curvature would be zero.

HOROWITZ: Flat space is an exact solution. In 26 dimensions, you can find other approximate solutions, but I think that's all one could expect. Classical space-time is supposed to be an approximation. So, I'm not upset that it is giving only approximate solutions.

UNRUH: By approximate, you mean that there is some condition that is a huge awful mess to write down such that, if the curvature is very small, it looks like $G_{ab} = 0$. There is some exact solution to which this is an approximation, and that solution has nonzero curvature.

HOROWITZ: That's right.

WALD: I'm very uneasy about the apparent disappearance of all structure in this, and I'm looking for a place where that structure may be hidden; a good candidate-place would seem to be the choice of this space of densities of weight 1/2 that you are working with in string field theory. And presumably this choice is equivalent to a choice of measure over the manifold you are dealing with. And there are zillions of inequivalent ones. The choice seems to determine the theory. You really have to pick one out. I can see where you can pick one out using the space-time structure that you started with; if you're going to throw that away, I guess I am concerned that there is either something that hasn't been specified in terms of this, or you really have to put back in the space-time structures, or equivalent structures, to get out the theory.

HOROWITZ: I certainly agree that that is a valid worry. I have thought a little bit about how to define mathematically the functions that I need to work with. And, really, the only thing I know at this point, as I mentioned during the talk, is this lattice regularization that Richard Woodard and Mark Srednicki proposed; and that, of course, does give us a way of defining everything precisely for a finite N, an arbitrarily large N, and things like that. I agree that it's not satisfactory. One has to look into it.

ANDERSON: Have you tried using some kinky topology on the target manifolds to see what effect that has on the solutions?

HOROWITZ: You mean, let it be a manifold but with any topology?

ANDERSON: Even the flat spaces. There is topology on them.

HOROWITZ: Yes.

ANDERSON: How does that affect things?

HOROWITZ: The arguments I gave apply to any manifold, so the results would be the same.

ASHTEKAR: A lot of people in the audience were asking this question. They were worried during the coffee break but don't want to ask this question, so I will. You gave this great presentation, and most of us felt for the first time that we were understanding the theory in detail, but...

ISHAM: That could be a dangerous move there... (laughter).

ASHTEKAR: What you did was, you started with this general framework; and then gave us some solutions, which seem to exist because, when we look around us, space-time has certain features, and so on. But there must be oodles and oodles of other solutions that also correspond to space-times. For example, suppose you had started with your candidate manifold of space-time as one in which the connection is different from the standard one: it admits torsion, as in the Einstein–Cartan theory. Again, one might have worked out something, and one might have found that there is a solution, say flat, and so forth. Another illustration of this is a solution with a cosmological constant. Is that true, first of all? If that is true, then one would require some further principle–minimizing energy or whatever–which will tell you that there is no torsion or cosmological constant, etc., etc.

HOROWITZ: Yes, I think it is true; in fact, various people have looked at torsion. There is an antisymmetric second rank tensor as one of the modes of the string, and you can put that on your background space-time and couple a string to it, adding a torsion to the connection. So I believe there are solutions with the torsion nonzero; this is all again relating it to ordinary four-dimensional relativity. Yes, one does need to show that not only is the natural ground state or whatever effectively four-dimensional, but that the background torsion is zero.

ASHTEKAR: O.K., then; so one could imagine what Roger Penrose likes to do, and consider complex self-dual solutions. (I am not suggesting that one do this, but I just want to consider a range of possibilities). And one can plug that in as the background structure and construct your Q from there, and say that the equations are again satisfied. At the moment the horizon is wide open, is that right?

HOROWITZ: These are complex, half-flat space-times—is that what you are talking about? I'm a little bit worried; I don't know if complex metrics would affect this argument. No calculations that I know of have looked into that. It might turn out that the argument that there is a solution for which the Ricci tensor vanishes, might break down if your metric somehow is intrinsically complex.

KUCHAŘ: Continuing in Abhay Ashtekar's vein, the fields that we know semi-classically propagate on a space-time background. Can you construct solutions that are based on the fields without actually using a metric? You did it for your ghost variables. Now you may conceive a similar procedure for some other field without really putting the metric into that solution. Wouldn't it give semi-classically something like a theory that isn't propagating on any metric background?

HOROWITZ: If I understand you, you are asking the question whether I can find solutions that, in some sense, have a background electromagnetic field but no space-time, no metric?

KUCHAŘ: Yes.

HOROWITZ: That's exactly the question that Bryce asked me right after my talk. My answer to him was: I think that's a great question to look at, and I plan to look at it, and I have never thought of it before. So, I don't know whether there are answers to that, but I don't think it will be that hard to analyze. Without a space-time metric, fields cannot have local dynamics. However, recently, topological field theories have been constructed with no metric, in which the solutions are based on the topology of the manifold. (For some simple examples and references to the literature see G. Horowitz, "Exactly Soluble Diffeomorphism Invariant Theories," ITP (1989) *Commun. Math. Phys.* 125: 417–437). If there are such solutions in string theory, what would one say? Well, I think it would be yet another example of this wide class of classical solutions that, one would have to argue, in the real solution—quantum solution, or whatever—just doesn't exist. Obviously, in the real ground state, we need a solution that has a space-time metric, fields and all of that. So, I think that's something which should be looked at.

KUCHAŘ: I am returning to the question of time. Suppose you talk about a ground state in ordinary quantum mechanics. You need a state with the lowest energy, which assumes that you have a time-independent Hamiltonian and that energy is canonically conjugate to time. Now, suppose you talk about the wave function of the universe. We know there is no time-independent frame, intrinsically, in quantum cosmology. Originally, that wave function was loosely called "ground state," though I think that Jim Hartle wouldn't call it ground state nowadays, but...

HARTLE: No, you can't tell what I would do.

KUCHAŘ: I want to say: you are coming to a theory which doesn't have any time, doesn't have any space-time, and you are still talking about ground states. What do these ground states mean operationally? Is it not a far-fetched extrapolation to take the formalism, which started firmly grounded in experience, and then pick up a structure, which relied on the other structure being present that gave the ground state its primary meaning, and apply it blindly without really knowing what it means?

HOROWITZ: Well, I completely agree, this is a far extrapolation from theories that we're used to. I'm using "gound state" only because I don't know other words. I don't know the appropriate language to describe the physics in this.

GELL-MANN: "Distinguished solution" would be much more suitable. Actually, it's much more likely to be associated with something like least action than it is with least energy, and in any case what you really mean is the distinguished solution. Superstring theory in general is a tremendous problem: the solutions proliferate. O.K., some people then say: If we look nonperturbatively and to all orders in the loop expansion, most of these may go away; well, maybe, that's one possibility. Another possibility is that there is some principle that selects a distinguished solution, which you can call ground state just as a metaphor. It's nothing more than a metaphor. Another possibility, less pleasant, is that the choice is a quantum-mechanical probabilistic choice: for this universe certain things are this way, and in other universes they would be other ways; that's the least pleasant alternative. They are all being discussed.

NEWMAN: Is there any reason why you can't hope for some Q's that work, that satisfy your conditions for a four-dimensional manifold?

HOROWITZ: You want a Q that works only in four dimensions? I don't know of any reason why there can't exist such Q's, so sure, one can hope.

SMOLIN: Along the lines of what I think was Karel Kuchař's question before, could you construct a BRST operator without a background metric, just from the P_μ and X^μ operators?

HOROWITZ: Yes.

SMOLIN: Would that have a physical interpretation?

HOROWITZ: The question would be something like: What's the spectrum of states if you perturb about that solution, look at linearized fields, and see what kind of states you get? Witten has discussed some related issues in his paper, "Topological Sigma Models."

ANDERSON: I would like to come back to the more technical issue of what path integral means, because you have this cubic action with no kinetic term in it. And usually, when I write down a path integral, I think there is a transition amplitude between some initial configuration and some final configuration. Now, since you don't have this structure, you are just summing over some configurations, right, just adding up all the different Φ's, to use a strong phrase, at an instant? O.K. So it's just a configuration. Right?

HOROWITZ: No. A field Φ would correspond to an ordinary field on space-time.

KUCHAŘ: On space-time, rather than space?

HOROWITZ: Space-time. It's definitely space-time.

ANDERSON: Φ is the field over the whole evolution of space-time.

HOROWITZ: Right. We're integrating over all Φ. It would be like doing the covariant functional integral, integrating over all Φ, subject to suitable asymptotic conditions.

ANDERSON: What I'm trying to get at is the following: I am supposed to treat this integral outside of the classical limit, making some semi-classical evaluation. You showed me that if I could find some Q's that were solutions to the equations of motion, I could perturb around them. Now, when I have the path integral and I do this, I generally would want to sum over all such stationary points of the action, so that presumably I am getting some interference between all backgrounds. Now let me accept what you said, that this is manifold independent, so I am getting a sum over all backgrounds. So we get a superposition of all possible universes. Isn't that imposed upon you?

HOROWITZ: So you want a semi-classical approximation?

ANDERSON: Yes.

HOROWITZ: Richard, do you want to answer that?

WOODARD: The functional formalism for string theory is highly background dependent. The reason why this is so is because it is not really true that $\Phi * \Phi = 0$ is a solution. You have to add a surface term to do the integration by parts; the surface term depends on what solution you want to come out. It is very highly background dependent and nobody knows how to deal with that. It's a problem. We should have a background-independent framework. But we don't.

HOROWITZ: I guess I would be a little more optimistic that the surface term will not be a serious problem.

Strings as Poor Relatives
of General Relativity

K. V. Kuchař and C. G. Torre

1. Summary

Canonical quantization of a closed bosonic string is studied as a model for canonical quantization of the gravitational field. Two quantizations are presented. First, the conventional Lorentz-covariant string quantization is reformulated to make world-sheet-diffeomorphism covariance manifest. This requires an enlargement of the usual string phase space to include the phase space corresponding to embeddings of Cauchy surfaces into the cylindrical space-time that is to be embedded as a string world-sheet in a d-dimensional Minkowski space. The factor ordering that allows for an anomaly-free representation of the Lie algebra of the diffeomorphism group is shown to be incompatible with conformal symmetry except in the critical dimension. The existence of the conformal anomaly implies a failure of the Dirac quantization scheme. Second, by breaking manifest Lorentz invariance, the embedding variables are extracted directly from the unextended string phase space. A factor ordering is found that again allows for general covariance of the quantum theory irrespective of the value for the target-space dimension. The conformal anomaly is still present, but this no longer prevents implementation of the Dirac quantization. The outstanding problem here concerns the status of the Lorentz group in the quantum theory.

2. Introduction

When one speaks of the canonical quantization of the gravitational field, one usually has in mind some variation of the Dirac quantization scheme (Dirac 1964), wherein the Hamiltonian and momentum constraints of geometrodynamics are viewed as operator restrictions on allowable wave functions. These restrictions are expected to impose both dynamics and general covariance on the quantum theory. Despite decades of effort, the conceptual and technical difficulties involved in implementing this scheme have proven to be too great to allow much progress to be made (see, however, the new results obtained in conjunction with Ashtekar's "new variables," described in this volume). Thus it is useful to study the quantization of simplified dynamical systems that can model the canonical structure of general relativity. The familiar prototype of such an approach is the relativistic point particle. There, the quantum Hamiltonian constraint amounts to the Klein–Gordon equation; it reflects the invariance of the theory under diffeomorphisms of the world-line of the particle. While this dynamical system provides a valuable conceptual workhorse for understanding quantum gravity, it is completely inadequate to model the intrinsically field-theoretic features that quantum general relativity should possess. We should therefore turn to the so-called "parametrized field theories," in which time and space coordinates (strictly speaking, embeddings) are adjoined to the phase space of, say, a scalar field on a fixed background, thereby rendering the canonical field theory generally covariant in much the same way as general relativity. Much can be learned by studying the quantization of such models (Kuchař 1989b). Still, as the kinematic space-time variables are introduced by hand, it is always too clear at each stage what are the internal time and space variables and what are the dynamical variables—a situation most definitely not encountered in canonical gravity.

One is naturally led to ask if there are dynamical systems that, like general relativity, are "already parametrized," and yet are simple enough to allow a detailed investigation of their canonical quantization. Probably the simplest of such systems is provided by the relativistic string (Green et al., 1987; Brink and Henneaux 1988), which is currently most popular not as a model for quantum gravity, but rather as an ultimate theory of quantum gravity! Nevertheless, in the present work, we will insist on relegating the string to the more humble role of a quantum-gravity paradigm. To see how this is possible, consider the action functional describing the string:

$$S[g,x] = -\frac{1}{2} \int_M \sqrt{-g}\, g^{ab} \eta_{\mu\nu} \nabla_a x^\mu \nabla_b x^\nu. \qquad (2.1)$$

Here, M is the two-dimensional space-time upon which the string evolves. We will choose $M = R \times S^1$, i.e., we consider a closed string. The

Lorentzian metric on M is g_{ab}. We will assume that the space-time (M, g) is globally hyperbolic with Cauchy surfaces diffeomorphic to S^1. The string coordinates, x^μ, $\mu = 0, \ldots, d-1$, are a set of scalar fields on M, and geometrically serve to embed M as a world-sheet in a d-dimensional Minkowski space M^d. The string equations of motion are obtained by varying both g_{ab} and x^μ; the action is thus invariant under diffeomorphic mappings of M onto itself. This unavoidably leads to constraints in the Hamiltonian formulation. If we define the momenta conjugate to x^μ in the usual way, the Hamiltonian form of (1.1) is

$$S[\vec{N}, x, p] = \int_{R \times S^1} (p_\mu \dot{x}^\mu - N\mathcal{H} - N^1 \mathcal{H}_1), \qquad (2.2)$$

where

$$\mathcal{H} = \frac{1}{2}(p_\mu p^\mu + x^\mu_{,1} x_{\mu,1}) \approx 0, \qquad (2.3)$$

$$\mathcal{H}_1 = p_\mu x^\mu_{,1} \approx 0 \qquad (2.4)$$

are the string Hamiltonian and momentum constraints. They are enforced by the lapse density N (of weight minus one) and the shift vector N^1 (a vector on S^1), which describe a foliation $X: R \times S^1 \to M$. Using the fundamental Poisson brackets

$$[x^\mu(\sigma), p_\nu(\sigma')] = \delta^\mu_\nu \delta(\sigma, \sigma')$$

($\sigma \in [0, 2\pi]$) is a coordinate on S^1), it is easy to verify that the constraint functions satisfy a two-dimensional "Dirac algebra." To write this out explicitly, define the "super-Hamiltonian"

$$H(N) := \int_{S^1} N\mathcal{H}, \qquad (2.5)$$

and the "supermomentum"

$$H(N^1) := \int_{S^1} N^1 \mathcal{H}_1. \qquad (2.6)$$

These dynamical variables satisfy the following Poisson bracket relations:

$$[H(N), H(M)] = H(L^1), \qquad (2.7)$$

$$[H(N), H(M^1)] = H(K), \qquad (2.8)$$

$$[H(N^1), H(M^1)] = H(J^1), \qquad (2.9)$$

where

$$L^1 = N\partial_1 M - M\partial_1 N,$$
$$K = N\partial_1 M^1 - M^1\partial_1 N,$$
$$J^1 = N^1\partial_1 M^1 - M^1\partial_1 N^1.$$

Note that the Dirac algebra for hypersurface deformations in a two-dimensional space-time is a true Lie algebra. It is in fact isomorphic to the Lie algebra $conf(M,g) \approx diff(S^1) \oplus diff(S^1)$ of the group of conformal isometries $Conf(M,g)$ of the metric g_{ab} on M.

The canonical structure of the string is of course quite familiar to those who have studied geometrodynamics. In fact, string dynamics defined by (2.1) can be viewed as the dynamics of d scalar fields coupled to gravity in two dimensions. What makes the string so useful as a model of quantum gravity is the relative simplicity of the corresponding canonically quantized theory. However, the usual quantization that is appropriate for the string considered as a framework for a "theory of everything" is not quite what one would envisage when using the string as a quantum gravity paradigm. Technically, when one tries to represent the algebra (2.7)–(2.9) on a (pseudo-) Hilbert space, one encounters an anomaly, i.e., one must be satisfied with a projective representation of $Conf(M,g)$. Thus, within the context of the usual string quantization, one can only consistently impose a (positive frequency) subset of the Hamiltonian and momentum constraints; it is not at all clear what has become of general covariance in the quantum theory. We will discuss this issue in more detail in Section 3.

Now, it is well known that, for field theories, unitarily inequivalent quantizations are possible. Is there a quantization of the string that respects general covariance and hence more closely models what one hopes to achieve in canonical quantum gravity? In what follows, we will try to answer this question in the affirmative. The key to doing this lies in making the general covariance (technically, covariance with respect to the diffeomorphism group $Diff(M)$) manifest in the canonical formulation, and then striving to preserve this in the quantization. This, in turn, is possible if we make the "already parametrized" nature of the string explicit. The basic strategy for accomplishing all of this, elaborated in the following sections, is as follows.

Begin classically, by finding phase space co-moments of $diff(M)$, the Lie algebra of $Diff(M)$. This has been done in detail (Kuchař and Torre 1989) by constructing homomorphisms from $diff(M)$ into the Poisson algebra of functions on various phase spaces for the string. The

$diff(M)$ co-moments, i.e., the diffeomorphism Hamiltonians, are then translated into well-defined operators on a suitable linear space. The groundwork for this step has been laid down by one of us (Kuchař 1989a). Here, one must overcome the difficulty that the simplest factor ordering of the quantum co-moments that renders them well-defined does not yield an anomaly-free representation of $diff(M)$. A consistent factor ordering *is* possible, and hence it is sensible to define *physical* wave functionals by requiring that they are $Diff(M)$ invariant,[1] i.e., annihilated by the diffeomorphism Hamiltonians. Having successfully completed these steps, we obtain a quantization of the string that explicitly incorporates general covariance. The lessons learned along the way give valuable insights into the corresponding issues in quantum gravity.

3. Quantization on an Extended Phase Space

There are essentially two different methods that one can adopt in trying to construct a canonical representation of $diff(M)$ in a generally covariant field theory like the string (Kuchař and Torre 1989). Both methods rely on one's ability to cast the theory into parametrized form. In this section, we use the approach wherein one explicitly enlarges the usual phase space to include the embeddings (i.e., the internal time variables) and their conjugate momenta in conjunction with a choice of "gauge." The results of this section can serve as a model for the quantum-mechanical treatment of space-time diffeomorphisms in general relativity (Isham and Kuchař 1985a, b).

The phase space of our system is given by

$$\Gamma' = T^*\left[Emb_\eta(S^1, M^d) \times Emb_g(S^1, M)\right].$$

The first factor in Γ' is the usual string phase space Γ, taken to be the cotangent bundle over the space of smooth space-like embeddings of a circle into M^d. The second factor in Γ' is the kinematic phase space: the cotangent bundle over embeddings of S^1 into M such that the embedded circle is space-like (hence a Cauchy surface). The metric with respect to which these latter embeddings are to be space-like is the metric g_{ab} fixed to the conformal gauge:

$$g_{ab} = e^{2\Omega}\eta_{ab}; \tag{3.1}$$

here, η_{ab} is a *fixed* flat metric on M, and $\Omega: M \to R^1$ is an arbitrary function. Given a coordinate chart X^α on M, the embeddings can be described parametrically as [2]

$$X^\alpha = X^\alpha(\sigma).$$

The momenta conjugate to $X^\alpha(\sigma)$ are denoted $P_\alpha(\sigma)$. Using $X^\alpha(\sigma)$ and $P_\alpha(\sigma)$ as fundamental canonical variables on $T^* Emb_g(S^1, M)$, we have the usual Poisson bracket relation

$$[X^\alpha(\sigma), P_\beta(\sigma')] = \delta^\alpha_\beta \delta(\sigma, \sigma').$$

The introduction of the embedding phase space has two related effects on the theory. First, it introduces new first-class constraints due to the nondynamical nature of the embeddings, namely

$$\begin{aligned} \mathrm{H}'_\alpha &:= P_\alpha - \gamma^{-\frac{1}{2}} n_\alpha \mathcal{H} + X^1_\alpha \mathcal{H}_1 \\ &\approx 0. \end{aligned} \tag{3.2}$$

The expression (3.2) has the following structure: The covector n_α (on M) represents the unit normal to the embedding and is defined via

$$X^\alpha_{,1} n_\alpha = 0, \tag{3.3}$$

$$g^{\alpha\beta}(X(\sigma)) n_\alpha n_\beta = -1. \tag{3.4}$$

The covector X^1_α is geometrically a projection operator that maps vectors on M (restricted to the embedding) into vectors on S^1. It is defined as

$$X^1_\alpha = g_{\alpha\beta}(X(\sigma)) \gamma^{-1} X^\beta_{,1}, \tag{3.5}$$

where $\gamma(\sigma)$ is the induced metric on the embedding:

$$\gamma = g_{\alpha\beta}(X(\sigma)) X^\alpha_{,1} X^\beta_{,1}. \tag{3.6}$$

It can be shown that H'_α is, in fact, independent of γ and is thus a functional of only the string variables and the embeddings. This result requires a gauge choice, such as (3.1), that respects Weyl invariance.

The second result of extending the string phase space is the reintroduction of diffeomorphism invariance, which is broken by the use of the conformal gauge. This can be seen by constructing the $diff(M)$ co-moments as follows. Begin by observing that the constraint functions in (3.2) satisfy an Abelian Poisson bracket algebra. Thus, for arbitrary smearing fields N^α and M^α, we have

$$[\mathbf{H}'(\vec{N}), \mathbf{H}'(\vec{M})] \equiv \left[\int_{S^1} N^\alpha \mathbf{H}'_\alpha, \int_{S^1} M^\beta \mathbf{H}'_\beta \right] \tag{3.7}$$
$$= 0.$$

Instead of using externally prescribed ("c-number") smearing fields, we can use functionals of the embeddings obtained by restricting a space-time vector field $\vec{V}$ to the embedding:

$$\mathbf{H}'(\vec{V}) = \int_{S^1} V^\alpha(X(\sigma)) \mathbf{H}'_\alpha. \tag{3.8}$$

The algebra (3.7) then becomes

$$[\mathbf{H}'(\vec{V}), \mathbf{H}'(\vec{W})] = \mathbf{H}'(-[\vec{V}, \vec{W}]), \tag{3.9}$$

where the bracket on the right-hand side of (3.9) is the commutator of vector fields. Thus, the functions (3.2), when smeared with the "q-number" fields as in (3.8), provide an anti-homomorphism from the Lie algebra of vector fields on M into the Poisson algebra of functions on Γ'. Using the standard anti-homomorphism between the algebra of (complete) vector fields $\vec{V}$ on M and elements $V \in diff(M)$, we see that the map

$$V \to \vec{V} \to \mathbf{H}'(\vec{V})$$

is a homomorphism from $diff(M)$ into the Poisson algebra of functions on Γ'. This is the precise way in which the canonical formalism based on Γ' can be considered to be generally covariant.

The diffeomorphism Hamiltonians (3.8) generate canonical transformations on Γ' that deform the embeddings according to the (infinitesimal) left action of $Diff(M)$. The deformation is accompanied by the corresponding dynamical evolution of the string data. Not all evolutions obtained in this manner are physically admissible; physical evolution begins from the constraint surface $\bar{\Gamma}' \subset \Gamma'$ obtained by imposing the constraints (3.2), (2.4) and (3.2).[3] It can be verified that these constraints are first class; thus, if the evolution begins on $\bar{\Gamma}'$, it remains there.

The quantum theory based on Γ' is obtained by the techniques developed in Kuchař 1989b. The representation space S' for the quantum

operators consists of wavefunctionals Ψ that are simultaneously functionals of the embeddings X and elements of a pseudo-Fock space $\mathcal{F}'$ for
the string degrees of freedom. The embedding momenta P_α (we will
now use the classical symbols to denote the quantum operators) act as
variational derivative operators. The string variables act via their creation
and annihilation operators in the usual way. The constant modes of x^μ
and p_μ act on Ψ as ordinary point-particle degrees of freedom. Note that
$\mathcal{S}'$ need not have the structure of a Hilbert space; the inner product is
only to be imposed on the solutions to the quantum-mechanical versions
of the constraints (2.3), (2.4) and (3.2). This is appropriate since, as we
shall see, the embedding coordinates play a role analogous to time in
the nonrelativistic Schrödinger formalism. Strictly speaking, there need
be no inner product on the space $\mathcal{F}'$ either. Nevertheless, it is instructive to allow this here. The physical states can then receive the inner
product induced from $\mathcal{F}'$; we are thus still very close to the usual string
quantization. In the next section, we will relax this assumption.

The operator transcription of the constraint functions is not entirely
straightforward: the Hamiltonian and momentum constraint functions
(which also appear in $\mathbf{H}'_\alpha$) have an ordering ambiguity. This ambiguity
must be resolved so that the resulting operators are well defined on $\mathcal{S}'$
and so that the diffeomorphism algebra (3.9) is preserved. As shown in
Kuchař 1989b, both criteria are satisfied if we appropriately order the
Hamiltonian and momentum constraint operators. Define

$$\mathcal{H}_\alpha := -\gamma^{-\frac{1}{2}} n_\alpha \mathcal{H} + X^1_\alpha \mathcal{H}_1 + \frac{d}{48\pi} A_\alpha[X]. \tag{3.10}$$

In (3.10), the Hamiltonian and momentum constraint operators are normal
ordered with respect to the time-like Killing vector for the flat metric
using in fixing g_{ab} to the conformal gauge. To do this, it is easiest to
begin in the Heisenberg picture, perform the normal ordering, and then
pass to the Schrödinger picture, which is being used here; see Kuchař
1989b. This ordering renders $\mathcal{H}_\alpha$ a well-defined operator on $\mathcal{S}'$; $\mathcal{H}_\alpha$ can
also be viewed as an (embedding-dependent) Hermitian operator on $\mathcal{F}'$.
The potential A_α is the result of an embedding-dependent re-ordering[4]
of the Hamiltonian and momentum constraint operators, and is given by

$$A_\pm = \left[\mp X^\pm_{,1} - \left((X^\pm_{,1})^{-1} \left(\frac{X^-_{,11}}{X^-_{,1}} - \frac{X^+_{,11}}{X^+_{,1}} \right) \right)_{,1} \right]. \tag{3.11}$$

The potential (3.11) is represented by its components in the preferred
null-coordinate basis associated with the conformal gauge fixing. It can

equally well be written in an arbitrary basis (Kuchař 1989b). We can also express $A_\pm$ explicitly as a covector on S^1. Define $\bar{\gamma}$ and $\bar{K}$ to be the first and second fundamental forms for the embedding $X^\pm = X^\pm(\sigma)$. These tensors on S^1 are defined with respect to the preferred flat metric specifying the conformal gauge. We can then rewrite (3.11) as

$$A_\pm = A_\pm^{(1)} + A_\pm^{(3)},$$

where

$$A_\pm^{(1)} = \mp X^\pm_{,1} \tag{3.12a}$$

$$A_\pm^{(3)} = -2\left((X^\pm_{,1})^{-1}\bar{\gamma}^{-\frac{1}{2}}\bar{K}\right)_{,1}. \tag{3.12b}$$

Note that the factor ordering of $\mathcal{H}_\alpha$ necessarily breaks Weyl invariance since one must refer to the metric η_{ab}. This breaking of Weyl invariance can be seen explicitly in $A_\alpha^{(3)}$, and is implicit in the integral kernel that normal orders the Hamiltonian and momentum constraint operators. As we shall see, this will spell doom for a true Dirac quantization of the string based on Γ'.

The additional operator ordering, represented by A_α, is needed so that, when we incorporate $\mathcal{H}_\alpha$ into the quantum-diffeomorphism generators

$$\mathbf{H}'_\alpha = P_\alpha + \mathcal{H}_\alpha,$$

the algebra (3.9) is preserved. Thus, it is consistent to impose

$$\begin{aligned}\mathbf{H}'_\alpha \Psi &= \frac{1}{i}\frac{\delta\Psi}{\delta X^\alpha} + \mathcal{H}_\alpha\Psi \\ &= 0\end{aligned} \tag{3.13}$$

as a condition on physical states. Notice that (3.13) represents a functional Schrödinger equation for the wave functional. Thus, the generators $\mathbf{H}'_\alpha$ serve both to reflect diffeomorphism invariance and to impose dynamics in the quantum theory.

The string wave functionals obtained through the Schrödinger equation (3.13) are not all allowable; we still must impose the conformal symmetry requirement

$$\mathcal{H}_\alpha\Psi = 0, \tag{3.14}$$

which, from (3.10), is tantamount to imposing the quantum Hamiltonian and momentum constraints. As suggested above, we will run into trouble here. The ordering that allows for general covariance in the quantum theory is incompatible with conformal symmetry. Indeed, if we compute the algebra of the operators $\mathcal{H}_\alpha$, we find an anomaly:

$$\frac{1}{i}[\mathcal{H}_\alpha(\sigma), \mathcal{H}_\beta(\sigma')] = C^\gamma_{\alpha\beta}(\sigma, \sigma')\left(\mathcal{H}_\gamma(\sigma) - \frac{d}{48\pi}A^{(3)}_\gamma(\sigma)\right)$$

$$\cdot \delta_{,\sigma}(\sigma, \sigma') - (\alpha\sigma \leftrightarrow \beta\sigma') - \frac{d}{48\pi}F^{(3)}_{\alpha\beta}(\sigma, \sigma'),$$

$$(3.15)$$

where

$$F^{(3)}_{\alpha\beta}(\sigma, \sigma') = \frac{\delta A^{(3)}_\beta(\sigma')}{\delta X^\alpha(\sigma)} - \frac{\delta A^{(3)}_\alpha(\sigma)}{\delta X^\beta(\sigma')}. \qquad (3.16)$$

The structure functions $C^\gamma_{\alpha\beta}$ are functionals of X obtained from the Poisson bracket version of (3.15). The anomaly as given in (3.15) comes in two parts. The term explicitly dependent on $A^{(3)}_\gamma$ is trivial in the sense that it can be removed from (3.15) by subtracting $A^{(3)}_\gamma$ from $\mathcal{H}_\alpha$. In the language of group homology, it is a trivial two-cocycle, i.e., a coboundary. Of course, we cannot remove $A^{(3)}_\gamma$ from $\mathbf{H}'_\alpha$ without introducing an anomaly into the algebra of the quantum-diffeomorphism Hamiltonians. The nontrivial part of the conformal anomaly resides in the "curvature" $F^{(3)}_{\alpha\beta}$. This term cannot be removed by a redefinition of $\mathcal{H}_\alpha$, i.e., it is a nontrivial two-cocycle. It is because of this term that the additional factor ordering beyond normal ordering was needed to represent $diff(M)$ without anomaly.

In the classical theory, the functions $\mathcal{H}_\alpha$ (without A_α), when smeared with conformal Killing vectors restricted to an embedding, are constants of motion. That is, they have vanishing Poisson brackets with the diffeomorphism Hamiltonians. This feature persists in the quantum theory. What changes is the algebra that the functions satisfy. Classically, these functions satisfy the conformal algebra $conf(M, g)$. Quantum mechanically, if we smear (3.15) with conformal Killing vectors restricted to an embedding, the curvature term represents a central extension of $conf(M, g)$. Thus, we see that diffeomorphism invariance can be preserved at the expense of using a projective representation of the conformal symmetry group. Because we obtain the centrally extended conformal algebra, there is only the trivial solution to (3.14). At best, we can impose a positive-frequency subset of these equations. We conclude that the quantization based on Γ' is simply a $diff(M)$ covariant treatment of the usual string quantization. We can make the connection with conventional string quantization more apparent if we enlarge the phase space further by the addition of the BRST ghosts. The BRST extensions of the diffeomorphism Hamiltonians, as well as the Hamiltonian and momentum constraint functions, have been derived in Kuchař and Torre 1989. We can turn these phase space functions into operators in a fashion analogous to that sketched here. One finds (Torre 1988) that the operator ordering that allows $diff(M)$ to be represented on the ghost-extended wave functionals again involves the potential A_α; but now the coefficient in front is $\frac{d-26}{48\pi}$. Thus, in the "critical dimension" $d = 26$, the quantum theory does carry a representation of both $diff(M)$ and $conf(M,g)$ if we include ghosts. Nevertheless, the unextended constraints (3.14) cannot be imposed since the anomaly remains in the string sector of the theory.

Now, one can trace the difficulties involved with imposing (3.14) back to our original insistence on a Hilbert space structure via the Fock space $\mathcal{F}'$ since, for example, the vacuum state will not satisfy (3.14). As suggested above, there is no compelling physical reason to require an inner product to be present except for states satisfying *all* of the constraints (3.13), (3.14). Indeed, the classical antecedent to (3.14) is supposed to express the fact that the embedding variables and their conjugate momenta are *already* present in the unextended string phase space Γ. As pointed out above, the embedding variables play the role of time; neither the embeddings nor their conjugate momenta need be represented as Hermitian operators on some Hilbert space. We should therefore do away with the artifice of adding the embeddings by hand, and instead strive to extract the "internal time" variables from Γ directly. In the next section, we shall see that this is indeed possible; this, in turn, will allow a Dirac quantization that respects general covariance.

4. Quantization on the Original Phase Space

Here, we quantize the string by making use of the fact that the string is "already parametrized." To achieve this, we must extract the embeddings and their conjugate momenta directly from Γ. In Kuchař and Torre 1989 we have done this in detail; here, we will merely sketch the results. As we shall see, the most natural choice for the embedding variables is such that the embeddings are expressed parametrically in a preferred conformal coordinate chart determined by the "light-cone gauge." One complication here is that, for the closed string, the light-cone gauge only fixes a conformal coordinate chart up to a choice of origin for the spatial coordinate. This remaining "gauge" degree of freedom is quite difficult to isolate, and we do not attempt to do so here. Instead, we follow the strategy of Section 3; add by hand this degree of freedom to the phase space at the expense of an additional constraint. Thus, we will work with a slight extension Γ^* of the usual string phase space Γ:

$$\Gamma^* = T^*\left[Emb_\eta(S^1, M^d) \times R^1\right].$$

Local coordinates on Γ^* are the string coordinates and momenta introduced previously, along with the canonical pair q, p,

$$[q, p] = 1.$$

The additional constraint that accompanies the enlarged phase space is simply

$$p \approx 0. \tag{4.1}$$

The reduction of Γ^* associated with this first class constraint leads us back to the original string phase space Γ.

Now, consider the following transformation on Γ^*:

$$X^\pm(\sigma) = -(p_-)_0^{-1}\left(x^+ \pm q \mp \int_0^\sigma dy\, p_-(y)\right), \tag{4.2}$$

$$P_\pm(\sigma) = -\frac{1}{2}(p_-)_0(p_+ \mp x^-_{,1} \pm p\delta(\sigma)), \tag{4.3}$$

$$(x^-)_0 = (p_-)_0^{-1}\int_{S^1}\left(p_- x^- - p_+ x^+ - \frac{qp}{2\pi}\right), \tag{4.4}$$

$$(p_-)_0 = \frac{1}{2\pi}\int_{S^1} p_-. \tag{4.5}$$

We will justify the reuse of the symbols $X^\pm$, $P_\pm$ shortly. In (4.2)–(4.5), we have used light-cone components for the string variables, which are defined in terms of a pair of (arbitrary, but fixed) covariantly constant null vectors k^μ, ℓ^μ in M^d, specified by

$$k^\mu k_\mu = 0 = \ell^\mu \ell_\mu,$$

and

$$k^\mu \ell_\mu = -1.$$

Specifically, we have

$$x^+ := -\ell_\mu x^\mu,$$
$$x^- := -k_\mu x^\mu,$$

and

$$p_+ := k^\mu p_\mu,$$
$$p_- := \ell^\mu p_\mu.$$

There are a couple of properties of the transformation worth mentioning. First and foremost, the transformation is *canonical* on Γ^* when complemented with the identity transformation for the remaining string variables x^i and p_i, $i = 1, \ldots, d-2$. This can be verified by computing the Poisson brackets between $X^\pm(\sigma)$, $P_\pm(\sigma)$ and $(x^-)_0$, $(p_-)_0$. The only nonvanishing brackets are

$$[X^\pm(\sigma), P_\pm(\sigma')] = \delta(\sigma, \sigma'),$$

$$[(x^-)_0, (p_-)_0] = 1.$$

Second, note that the momenta $P_\pm$ are in fact distributions on S^1; strictly speaking, all expressions involving $P_\pm$ should be smeared with suitable test functions. This feature of the formalism can be avoided if we allow for an externally prescribed measure on the circle. For simplicity, we will omit this improvement and retain the delta function "measure."

The canonical transformation (4.2)–(4.5) is a smooth bijective map between the variables (q, p, x^μ, p_μ) and $((x^-)_0, (p_-)_0, X^\pm, P_\pm, x^i, p_i)$. We can interpret it as a map between Γ^* and

$$T^*\left[R^1 \times Emb(S^1, M) \times C^\infty(S^1, R^{d-2})\right],$$

where T^*R^1 is the phase space of the "$-$" direction zero modes, and $T^*C^\infty(S^1, R^{d-2})$ is the phase space of the transverse string variables. Firstly, using (4.2)–(4.5), the Hamiltonian and momentum constraints (2.3), (2.4) become

$$\mathcal{H} = P_+ X^+_{,1} - P_- X^-_{,1} - \frac{1}{2}(p_-)_0(X^+_{,1} + X^-_{,1})\delta(\sigma)\int_{S^1}(P_+ - P_-)$$

$$+ \frac{1}{2}(p_i p^i + x^i_{,1} x_{i,1}) \approx 0, \tag{4.6}$$

$$\mathcal{H}_1 = P_+ X^+_{,1} + P_- X^-_{,1} - \frac{1}{2}(p_-)_0(X^+_{,1} - X^-_{,1})\delta(\sigma)\int_{S^1}(P_+ - P_-)$$

$$+ p_i x^i_{,1} \approx 0. \tag{4.7}$$

We also have the constraint (4.1), which can be written as

$$\int_{S^1} (P_+ - P_-) \approx 0 \tag{4.8}$$

in terms of the new variables. Secondly, we replace the constraints (4.6)–(4.8) by an equivalent system of constraints

$$\tilde{\mathcal{H}} := P_+ X^+_{,1} - P_- X^-_{,1} + \frac{1}{2}(p_i p^i + x^i_{,1} x_{i,1}) \approx 0, \tag{4.9}$$

$$\tilde{\mathcal{H}}_1 := P_+ X^+_{,1} + P_- X^-_{,1} + p_i x^i_{,1} \approx 0 \tag{4.10}$$

and

$$R := \int_{S^1} [(X^+_{,1})^{-1} \tilde{\alpha}_i \tilde{\alpha}^i + (X^-_{,1})^{-1} \alpha_i \alpha^i] \approx 0, \tag{4.11}$$

where we have introduced

$$\tilde{\alpha}^\mu := \frac{1}{2}(p^\mu + x^\mu_{,1}),$$

$$\alpha^\mu := \frac{1}{2}(p^\mu - x^\mu_{,1}).$$

It can be shown (Kuchař 1978; Kuchař and Torre 1989) that (4.9) and (4.10) are respectively the super-Hamiltonian and supermomentum constraints in the parametrized formalism for a set of $d - 2$ free massless scalar fields x^i. This interpretation of (4.9), (4.10) has the variables $X^\pm(\sigma)$ playing the role of embeddings registered in a null coordinate system $X^\pm$. The null coordinate system used to describe the embeddings can be shown to be that associated with the light-cone gauge:

$$X^\pm = \tau \pm \sigma,$$

where τ is the evolution parameter on $R \times S^1$. It is fairly well known that the light-cone gauge is not completely admissible (Patrascioiu 1974; Kuchař and Torre 1989). That is to say, the transformation from arbitrary coordinates on M to light-cone gauge coordinates can be singular. It turns out that the above formal identification of the string as a parametrized system breaks down precisely when the light-cone gauge fails. The portion of Γ where this gauge *is* admissible is a dense subset of Γ. We will assume in what follows that we restrict our attention to this open subspace. This will justify, among other things, our implicit assumption that $(p_-)_0 \neq 0$.

To summarize, insofar as the light-cone gauge is admissible, the classical dynamics of the closed string is canonically equivalent to the dynamics of a parametrized set of $d - 2$ massless scalar fields with the subsidiary constraint (4.11). This constraint is an invariant version of the so-called "shift of origin" constraint that remains when one imposes the light-cone gauge for the closed string. It is an intrinsic feature of the closed string and has no counterpart in parametrized theories.

Having rewritten the string as a parametrized theory, it is straightforward to construct representatives of $diff(M)$ using the ideas developed in Section 3. Thus, the counterparts of the constraints (3.2) take the form

$$\mathbf{H}_+ = P_+ + (X^+_{,1})^{-1}\tilde{\alpha}_i\tilde{\alpha}^i \approx 0, \qquad (4.12a)$$

$$\mathbf{H}_- = P_- - (X^-_{,1})^{-1}\alpha_i\alpha^i \approx 0. \qquad (4.12b)$$

These constraints, however, no longer appear *in addition* to the Hamiltonian and momentum constraints, but (*modulo* (4.11)) are *equivalent* to them. As before, they satisfy an abelian Poisson algebra among themselves as well as with the constraint (4.11). If the functions in (4.12) are projected perpendicularly and parallel to a given embedding, they satisfy the Dirac algebra (2.7)–(2.9). To construct the diffeomorphism Hamiltonians, proceed as follows. Fix a set of coordinates $X^\pm$ on M by letting $X^\pm \in (-\infty, \infty)$ with the identification

$$X^+ - X^- \sim X^+ - X^- + 4\pi.$$

We can consider $X^\pm(\sigma)$ as an embedding of S^1 into M expressed parametrically in the coordinates $X^\pm$. For a given solution to the string equations of motion, these coordinates will be null coordinates relative to the metric g_{ab} as well as with respect to the induced metric on the world-sheet. Given a vector field $\vec{V}$ on M representing $V \in diff(M)$, we take its components in the coordinate basis provided by $X^\pm$ and pull the resulting functions $V^\pm$ back to S^1 along the embedding. The resulting functions on S^1 (functionals of the embeddings) are used to smear the constraint functions (4.12):

$$\mathbf{H}(\vec{V}) = \int_{S^1} (V^+\mathbf{H}_+ + V^-\mathbf{H}_-). \qquad (4.13)$$

The corresponding map

$$V \to \vec{V} \to \mathbf{H}(\vec{V})$$

is a homomorphism from $diff(M)$ into the Poisson algebra of functions on Γ^*:

$$[\mathbf{H}(\vec{V}), \mathbf{H}(\vec{W})] = \mathbf{H}(-[\vec{V}, \vec{W}]). \qquad (4.14)$$

The algebra $conf(M,g)$ can be represented in a variety of ways. Of particular interest is the representation of this algebra as a symmetry algebra for the transverse string variables. To obtain this representation, observe that $v \in conf(M,g)$ corresponds to a conformal Killing vector $\vec{v}$ on M. In the null coordinate basis, the solutions to the conformal Killing equation take the form

$$v^{\pm} = v^{\pm}(X^{\pm}).$$

Once again, we restrict such vectors to an embedding, and use the resulting functions on the circle as smearing fields to obtain

$$\mathbf{h}(\vec{v}) := \int_{S^1} \left[v^+(X^+{}_{,1})^{-1}\tilde{\alpha}_i\tilde{\alpha}^i - v^-(X^-{}_{,1})^{-1}\alpha_i\alpha^i \right]. \qquad (4.15)$$

The functions $\mathbf{h}(\vec{v})$ are constants of motion i.e., they have vanishing Poisson brackets with the diffeomorphism Hamiltonians. The symmetry group that they generate is, of course, $Conf(M,g)$:

$$[\mathbf{h}(\vec{v}), \mathbf{h}(\vec{w})] = \mathbf{h}([\vec{v}, \vec{w}]).$$

Thus, the map

$$v \to \vec{v} \to \mathbf{h}(\vec{v})$$

is an anti-homomorphism from $conf(M,g)$ into the Poisson algebra of functions on Γ^*.

The quantization now proceeds in close analogy to what was done in Section 3. The representation space $\mathcal{S}$ for the quantum operators consists of elements Ψ which are functionals of the embeddings as well as elements of a Fock space $\mathcal{F}$ for the transverse string variables. The operators $(x^-)_0$ and $(p_-)_0$ (as well as the homogeneous modes of the transverse string variables) act on Ψ in the usual way for point-particle degrees of freedom. The diffeomorphism generators $\mathbf{H}_\alpha$ as well as the operator R are turned into well-defined operators on $\mathcal{S}$ by normal ordering the expressions that involve the transverse string variables. Here, the positive frequency part of an operator is defined in terms of the preferred time coordinate $T := \frac{1}{2}(X^+ + X^-)$. Given a solution to the string equations of motion, this time parameter is that associated with the time-like Killing vector for the flat metric, which, in the light-cone gauge, is conformal to the induced world-sheet metric. As before, the algebra (4.14) can be represented without anomaly only if we perform an embedding-dependent reordering of the conformal symmetry generators. Thus, we define

$$\mathbf{h}_+ := (X^+_{,1})^{-1} \tilde{\alpha}_i \tilde{\alpha}^i + \frac{d-2}{48\pi} A_+[X], \qquad (4.16a)$$

$$\mathbf{h}_- := -(X^-_{,1})^{-1} \alpha_i \alpha^i + \frac{d-2}{48\pi} A_-[X], \qquad (4.16b)$$

where normal ordering is implicit in the terms quadratic in $\tilde{\alpha}^i$ and α^i. The potential A_α is again given by (3.11) or (3.12), but now with the identification (4.2):

$$A^{(1)}_+ = 2(p_-)_0^{-1} \tilde{\alpha}^+, \qquad (4.17a)$$

$$A^{(1)}_- = 2(p_-)_0^{-1} \alpha^+, \qquad (4.17b)$$

$$A^{(3)}_+ = -(p_-)_0 \partial_1 \left[(2\tilde{\alpha}^+)^{-1} \left(\frac{\alpha^+_{,1}}{\alpha^+} - \frac{\tilde{\alpha}^+_{,1}}{\tilde{\alpha}^+} \right) \right], \qquad (4.17c)$$

$$A^{(3)}_- = (p_-)_0 \partial_1 \left[(2\alpha^+)^{-1} \left(\frac{\alpha^+_{,1}}{\alpha^+} - \frac{\tilde{\alpha}^+_{,1}}{\tilde{\alpha}^+} \right) \right]. \qquad (4.17d)$$

Since R is just a special case of the conformal symmetry generator (corresponding to a conformal isometry generated by the vector field tangent to the $T = constant$ circles), the reordering should in principle also be done for R. However, because

$$\int_{S_1} (A_+ - A_-) = 0,$$

R is unaffected by this additional factor ordering.

Having obtained a consistent factor ordering for the diffeomorphism Hamiltonians, we can select physical wave functionals via the functional Schrödinger equation:

$$\mathbf{H}_+ \Psi = \frac{1}{i} \frac{\delta \Psi}{\delta X^+} + \mathbf{h}_+ \Psi = 0, \qquad (4.18a)$$

$$\mathbf{H}_- \Psi = \frac{1}{i} \frac{\delta \Psi}{\delta X^-} + \mathbf{h}_- \Psi = 0, \qquad (4.18b)$$

along with

$$R\Psi = 0. \tag{4.19}$$

These equations are just the usual "many-fingered time" evolution equations for $d - 2$ free scalar fields (with a quantum-corrected energy-momentum density given by A_α), along with a subsidiary constraint. The conditions (4.18), (4.19) are the only conditions to be imposed on the wave functional. Moreover, the operators $\mathbf{H}_\pm$, R are commuting on S thanks to the factor ordering adopted in (4.16). We conclude that a true Dirac quantization can be implemented consistently irrespective of the value of d. This is not to say that the conformal anomaly is not present; it is. If we drop the potentials $A_\pm$ from $\mathbf{h}_\pm$, these functions are still constants of motion when smeared with conformal Killing vectors restricted to an embedding. The symmetry algebra generated by these operators, as before, is a central extension of $conf(M, g)$. This however poses no problem for a quantization à la Dirac.

5. Discussion

In the work we have described, our goal has been to gain insight into issues that can be expected to arise in any canonical approach to quantum gravity. We have placed considerable emphasis on the (un-projected) Abelian constraints of the string rather than, as is usually done, on the (projected) non-Abelian super-Hamiltonian and supermomentum constraints. To an extent this emphasis undoes decades of work that led to the viewpoint that, for generally covariant theories, the projected constraints are preferred both geometrically and physically. Nevertheless, there is considerable advantage in avoiding direct use of the super-Hamiltonian and supermomentum constraints, especially in the quantum theory. The conceptual advantage is that, by working with the unprojected constraints, the status of the space-time diffeomorphism group, which serves both as a symmetry group and as a dynamical group, is always manifest. From a technical point of view, a consistent factor ordering of the quantum constraints is most easily obtained in terms of the diffeomorphism Hamiltonians. Of course, one can always project these Hamiltonians along directions perpendicular and tangent to the Cauchy surfaces, thereby recovering the corresponding factor ordering of the Hamiltonian and momentum constraint operators. If this is done, one finds this factor ordering to be quite complicated; it is hard to imagine arriving at this operator ordering by working directly with the projected constraint operators, i.e., without appeal to the operators representing the space-time diffeomorphisms. As a bonus, by defining physical states of the theory in terms of $diff(M)$-invariant wave functionals, we find that they satisfy a functional Schrödinger equation, and thus the probability interpretation for the physical states is straightforward to obtain.

A prerequisite for making manifest the $diff(M)$ covariance of a generally covariant theory within the confines of the canonical formalism is an identification of the internal time and space variables, i.e., the embeddings. These variables, representing a "many-fingered time," are not to be treated as Hermitian operators on some Hilbert space. This structure is reserved for the remaining dynamical variables. The consequences of ignoring this physical consideration are seen in Section 3. By treating all the string variables on the same footing, we have the elegance of manifest Lorentz invariance, but we lose the ability to impose consistently all of the constraints of the theory due to operator-ordering problems. When we are able to extract the internal time variables and then treat them in the appropriate matter,[5] a $diff(M)$-covariant quantization is possible that incorporates all of the constraints. This way of quantization, of course, no longer treats all of the variables on the same footing; this implies a loss of manifest Lorentz invariance for the string. In particular, a $diff(M)$-covariant treatment in the quantum theory implies, in the language of geometric quantization, a rather different *polarization* (see (4.2)–(4.5)) than that one might normally use, based on considerations of Lorentz symmetry. However, while the emphasis on manifest covariance leads to a preferred polarization, there still remains the rather deep question concerning the dependence of the theory on this choice. In particular, there is no reason that one cannot obtain other, equally admissible, identifications of the kinematic space-time variables from the original string phase space. In this case, variables that we treated in the original quantization as dynamical are now kinematical and *vice versa*. It is not at all clear how these various quantizations will be related. In the context of the string quantization given in Section 4, this question is closely related to the status of the Lorentz symmetry, which is no longer manifest in the theory. Classically, this symmetry is still present; one simply uses the transformation (4.2)–(4.5) to translate the usual Lorentz group co-moments. In the quantum theory, however, the operator ordering associated with the potentials A_α makes it difficult to see how Lorentz symmetry can be maintained since these potentials are rather complicated functionals of the light-cone string variables (see (4.17)). If Lorentz invariance were shown to hold in the quantum theory, we would have a rather interesting string quantization away from the critical dimension.

The question as to the status of the Lorentz group for the string quantized as in Section 4 has important consequences even in quantum gravity. Of course, there is no analog of this symmetry group in geometrodynamics. In fact, it has been shown that geometrodynamics possesses no symmetries whatsoever (Kuchař 1981). What we have in mind is, rather, the following. The action of the Lorentz group on the string variables has the effect of mixing the dynamical variables with the kinematical variables. If this symmetry group does in fact act in the quantum string theory, it serves to blur the distinction between these two rather different types of variables. Thus, the string provides us with an excellent model for exploring the quantum-mechanical role of time in a generally covariant theory. Whether the Lorentz group acts in the quantum theory is not yet known. Within the framework of light-cone gauge quantization of the string, it is known that this symmetry is only present in the critical dimension. It may be that this is also true for the quantization advocated in Section 4. However, this is far from obvious. The demand for $diff(M)$ covariance has necessitated both a nonstandard polarization and a nonstandard factor ordering relative to traditional string quantization. We hope to address the question of Lorentz invariance in a future publication.

Acknowledgments. We thank the organizers of the Osgood Hill Conference for their support. This work was also supported in part by NSF grant PHY 85-03653 to the University of Utah.

NOTES

[1] Strictly speaking, invariant under the diffeomorphisms that are in the connected component of the identity.

[2] Greek letters from the beginning of the alphabet denote coordinate indices on M. We reserve the latter part of the alphabet to denote target space Lorentz indices on M^d.

[3] Here, the Hamiltonian and momentum constraint functions serve to generate a conformal *symmetry* group for the string.

[4] From the point of view of the string variables, A_α is just an elaborate (time-dependent) normal-ordering constant.

[5] Of course, this is an outstanding problem in geometrodynamics.

REFERENCES

Brink, Lars, and Henneaux, Marc (1988). *Principles of String Theory.* New York: Plenum.

Dirac, Paul A. M. (1964). *Lectures on Quantum Mechanics.* Belfer Graduate School of Science Monographs, no. 2. New York: Belfer Graduate School of Science, Yeshiva University.

Green, M. O.; Schwarz, J.; and Witten, Edward (1987). *Superstring Theory.* Cambridge: Cambridge University Press.

Isham, Christopher J., and Kuchař, Karel V. (1985a). "Representations of Spacetime Diffeomorphisms. I. Canonical Parametrized Field Theories." *Annals of Physics* 164: 288–315.

———(1985b). "Representations of Spacetime Diffeomorphisms. II. Canonical Geometrodynamics." *Annals of Physics* 164: 316–333.

Kuchař, Karel V. (1978). "On Equivalence of Parabolic and Hyperbolic Super-Hamiltonians." *Journal of Mathematical Physics* 19: 390–400.

———(1981). "General Relativity: Dynamics without Symmetry." *Journal of Mathematical Physics* 22: 2640–2654.

———(1989a). "Parametrized Scalar Field on $R \times S^1$: Dynamical Pictures, Spacetime Diffeomorphisms and Conformal Isometries." *Physical Review D* 39: 1579.

———(1989b). "The Dirac Constraint Quantization of a Parametrized Field Theory by Anomaly-Free Operator Representations of Spacetime Diffeomorphisms." *Physical Review D* 39: 2263.

Kuchař, Karel V., and Torre, C. G. (1989). "World Sheet Diffeomorphisms and the Canonical String." *Journal of Mathematical Physics* 30: 1769–1793.

Patrascioiu, A. (1974). "Quantum Dynamics of a Massless Relativistic String. II." *Nuclear Physics B* 81: 525–546.

Torre, C. G. (1988). Unpublished typescript.

Discussion

WOODARD: So you're just rotating the Ψ, not X_0?

TORRE: Yes, that's right. Traditionally what people do is they rotate everything. So, the action really is positive when anybody does a calculation in string theory. But, to make an analogy with what one does in general relativity, one should rotate the space-time and ignore what the internal time is doing, in some sense.

WOODARD: There is, however, a totally trivial way to get a positive action.

TORRE: By trivial, do you mean simply by rotating the time variable?

WOODARD: X_0.

TORRE: I quite agree.

DEWITT: But you can't possibly rotate one without the other. There are no solutions. And in trying to get at the positive-definite metric, you rotate the string. You're making it positive-definite, and yet it's embedded in, it's going in, the time direction in the embedding space. And one of the equations identifies the external and internal induced metrics.

TORRE: I agree with that geometrical interpretation. Just take a matter field on a two-dimensional background. You can embed it any way you want. I quite agree. Physically, what should happen is, you should use all Lorentzian or all Euclidean time. This is something like what happens in general relativity, where we don't have this target space to give us any physical intuition about it. The choice of slice determines what the ordering and the effect of it is, which is to add on a potential term.

WOODARD: Is it correct to extrapolate it, in that statement, the consequence of your saying that there is no anomaly, or that you've found a set of states where there is no anomaly?

TORRE: I probably want to come back to that, but I'll say yes and no. I'm not saying that I've quantized the string in a manner that is anomaly free. That is certainly not true. In fact, I'm about to show that this quantization scheme produces conventional string results. I would say I do not have an anomaly in the representation of the two-dimensional diffeomorphism group, but only in the subgroup of conformal isometries.

ASHTEKAR: For the trivial embedding, it is zero?

TORRE: That's right. If you go to the trivial foliation, everything just becomes ordinary quantum field theory. It's a conformally-invariant scalar field propagating in a two-dimensional space-time.

JACOBSON: Isn't there another real difference in what you've done, in that you have another variable in your wave equation?

TORRE: Well, in this second version, those variables are functions of the original string phase-space variables. So, in fact, this is very much tied into light-cone gauge quantization. The variables that are serving as the many-fingered time are just the light-cone variables of the string.

WOODARD: Is the Lorentz group a symmetry of the quantum theory?

TORRE: It's not clear, it may not be. It may be that it is, in some critical dimension. Or it may turn out that, because this embedding potential is a highly nontrivial functional of a non-Lorentz-invariant combination of the string phase-space variables, certain generators of the Lorentz group will not even be conserved.

PART IV

Approaches to the Quantization of Gravity

Canonical Groups and the Quantization of Geometry and Topology

C. J. Isham

1. Introduction

At a heuristic level, the concept of the "canonical" quantization of a relativistic field theory is understood reasonably well. A specific choice of Lorentzian time label enables the fixed Minkowski space-time to be foliated into a one-parameter family of three-dimensional space-like planes. For a scalar field theory, the basic configuration variable is the field φ evaluated on a selected three-plane (corresponding to a particular choice of "initial" time t_0), which, together with its canonical conjugate π, satisfies the formal canonical commutation relations (CCR)

$$[\varphi(x), \varphi(y)] = 0 \tag{1.1}$$
$$[\pi(x), \pi(y)] = 0 \tag{1.2}$$
$$[\varphi(x), \pi(y)] = i\delta(x, y), \tag{1.3}$$

or, in smeared form,

$$[\varphi(f_1), \varphi(f_2)] = 0 \tag{1.4}$$
$$[\pi(g_1), \pi(g_2)] = 0 \tag{1.5}$$
$$[\varphi(f), \pi(g)] = i(f, g) := i \int d^3x \, f(x)g(x), \tag{1.6}$$

where the real test-functions f and g belong to some topological vector space $\mathcal{T}$. Self-adjoint representations of this algebra are best studied in terms of unitary operators $U(f)$, $V(g)$ that are weakly continuous in f and g and obey the Weyl–Heisenberg relations

$$U(f_1)U(f_2) = U(f_1 + f_2) \tag{1.7}$$

$$V(g_1)V(g_2) = V(g_1 + g_2) \tag{1.8}$$
$$U(f)V(g) = e^{-i(f,g)}V(g)U(f). \tag{1.9}$$

These reproduce (1.4–6) if the operators $\varphi(f)$ and $\pi(g)$ are identified as the self-adjoint generators of $U(f)$ and $V(g)$ in the form $U(f) = \exp[-i\varphi(f)]$, $V(g) = \exp[-i\pi(g)]$.

With the aid of a suitable topology on the real vector space T (typically it must be nuclear), the spectral theorem applied to the representation (1.7) of the Abelian group T shows that it is unitarily equivalent to one in which $\varphi(f)$ is a "diagonal" operator. That is, the Hilbert space of the representation can be written as $L^2(T', d\mu)$, where μ is some measure on the topological dual T' of T, and $U(f)$ acts on a state functional $\Psi \in L^2(T', d\mu)$ as

$$(U(f)\Psi)[\kappa] = e^{-i\kappa(f)}\Psi[\kappa], \quad \kappa \in T', \tag{1.10}$$

or, more heuristically,

$$(\varphi(x)\Psi)[\kappa] = \kappa(x)\Psi[\kappa]. \tag{1.11}$$

The measure μ is quasi-invariant under translations of T' by elements of T (identified as a subspace of T via the inner product (f,g) on T), and $V(g)$ acts as

$$(V(g)\Psi)[\kappa] = \sqrt{\frac{d\mu(\kappa - g)}{d\mu(\kappa)}}\Psi[\kappa - g], \tag{1.12}$$

where $\frac{d\mu(\kappa-g)}{d\mu(\kappa)}$ denotes the Radon–Nikodym derivative of the translated measure. Equation (1.12) implies formally that the conjugate field $\pi(x)$ acts as a functional derivative:

$$(\pi(x)\Psi)[\kappa] = -i\frac{\delta}{\delta\kappa(x)}\Psi[\kappa] + \rho_\kappa(x)\Psi[\kappa] \tag{1.13}$$

for some function $\rho_\kappa(x)$. Note that the ρ_κ term only vanishes if the measure μ is actually *invariant* under translations, rather than being merely quasi-invariant. In practice, this never happens and the extra term in (1.13) is always present.

The discussion above is concerned with the "kinematics" of quantization. Dynamics is introduced with the choice of a Hamiltonian function

H of φ and π, which determines the evolution of a state vector via the differential equation

$$i\frac{d}{dt}\Psi_t[\kappa] = (H\Psi_t)[\kappa]. \qquad (1.14)$$

Alternatively, one can use the Heisenberg picture, in which the time evolution is carried by the operator fields:

$$\varphi(x,t) = e^{-itH}\varphi(x)e^{itH}. \qquad (1.15)$$

This provides a straightforward relation between the space-time fields of the "covariant" picture and the canonical picture with its fields defined on a single spatial three-plane.

Unfortunately, this simple description of a quantum field theory breaks down dramatically for interacting systems, and ultra-violet divergences render the naive CCR irrelevant for a nonlinear theory in space-times of dimension greater than two. The ensuing need to perform renormalization in a Lorentz-invariant way has led to a general rejection of canonical quantization as anything other than a formal device with which to "validate" covariant methods such as, for example, the modern BRS/functional-integral perturbative quantizations of non-Abelian gauge theories.

However, the situation in quantum gravity is somewhat different. Even at the classical level, the connection between the canonical view of general relativity and the full space-time picture is rather subtle, and there is no a priori reason for expecting a canonical quantization to be necessarily equivalent to a "covariant" one, or vice versa. The heart of the matter is the absence in general relativity of any analogue of the pre-ferred family of Lorentzian foliations of Minkowski space-time; instead, one must allow for arbitrary decompositions involving all possible one-parameter families of space-like hypersurfaces. Furthermore, the concept of "space-like" itself depends on the very four-metric one is attempting to construct with the aid of the Hamiltonian method. These features are well understood in the classical theory, but they have far-reaching implications for the quantum gravity program.

The canonical analysis of general relativity (Dirac 1958a, 1958b, 1959; Arnowitt, Deser and Misner 1962) shows that the natural config-uration variables of the theory are the set $Riem\,\Sigma$ of all Riemannian metrics on some three-manifold Σ. This suggests that the starting point for canonical quantization should be commutation relations analogous to (1.1–3) in the form

$$[g_{ab}(x), g_{cd}(y)] = 0, \tag{1.16}$$

$$[\pi^{ab}(x), \pi^{cd}(y)] = 0, \tag{1.17}$$

$$[g_{ab}(x), \pi^{cd}(y)] = i\delta_a^{(c}\delta_b^{d)}\delta(x, y). \tag{1.18}$$

If desired, the operators can be smeared to yield CCR analogous to those of (1.4–6). If one wishes to keep to the unsmeared form, it is useful to define the components of the metric with respect to some globally-defined triad θ^a, $a = 1, 2, 3$ of one-forms on Σ, i.e., $g_x = g_{ab}(x)\theta^a(x) \otimes \theta^b(x)$ (such a set exists on any orientable three-manifold). This guarantees that the six classical functions g_{ab} are defined *globally* on Σ, which would not generally be the case if a coordinate frame was used. Wherever appropriate, we shall assume this has been done.

I shall return later to the question of whether (1.16–18) are in fact justified in the present context. If we assume for the moment that they are, then, as in the case of the scalar field, it is natural to seek representations in which $g_{ab}(x)$ is a diagonal operator acting on state functionals $\Psi[\kappa]$ as $(g_{ab}(x)\Psi)[\kappa] = \kappa_{ab}(x)\Psi[\kappa]$. However, this is not the end of the story since the invariance of classical general relativity under space-time diffeomorphisms implies that the canonical theory should be invariant under the action of the spatial diffeomorphism group $Diff\,\Sigma$. There are various ways in which this might be understood at the quantum level, but the usual approach is to suppose that the Hilbert space carrying the representation of the CCR (1.16–18) also carries a unitary representation of $Diff\,\Sigma$ that intertwines correctly with these canonical operators. A state Ψ is then deemed to be physical only if it is invariant under this action of the diffeomorphism group. If the domain space of Ψ were the set $Riem\,\Sigma$ of all *smooth* metrics on Σ (rather than "distributional" metrics– see below), physical states could then be identified with functionals on the orbit space $Riem\,\Sigma/Diff\,\Sigma$ of equivalence classes of these metrics.

There are many "ifs" and "buts" about this approach, some of which will be discussed later. However, proceeding formally, one notices a peculiar feature of the resulting state function: to specify an element in $Riem\,\Sigma/Diff\,\Sigma$ requires $6 - 3 = 3$ functions per spatial point, whereas the correct count for the number of canonical degrees of freedom for the gravitational field is two. The well-known answer to this "paradox" is that the canonical content of the invariance under space-time diffeomorphisms is not exhausted by requiring invariance under $Diff\Sigma$. There should be an additional invariance, under what would be deformations of the spatial hypersurface normal to itself were the system to be viewed from the perspective of an enveloping space-time. The canonical generator of such deformations is the well-known function $\mathcal{H}(g, \pi)$ of the canonical variables,

$$\mathcal{H}(g, \pi) = 2\kappa\, g^{-1/2} \left(g_{ab}g_{cd} - \frac{1}{2}g_{ac}g_{bd} \right) \pi^{ab}\pi^{cd} - \frac{1}{2\kappa}g^{1/2}\, {}^3R(g),$$

(1.19)

where $g := \det g_{ab}$, ${}^3R(g)$ is the scalar curvature of g on Σ, $\kappa = 8\pi G$ and G is Newton's constant. The requirement that this generator annihilate physical quantum states,

$$\mathcal{H}(g, \pi)\Psi = 0,$$

(1.20)

is the famous Wheeler–DeWitt equation.

This equation features in many of the contributions to this volume, and lies at the heart of the canonical quantization of gravity. In many respects it plays the role of the Schrödinger equation (1.14), with the absence of the d/dt term reflecting the absence in the theory of any preferred time label. Instead, "time" is to be understood as the "extra" third variable mentioned above, and which must therefore be identified as some function of the gravitational field variables themselves. After making such an identification, one can attempt to interpret the Wheeler–DeWitt equation as describing the evolution of the remaining two physical degrees of freedom with respect to this particular choice of time. There is of course no unique way of doing this (that is what is meant by saying that the theory is invariant under deformations of the spatial hypersurface), and the question of how different choices are related, and what this implies for the quantum theory, is complex and by no means fully understood.

For our present purposes, it suffices to emphasize that what is regarded in the classical theory as *space-time* only emerges *after* an identification of internal time has been made. Furthermore, there is a good case for arguing that the interpretation of this internal variable as "time" will only be meaningful in, at best, a semi-classical approximation to the quantum theory. That is, it is only in this limit that the Wheeler–DeWitt equation can be made to look like a genuine Schrödinger equation. This, in turn, implies that the concept of "space-time" (as opposed to "space") is one that cannot be sustained in a full canonical theory of quantum gravity. Note, however, that this does not mean that four-dimensional manifolds cannot be used in such a theory; for example, they are employed by Hartle and Hawking (1983) in their functional integral technique for generating a specific "wave-function-of-the-universe" solution to the Wheeler–DeWitt equation. This example might be thought inappropriate since their metrics have a *Riemannian* (rather than Lorentzian) signature. But this misses the crucial point that the "phenomenological" four-dimensional (Lorentzian) space-time that is reconstructed from the canonical state Ψ is not necessarily related, either metrically or topologically, to any four-dimensional manifold that happens to be used in the construction of the state. Indeed, if the concept of "time" is only semi-classical, it is incorrect to talk at all about a four-dimensional physical manifold at the quantum level. The canonical theory suggests instead that the fundamental entity is the *three*-dimensional space Σ, with a four-dimensional space-time (topologically of the form $\Sigma \times \mathbb{R}$) appearing only in the semi-classical regime.

Clearly, there are radical differences between a canonical theory of quantum gravity and any approach that contains a four-dimensional manifold $\mathfrak{M}$ that is deemed to have a physical significance beyond just the semi-classical limit (this is essentially my *definition* of a "covariant" theory). But what would such a theory look like? In spite of much "covariant-canonical" debate over the years, I do not think we have a clear answer to this. For example, no one has ever constructed a complete set of axioms that could play the same role for covariant quantum gravity as do the Wightman axioms for conventional relativistic quantum field theory. A minimal requirement would presumably be the association with each open subset O of $\mathfrak{M}$ of an operator algebra $\mathfrak{A}(O)$, which could be thought of as being generated by the quantum fields localized in this region. The role of the diffeomorphism group $Diff\,\mathfrak{M}$ becomes rather problematical at this point since, regarded as an active transformation group of $\mathfrak{M}$, it maps one such region into another. Thus, like space-time points, open regions have no fundamental ontological status in a $Diff\,\mathfrak{M}$-covariant theory. But setting this aside, what sort of structure might be exhibited by these local algebras? In conventional quantum field theory, a crucial role is played by Einstein causality, whereby the elements of algebras associated with space-like separated regions commute. For example, this justifies the vanishing of the canonical equal-time commutators (1.1–3) when the spatial points x and y are not equal.

However, as emphasized by Haag et al. (1984), in a quantum theory of gravity all possible space-time metrics are presumably "available" (at least virtually) to the system. But, given any pair of points in $\mathfrak{M}$

that are sufficiently close in a topological sense, there will always exist Lorentzian metrics in which they are *not* space-like separated, which suggests that the local algebras should never commute with each other, and that the causal properties of the theory are therefore carried entirely by the quantum *states* of the system rather than by the algebra of observables. This implies that a covariant/space-time theory of quantum gravity might be strikingly different from anything encountered so far in conventional quantum field theory. It might also seem at variance with the imposition of the commutation relations (1.16–18), which imply that Σ is always space-like, irrespective of the quantum state. One would probably argue that this is correct, but that it means that Σ cannot be regarded as a fixed subspace of a fixed space-time. Rather, the "semi-classical" space-time must be constructed from the quantum theory *before* saying which hypersurface Σ is supposed to represent. However, the question is an interesting one and deserves further study. At the very least, it serves to emphasize that, in choosing between a canonical or a covariant approach, one is taking a step that cannot easily be reversed. This is reminiscent of the long-running argument about whether the metrics in a space-time functional integral should be Lorentzian or Riemannian: Once you have committed yourself, it is difficult to make much contact with the opposite camp!

With these cautionary remarks in mind, the rest of the paper will be concerned with various issues relating to the quantization of gravity via the imposition of something like conventional canonical commutation relations. Since the conference was concerned with *conceptual* problems in quantum gravity, I have placed the emphasis firmly on general ideas rather than detailed mathematical formalism. Section 2 contains a discussion of a variety of questions that arise in the context of canonical quantum gravity, some of which are resolved in Section 3 with the aid of a group-theoretical approach to quantization. I have been "pushing" this point of view for some time, and it is relatively uncontroversial. More speculative, however, is the material in Section 4, which contemplates the idea of applying canonical methods to the *topological* structure of Σ, not just to the Riemannian metrics which it carries. The possibility is considered of quantizing all topologies on a given set X, and also (and less exotically) the set of all metric topologies on X. The driving motivation behind such speculations is the desire to escape from the trammels of differential geometry, which I feel belongs at most to the semi-classical limit of the true quantum theory of gravity.

2. Some General Questions in Canonical Quantum Gravity

2.1. THE A PRIORI ROLE OF SPACE-TIME CONCEPTS

The two major mathematical ingredients in the classical theory of general relativity are: (i) the differentiable manifold, which represents space-time (or space); and (ii) the Lorentzian (or Riemannian) metric defined on this space, which represents the gravitational field. It is clear from the discussion above that in any approach to quantum gravity, a crucial question is: "what is the status at the quantum level of these classical concepts of space and time?"

The type of answer given to this central question will depend strongly on the approach to quantum gravity that is adopted. Throughout the history of the subject, there have been two contrasting schools of thought. The first sees the construction of a quantum theory of gravity as involving primarily a move from classical to quantum physics; the second sees a move in the opposite direction. Within the first category lie all those approaches whose primary ingredient is classical general relativity, which is then "quantized" in some way. These include most of the schemes in which attempts are made to replace the classical metric by a quantum operator, or to perform a functional integral over such metrics. In this context, it should be noted that "quantize" is a verb: it is something one "does" to a classical theory to improve it in some way (of course, this is how most elementary quantum mechanics is taught: we "quantize" the hydrogen atom, the simple harmonic oscillator, etc.). It is not surprising that, in approaches of this type, the basic epistemological and conceptual frameworks are derived from the classical theory. In particular, the fundamental question concerning the quantum status of space and time is usually interpreted to mean "to what extent are the classical space-time concepts *preserved* (or *modified*) in the transition from the classical to the quantum world?"

The second school of thought is more radical. It seeks a theory that is fundamentally quantum mechanical, with the classical theory of general relativity only emerging in some "low-energy," or "semi-classical," limit. Currently, the most active research area of this type is superstring theory; earlier instances were the various "induced gravity" programs. Note that in both examples one does in fact start with *some* classical theory, which is then "quantized" in the time-honored fashion. However, the critical point is that this starting theory is *not* general relativity, but something quite different. This is particularly true in the case of string theory.

The implications of this line of thought are quite striking. It suggests that any attempt to assign fundamental status to a quantized metric tensor is misguided. Such an operator should at best be regarded as a convenient technical tool with which to probe certain low-energy gravitational phenomena, just as the (non-renormalizable) nonlinear σ-model can be used to give low-energy predictions for the more basic theory of QCD. Since much of the interest in quantum gravity lies in very *high* energy phenomena (i.e., Planck-length effects) this means that most past (and present) activity was (and is) doomed to inevitable failure.

I suspect that most people working today in quantum gravity would prefer to adopt a "quantum-to-classical" path, but the problem is knowing where to start. The fundamental question concerning the quantum status of space-time concepts becomes: "how are these concepts to be *recovered* in the appropriate limit?" with the inevitable concomitant: "from *what* are these concepts recovered?" i.e., "What is the appropriate conceptual framework in the truly nonclassical regime?" This question is considerably more open-ended than simply asking how classical concepts are modifed in a "classical-to-quantum" approach. Even the status of the space-time manifold itself becomes problematical. Indeed, if the metric tensor is deemed to be only a low-energy effect, why should not the same be true of the continuum space on which it is defined? Perhaps the entire paraphernalia of differential geometry is only appropriate at scales greater than the Planck length. But if so, with what should it be replaced: spaces that are not smooth manifolds, but which still carry a metric topology; nonmetrizable topological spaces; or discrete sets, with an underlying approach to physics that involves combinatorics rather than continuity or differentiation? The idea that physics should not involve the continuum at a fundamental level is one to which I have become increasingly attracted, even though it poses many difficulties vis-a-vis the more traditional approaches to quantum gravity. For example, what is the status of the diffeomorphism group of space-time (or space) if the manifold structure is only meaningful in some semi-classical sense? In a conventional "gauge" theory, the imposition of gauge invariance at *all* levels is a fundamental ingredient in any quantization scheme.

It is clear that, in any approach to quantum gravity, one should strive actively to be aware of the origin and validity of any space-time concepts that are employed. Unfortunately, it is all too easy to slip them in "under the carpet" and then to recover them with the proud claim that the structure of space-time has "emerged" from the theory!

Space-time concepts are often derived from the use of some background space and/or metric. For example, in the particle-physics approaches to covariant quantum gravity, the metric tensor $g_{\mu\nu}(x)$ is usually expanded about a background Minkowskian metric $\eta_{\mu\nu}$ in the form

$$g_{\mu\nu}(x) = \eta_{\mu\nu} + h_{\mu\nu}(x), \tag{2.1}$$

with $h_{\mu\nu}(x)$ interpreted as the Poincaré-covariant quantum field describing a spin-two massless graviton. Note that both the background *metric* and the background *manifold/topology* are prescribed in this formalism. A similar remark would apply to any approach to canonical quantization in which the metric tensor $g_{ab}(x)$ on the three-manifold Σ was expanded as $g_{ab}(x) = \delta_{ab} + h_{ab}(x)$ with CCR being satisfied by $h_{ab}(x)$ and its canonical conjugate.

The imposition of a background structure of this particular type has been criticized many times over the years, and justifiably so. However, it by no means easy to construct, or indeed to interpret, a formalism in which no background appears at all. It is relatively straightforward to conceive a theoretical framework with no background *metric* (so that, for example, the concept of "causality" can only be extracted in a state-dependent way), although even here it is difficult to know a priori how such a theory should be interpreted: Bohr's insistence on the primary role of classical ideas in quantum theory was not without its point. For example, it is not trivial to formulate a well-defined "measurement theory" that does not employ some sort of fixed reference system for its rods and clocks. Traditional "Copenhagen" instrumentalism seems especially reliant on such a classical background, but attempts to replace it with a more realist interpretation produce difficulties of their own. This problem is particularly acute in quantum cosmology, and involves the whole question of how quantum theory should be interpreted in such situations—an issue that is still much debated (Gell-Mann and Hartle 1990, Hartle 1991, Penrose 1987).

It is, however, even harder to envisage a mathematical structure in which there is no background *manifold*. Indeed, one encounters at once the difficult question of what should take its place. Note that, notwithstanding occasional claims to the contrary, most current formulations of superstring theory do not assist in resolving this problem. Any discussion of strings as extended objects moving *in* a space-time has already introduced that space-time as part of the fixed background. Of course, workers in this area are well aware of the problem, and have attempted to address it in various ways. For example, the Friedan–Shenker (1986, 1987) scheme avoids any discussion of space-time as such. Another approach (Horowitz and Witt 1987, Horowitz 1991) involves summing over all space-time manifolds, which averts the need to give primary status to any particular one (although even here it is assumed implicitly that space-time is a smooth manifold rather than, for example, a more general topological space).

Being acutely conscious of these problems, it would be nice to be in a position to write down an a priori quantum structure and derive the classical space-time concepts in some appropriate limit. However, lacking any deep insight into what such a fundamental conceptual framework should be, I have adopted the more cautious stance of taking seriously the "quantization" of a classical system. Thus, the rest of the paper is concerned with the classical-to-quantum approach to the canonical quantization of gravity, with the hope that the ensuing changes in the ideas of space and/or time may afford a genuine glimpse of what quantum physics at the Planck length might look like. But a "glimpse" is probably all one can hope for, and I would not like to claim that this, or any other, approach to canonical quantization has any more fundamental validity than, for example, does the nonlinear σ-model as an approximation to QCD.

2.2. THE A PRIORI STATUS OF THE METRIC COMMUTATION RELATIONS

Let us consider what can be said in general about quantizations of general relativity that involve the imposition of canonical commutation relations like (1.16–18). We are not concerned now with the conceptual issues raised in the introduction, but with the more technical matters that arise in attempts to find irreducible self-adjoint operator representations of these commutation relations. For example:

(i) *Why Should the Representation be Irreducible?*

In a conventional, canonically-quantized theory, the basic canonical variables have the property that, classically, every observable can be written as a function of elements of this set. The requirement that the representation be irreducible (i.e., no "superselection sectors") is the operator analogue of this condition. Physically, it means that there are no observables in the quantum theory other than "quantizations" of classical observables. In the case of quantum gravity, there is no strong justification for this assumption, since it is quite feasible that there could be quantities relevant at the Planck length that have no direct equivalent in the classical theory of general relativity. In so far as any reducible representation can be decomposed into a direct sum/integral of irreducible representations, it is not unreasonable to start by concentrating on the latter. However, if some "naturally occurring" representation turns out not to be irreducible, one should keep in mind the possibility that this may reflect some new physical phenomenon of genuine interest.

(ii) *Why Should the Operators be Self-Adjoint?*

The point has been made frequently (e.g., Komar 1970; Kuchař 1986, 1987) that four of the six degrees of freedom per space point of the metric tensor $g_{ab}(x)$ are "gauge" variables, and there is no justification for requiring these nonphysical parts of $g_{ab}(x)$ to be represented by operators that are self-adjoint. An analogous argument applies to any gauge theory, and is closely connected to the question of whether the operator algebra should be represented on an indefinite inner-product space rather than a genuine Hilbert space. This is a valid point, although it is complicated by the absence of any preferred decomposition of the field variables into those that are "physical" and those that are not; indeed, this is precisely what is meant by saying that the theory is gauge invariant. It is also very difficult technically to handle operators that are not self-adjoint. In particular, one loses the possibility of studying the CCR representation via a unitary representation of the corresponding group whose self-adjoint generators are the canonical quantum fields. It is mainly for this reason that I will restrict my attention in what follows to genuinely self-adjoint representations.

(iii) *What is the Role of the Diffeomorphism Group $Diff\ \Sigma$?*

This question is related to the previous one since three of the four "gauge" degrees of freedom are associated with the diffeomorphism group of the spatial three-manifold Σ. One possibility would be to quantize directly on the space $Riem\ \Sigma/Diff\ \Sigma$ of equivalence classes of metrics. This is an attractive idea, but it has never been implemented mainly because of the absence of any natural coordinates on this space. The obvious approach is to fix a gauge for the spatial diffeomorphisms and then to quantize the remaining variables. Unfortunately, this procedure is not globally defined since, in practice, it means choosing a fixed coordinate system on Σ. A proper gauge choice for $Diff\ \Sigma$ would involve finding a cross-section for the principal bundle $Diff\ \Sigma \rightarrow Riem\ \Sigma \rightarrow Riem\ \Sigma/Diff\ \Sigma$. But this $Diff\ \Sigma$-bundle is nontrivial, and so no such cross-sections exist, i.e., there is a Gribov effect (Gribov 1978; Singer 1978).

For these reasons most approaches have adopted the Dirac procedure mentioned earlier. Thus, the Hilbert space carrying the CCR representation is also required to carry a representation D of $Diff\,\Sigma$ with the property of reproducing correctly the transformations of the canonical fields:

$$D(\alpha)g_{ab}(x)D^{-1}(\alpha) = J_{\alpha a}{}^{c}(x)\,J_{\alpha b}{}^{d}(x)\,g_{cd}(\alpha(x)), \qquad (2.2)$$

where the Jacobian matrix $J_{\alpha a}{}^{c}(x)$ of $\alpha \in Diff\,\Sigma$ is defined with respect to the global frame θ^{a} of one-forms on Σ by $(\alpha^{*}\theta^{a})_{x} = \theta_{x}^{b}J_{\alpha b}{}^{a}(x)$. The physical states are then required to satisfy

$$D(\alpha)\Psi = \Psi \qquad \text{for all} \quad \alpha \in Diff\,\Sigma. \qquad (2.3)$$

This procedure is not entirely satisfactory since the operators $D(\alpha)$ will typically have spectra that are continuous. Hence, states satisfying (2.3) will not lie in the original Hilbert space $\mathfrak{H}$, but must instead be regarded as generalized (i.e., "δ-function normalized") eigenstates. This means that the inner product on $\mathfrak{H}$ cannot be used directly to give an inner product on the set of physical states—an important consideration when considering the probabilistic interpretation of the theory.

An alternative approach is to concentrate on observables rather than states, and to define physical observables to be those operators on $\mathfrak{H}$ that commute with the unitary operators $D(\alpha)$. Note that even if the representation of the CCR is irreducible, the representation of $Diff\,\Sigma$ must be very reducible if a reasonable number of physical observables is to be obtained. What do physical states and/or observables look like? Contrary to naive expectations, scalar functions of the metric, such as for example the Ricci scalar $R(g)$, do not qualify since

$$D(\alpha)R(g(x))D^{-1}(\alpha) = R(g(\alpha(x))) \neq R(g(x)). \qquad (2.4)$$

The simplest observables are obtained by integrating such scalars over Σ with the aid of the volume element constructed from g (if Σ is noncompact, we will restrict our attention to diffeomorphisms of compact support so that boundary terms do not appear). Observables of this type (or the analogous states) are very nonlocal, but this seems to be an intrinsic feature of general relativity in the absence of any background reference frame. To illustrate this further, suppose that the theory includes a field that is a one-form $A(x)$. Then the integral of A along any path in Σ

could (at least, in principle) be a well-defined operator. But, if Ω is some physical state, the new state

$$\Psi_P := \int_P A_a(x)dx^a\Omega \qquad (2.5)$$

will not be physical since $D(\alpha)\Psi_P = \Psi_{\alpha(P)}$, where $\alpha(P)$ is the path obtained by transforming P with the diffeomorphism α. Genuine physical states could be defined as equivalence classes of such states, and would therefore be labelled by elements of the quotient space $Path\,\Sigma/Diff\,\Sigma$. This is one way in which the currently fashionable theory of knots could arise in quantum gravity; an explicit example is provided by the intriguing work of Smolin and Rovelli reported in this volume. Note once again that the states thus obtained are highly nonlocal.

In general, quantum states might be labelled by the orbits of $Diff\,\Sigma$ on any space X, on which it acts as a group of transformations. In the case above, X is the path space of Σ; another example would be the space of two-dimensional submanifolds of Σ, in which two submanifolds of different genii would necessarily lie in inequivalent classes. The simplest example is to take X to be Σ itself. But $Diff\,\Sigma$ acts transitively on Σ and hence there is just one orbit in this case. This is effectively what happens in the example of the Ricci scalars $R(g(x))$ considered above: all these operators (i.e., one for each point in Σ) combine to give a single physical observable.

(iv) *What is the Role of the Background Manifold Σ?*

There should be no problem in discussing representations of the CCR (or similar ones) in the absence of a background three-metric, but the very formulation of the CCR supposes that we have a background *manifold* Σ, on which the canonical fields are defined. It would be possible to extend the idea of the CCR to include fields defined on a disjoint union of different three-manifolds, but the basic fact remains that the manifold structure is given and fixed. One interesting question is whether the topological structure of Σ plays any role in the representation theory. We shall see in the next section that this is indeed the case once the correct identification of the canonical group has been made. A much harder question is whether a theory can be constructed that involves a simultaneous quantization of both the metric and the manifold on which it is defined. An obvious difficulty is that if, as seems likely, quantization of topology involves moving away from differentiable structures, then it is quite unclear what should replace the Riemannian metric on those spaces Σ that are topological spaces but not smooth manifolds. This is part of the very general, and highly nontrivial, question of whether the idea of "quantum topology" is really meaningful, and, if so, how it fits into a quantum theory of gravity.

(v) *Is a Quantized Metric Compatible With Classical Riemannian Geometry?*

An intriguing and nontrivial question in the context of canonical quantization is the extent to which a quantized metric field $\hat{g}_{ab}(x)$ satisfying CCR like (1.16–18) is compatible with the classical ideas of Riemannian geometry. There is an analogous problem in the nonlinear σ-model, in which the configuration space is the infinite-dimensional nonlinear manifold $C(\Sigma, M)$ of all smooth maps from physical space Σ into some target manifold M. The question then is what meaning, if any, can be associated to converting a classical function $\varphi \in C(\Sigma, M)$ into a quantum operator $\hat{\varphi}$. Both these examples are special cases of a very general and important problem: given a classical theory with configuration space Q, how many properties of Q can be preserved in the quantum theory?

We do not expect a quantized metric to possess analogues of *all* the classical properties of elements of $Riem\,\Sigma$ — were that the case, there would be little of interest in the subject of quantum gravity! The problem, however, is to distinguish genuine quantum effects from pathological behavior that merely results from choosing a bad method of quantization in the first place. For example, a "weak-field" approach to the canonical commutation relations might involve expanding the metric about flat three-space in the form $g_{ab}(x) = \delta_{ab} + h_{ab}(x)$, and then using this background metric δ_{ab} to construct a standard Fock space representation for $h_{ab}(x)$ and its conjugate momenta. It is well-known that, in any (exponentiable) representation of the conventional CCR, the spectra of the operators take on arbitrarily large positive and negative values. But it is clear that only relatively small variations in h_{ab} are allowed if the sum $\delta_{ab} + h_{ab}(x)$ is to maintain the positive definiteness required of a Riemannian metric. An advocate of this scheme might try to argue that this shows the dramatic way in which Riemannian geometry is lost by quantization. But a more realistic reaction is that the loss of this structure is the result of choosing an expansion of the metric that is incompatible with the quantization program used thereafter.

However, this problem is not simply a result of the weak-field expansion, but is endemic in the CCR (1.16–18) themselves, which are fundamentally incompatible with the condition that the tensor $g_{ab}(x)$ be positive definite. This is closely analogous to the well-known problem of quantizing a particle moving on the positive real line. When exponentiated, the conventional CCR $[x, p] = i$ becomes

$$e^{-iap} \, x e^{iap} = x - a, \tag{2.6}$$

which, since a can be any real number, shows at once that the CCR are incompatible with the condition $x > 0$. Note that constraints of this type are "open," i.e., they are defined by *inequalities,* rather than the "closed" type, in which the constraint is the vanishing of some expression: for example, a particle moving on the unit circle in the $x - y$ plane is subject to the constraint $x^2 + y^2 - 1 = 0$. Systems of closed constraints are often handled with the aid of Dirac brackets, but open constraints require a different approach, which will be discussed in the next section.

(vi) *Is Riemannian Geometry Compatible With the Singular Nature of Quantum Fields?*

For a conventional scalar field satisfying the CCR (1.4–6), the measure μ used in the construction of the Hilbert space is usually concentrated on elements of the dual test function space T' that do not belong to the space T. Thus, the κ appearing in (1.11) is typically a *distribution,* not a smooth function. This is a reflection of the familiar ultraviolet divergence problem since (1.11) implies that

$$((\varphi(x))^2 \Psi)[\kappa] = (\kappa(x))^2 \Psi[\kappa]. \tag{2.7}$$

and the right hand side may not be defined if it involves the square of a distribution. In the case of quantum gravity, this suggests that the objects g appearing as arguments in the state vector $\Psi[g]$ may be *distributional* rather than smooth metrics. However, it is not at all clear what is meant by a "distributional" metric, and how this would affect the usual ideas of Riemannian geometry. Any careful analysis of canonical quantum gravity will need to address this issue and the related question of whether the metric operator can be renormalized without totally destroying its geometrical meaning. It is clear that a similar question can be asked of the nonlinear σ-model, viz., what is the appropriate definition of a distributional dual for the infinite-dimensional *nonlinear* manifold $C(\Sigma, M)$?

Another question in this context is the fate of the effects that are expected to be associated with the nontrivial topological structure of the classical configuration space $Riem\,\Sigma/Diff\,\Sigma$. This space has non-vanishing fundamental group $\pi_1(Riem\,\Sigma/Diff\,\Sigma) \approx \pi_0(Diff\,\Sigma)$, and hence, according to the usual analysis (Friedman and Sorkin 1980, Isham 1981, Friedman 1991) of a quantization that takes $Riem\,\Sigma/Diff\,\Sigma$ as the classical configuration space, the most general state vector will not be simply a function on $Riem\,\Sigma/Diff\,\Sigma$ but rather a cross-section of a flat vector bundle on this space. Equivalence classes of such bundles are in one-to-one correspondence with homomorphisms from the group $\pi_1(Riem\,\Sigma/Diff\,\Sigma)$ into the unitary group $U(n)$ for any n, and the ensuing "θ-angle" (i.e., the specific choice of homomorphism) may have important physical effects. However, this analysis is based on the idea that quantum state vectors are essentially functions on the classical configuration space Q, and must be reconsidered carefully in this infinite-dimensional case if a more appropriate choice is a "distributional" version of Q.

3. A Group-Theoretic Approach To Canonical Quantum Gravity

3.1. THE GENERAL THEORY OF CANONICAL QUANTIZATION

The question of the compatibility of a quantized metric with the ideas of Riemannian geometry and of the appropriate concept of a distributional metric can be approached with the aid of a group-theoretic approach to canonical quantization. Since the details have been presented at length elsewhere (Isham 1984), only a very brief description will be given here.

Let us start by considering the problem of constructing a quantum theory for a classical system whose phase space is some general symplectic manifold $\mathfrak{S}$. The conventional approach would be to replace the classical observables (real-valued functions on $\mathfrak{S}$) with self-adjoint operators on a Hilbert space in such a way that Poisson brackets go into operator commutators. This means that quantization consists of finding an (irreducible) self-adjoint representation of the Lie algebra of the full group of symplectic/canonical transformations of $\mathfrak{S}$; a refinement would be to seek some $\hbar$-dependent deformation of this classical Poisson bracket algebra. Unfortunately, operator-ordering problems (Groenewold 1946; Van Hove 1951) render this program impossible, and in practice the scheme can only be implemented for small subsets of observables. But this raises the critical question of how such preferred subsets are to be chosen. In the familiar example where $\mathfrak{S}$ is a vector space, the natural choice is the set of globally-defined canonical coordinates $(x^1, x^2, \ldots x^n; p_1, p_2, \ldots p_n)$ (with appropriate care, n may be infinite) satisfying the usual CCR

$$[x^i, x^j] = 0, \quad [p_i, p_j] = 0, \quad [x^i, p_j] = i\hbar\delta^i_j. \tag{3.1}$$

The operator representation of this algebra is required to be irreducible, this being the quantum analogue of the property that every classical observable can be written as a function of this basic set of observables.

For a general phase space $\mathfrak{S}$, there will be no such globally-defined coordinate functions, and it is not clear a priori how to proceed. Of course, in any specific theory, a particular set of preferred observables may suggest itself as being especially natural; a good example are the loop-space observables currently being investigated (Rovelli and Smolin 1990; Smolin 1991) in the context of Ashtekar's formulation of canonical general relativity (Ashtekar 1988). In general, the nearest analogue to the coordinate functions $(x^1, x^2, \ldots x^n; p_1, p_2, \ldots p_n)$ would be a set of classical observables $f_A \in C(\mathfrak{S}, \mathbb{R})$ that are labelled by the elements A of a Lie algebra $\mathfrak{L}$ and are such that:

(i) $\{f_A, f_B\} = f_{[AB]}$, where $[AB]$ denotes the Lie bracket of a pair of elements $A, B \in \mathfrak{L}$, i.e., the Poisson bracket algebra of $\{f_A \mid A \in \mathfrak{L}\}$ is isomorphic to $\mathfrak{L}$.

(ii) The set of observables $\{f_A \mid A \in \mathfrak{L}\}$ is *small* enough to avoid the Van Hove phenomenon. This allows a consistent quantization, in which Poisson brackets are replaced by commutators; i.e., one seeks self-adjoint operator representations of the Lie algebra $\mathfrak{L}$.

(iii) The set of observables $\{f_A \mid A \in \mathfrak{L}\}$ is *large* enough that every observable can be expressed as a function of members of this set. This makes it meaningful to look for operator representations of $\mathfrak{L}$ that are *irreducible*.

In a general symplectic manifold $\mathfrak{S}$, there is no guarantee that such a preferred set of variables will exist. One approach to finding such a set is to look for a subgroup $\mathcal{G}$ of the group of all symplectic transformations of $\mathfrak{S}$ whose Lie algebra can serve as a model for $\mathfrak{L}$ (Isham 1984). The critical step is to find a group that is big enough so that every observable can be written as a function of its generators but which is small enough to avoid the Van Hove phenomenon. The first condition will be satisfied if the group G acts transitively on $\mathfrak{S}$, and a sufficient condition for the second is that G contains no proper subgroups that act transitively. If these requirements can be satisfied, then (modulo certain global problems involving $\pi_1(\mathfrak{S})$, which are not relevant to our present concerns) $\mathfrak{L}$ can be chosen to be either the Lie algebra $\mathfrak{L}(G)$ of G, or a central extension of it (this depends on the fine details of the G-action on $\mathfrak{S}$).

Thus, the steps in quantizing the system are as follows.

(i) Find a suitably small, transitively-acting, group G of symplectic transformations of $\mathfrak{S}$.

(ii) Decide if $\mathfrak{L}(G)$ can be used, or whether a central extension is needed.

(iii) Look for irreducible *unitary* representations of the Lie group G (or, if appropriate, of a central extension of G). The self-adjoint generators of this representation will be the preferred quantum observables.

Thus, we will end up with a set $\{\hat{f}_A \mid A \in \mathfrak{L}\}$ of self-adjoint operators satisfying

$$[\hat{f}_A, \hat{f}_B] = i\hat{f}_{[AB]} \quad \text{(plus a possible central term)}, \qquad (3.2)$$

and which constitute the kinematical aspects of the quantization. In effect, we have replaced the familiar Weyl–Heisenberg group of ordinary wave mechanics (i.e., when $\mathfrak{S} = \mathbb{R}^{2n}$) with a canonical group G that is more adapted to the global structure of the new phase space. For many symplectic spaces there will be no such group, in which case one is forced to think again. Or there may be more than one group that satisfies the necessary conditions. This will produce genuinely inequivalent quantizations of the same classical system, corresponding to different choices for the basic set of preferred observables. In the context of quantum gravity, the crucial claim is that suitable groups *do* exist and that a group-theoretic approach of this type can be used to handle some of the problems concerning the compatibility of classical geometrical concepts with a quantized metric field. But first let us give a few simple examples where the scheme sketched above can be implemented successfully.

3.2. Some Examples of the Group-Theoretic Quantization Scheme

(a) The simplest case is when the configuration space Q is just the trivial vector space $\mathbb{R}^n$. Then $\mathfrak{S} = T^*\mathbb{R}^n \approx \mathbb{R}^{2n}$, and the obvious choice for the transitively acting symplectic group is the Abelian group $\mathbb{R}^{2n}$. This turns out to be one of the (rather rare) cases in which a central extension is needed, and the final group is just the Weyl–Heisenberg group. Thus we reproduce the usual quantization of a system with configuration space $\mathbb{R}^n$.

(b) A far more interesting example is when Q is a homogeneous space G/H for some pair of finite-dimensional Lie groups G and H so that $\mathfrak{S} = T^*(G/H)$. It can be shown (Isham 1984) that a suitable canonical group can be obtained by first finding a vector space W that carries a linear representation of G with the property that at least one of the G-orbits is diffeomorphic to G/H (i.e., there is a vector in W such that the isotropy group of the G-action is H). Then, an appropriate set of preferred observables is given by the generators of G and the elements of W^* (considered as functions on the copy of Q embedded in W), and the canonical group is the semi-direct product $W^* \circledS G$. Vis-a-vis the more familiar quantization with the Weyl–Heisenberg group, it should be noted that: (i) the group G of "momentum" variables is typically *not* abelian; (ii) there is no reason why the number of "q" variables and "p" variables should be the same (i.e., $\dim W^* \neq \dim G$ in general); (iii) since $\mathcal{G}$ is a semi-direct product its representations can be studied with the aid of Wigner/Mackey induced representation theory; (iv) topological effects that are sometimes anticipated in the quantum version of a classical system (such as the use of flat vector bundles mentioned earlier) should be reflected in the representation theory of the group.

It is important to note that this scheme can also be applied to cases in which the configuration space is *infinite* dimensional, including all non-linear σ-models with a homogeneous target space, as well as canonical quantum gravity where $Q = Riem\,\Sigma$. A careful analysis of these infinite-dimensional systems shows that an appropriate choice for W^* is not the algebraic dual (which is much too big), but an L^2-dual. The resulting canonical group $\mathcal{G}$ is typically of the form $V \circledS G$, where V is a topological vector space of smooth functions and G is an infinite-dimensional Lie group. A spectral theory analysis applied to the Abelian group V shows that typical states will involve functions (or cross-sections of vector bundles) defined on an orbit (or collection of orbits if the action is strictly ergodic) of the G-action on the topological dual space V'. However, if V is a space of smooth functions, V' typically contains *distributions,* and this is the means whereby one can obtain an appropriate definition of what is meant by a distributional version of a nonlinear σ-model map or, in quantum gravity, a distributional metric.

(c) The one-dimensional system with $Q = \mathbb{R}_+$ (the set of all positive real numbers) was considered many years ago by Klauder (1970) as a model for the positive-definite restriction on the metric tensor in canonical quantum gravity. As remarked in the context of (2.6), conventional CCR are incompatible with this system, and Klauder suggested replacing $[x, p] = i$ with the "affine" relations

$$[x, \pi] = ix \tag{3.3}$$

on the grounds that, in the two (nontrivial) irreducible representations of (3.3), the spectrum of the operator x is strictly positive and strictly negative respectively. Thus, by choosing the appropriate representation, it is possible to quantize the system in a way that is compatible with the classical constraint $x > 0$. From the group-theoretical perspective espoused above, it is easy to see the origin of (3.3). The configuration space $Q = \mathbb{R}_+$ can be written as a homogeneous space G/H by letting G be the multiplicative group $\mathbb{R}_+$ of positive numbers, which acts transitively on Q in the obvious way: $(a, x) \rightarrow ax$, where $a \in G$ and $x \in Q$ (note that H is trivial in this case). The simplest choice for a vector space W carrying a linear G representation with this orbit is $\mathbb{R}$ (with the linear action $(a, x) \rightarrow ax$, where $a \in G \approx \mathbb{R}_+$ and $x \in W \approx \mathbb{R}$). Thus, the canonical group $\mathcal{G}$ is just $\mathbb{R}^* \circledS \mathbb{R}_+ \approx \mathbb{R} \circledS \mathbb{R}_+$, whose Lie algebra relations are precisely (3.3).

(d) A more realistic model for quantum gravity is given by the infinite-dimensional system whose configuration space is $Q = C(\Sigma, \mathbb{R}_+)$ — the set of all smooth, positive real functions defined on the spatial manifold Σ. The natural choice for G is the group $C(\Sigma, \mathbb{R}_+)$, in which multiplication is defined point-wise by $(\Lambda_1 \Lambda_2)(x) := \Lambda_1(x)\Lambda_2(x)$ for all $\Lambda_1, \Lambda_2 \in C(\Sigma, \mathbb{R}_+)$ and $x \in \Sigma$. The action λ_Λ of $\Lambda \in C(\Sigma, \mathbb{R}_+)$ on a function φ in Q is similarly defined as $(\lambda_\Lambda \varphi)(x) := \Lambda(x)\varphi(x)$, and the vector space W carrying the G-representation is $C(\Sigma, \mathbb{R})$. Thus (using an L^2-dual), the canonical group $\mathcal{G}$ is the semi-direct product $C(\Sigma, \mathbb{R}) \circledS C(\Sigma, \mathbb{R}_+) \approx C(\Sigma, \mathbb{R} \circledS \mathbb{R}_+)$, whose basic nonvanishing Lie bracket is (cf. (3.3))

$$[\varphi(f), \pi(h)] = i\varphi(fh), \tag{3.4}$$

where $f \in \mathfrak{L}(C(\Sigma, \mathbb{R})) \approx C(\Sigma, \mathbb{R})$ and $g \in \mathfrak{L}(C(\Sigma, \mathbb{R}_+)) \approx C(\Sigma, \mathbb{R})$. Hence, the present system can be regarded as a "current algebra" version of the theory with $Q = \mathbb{R}_+$.

In the context of our interest in the role of the topology of Σ, there is an interesting difference between the affine relations (3.4) and the more conventional scalar field CCR (1.6). In order to define the latter, it is first necessary to specify the inner product (f, g) between a pair of functions on the spatial manifold Σ. Typically, this involves defining a measure on Σ, which is likely to depend on the topology on Σ. On the other hand, the affine relations (3.4) do not invoke the topology of Σ at all. Indeed, it is not the Hilbert space structure of the test functions that is relevant (as in the case of the conventional CCR) but rather their *ring* structure. This is of some interest since the topological structure of a compact Hausdorff space can be uniquely reconstructed from the algebraic structure of its ring of continuous functions. Of course, to "recover" the field $\varphi(x)$ at a point from the smeared operators $\varphi(f)$, heuristically one thinks of taking a sequence of functions f that converges (formally) to a δ-function, and this notion of convergence employs a topology on the space of functions, which usually involves the topology on Σ (although note that even then it is a Banach algebra, rather than Hilbert space, structure that is naturally suggested by the algebraic form of (3.4)). However, the fact that affine relations can be defined independently of the topological structure on Σ could be significant if one were contemplating the construction of a quantum field theory in which the topology of Σ was itself subject to quantum fluctuations.

3.3. QUANTIZING THE METRIC TENSOR

Finally, we can consider the application of the group-theoretical approach to quantum gravity, in which the configuration space is the set $Riem\,\Sigma$ of all Riemannian metrics on Σ. With the aid of the global triad θ^a of one-forms introduced earlier, a one-to-one correspondence is established between metrics and functions taking their values in the set $Sym_+(3, \mathbb{R})$ of real, symmetric, positive-definite 3×3 matrices. Thus the configuration space can be taken to be the space of smooth functions $C(\Sigma, Sym_+(3, \mathbb{R}))$. The analogy with the progression above from $Q = \mathbb{R}_+$ to $Q = C(\Sigma, \mathbb{R}_+)$ suggests that it might be instructive to consider first the finite-dimensional model with $Q = Sym_+(n, \mathbb{R})$. The critical point is that the naive CCR

$$[S_{ab}, \pi^{cd}] = i\delta_a^{(c}\delta_b^{d)} \tag{3.5}$$

are incompatible with the positive-definiteness of the matrices S in $Sym_+(n, \mathbb{R})$, just as $[x, p] = i$ is incompatible with the constraint $x > 0$.

The first step in constructing the correct quantization is to find a group G that acts transitively on $Sym_+(n,\mathbb{R})$. The obvious choice is $GL^+(n,\mathbb{R})$ with the left action $\lambda_g(S) := gSg^t$, where g^t denotes the transpose of the matrix $g \in GL^+(n,\mathbb{R})$. Clearly, this preserves the symmetry and positive-definiteness of $S \in Sym_+(n,\mathbb{R})$, and transitivity follows from the fact that every symmetric matrix can be diagonalized by an orthogonal group operation, i.e., for each $S \in Sym_+(n,\mathbb{R})$ there exists an element $O \in SO(n,\mathbb{R}) \subset GL^+(n,\mathbb{R})$ such that $S = ODO^t$, where D is a diagonal matrix. Since S is positive-definite, the diagonal elements of D are positive real numbers, and hence a positive square root $D^{1/2}$ is well defined. Thus, we can write $S = (OD^{1/2})(OD^{1/2}) = \lambda_{(OD^{1/2})}(1)$, where 1 is the unit $n \times n$ matrix. Hence, the $GL^+(n,\mathbb{R})$ action is transitive.

The isotropy group of this action at 1 is clearly $SO(n,\mathbb{R})$ so that $Sym_+(n,\mathbb{R})$ is diffeomorphic to the homogeneous space $GL^+(n,\mathbb{R})/SO(n,\mathbb{R})$. The next step is to find a vector space W on which $GL^+(n,\mathbb{R})$ acts linearly such that the isotropy group at some vector is $SO(n,\mathbb{R})$. An obvious choice is the set $Sym(n,\mathbb{R})$ of *all* $n \times n$ real, symmetric matrices, with the (linear) action of $g \in GL^+(n,\mathbb{R})$ taking $M \in Sym(n,\mathbb{R})$ into gMg^t. This provides an equivariant embedding of $Sym_+(n,\mathbb{R})$ as an open subset of the vector space $Sym(n,\mathbb{R})$ and leads to the canonical group $\mathcal{G} = Sym(n,\mathbb{R}) \circledS GL^+(n,\mathbb{R})$. ($Sym(n,\mathbb{R})$ can be identified with its dual by associating $A \in Sym(n,\mathbb{R})$ with the linear map $B \rightarrow \text{Trace}(AB)$). The Lie algebra of this group is

$$[S_{ab}, S_{cd}] = 0, \tag{3.6}$$

$$[\pi_a{}^b, \pi_c{}^d] = i(\delta_a^d \pi_c{}^b - \delta_c^b \pi_a{}^d), \tag{3.7}$$

$$[S_{ab}, \pi_c{}^d] = i\delta_{(a}^d S_{b)c}. \tag{3.8}$$

This provides a concrete illustration of several general points made earlier. For example, (i) the number of "q"-variables is $\frac{1}{2}n(n+1)$, which is not the same as the number n^2 of "p-variables;" (ii) the "p-group" is $GL^+(n,\mathbb{R})$, whose Lie algebra (3.7) is non-Abelian; and (iii) the canonical group is a semi-direct product, whose representations can be studied with the aid of induced representation theory. For a general semi-direct product $V \circledS G$, this involves constructing vector bundles over the orbits of G in the dual of V. One particular orbit in $Sym(n,\mathbb{R})$ is, of course, $Sym_+(n,\mathbb{R})$ itself, with the little group $SO(n,\mathbb{R})$. Thus, a typical state vector will be a vector-valued function (more precisely, a cross-section) $\psi^a(S)$, in which the index a corresponds to some representation of this little group. Thus, a generic state involves an $SO(n, \mathbb{R})$ "internal spin" as well as the "orbital motion" associated with ψ^a being a function on the classical configuration space $Sym_+(n, \mathbb{R})$. Note, however, that there are other orbits in $Sym(n, \mathbb{R})$ composed of degenerate matrices whose rank is less than n. These could be relevant to the study of the effect of quantum theory on the gravitational singularities that arise in classical general relativity.

This finite-dimensional model can be generalized to the case of canonical quantum gravity just as the quantization of $Q = \mathbb{R}_+$ was generalized to $Q = C(\Sigma, \mathbb{R}_+)$. The configuration space is now $Q = C(\Sigma, Sym_+(3, \mathbb{R}))$, and the natural choice for G is $C(\Sigma, GL^+(3, \mathbb{R}))$, with the action of $\Lambda \in C(\Sigma, GL^+(3, \mathbb{R}))$ on $g \in C(\Sigma, Sym_+(3, \mathbb{R}))$ being defined as

$$(\lambda_\Lambda g)_{ab}(x) := \Lambda_a{}^c(x)\Lambda_b{}^d(x)g_{cd}(x). \tag{3.9}$$

The vector space W can be taken to be $C(\Sigma, Sym(3, \mathbb{R}))$, and (using an L^2-dual) the final canonical group is (Isham 1984; Isham and Kakas 1984a, 1984b):

$$\mathcal{G} = C(\Sigma, Sym(3, \mathbb{R})) \circledS C(\Sigma, GL^+(3, \mathbb{R}))$$
$$\approx C(\Sigma, Sym(3, \mathbb{R}) \circledS GL^+(3, \mathbb{R})),$$

with the (unsmeared) commutation relations

$$[g_{ab}(x), g_{cd}(y)] = 0, \tag{3.10}$$
$$[\pi_a{}^b(x), \pi_c{}^d(y)] = i(\delta_a^d \pi_c{}^b(x) - \delta_c^b \pi_a{}^d(x))\delta(x, y), \tag{3.11}$$
$$[g_{ab}(x), \pi_c{}^d(y)] = i\delta_{(a}^d g_{b)c}(x)\delta(x, y). \tag{3.12}$$

A comparison with (3.6–8) makes clear the sense in which the kinematical aspects of canonical quantum gravity can be viewed as a (noncompact) current algebra version of the quantization of a system whose configuration space is $Sym_+(3, \mathbb{R})$. The algebra (3.10–12) is geared to maintaining as much as possible of the geometrical structure of Riemannian geometry, and was first introduced in this context by Klauder (1970) and Pilati (1982, 1983). There is an obvious generalization to triads (rather than metrics) and the formalism can also be extended to the case where the manifold Σ has a dimension other than three. If Σ is not parallelizable, there will not be any globally-defined frame θ^a, and therefore group transformations cannot be used on the indices of the metric tensor as in

(3.9). Instead, the group G becomes the group of automorphisms of the frame bundle of Σ, and the vector space W is the set of all symmetric covariant tensor fields on Σ.

A number of interesting things can be said about the affine commutation relations (3.10–12). For example, $\pi_a{}^b(x)\pi_b{}^a(x)$ is a Casimir function for the group $C(\Sigma, GL^+(3, \mathbb{R}))$, and is a natural "free-field" Hamiltonian for the system. But this is essentially what is obtained from the quantum gravity "Hamiltonian" (1.19) in the limit as Newton's constant $G \to \infty$. The use of such a strong-coupling perturbation theory in quantum gravity (Isham 1976; Teitelboim 1980) would probe a totally different regime from that of the weak-field perturbative techniques of conventional particle-physics approaches to the subject.

However, any real development in this area requires a deep understanding of the representation theory of the canonical group $C(\Sigma, Sym(3, \mathbb{R}))$ ⑧ $C(\Sigma, GL^+(3, \mathbb{R}))$. As was remarked earlier, the orbits of $C(\Sigma, GL^+(3, \mathbb{R}))$ on the topological dual of $C(\Sigma, Sym(3, \mathbb{R}))$ would be closely involved in answering questions about the role in quantum gravity of distributional Riemannian geometry. Some simple representations were discussed in Klauder and Aslaken 1970, Pilati 1982, 1983, and Isham and Kakas 1984b, but little is known in general. It seems very likely that any really useful representations will involve strictly ergodic actions of $C(\Sigma, GL^+(3, \mathbb{R}))$ on the dual of $C(\Sigma, Sym(3, \mathbb{R}))$, and it would be of considerable interest if concrete examples of this type could be found. The role of the diffeomorphism group $Diff\,\Sigma$ is also rather intriguing in this context, especially in relation to the expected occurrence of physical observables labelled by equivalence classes of geometrical objects in Σ.

Another important question concerns the appropriate regularization of formally divergent expressions. Klauder (1973) has argued strongly that a multiplicative renormalization is most appropriate for systems with affine-type commutation relations, but this poses peculiar problems in the case of quantum gravity. In particular, it turns out to be rather difficult to maintain the correct $Diff\,\Sigma$ behavior of a regularized object of this type (Pilati 1983; Isham and Kakas 1984b). A good example is the inverse $g^{ab}(x)$ of the Riemannian metric. In order to get the correct diffeomorphism group behavior, it seems to be necessary to introduce an extra scalar field φ into the theory, and then the regularized inverse and the original metric operator satisfy the relations

$$\left(g_{reg}^{ab}(x)\, g_{bc}(x)\right)_{reg} = \delta_c^a \varphi(x). \qquad (3.13)$$

This suggests that there might be two natural phases in the theory. One would be a weak-field phase in which $\langle \varphi(x) \rangle$ is a nonzero constant (or function), corresponding presumably to the presence of some nonvanishing background metric $\langle g_{ab}(x) \rangle$, about which the fields can be expanded perturbatively; in particular, an inverse of the metric could be defined (at least formally) in this way. The second phase would be associated perhaps with some strong-field limit and would be signalled by the vanishing of $\langle \varphi(x) \rangle$. The expectation value of $g_{ab}(x)$ could also be expected to vanish in this limit, corresponding to the absence of any background metric. Thus, it might be impossible to define the inverse metric (and associated geometrical objects) in this "confined" phase. Speculations of this type are intriguing, and are reminiscent of the interpretation advanced by Witten (1988) of his "topological quantum gravity." However, any further advance in this direction is conditional on finding out more about the representations of the algebra (3.10–12).

Finally, let us note the existence of some potentially interesting topological properties of the canonical group $\mathcal{G} = C(\Sigma, Sym(3, \mathbb{R})) \circledS C(\Sigma, GL^+(3, \mathbb{R}))$. Firstly, it is not connected, i.e., $\pi_0(\mathcal{G}) \neq \{0\}$. To see this, we can exploit the fact that this group space is homotopic to $C(\Sigma, SO3)$—a space that has been studied in some detail in the past (Shastri et al. 1980; Isham 1983). It follows from these results that the group $\pi_0(\mathcal{G})$ fits into a short exact sequence of the form

$$0 \longrightarrow H^3(\Sigma, \mathbb{Z}) \longrightarrow \pi_0(\mathcal{G}) \longrightarrow H^1(\Sigma, \mathbb{Z}_2) \longrightarrow 0, \qquad (3.14)$$

and is therefore determined by the cohomology groups (and hence the topology) of Σ. If Σ is any compact three-manifold, $H^3(\Sigma, \mathbb{Z}) \approx \mathbb{Z}$, and hence the different components of the canonical group are labelled by an integer (a "large" winding-number) plus a torsion element from $H^1(\Sigma, \mathbb{Z}_2) \approx Hom(\pi_1(\Sigma), \mathbb{Z}_2)$. However, the sequence (3.14) does not split into a direct sum of $H^3(\Sigma, \mathbb{Z})$ and $H^1(\Sigma, \mathbb{Z}_2)$; instead, there is the more complicated result:

$$\pi_0(\mathcal{G}) \approx \{(\alpha, \beta) \in H^1(\Sigma, \mathbb{Z}_2) \oplus H^3(\Sigma, \mathbb{Z}) \mid \alpha \cup \alpha \cup \alpha = \beta \bmod 2\}, \qquad (3.15)$$

which identifies $\pi_0(\mathcal{G})$ with a specific subgroup of $H^1(\Sigma, \mathbb{Z}_2) \oplus H^3(\Sigma, \mathbb{Z})$.

Another significant topological property is that the canonical group is not simply-connected. Such a phenomenon cannot occur for the Weyl group (which is topologically trivial) and is potentially one of the most interesting properties of the group-theoretical quantization of general phase spaces. In the present case, note that the phase space $T^*Riem\,\Sigma$ is contractible. Thus, the effect is a genuine property of the canonical group, not of the phase space on which it acts. The full expression for $\pi_1(\mathcal{G})$ is rather complicated, but, for example, if $\Sigma \approx S^3$, then $\pi_1(\mathcal{G})$ contains a part that comes from the relation $\pi_4(SO3) \approx \mathbb{Z}_2$. This means there could exist *spinorial*-type representations of the affine commutation relations (3.10–12) associated with faithful representations of the universal covering group of the canonical group $\mathcal{G}$. It would be extremely interesting to find a concrete example of a representation of this type.

4. Quantum Topology

4.1. THE EUCLIDEAN, CO-BORDISM APPROACH

The idea that the topological structure of space and/or space-time should be quantized has recurred frequently in studies of quantum gravity. The first advocate was Wheeler (1964) who, arguing from within the framework of the canonical formalism, claimed that large quantum fluctuations in the gravitational field would lead to associated changes in the topology of the three-manifold. More recently, Hawking (1984, 1987, 1988) has developed a similar idea within the context of his Euclidean approach to quantum gravity, leading to a picture of "virtual black-holes," "baby-universes," and an associated foam-like structure for the four-dimensional space-time. When employing the canonical formalism, it is tempting to talk, as above, of "changes" in the topology; but this could be very misleading if by this is meant changes with respect to some continuous time parameter. We have argued already that the notion of "time" in quantum gravity has at best a phenomenological and semi-classical significance, and it would be even more strange to talk of conventional "changes" if the topology of space were involved.

One advantage of the Euclidean approach over the canonical formalism is that it is harder to fall into this trap. For each four-manifold $\mathfrak{M}$, whose boundary is the disjoint union of a pair of three-manifolds Σ_1 and Σ_2, there is an associated "transition" amplitude defined by

$$K_{\mathfrak{M}}(g_1, \Sigma_1; g_2, \Sigma_2) = \int d\mu(g_{\mathfrak{M}}^{(4)}) \exp[-S(g_{\mathfrak{M}}^{(4)})], \qquad (4.1)$$

where the functional integral is over all Riemannian metrics $g_{\mathfrak{M}}^{(4)}$ on $\mathfrak{M}$ whose restrictions to Σ_1 and Σ_2 are g_1 and g_2 respectively. The complete transition amplitude is the sum over all such four-manifolds:

$$K(g_1, \Sigma_1, g_2, \Sigma_2) = \sum_{\mathfrak{M}} \nu(\mathfrak{M}) \, K_{\mathfrak{M}}(g_1, \Sigma_1; g_2, \Sigma_2), \qquad (4.2)$$

in which ν is some weight factor (matter fields can be included in a similar way).

However, although the word "transition" has been used, a four-manifold that interpolates between topologically inequivalent three-spaces cannot be foliated globally into a one-parameter family of three-manifolds, and hence there is no global time with respect to which the change could be said to take place (a feature that becomes even more pronounced after summing over the interpolating manifolds). Cobordisms of this type are not possible in classical general relativity, but this is no contradiction since the four-manifolds in the quantum theory carry metrics with a Euclidean, rather than Lorentzian, signature. This is consistent with a view of the topology change as being due to "quantum tunnelling," and therefore with no analogue in the classical world. If, as in the Hartle–Hawking scheme for quantum cosmology, (4.1) is viewed as a way of generating a solution to the Wheeler–DeWitt equation (or a Green function—see Halliwell 1988, 1991) at the three-dimensional boundaries, then the problem of "time" reverts to the one mentioned earlier, i.e., the recovery of an internal time from the wave functional of the three-metrics. Of course, something rather odd must happen to the Wheeler–DeWitt equation if (4.1) is studied close to a bifurcation of topology, but this is to be expected, and is not a criticism of the Euclidean formalism per se.

The question of whether Riemannian or Lorentzian metrics should be employed in quantum gravity has been debated at length, and the argument will doubtless continue. The use of Lorentzian metrics in discussions of topology change cannot be rule out *ab initio* provided one is prepared to accept degenerate points or closed time-like curves. This has been discussed by Sorkin (1988), and was used by Anderson and DeWitt (1986) in their stimulating analysis of the "pair-of-pants" problem. However, my main objection to expressions like (4.1–2) (in either Euclidean or Lorentzian form) is that they are heavily oriented towards the use of smooth manifolds and continuum physics. If quantizing fields leads naturally to distributional objects, one might wonder whether something similar is also true for topology, in which case (4.2) should be a sum/integral over spaces that are more singular than manifolds. The problem then is one of knowing how to select: (i) this class of spaces, and (ii) the measure ν defined on them (which is related to the issue of whether "dynamical laws" should be assigned to the topology changes). These questions appear rather naturally in the Regge-calculus approximation to Euclidean quantum gravity, and have been discussed in this context by Hartle (1985).

The usual response to this worry is that one is taking too seriously equations (4.1–2), which are merely a low-energy approximation to the true theory of quantum gravity. That may well be so. But then it is clear (Coleman 1988) that topological entities like wormholes can only be discussed meaningfully in this framework if their size is considerably larger than the Planck scale. If one is interested (as I am) in real Planck length effects (i.e., where the theory based on (4.1) breaks down), then it seems unlikely that the "quantum topology" sum (4.2) has much relevance. The heart of the matter is that expressions like (4.1–2) are grounded firmly in the classical-to-quantum route to quantum gravity, whereas quantum topology seems to cry out for a quantum-to-classical treatment.

4.2. CANONICAL QUANTUM TOPOLOGY

For the reason outlined above, Noah Linden and I have been looking recently at the possibility of constructing a genuine theory of quantum topology that does not assume too many continuum concepts a priori. The approach is "canonical" in the sense that we are, quite literally, interested in things like wave functions $\psi(\tau)$ of topologies τ, but the word does not imply a commitment to any specific view of time-development and topologies "changing" in time; that would require a complete picture of quantum gravity, which we certainly do not have.

Many questions arise in contemplating an enterprise of this type. For example:

(i) Can we talk about "all" topologies in a meaningful way?

(ii) Do they form a "nice" space? For example, does the family of all topologies form a topological space in its own right?

(iii) What is special about "topologies" rather than other types of mathematical structure?

(iv) Is the use of topological spaces that are *not* manifolds compatible with the basic ideas of quantum theory?

Let us consider these briefly in turn. The answer to the first question is "no," in the sense that the family of all topologies forms a class, not a set. This is not unrelated to the Russell-type paradox lurking in question (ii).

One way of resolving this problem is to consider just the family $\tau(X)$ of all topologies on a particular set X. Thus, each member of this family is a set θ of subsets of X that satisfy the axioms for a topology, i.e., θ contains X and the empty set Φ, and is closed under finite intersections and arbitrary unions. The collection $\tau(X)$ of all such θ *is* a set, and so it is meaningful to ask whether it carries a natural topological structure in its own right. However, we should consider first how this master set X is to be chosen. In some primordial sense, its elements represent the points of physical space, and therefore its most fundamental property is perhaps its *cardinality,* i.e., the "number" of points it contains. I would like to argue that this is the *only* property that is relevant at this stage, and therefore it should not matter if X is replaced by any other set of the same cardinality. In particular, any topology placed on $\tau(X)$ should be invariant under the action on the subsets of X induced by the group $Perm(X)$ of all bijections of X onto itself; thus, $Perm(X)$ is a sort of gauge-group for the theory. Note that fixing the cardinality of X still leaves a large class of spaces. For example, if X has the cardinality of the real numbers, then every differentiable manifold of any dimension is included, as is the great host of spaces with this cardinality that do not admit a differentiable structure. One anticipates that a suitable topology on $\tau(X)$ would have the property that the special points in $\tau(X)$ corresponding to differentiable manifolds can serve in a natural way as the classical limits of a quantum theory of topology.

From a mathematical perspective, there are several natural ways of imparting a topological structure to $\tau(X)$, although which of them, if any, is relevant to a quantum theory is debatable. One possibility is to exploit the fact that $\tau(X)$ is a partially ordered set, in which $\tau_1 < \tau_2$ means that τ_1 is strictly weaker than τ_2 (i.e., it has fewer open sets). A topology can be induced on $\tau(X)$ by choosing a sub-basis to be the set of all subsets of $\tau(X)$ of the form $O_{\tau_1 \tau_2} := \{\tau \in \tau(X) \mid \tau_1 \leq \tau \leq \tau_2\}$ (just as the topology on the real line is generated by its closed intervals). Some variations on this theme would be to use open, or semi-infinite, chains. Note that the action of $Perm(X)$ preserves the partial ordering on $\tau(X)$, and hence these topologies satisfy the requirement of $Perm(X)$-invariance.

Other possibilities stem from the natural "Vietoris type" topologies carried by the closed sets $\mathfrak{F}(Y)$ of any topological space Y (Michael 1951; Marjanovic 1966; Beer et al 1987). For example, if U is any subset of Y, define

$$A^- := \{F \in \mathfrak{F}(Y) \mid F \cap A \neq \Phi\}, \quad A^+ := \{F \in \mathfrak{F}(Y) \mid F \subseteq A\}.$$
(4.3)

Then the basic Vietoris topology on $\mathfrak{F}(Y)$ is defined to be the one with a sub-base that is all sets of the form U^- and U^+, where U is open in Y. Now, any topology θ on X is a subset of $P(X)$—the set of all subsets of X. Thus, if $P(X)$ can be equipped in some way with a topology such that each θ is a *closed* subset of $P(X)$, then the set $\tau(X)$ of all topologies on X could be given a Vietoris-type topology with respect to $P(X)$. The power set $P(X)$ might be topologized in various ways (for example, it too is a complete lattice), and the question of whether a generic topology θ on X is a closed subset of $P(X)$ is significantly related to the intriguing question: "what is special about topologies?"

For example, if any particular θ is not closed, its closure could be taken and then considered as the basic entity. But the closure of θ may well be a family of subsets of X that do not themselves satisfy the axioms for a topological space, thus providing a route whereby mathematical structures other than topology could enter. Certainly, it is difficult to give any general answer to the question, "why topologies?". There are many mathematical structures that are similar to topology and with which a differentiable manifold is automatically equipped (e.g., limit spaces, uniform spaces, proximity spaces, measure spaces...). Although we tend to talk of quantum "topology," there is no a priori reason for preferring one of these structures to another once we have escaped from the category of differential geometry. It is important therefore to see if certain types of structure can be more readily topologized than others, and hence perhaps to get a handle on what the domain space of quantum "topology" might really be. (For another recent approach to this question see Isham, 1989).

The final question: "are nonmanifold structures compatible with quantum theory?" can be read in a variety of ways. At a technical level, we might note that, whatever topology is finally chosen for $\tau(X)$, it will be precisely that: a topology, but *not* a differentiable manifold. Thus, we have a special case of a rather more general question: "can one quantize a system whose configuration space Q is a topological space but not a smooth manifold?" Of course, in a sense, the use of the verb "quantize" is particularly inappropriate here. If Q is not a manifold, there is no phase space T^*Q, and hence no classical system to quantize in the first place. However, there are various ways in which standard quantum theory might be extended to include this case. For example, state vectors could be defined to be functions on Q, which at least would allow the continuous functions $C(Q)$ on Q to be afforded the status of quantum operators. A crucial problem then would be to decide on an appropriate measure μ on Q, so that the quantum state space is $L^2(Q, d\mu)$. However, it is salutary to recall that such a strategy breaks down already for conventional quantum field theory, where one is obliged to employ the distributional dual of Q. It is intriguing to wonder if there is a "distributional" version of $\tau(X)$, and if that is relevant to quantum topology!

Another key question concerns the correct choice of operators (the "momenta") to supplement the Abelian algebra $C(Q)$ of the configuration variables. It is no good saying that every self-adjoint operator on $L^2(Q, d\mu)$ must be *some* quantum observable. Even if that were true, it wouldn't help much: the problem is knowing how to *label* the operators in a way that gives them physical meaning.

In conventional quantum theory, the group-theoretical approach is well equipped to handle this problem. For a system with classical phase space T^*Q, the biggest natural group of symplectic transformations is $C(Q) \circledS Diff(Q)$, and the quantization scheme discussed briefly in Section 3.2 involves finding a small subgroup $V \circledS G$ of this group that still acts transitively on T^*Q and whose generators can be identified with specific physical observables. In the case of a system whose configuration space Q is not a manifold, one might proceed by analogy and study various subgroups of $C(Q) \circledS Hom(Q)$, where $Hom(Q)$ is now the group of homeomorphisms of Q. If Q happened to be a homogeneous space G/H (where G and H are topological, but not Lie, groups), then a natural starting point would be the unitary representations of the subgroup $C(Q) \circledS G$. Note that, as a consequence of Stone's theorem, a unitary, weakly continuous representation of G will yield a self-adjoint operator associated with each continuous path in G, even though G is not actually a Lie group (thus the arc-wise connectivity properties of Q and G are of some importance). By this means a sort of "quasi phase-space" structure might be induced on Q.

This discussion is probably hypothetical since it seems unlikely that $\tau(X)$ is a homogeneous space. This problem might possibly be tackled by arguing that we should be constructing a quantum *field* theory over $\tau(X)$, rather than a quantum theory in which $\tau(X)$ is the "configuration" space (the former is actually easier!). This would be consistent with recent work in quantum cosmology, in which the Wheeler–DeWitt function $\Psi[g]$ is regarded as a quantum field (for example, Giddings and Strominger 1988).

However, there is a more fundamental objection to the concept of "quantum topology." The Hilbert space of a quantum theory employs the real/complex numbers with their conventional topology, but does this not already assume implicitly a differentiable picture for space and/or space-time? The real numbers are used in quantum theory because they model the results of making measurements, and all measurements ultimately involve measurements of distances in space, or intervals of time. Hence, it could be argued, conventional quantum theory is only valid in regimes where the representation of space-time by a differentiable manifold is valid. In particular, one must keep well away from the Planck length, in which case to talk of "quantum" effects, topological or otherwise, at that scale would be totally misguided. I must confess that I am very sympathetic to this view, even though it does render most of this paper redundant!

4.3. RIEMANNIAN-METRIC-DRIVEN QUANTUM TOPOLOGY

It is clearly going to be very difficult to invent a credible theory of full quantum topology. But perhaps it is still possible to follow the classical-to-quantum path, and ask a more modest question: "within the context of canonical quantum gravity on a background manifold Σ, is there any mechanism for producing quantum fluctuations of the topology away from that of Σ?" Presumably this is how Wheeler first saw the question of quantum topology, and even if it does not satisfy the desire to construct a theory free of a priori continuum concepts, it may at least afford some insight into the structure of the appropriate space of topologies (or whatever) in the neighborhood of the manifold Σ.

A natural starting point for such an investigation is the familiar expression in classical differential geometry for the distance function $d(x, y)$ between a pair of points x, y on the Riemannian manifold Σ:

$$d(x, y) = \inf_{\gamma} d_\gamma(x, y), \tag{4.4}$$

$$d_\gamma(x, y) = \int_{t_x}^{t_y} (g_{\gamma(t)}(\dot{\gamma}, \dot{\gamma}))^{1/2} dt \equiv \int_{t_x}^{t_y} (g_{ab}(\gamma(t))\dot{\gamma}^a\dot{\gamma}^b)^{1/2} dt, \tag{4.5}$$

in which the *infimum* is taken over all curves γ whose end-points are x and y. Here, $\dot{\gamma}$ is the tangent vector to γ and $\dot{\gamma}^a$ denotes the components of $\dot{\gamma}$ in a frame dual to the frame θ^a of one-forms. This distance function satisfies the basic inequalities that define a metric topology:

$$d(x, y) = d(y, x) \qquad \text{for all } x, y \in \Sigma, \tag{4.6}$$

$$d(x,y) \geq 0 \qquad\qquad \text{for all } x, y \in \Sigma$$
$$= 0 \qquad\qquad \text{if and only if } x = y, \tag{4.7}$$

$$d(x,z) \leq d(x,y) + d(y,z) \quad \text{for all } x, y, z \in \Sigma. \tag{4.8}$$

Of course, the metric function $d(,)$ depends explicitly on the Riemannian metric g, but it is a standard result that the topologies induced by *all* Riemannian metrics are homeomorphic to each other and to the underlying topology on the manifold Σ.

Now suppose we have a canonical quantum theory, in which $g_{ab}(x)$ is a self-adjoint operator. Then (4.4–5) implies that, at least in principle, the distance function also becomes an operator. "Quantum topology" could then be interpreted to mean, for example, that the distance function $d^\psi(x, y)$ defined as the expectation values $\langle \psi, d(x, y)\psi \rangle$ may induce a topology on Σ that depends on the quantum state ψ. If the matrix elements of the operator $g_{ab}(x)$ were genuine Riemannian metrics, then the associated topologies would all be homeomorphic, and nothing new would emerge. However, as we have argued earlier, a quantized metric is likely to be more singular than a classical one, which could have dramatic effects on the quantized distance function. For example, in the representations of the affine algebra (3.10–12) discussed by Isham and Kakas (1984b), there exist states ψ for which $\langle \psi, d(x, y)\psi \rangle = 0$ for pairs of points x, y with $x \neq y$. The effective topology induced on Σ in such a state would be non-Hausdorff, corresponding to the fact that d^ψ is a pseudo-metric rather than a metric. One might argue that points zero distance apart are physically equivalent, in which case they should be identified. This leads to a topology on Σ that is once again Hausdorff, but which is unlikely to be equivalent to that on the original (or any other) manifold. Thus, although we started with a background space that was a three-manifold Σ, "quantum fluctuations" in the topology would take us outside the category of manifolds into the much wider class of general, metrizable topological spaces. A theory of this sort could be said to involve "Riemannian-metric-driven topology change" since the effects stem from quantizing the Riemannian metric on the original manifold Σ.

There are several ways in which we might try to make these heuristic ideas more precise. The first would be to take (4.4–5) literally, and attempt to give some proper meaning to these equations when $g_{ab}(x)$ is an operator. Taking the square root of $g_{ab}(\gamma(t))\dot{\gamma}^a(t)\,\dot{\gamma}^b(t)$ should pose no real problem since part of the motivation for (3.10–12) is its compatibility with the relation $g_{ab}(x)v^a(x)v^b(x) \geq 0$ for any vector field **v**. Similarly, no major difficulty should arise in taking the *infimum* in (4.4) of the collection of unbounded, but positive semi-definite, operators $d_\gamma(x, y)$ (perhaps, after smearing in x and y). The main problem is that, if the integral in (4.5) is construed to mean the operator g smeared with a certain test-function, then the "function" involves a δ-function concentrated on the curve $t \to \gamma(t)$, and presumably this is too singular for the purpose. Thus, one needs to embark on a careful study of the regularization of (4.4), which requires a detailed knowledge of the representations of the commutation relations (3.10–12).

A second possibility, which is very much in the spirit of the group-theoretic approach to canonical quantization, is to construct a *new* canonical algebra that contains $\{d_\gamma(x, y), \gamma \in Path\, \Sigma\}$ as an Abelian subset of operators. Thus, the primary configuration variable would no longer be the Riemannian metric $g_{ab}(x)$ but instead the distance between pairs of points measured along a particular curve γ; however, the conjugate momenta would continue to be something like the existing local variable $\pi_a{}^b(x)$. This asymmetric treatment, in which configuration variables and their conjugates are associated respectively with curves (or, more generally, chains) and points in Σ, is feasible because of a feature of the affine relations that can be seen already in the scalar field example (3.4). We remarked earlier that these relations are meaningful if f, h belong to an arbitrary ring $\mathfrak{A}$, but note that they are also well-defined if the smearing function f for the canonical variable φ is restricted to some *ideal* $\mathfrak{J} \subset \mathfrak{A}$, or if f belongs to a linear space that is a *module* over the vector space generated by the functions h that smear the conjugate variable π. In particular, the set of δ-functions concentrated on paths in Σ generate a module over the ring of smooth functions on Σ. It is this property of the affine relations that allows us to contemplate an asymmetric structure for the canonical variable and its conjugate.

The main problem is to construct an appropriate canonical algebra. The obvious tactic is to use the affine relations (3.10–12) to compute the commutators between $\pi_c{}^d(y)$ and $d_\gamma(x, y)$ (defined heuristically by (4.5)), and then to "axiomatize" these as the relations we are seeking (augmented by (3.11) and the Abelian structure of $\{d_\gamma(x, y) \mid \gamma \in Path\, \Sigma\}$). However, this tactic does not work as it stands because the commutation relations do not close. To see this, note that (3.10–12) imply that the unitary representation operators $U(\Lambda)$ of $\Lambda \in C(\Sigma, GL^+(3, \mathbb{R}))$ intertwine with $g_{ab}(x)$ as

$$U^{-1}(\Lambda)g_{ab}(x)U(\Lambda) = \Lambda_a{}^c(x)\Lambda_b{}^d(x)\, g_{cd}(x), \qquad (4.9)$$

and hence

$$
\begin{aligned}
U^{-1}(\Lambda)d_\gamma(x,y)U(\Lambda) \\
= \int_{t_x}^{t_y} \left(g_{cd}(\gamma(t))\Lambda_a{}^c(\gamma(t))\Lambda_b{}^d(\gamma(t))\dot{\gamma}^a(t)\dot{\gamma}^b(t)\right)^{1/2} dt.
\end{aligned} \tag{4.10}
$$

However, the right hand side of (4.10) is not of the form $d_{\gamma'}(x,y)$ since, although $t \to \dot{\gamma}^a(t)\Lambda_a{}^c(t)$ is a path of tangent vectors, it is not in general tangent to any curve γ' in Σ. One way of resolving this difficulty is to include *all* paths in $T\Sigma$ by defining an extended class of objects

$$
d_c := \int g_{\pi\circ c(t)}(c, c)dt \equiv \int g_{ab}(\pi(c(t)))\,c^a(t)\,c^b(t)dt, \tag{4.11}
$$

where c is any curve in the tangent bundle $T\Sigma$, and $\pi: T\Sigma \to \Sigma$ is the projection. We now have closure in the sense that

$$
U^{-1}(\Lambda)d_cU(\Lambda) = d_{c\Lambda}, \tag{4.12}
$$

where $(c\Lambda)^a(t) := c^b(t)\Lambda_b{}^a(\pi(c(t)))$.

The properties of the associated canonical group are currently being investigated and will be reported elsewhere. This new structure is clearly well adapted to discussions of "Riemannian-metric"-driven quantum topology and, in its use of a path space, is reminiscent of the work by Rovelli and Smolin (1990), whose variables include integrals of the connection along paths in Σ. It would be interesting to see if these two approaches could be combined so that all the basic canonical variables in quantum gravity are associated with paths in Σ or $T\Sigma$.

4.4. QUANTUM THEORY ON $Q = Metric(X)$

A far more ambitious approach to quantizing the distance functions is to deal directly with the set $Metric(X)$ of all metrics (not to be confused with Riemannian metrics) on a set X; i.e., the set of all functions $d: X \times X \to \mathbb{R}$ satisfying the basic relations (4.6–4.8). Two metrics on X are *equivalent* if they induce the same topology on X, but if X is anything other than a finite set it will carry many inequivalent topologies of this type, including all manifolds if its cardinality is that of the real numbers. This takes us away from "quantum fluctuations" around a background manifold topology Σ into the domain of quantum topology proper. This seems to me to be the most promising starting point for constructing a genuine quantum theory of nonmanifold topologies. Metric topologies are a significant generalization of smooth manifolds, and yet the real numbers are still included in a fundamental way (the distance function is real-valued). This may help to allay the fears expressed earlier about a possible basic inconsistency between the use of the continuum of numbers in quantum theory and the desire to escape from differential geometry at the scale of the Planck length.

Thus, the task is to construct a quantum theory associated with the set $Metric(X)$. This would be a scalar quantum field theory, in which the fields $d(x, y)$ are defined on *pairs* of points in X, and which is compatible in some way with the classical inequalities (4.6–8). Note that, if X is a finite set of N points, these are a subset of the relations applied to a simplicial complex with N vertices in Regge calculus. However, the interpretation there is different from the one we have in mind. In the former, the N vertices are "filled" in with simplices of the appropriate dimension, whereas in our case all spatial points are already contained in the set X. Note also that a set of vertices satisfying the *equality* in (4.8), $d(x, z) = d(x, y) + d(y, z)$, corresponds to a *singular* simplex, whereas nothing in particular is attached to this situation in the metric interpretation.

One immediate problem is that we might like to allow for the possibility that the quantum distance between points in X is "distributional," and hence to smear the field d with a pair of test functions. But then the question arises as to the space $\mathfrak{L}$ to which these functions should belong. Quantum fields are normally smeared with continuous (or smooth) functions of compact support, but this is inappropriate here since notions like "continuous" or "compact" are only meaningful if there is a background topology on X, which we wish to avoid. A similar problem would arise if we were trying to define a normal scalar quantum field φ on X (this was touched upon briefly in the discussion following the affine commutation relations (3.4)). One possibility for a "universal" test-function space (i.e., one that is independent of any background topology on X) stems from the observation that any continuous compact-support function on a topological space is necessarily bounded. This suggests that the set $\mathfrak{F}_b(X)$ of *all* bounded functions on X might be a suitable candidate. Note that this forms a Banach algebra with the norm

$$\|f\| := \sup_{x \in X} |f(x)|. \tag{4.13}$$

The heart of a quantum theory on $Metric(X)$ lies in the relations (4.6–8). The smeared form of inequality (4.8) appears to require an integral I for its definition as

$$I(g)\, d(f,h) \leq I(h)\, d(f,g) + I(f)\, d(g,h) \qquad (4.14)$$

for all positive test-functions f, g, h. The significance of the choice of I is not clear, and it seems advisable to go back to the relations (4.6–8) and ask *ab initio* how they can be incorporated into a quantum scheme. The first question concerns the topology to be placed on $Metric(X)$. One possibility might be to exploit the recent work of Gromov on exactly this topic. (Gromov 1981a, 1981b; Epstein 1987). This would be particularly appropriate if one were interested in metric spaces arising as limits of sequences of manifolds.

From the viewpoint of the group-theoretic approach to quantization, a more natural path would be to choose the configuration space to be the set $Metric_b(X)$ of bounded metrics on X. Nothing essential is lost thereby since every metric topology can be derived from a bounded metric, and the advantage is that $Metric_b(X)$ can be embedded in the space of all bounded maps from $X \times X$ into $\mathbb{R}$, which can be topologized in several natural ways. Ideally, there would be a topological group acting on this space with $Metric_b(X)$ as one of its orbits (i.e., it preserves the relations (4.6–8)). This would be a precise analogue of the way $Riem\,\Sigma$ is embedded in $C(\Sigma, Sym(3, \mathbb{R}))$ as an orbit of the group $C(\Sigma, GL^+(3, \mathbb{R}))$. Unfortunately, I have no idea whether such a group exists. If it does, its construction will be quite different from that of $C(\Sigma, GL^+(3, \mathbb{R}))$ since, by their very nature, the defining relations (4.6–8) are highly nonlocal, whereas the conditions for $g_{ab}(x)$ to be a Riemannian metric are imposed independently at each $x \in \Sigma$.

An alternative to embedding $Metric_b(X)$ in the space of all functions from $X \times X$ into $\mathbb{R}$, is to consider metrics on X derived from the set of maps from X into some fixed Banach space $\mathfrak{B}$, i.e., to each $\iota : X \to \mathfrak{B}$, there corresponds a distance function $d_{(\iota)}$ defined by

$$d_{(\iota)}(x, y) := \|\iota(x) - \iota(y)\|, \qquad (4.15)$$

which automatically satisfies the defining conditions (4.6–8), except that $d_{(\iota)}(x, y) = 0$ does not necessarily imply that $x = y$ (i.e., $d_{(\iota)}$ is a pseudo-metric rather than a metric). This could be avoided by restricting attention to the set of *injections* of X into $\mathfrak{B}$. Or, it could be argued, we should start with the set of pseudo-metrics on X, with any non-Hausdorff topologies that emerge being made Hausdorff by an identification of degenerate points. Note that many different functions ι produce the same distance function (a sort of "gauge-invariance"); in particular, $\iota_v(x) := \iota(x) + v$ is equivalent to $\iota(x)$ for all vectors **v** in $\mathfrak{B}$. If desired, this translation degeneracy can be removed from the outset by requiring ι to satisfy the partial "gauge-condition" $\iota(x_0) = 0$, where x_0 is some arbitrary, but fixed, basepoint in X.

A crucial question is how many metrics can be written in the form (4.15). Clearly, the answer depends on the Banach space $\mathfrak{B}$, and for any particular set X, it is important to know if there is a choice of $\mathfrak{B}$ with the property that *all* metrics on X can be expressed in this way. One candidate is the space $\mathfrak{F}_b(X)$ of all bounded functions on X with the norm in (4.13). To see that this works, let d be any metric on X and define the map $\iota_d: X \to \mathfrak{F}_b(X)$ by $(\iota_d(x))(z) := d(x, z) - d(x_0, z)$, where x_0 is a basepoint in X. Now, (4.8) implies that $|d(x, z) - d(y, z)| \le d(x, y)$ for all $x, y, z \in X$, and hence $|(\iota_d(x))(z)| \le d(x, x_0)$, i.e., $\iota_d(x)$ is a bounded function on X. Furthermore,

$$\sup_{z \in X} |d(x, z) - d(y, z)| \le d(x, y). \tag{4.16}$$

However, $|d(x, x) - d(y, x)| \equiv d(x, y)$, and hence the equality holds in (4.16). Thus

$$d(x, y) = \sup_{z \in X} |d(x, z) - d(y, z)| = \sup_{z \in X} |(\iota_d(x))(z) - (\iota_d(y))(z)|$$
$$= \|\iota_d(x) - \iota_d(y)\|, \tag{4.17}$$

as required.

The next step would be to find a group G of transformations on the maps ι that acts transitively and preserves the equivalence relation $\iota_1 \equiv \iota_2$ if ι_1 and ι_2 induce the same metric via (4.14). One natural group in the system is the set of all continuous linear transformations of the Banach space $\mathfrak{B}$. However, this does not preserve the equivalence relation, and passing from here to a group that does is likely to be as difficult as finding a group that acts on $Riem\,\Sigma / Diff\,\Sigma$ given that $C(\Sigma, \mathrm{GL}^+(3, \mathbb{R}))$ acts transitively on $Riem\,\Sigma$.

The possibility of finding a suitable group for a nontrivial X is currently under investigation, but, to show that embedding techniques are not totally intractable, let us conclude by considering a very elementary example (albeit of a Hilbert, not Banach, carrier space). The simplest nontrivial case is when X is a finite set $\{0, 1\}$ of just two points. Then $Metric(X) \approx \{r \in \mathbb{R} \mid r \ge 0\}$, which is the disjoint union of the positive real numbers and zero. Quantization on the former was discussed in the context of (3.3), and the latter is trivial. The next simplest example is when X is a finite set $\{0, 1, 2\}$ of three points. Thus, denoting the distances between these points by $\alpha = d(0, 1)$, $\beta = d(0, 2)$ and $\gamma = d(1, 2)$, we are interested in all triples (α, β, γ) of positive real numbers satisfying the triangle inequalities

$$\alpha \leq \beta + \gamma, \quad \beta \leq \alpha + \gamma, \quad \gamma \leq \alpha + \beta. \tag{4.18}$$

Any such triple can always be realized by embedding $\{0, 1, 2\}$ into the vector space $\mathbb{R}^2$ equipped with its usual distance function (note that this is a "Regge calculus" embedding in a Hilbert space; it is *not* a special case of the $\mathfrak{F}_b(X)$ embedding discussed above). The set of all such triples forms a submanifold of $\mathbb{R}^2$, whose boundaries correspond to the special cases where one or more equalities hold in (4.18). It might be possible to patch together a quantum theory defined on these one-dimensional boundaries with one on the two-dimensional interior; or one might argue that only the open interior has any significant effect. In any event, we will concentrate here on the latter, and restrict our attention to the case where strict inequalities hold. Of course, the topology induced on the finite set $\{0, 1, 2\}$ is always the discrete one, so this is not an example of "quantum topology" proper, but only of the simpler idea of quantizing metrics that induce the same topology.

Many embeddings yield the same triplet of distance functions, and this nonuniqueness must be removed by choosing a "gauge." If the images of the three points 0,1,2 are the vectors x_0, x_1, x_2, respectively, then we can choose the unique embedding with $x_0 = 0$ (the null vector), $x_1 = \alpha e_1$, $x_2 = \mu e_1 + \nu e_2$ where $\{e_1, e_2\}$ is an orthonormal base for $\mathbb{R}^2$ with $\alpha, \nu > 0$. It is easy to show that a transitive group of "gauge-preserving" transformations on $Metric\{0, 1, 2\}$ is induced by the linear transformations on $\mathbb{R}^2$:

$$\begin{pmatrix} x_1 \\ x_2 \end{pmatrix} \rightarrow \begin{pmatrix} a & 0 \\ b & c \end{pmatrix} \begin{pmatrix} x_1 \\ x_2 \end{pmatrix}, \quad a > 0, \ c > 0, \tag{4.19}$$

in the form

$$\begin{pmatrix} \alpha^2 \\ \beta^2 \\ \gamma^2 \end{pmatrix} \rightarrow \begin{pmatrix} a & 0 & 0 \\ b(b+c) & c(b+c) & -bc \\ (a-b)(a-b-c) & -c(a-b-c) & c(a-b) \end{pmatrix} \begin{pmatrix} \alpha^2 \\ \beta^2 \\ \gamma^2 \end{pmatrix}, \tag{4.20}$$

which, it should be noted, is linear on the *squares* of the distances. Thus, in this case, the choice for the group G is the set of all triangular matrices $\begin{pmatrix} a & 0 \\ b & c \end{pmatrix}$ with $a > 0$ and $c > 0$. The vector space W is chosen to be $\mathbb{R}^3$ with the G-action (4.20), in which the set of metrics (embedded via the squares of the lengths) on $X = \{0, 1, 2\}$ is clearly an orbit. Thus, the canonical group for a system with this configuration space would be the semi-direct product $\mathbb{R}^3 \circledS G$, and its unitary representations will yield the associated quantum theory.

This example can be generalized in an obvious way to any finite set of points $X = \{0, 1, 2, \ldots, N\}$. Once again, there is no real topology change (since all the metric topologies on a finite set are discrete), but at least one can see the beginning of a general scheme. Note that if one tried to emulate this procedure for a set X whose cardinal number is that of the real numbers (the case of interest in quantum gravity?), a well-ordering of X would be needed to imitate the "gauge-fixing" above. (For a recent development of these ideas see Isham, Kubyshin and Renteln (1990))

5. Conclusions

One of the major points of debate in quantum gravity has always been whether the construction of a successful theory is possible within the general framework of existing physics, or whether it necessarily entails a radical reappraisal of the fundamental concepts of space, time and matter. Most classical-to-quantum schemes assume the former, albeit with an expectation that a modest shift in basic ideas may emerge at the end of the day. On the other hand, many advocates of the quantum-to-classical approach are rather equivocal about the status of our current modes for analyzing the world, and are really seeking a "?-to-quantum-to-classical" framework, in which the unknown "?" would probably be as different from quantum physics as is the latter from classical physics. Thus, the apparent primacy of quantum ideas in a quantum-to-classical scheme is of provisional significance only, and merely reflects the belief that the structure "?" is more likely to be found by reflecting on the nature of quantum physics than on the classical world of Newton and Einstein.

The great difficulty in pursuing such a path is the absence of any definitive understanding of quantum physics that does not assign some fundamental status to the conceptual framework of the classical world, which it is intended to supplant. Such a Janus-faced starting point is especially inappropriate when dealing with quantum gravity, in which space-time—that paradigmatic feature of classical physics—is itself called into question. This is why most attempts in this direction seem ad hoc and, as a consequence, have rarely succeeded in holding the attention of the scientific world. It is for this reason that I have concentrated upon the classical-to-quantum path, with the hope that a careful investigation of the "quantization" of general relativity may yield some genuine insight into physics at the scale of the Planck length or, perhaps better, the extent to which the current conceptual framework becomes inapplicable at this scale. In particular, I feel there is still much to learn from the use of nonstandard choices of canonical variable such as the affine relations (3.10–12), the d_γ-variables in Section 4, or the loop-variables of Rovelli and Smolin.

However, on moving from quantum geometry to quantum topology, the situation becomes less clear. When working in such an obscure area, there is always the danger of falling into what Whitehead perspicaciously called the "fallacy of misplaced concreteness": the general scientific tendency to reify and objectify what are really highly abstract ideas. "Space" and "time" are two prime examples of this type, and one should be cautious therefore about taking too literally ideas like "virtual black holes," "Planck length wormholes," "topology on $\tau(X)$" and similar ideas.

This should not be misconstrued as implying that all work on quantum topology is misguided. Even in such an adventurous realm as modern quantum cosmology, the use of semi-classical language may well be valid as long as one keeps away from the Planck length scale, at which it is called into doubt. But the Planck length itself remains an elusive mystery. Perhaps nonperturbative formulations of superstring theory will indeed produce the major breakthrough that their votaries expect. However, my conviction is that the use of real numbers in physics is a critical example of "misplaced concreteness," and this makes me uneasy about any scheme in which differential geometry (be it real or complex) plays a central role. Similar reasons underlie my feeling that the use of $Metric(X)$ is as far as one can reasonably go in attempting to formulate a quantum theory of topology that employs conventional quantum ideas.

I have argued in the text that the main property of X is its cardinality, but it is not clear what this should be. The obvious choices are finite, countable or continuum. It seems unlikely that a finite set could yield sufficient structure (unless perhaps McGoveran's (1988) ideas on "finite" topology can be brought into play), and countable sets are difficult to use because the associated metric topologies are highly disconnected. Thus, the continuum case might seem to be the most appropriate, although it could be argued that this is already granting physical status to a mathematical idea that is too abstract. Another option is to restrict attention to compact metric spaces. Every such space has a countable dense subset and, therefore, within this class, it suffices to define the metrics on a countable set and then to take the appropriate completion. There are a number of intriguing possibilities here, and I am sure that a study of quantum theory on $Metric(X)$ will be rewarding. However, it remains to be seen if it will be sufficient to enable us finally to dispense with the beautiful shackles of differential geometry!

Acknowledgments. I have enjoyed many stimulating discussions on these matters with Noah Linden during the last year. I would also like to thank the Theoretical Physics Group at MIT, the Boston University Center for Einstein Studies, and the Relativity Group at Maryland University for financial support during my visit to the USA to attend the Osgood Hill meeting.

REFERENCES

Anderson, Arlen, and DeWitt, Bryce S. (1986). "Does the Topology of Space Fluctuate?" *Foundations of Physics* 16: 91–105.

Arnowitt, R., Deser, S., and Misner, Charles W. (1962). "The Dynamics of General Relativity." In *Gravitation: An Introduction to Current Research.* Louis Witten, ed. New York: Wiley, 227–265.

Ashtekar, Abhay, ed. (1988). *New Perspectives in Canonical Gravity.* Naples: Bibliopolis.

———(1991). "Old Problems in the Light of New Variables." See this volume, pp. 401–426.

Beer, G.A., et al. (1987). "The Locally Finite Topology on 2^x." *American Mathematical Society. Proceedings* 101: 168–172.

Coleman, S. (1988). "Why there is Nothing Rather than Something: A Theory of the Cosmological Constant." *Nuclear Physics B.* 310: 643–668.

Dirac, Paul A. M. (1958a). "Generalized Hamiltonian Dynamics." *Royal Society of London. Proceedings A* 246: 326–332.

———(1958b). "The Theory of Gravitation in Hamiltonian Form." *Royal Society of London. Proceedings A* 246: 333–343.

———(1959). "Fixation of Coordinates in the Hamiltonian Theory of Gravitation." *Physical Review* 114: 924–930.

Epstein, D. B. A. (1987). "Complex Hyperbolic Geometry." In *Analytical and*

Geometric Aspects of Hyperbolic Space. London Mathematical Society Lecture Notes Series III. D. B. A. Epstein, ed. Cambridge: Cambridge University Press.

Friedan, Daniel, and Shenker, Stephen (1986). "The Integrable Analytic Geometry of Quantum String." *Physics Letters B* 175: 287–296.

———(1987). "The Analytic Geometry of Two-Dimensional Conformal Field Theory." *Nuclear Physics B* 281: 509–545.

Friedman, John L. (1991). "Spacetime Topology and Quantum Gravity." See this volume, pp. 539–572.

Friedman, John L., and Sorkin, Rafael D. (1980). "Spin 1/2 from Gravity." *Physical Review Letters* 44: 1100–1103.

Gell-Mann, Murray and Hartle, James B. (1990). "Quantum Mechanics in the Light of Quantum Cosmology." In *Complexity, Entropy and the Physics of Information.* Santa Fe Institute Studies in the Sciences of Complexity, vol. VIII, Wojciech H. Zurek, ed. Reading, Mass.: Addison Wesley; also in *Proceedings of the Third International Symposium on Foundations of Quantum Mechanics in the Light of New Technology.* H. Ezawa, S. Kobayashi, Y. Murayama and S. Nomura, eds. Tokyo: Japan Physical Society.

Giddings, S. B., and Strominger, Andrew (1988). "Baby Universes, Third Quantization and the Cosmological Constant." Harvard University pre print, HUTP-88/A036.

Gribov, V. N. (1978). "Quantization of Non-Abelian Gauge Theories." *Nuclear Physics B* 139: 1–19.

Groenewold, H. J. (1946). "On the Principles of Elementary Quantum Mechanics." *Physica* 12: 405–460.

Gromov, Mikhael (1981a). "Groups of Polynomial Growth and Expanding Maps." *Institut des Hautes Etudes Scientifiques. Publications Mathématiques* 53: 53–73.

———(1981b). *Structures Métriques pour les Variétés Riemanniennes.* Paris: Fernand Nathan.

Haag, Rudolf; Namhofer, Heide; and Stein, Ulrich (1984). "On Quantum Field Theory in Gravitational Background." *Communications in Mathematical Physics* 94: 219–238.

Halliwell, Jonathan J. (1988). "Derivation of the Wheeler–DeWitt Equation from a Path Integral for Minisuperspace Models." University of Santa Barbara preprint, NSF-ITP-88-25.

———(1991). "The Wheeler–DeWitt Equation and the Path Integral in Minisuperspace Quantum Cosmology." See this volume, pp. 75–115.

Hartle, James. B. (1985). "Unruly Topologies in Two-Dimensional Quantum Gravity." *Classical and Quantum Gravity* 2: 707–720.

———(1991). "Time and Prediction in Quantum Cosmology." See this volume, pp. 172–203.

Hartle, James B., and Hawking, Stephen W. (1983). "Wave Function of the Universe." *Physical Review D* 28: 2960–2975.

Hawking, Stephen W. (1984). "Non-Trivial Topologies in Quantum Gravity." *Nuclear Physics B* 244: 135–146.

———(1987). "Quantum Coherence down the Wormhole." *Physics Letters B* 195: 337–343.

———(1988). "Wormholes in Spacetime." *Physical Review D* 37: 904–910.

Horowitz, Gary T. (1991). "String Theory without Spacetime." See this volume, pp. 299–325.

Horowitz, Gary T., and Witt, D. (1987). "Toward a String Field Theory Independent of Spacetime Topology." *Physics Letters B* 199: 176–182.

Isham, Christopher J. (1976). "Some Quantum Field Theory Aspects of the Superspace Quantization of General Relativity." *Royal Society of London. Proceedings A* 351: 209–232.

———(1981). "Topological θ-Sectors in Canonically Quantized Gravity." *Physics Letters B* 106: 188–192.

Isham, Christopher J. (1983). "Vacuum Tunneling in Static Space-Times." In *Old and New Questions in Physics, Cosmology, Philosophy and Theoretical Biology: Essays in Honor of Wolfgang Yourgrau.* Alwyn van der Meave, ed. New York: Plenum Press, 189–211.

Isham, Christopher J. (1984). "Topological and Global Aspects of Quantum Theory." In *Relativity, Groups and Topology II*, Bryce S. DeWitt and Raymond Stora, eds. New York: North Holland, 1059–1290.

Isham, Christopher J. (1989). "Quantum Topology and Quantization on the Lattice of Topologies." *Classical and Quantum Gravity* 6: 1509–1534.

Isham, Christopher J., and Kakas, A. C. (1984a). "A Group Theoretical Approach to the Canonical Quantization of Gravity: I. Construction of the Canonical Group." *Classical and Quantum Gravity* 1: 621–632.

———(1984b). "A Group Theoretical Approach to the Canonical Quantization of Gravity: II. Unitary Representations of the Canonical Group." *Classical and Quantum Gravity* 1: 633–650.

Isham, Christopher J., Kubyshin Yuri and Renteln, Paul (1990). "Quantum Norm Theory and the Quantization of Metric Topology." *Classical and Quantum Gravity* 7: 1053–1074.

Klauder, John R. (1970). "Soluble Models of Quantum Gravitation." In *Relativity: Proceedings of the Relativity Conference in the Midwest, Cincinnati, Ohio, June 2–6, 1969.* Moshe Carmeli, Stuart I. Fickler, and Louis Witten, eds. New York: Plenum, 1–17.

———(1973). "Functional Techniques and their Application in Quantum Field Theory." In *Mathematical Methods in Theoretical Physics.* Wesley E. Britten, ed. Boulder: Colorado Associated University Press, 329–421.

Klauder, John R., and Aslaken, Erik W. (1970). "Elementary Model for Quantum Gravity." *Physical Review D* 2: 272–276.

Komar, Arthur (1970). "Consistent Factor Ordering of General-Relativistic Constraints." *Physical Review D* 20: 830–833.

Kuchař, Karel V. (1986). "Covariant Factor Ordering of Gauge Systems." *Physical Review D* 34: 3044–3057.

———(1987). "Consistent Factor Ordering of Constraints May be Ambiguous." *Physical Review D* 35: 596–599.

Marjanovic, M. (1966). "Topologies on Collections of Closed Subsets." *Institut Mathématiques* (Belgrade). *Publications* 20: 125–130.

McGoveran, D. (1988). "Foundations of a Discrete Physics." Stanford University preprint.

Michael, Ernest (1951). "Topologies of Spaces of Subsets." *American Mathematical Society. Transactions* 71: 152–182.

Penrose, Roger (1987). "Newton, Quantum Theory and Reality." In *Three Hundred Years of Gravitation.* Stephen W. Hawking and W. Israel, eds. Cambridge and New York: Cambridge University Press, 17–49.

Pilati, Martin (1982). "Strong-Coupling Quantum Gravity: I. Solution in a Particular Gauge." *Physical Review D* 26: 2645–2663.

———(1983). "Strong-Coupling Quantum Gravity: II. Solution Without Gauge Fixing." *Physical Review D* 28: 729–744.

Rovelli, Carlo, and Smolin, Lee (1990). "Loop Representation for Quantum General Relativity." *Nuclear Physics B.* 331: 80–152.

Shastri, A. R.; Williams, J. G.; and Zvengrowdki, P. (1980). "Kinks in General Relativity." *International Journal of Theoretical Physics* 19: 1–23.

Singer, I. M. (1978). "Some Remarks on the Gribov Ambiguity." *Communications in Mathematical Physics* 60: 7–12.

Smolin, Lee (1991). "Nonperturbative Quantum Gravity via the Loop Representation." See this volume, pp. 440–489.

Sorkin, Rafael D. (1988). "Spinors, Twistors and Complex Manifolds in General Relativity." *Talk given at the 1988 Durham Symposium on "Spinors, Twistors and Complex Manifolds in General Relativity."*

Teitelboim, Claudio (1980). "The Hamiltonian Structure of Space-Time." In *General Relativity and Gravitation: One Hundred Years after the Birth of Albert Einstein.* A. Held, ed. New York: Plenum, 195–225.

Van Hove, L. (1951). "On the Problem of the Relations between the Unitary Transformations of Quantum Mechanics and the Canonical Transformations of Classical Mechanics." *Académie Royale de Belgique. Classe des Sciences. Bulletin* 37: 610–620.

Wheeler, John A. (1964). "Geometrodynamics and the Issue of the Final State." In *Relativity, Groups and Topology.* Lectures delivered at les Houches during the 1963 Session of the Summer School of Theoretical Physics, University of Grenoble, Cecile DeWitt and Bryce S. DeWitt, eds. New York and London: Gordon and Breach, 316–520.

Witten, Edward (1988). "Topological Quantum Gravity." Princeton University preprint.

Discussion

HOROWITZ: Chris, people were confused as to why your quantum operators preserve classical properties.

ISHAM: Oh! I am not *demanding* it, quite the opposite. What I want to see is the maximum degree to which they can be preserved. But my honest belief about quantum gravity is that, at the Planck-length scale, classical differential geometry is simply incompatible with quantum theory.

KUCHAŘ: What happens if you use variables that take the positivity into account, like triads for example?

ISHAM: Well, in fact, triads do not really do that. If you have a triad of vectors it is necessary to impose the additional requirement that they are a linearly independent set of vectors. And, for most purposes, that is just as bad as saying that the metric is positive definite.

HARTLE: Would you comment on the work of Haag et al?

ISHAM: There are a lot of interesting ideas in that paper but they are looking at things from a quite different perspective and I cannot really see much connection with what I have been talking about. Neither do I agree with all that they say.

JACOBSON: What are they saying?

ISHAM: They are trying to strip away almost everything and essentially end up with just quantization at each space-time point. Their approach is a covariant one, but I don't think they have really grappled with the implications of the differential-geometric structure of the space-time manifold.

DEWITT: It looks like it is connected with Eckenberg's approach.

ISHAM: Yes, but they haven't done the patching, that's the trouble. You can always quantize at a point. I mean that, if you have a manifold, you can always take a tangent plane and quantize on that, which is basically what they are doing. But the problem then is to patch together the different tangent space quantizations in a way that is meaningful, and this they haven't done.

ASHTEKAR: Another thing is that they never discuss dynamics.

ROVELLI: In the representations of the semidirect-product group of the quantized topological metrics, is there anything that has the memory of the original topology? How do you reconstruct it?

ISHAM: I only finished that particular piece of work very recently and I haven't had time yet to look at the representations. Imperial College is a hectic place in which to work these days!

JACOBSON: In choosing between the different possible representations, one thing that you would demand would be agreement with observations, although of course there might be unexpected effects. For example, in the early days of quantum mechanics there was the concept of the internal spin of the electron. But, in your case, such effects would have to be things that we don't see at a macroscopic scale. The only thing I can think of is that either we pretty much have the trivial representation, or, maybe, your variables sometimes label things that are different from what we would, at the macroscopic level, now call fields.

ISHAM: Well, in a sense that's true, it's a very good point. When I said that $g_{ab}g^{ab} = \varphi$, I didn't mean that φ is an extra field in the classical theory. I don't quite understand the significance of this, but there's a certain sense in which these other fields (like φ) tend to arise from the kinematical structure of the quantum theory.

SMOLIN: In this last topological quantization, is there some sense in which, when all those points lie in a one-dimensional subspace, there exists a subgroup that is the one-dimensional diffeomorphism group?

ISHAM: Diffeomorphism group?

SMOLIN: Well, because in some sense the transformations, homeomorphisms are...

ISHAM: Ah, but you see *that* would break the gauge. The set of vectors I have chosen is constrained to lie in what is really a *flag space* of the vector space, and any permutation would take you out of that. If you like, this is a gauge-fixed quantization, so I can't ask your question; the gauge has been fixed via the flag space before doing the quantization.

SMOLIN: Is there some sense in which diffeomorphism groups should emerge out of these more general topological things?

ISHAM: In ordinary canonical quantum gravity, it is not the case that the diffeomorphism group has to be added, so that the complete quantizing group (i.e., the group whose *irreducible* representations define the Hilbert space of the quantum theory) is the (semidirect) product of the canonical group and the diffeomorphism group. The canonical group alone is sufficient. That is to say, the generators of the diffeomorphisms are *functions* of the canonical generators, as, for example, is the Hamiltonian. The correct statement is only that the irreducible representation that you choose for the canonical group must be extendible to the semidirect product with the diffeomorphism group.

MASON: One critical problem is how to select a particular representation of your canonical group. In the construction of the quantum theory, is there any reason for supposing that any one representation has preference over the others?

ISHAM: It is true that, even for finite-dimensional systems, there may be many irreducible representations, classified for examples by the values of the Casimir operators. You have to decide which is relevant for any particular physical system. For example, consider the case of a system whose configuration space is a two-sphere and someone asks what do you mean by quantum mechanics on a two-sphere? I do not think there is any absolute God-given answer to that. All I can say is that, in this particular approach, it is suggested very strongly that all the representations should be treated on an equal footing since they carry information of physical significance. In the case of the two-sphere, the different representations of the canonical group $R^3 \circledS SO3$ are classified by the integer that specifies the possible irreducible representations of the $SO2$ subgroup of the $SO3$ group that acts transitively on the sphere. This integer has a direct physical interpretation as the intrinsic magnetic monopole charge of the system.

ASHTEKAR: In the beginning you emphasized that wave functions have support on distributional fields. Two remarks. The first is that this is true if you work in the Schrödinger representation. However, if you work with a Kähler polarization of holomorphic functions (i.e., a Bargmann representation), the wave functions are just holomorphic functionals of one-particle states, and the one-particle states are just respectable functions. There are no distributions there. So in that case, one doesn't have this problem. My second remark is that, even in the Schrödinger representation, the way that one normally constructs states—goes about doing it as opposed to looking at the finished quantum picture—is by using field configurations at an initial instant of time that are fields on R^3 and that have some fall off so that they are normalizable with respect to some energy norm or other. So it's a respectable function space. Construct a Gaussian measure on that. Then, if you look at the square-integrable functions, you see that all these nontrivial things actually arise in the Cauchy completion. So, although it's true that, in the finished picture, the measure is concentrated in some distributional space, in an actually constructive approach, one doesn't have to start there.

ISHAM: What you say is true, but I think it is only really useful for Gaussian/Wiener measures. That's the trouble with the use of the triplet of spaces you have in mind. But it is only a very special situation in which one gets a simple Gaussian measure.

ASHTEKAR: Let's look at a theory in a lower dimension. I think it may be true there for a wider class of measures. In lower dimensions, for exactly soluble models in, say, a box, where we don't have to worry about infrared problems, it is indeed true that one can again repeat that kind of construction.

ISHAM: Oh, that's true here too. The singularity structure of the distributions is very dimension dependent. Of course, that's quite right: in lower dimensions you will find such representations. But I am mainly interested in three dimensions and above.

ASHTEKAR: Let me ask another question. What is the general status if you incorporate the constraints in this formalism?

ISHAM: Well, one could proceed in any of the several standard ways. But, to be honest, I don't think you can really quantize gravity. I am not trying to save the situation by doing this. I genuinely believe that one will not be able to use differential geometry in the true quantum-gravity theory. What I'm trying to do is to get some idea of the type of space and mathematical objects that one might encounter in such a theory.

Old Problems in the Light of New Variables

Abhay Ashtekar

1. Introduction

It is well known that attempts to use perturbative, field-theoretic techniques to obtain a quantum version of general relativity fail by their own criteria. Introduction of higher derivative terms and extension to supergravity have not, in the final analysis, altered this situation. In the last couple of years, there were hopes that the superstring theory may change this picture dramatically. Indeed, there are now claims that the theory is *finite* in a technical sense. However, the terminology is perhaps misleading to non-experts: it is only the individual terms in the perturbation series that are claimed to be finite. The series, when summed, *does diverge* and does so rather badly.[1] Hence, even in string theory, perturbative techniques ultimately fail. The problem, it seems, lies in the basic assumption of the perturbation theory that the space-time geometry continues to be smooth even at short distances, where, by simple dimensional arguments, quantum-gravity effects should be crucially important. Thus, in spite of imaginative attempts to prolong the comfortable regime of perturbation theory, we are now pretty much forced to abandon it as the main avenue to quantum gravity. We are back to the grand old question: *Does quantum gravity makes sense nonperturbatively?*

Once one accepts the premise that the problem of quantum gravity should be faced nonperturbatively, the argument for abandoning general relativity as the starting point loses much of its force. It is again of considerable interest to find out if quantum general relativity makes sense. Rather than feeding in from outside a model of the micro-structure of space-time as in perturbative attempts, one now wants the theory itself to tell us what this structure is. Of course, there is a serious possibility that quantum general relativity may not be the correct theory because it may fail to incorporate important features of the short-distance dynamics of the real physical world. Nonetheless, to investigate many of the *conceptual* problems of quantum gravity, at this stage general relativity is as good a starting point as any. The hope is that, in spite of our ignorance of the short-distance dynamics, by using techniques that are reasonably "robust" to changes in the details of the theory, one may be able to extract at least some qualitative features of the Planck-scale physics.

The canonical approach offers a natural avenue to venture into non-perturbative quantum gravity. For, in this approach, one does not need to fix a background geometry or build strong fields by superposing weak fluctuations; one can deal directly with highly curved geometries. Let us therefore begin by recalling the mathematical problems that this program must face. Can one solve the *quantum* constraint equations exactly? Is there a "general" solution? These solutions would represent the physical states of the theory. Is there a natural inner product on this space? What is the role of time? Can one reinterpret the Hamiltonian constraint as the quantum "evolution equation" by identifying, prior to the imposition of constraints, one of the arguments of the wave function with "time"? Can this be done exactly? Or, is the notion of "time-evolution"—and the corresponding unitarity of dynamics—only an approximate one? Can one introduce suitably-regularized quantum observables? More generally, is there a well-defined regularization scheme that one can use in the passage from the classical phase-space description to the quantum one?

Then there is a second set of questions concerning the relation of the canonical approach to other approaches and theories. Over the years, Newman, Penrose and Plebanski have developed a detailed mathematical framework to treat self-dual solutions to Einstein's equations (see, e.g., Ko et al. 1981). Penrose has suggested that a restricted class of these solutions should be interpreted as "nonlinear gravitons." Can one see the significance of these solutions in the canonical program? Can they indeed be interpreted as "particle-like states" from a more traditional point of view? More generally, does "self-duality" have a role to play in canonical quantization? Is there a mathematically consistent *quantum* theory of only the self-dual fields? Next, recall that in the three decades since the canonical program was first introduced in quantum gravity by Dirac, Bergmann, Arnowitt–Deser–Misner, and others, gauge theories have proved themselves to be very successful in electroweak and strong interactions. Can one reformulate canonical quantum gravity in a manner that treats gravity in the same way as other interactions, at least mathematically? Can one import techniques from, say, QCD into quantum gravity? One can add terms of topological origin to the Yang–Mills action, which, in quantum theory, lead to CP violation. Does a similar phenomenon occur in quantum gravity? Are there other general predictions of quantum gravity that may have ramifications in particle physics?

Finally, there is a set of deeper, conceptual questions. What is the micro-structure of space-time like? How does the continuum picture arise? What are the appropriate concepts for physical theories in the absence of a space-time continuum? In Minkowskian physics, (infinite) renormalization of quantum field theories can be considered as a "short-cut" that enables one to compute physical quantities without having to worry about the small-scale structure of space-time. What is it about the micro-structure of space-time that makes renormalization "work"? Finally, there are the questions relating to the measurement theory and the issue of interpretation of quantum mechanics, discussed in the first part of this volume.

The program that I want to discuss here is aimed at answering these three groups of questions.

A major obstacle to progress in the canonical approach has been the complicated nature of the field equations in the traditional canonical variables, (q_{ab}, p^{ab}). For example, since the constraint equations are nonpolynomial in q_{ab}, it has not been possible to solve them in the (full) quantum theory. This obstacle was removed recently by the introduction of new canonical variables. In terms of these, *all* equations of the theory become polynomial, in fact, at worst quartic (Ashtekar 1986a, 1986b, 1987). Furthermore, the use of these variables brings out a hidden relation between Yang–Mills theory and general relativity, enabling one to borrow for general relativity and quantum gravity ideas from Yang–Mills theory and QCD. These two features—polynomialization of the equations and the close relation to Yang–Mills theory—seem to be quite robust. They continue to exist if one allows for a nonzero cosmological constant, introduces matter sources (Klein–Gordon, Dirac and Yang–Mills fields), or considers a supersymmetric extension of the theory (Ashtekar et al. 1988b; Jacobson 1988a, 1988b). As a result, there has been a significant resurge of activity in nonperturbative, canonical quantum gravity.

The purpose of this article is to reexamine the old problems of quantum gravity, discussed above, in the light of these developments. While only a few of these problems have been fully solved by now, substantial progress has been made on many. The current status may be summarized as follows. In the first two of the three groups of questions, the use of new variables has provided fresh directions and new tools—both, mathematical and conceptual—which, in turn, have yielded concrete results. These results have provided a platform from which one can now address the deeper, conceptual problems listed in the third group. However, here the work has barely begun.

Many of these developments have already appeared in print and the results obtained prior to January 1988 are discussed in detail in a book (Ashtekar et al. 1988a). Therefore, in this article I will only provide a global picture of what has been achieved and discuss in some detail only the issue of "time," which has not been treated elsewhere. The organization of this article is as follows: Section 2 contains the basic framework; Section 3 summarizes the current status of the questions raised above; and Section 4 discusses the issue of time.

2. The General Framework

In this section, I will outline the quantization program. For simplicity, I will restrict myself to source-free general relativity. (For inclusion of sources and extension to supergravity, see Ashtekar et al. 1988b and Jacobson 1988a, 1988b.)

It is convenient to begin with complex general relativity and then, at the end, take the "real section" of the resulting phase-space by imposing suitable reality conditions. Thus, we wish to first consider the system of complex, Ricci-flat metrics $g_{\mu\nu}$ on a *real* four-manifold M. For the phase-space formulation, we begin by fixing a three-manifold Σ; M will have the topology $\Sigma \times \mathbb{R}$. The new canonical pair $(E^a_i, A_a{}^i)$ consists of complex fields on Σ : E^a_i is a triad with density weight one, while $A_a{}^i$ is a $SO(3)$-connection one-form on Σ, where "a" is the vector index and "i" the $SO(3)$, internal index. The geometrical interpretation of these fields is as follows: E^a_i is the "square-root" of the three-metric while, in any solution to the field equations, $A_a{}^i$ turns out to be the potential of the anti self-dual part of the Weyl tensor. More precisely, the relation to the more familiar variables, the three-metric q_{ab} and the extrinsic curvature K^{ab}, is the following:

$$E^a_i E^{bi} = q(q^{ab}) \qquad \text{and} \qquad G A_a{}^i = \Gamma_a{}^i - i K_{ab} e^{bi}. \qquad (2.1)$$

Here, q is the determinant of q_{ab}, $e^b_i = (\sqrt{q})^{-1} E^b_i$ is the unweighted triad, and $\Gamma_a{}^i$ is the $SO(3)$-connection on internal indices, compatible with the triad $e^a{}_i$, given by

$$0 = D_a e^b{}_i \equiv \partial_a e^b{}_i + \epsilon_{ijk} \Gamma_a^j e^{bk} - \Gamma_{ac}^b e^c{}_i,$$

where $\Gamma^b{}_{ac}$ are the Christoffel symbols defined by q_{ab}. If the three-manifold Σ is noncompact, the pair $(E^a{}_i, A_a{}^i)$ is subject to asymptotically-flat boundary conditions. Note that the relation between the new and the old canonical variables is nonpolynomial. The relation is precisely such that, as we shall see, the field equations become polynomial in the new pair. Note also that the connection one-form $A_a{}^i$ has information about both the spatial and the time derivatives of the triad. In fact, it contains precisely that combination of derivatives that makes it a potential for the anti-self-dual part of the Weyl tensor in any solution to the field equations:

$$GF_{ab}{}^i e^c{}_i \epsilon^{abd} = E^{cd} - iB^{cd}, \tag{2.2}$$

where E^{ab} and B^{ab} are, respectively, the electric and the magnetic parts of the Weyl tensor. Finally, the basic (nonvanishing) canonical commutation relation is simply:

$$\{E^a{}_i(x), A_b{}^j(y)\} = -i\delta^a_b \delta^j_i \delta(x, y). \tag{2.3}$$

The canonical pair is subject to three constraint equations:

$$\mathcal{D}_a E^a_i \equiv \partial_a E^a_i + G\epsilon_{ijk} A_a^j A^{ak} = 0$$

$$F^i_{ab} E^a_i \equiv \vec{B}_i \times \vec{E}^i = 0 \tag{2.4}$$

$$\epsilon^{ijk} F_{abk} E^a_i E^b_j \equiv \epsilon^{ijk} \vec{B}_i \cdot \vec{E}_j \times \vec{E}_k = 0,$$

where $\vec{B}_i$ is the magnetic field constructed from A_{ai}. These constraints form a first-class system. The first equation is just the Gauss constraint, which generates rotations on internal indices that are asymptotically identity. A linear combination (with q-number coefficients) of the first two constraints generates spatial diffeomorphisms (Ashtekar et al. 1987) that tend to identity at spatial infinity, while the third generates asymptotically identity time-evolutions. It is manifestly clear from their expressions that all constraints are polynomial in the basic variables.

The notation also makes it clear that the above gravitational phase-space is naturally isomorphic to the standard phase-space of Yang–Mills theory; $A_a{}^i$ can be regarded as the Yang–Mills potential and E^a_i can be identified with the Yang–Mills electric field. What is the situation with the constraints? The first constraint can now be identified with the Yang–Mills Gauss-law constraint. The last two, on the other hand, do not have a role to play in Yang–Mills theory. Thus, what we have is a proper imbedding of the constraint surface of Einstein's theory into that of Yang–Mills. It is this imbedding that enables one to relate the two theories. Note that the degrees of freedom match. Yang–Mills theory has six degrees of freedom (2 (helicity states) × 3 (internal degrees)), so that after the imposition of the four equations contained in the last two constraints, we are left with two true gravitational degrees of freedom. Finally, note that the last two constraints also have rather simple expressions in terms of Yang–Mills field strengths and that, regarded as equations on the Yang–Mills phase-space, they *do not* depend on any auxiliary structure such as a background space-time metric. (In fact, there are exactly five such expressions one can construct subject to the condition that they be at worst quadratic in the electric and magnetic fields. Four of these turn out to be the gravitational constraints and the fifth arises in the analysis of the gravitational θ-vacua!) These features of the gravitational constraints have been exploited by Renteln and Smolin (1989) to carry over ideas from the Hamiltonian lattice QCD to quantum gravity.

While the constraints of the two theories are similar, the dynamics is entirely different. Given a lapse[2] N and a shift N^a, the Hamiltonian can be written as:

$$
H(E, A) = i \int_{\Sigma} d^3x (N^a E^b_{\mathbf{i}} F_{ab}{}^{\mathbf{i}} - \frac{i}{2} N E^a_{\mathbf{i}} E^b_{\mathbf{j}} F_{abk} \epsilon^{\mathbf{ijk}})
$$
$$
- \oint_{\partial\Sigma} d^2 S_a (N E^a_{\mathbf{i}} E^b_{\mathbf{j}} A_{bk} \epsilon^{\mathbf{ijk}} + 2i N^{[a} E^{b]}{}_{\mathbf{i}} A_b{}^{\mathbf{i}}).
$$

(2.5)

As usual, the volume terms are linear combinations of constraints while the surface terms yield the ADM four-momentum. In particular, if we set the shifts N^a equal to zero on the constraint hypersurface, the resulting surface term can be reexpressed entirely in terms of the triad vectors $E^a_{\mathbf{i}}$ and is real and positive if we restrict ourselves to the real section of the phase space (see below). Finally, the equations of motion are:

$$
\dot{A}_b{}^{\mathbf{i}} = i N E^a_{\mathbf{j}} F_{abk} \epsilon^{\mathbf{ijk}} + N^a F_{ab}{}^{\mathbf{i}},
$$
$$
\dot{E}^b_{\mathbf{i}} = -i D_a (N E^a_{\mathbf{j}} \tilde{E}^b_{\mathbf{k}}) \epsilon_{\mathbf{ijk}} + 2 D_a (N^{[a} E^{b]}_{\mathbf{i}}).
$$

(2.6)

Thus, there is no simple relation between the evolution equations of the Einstein and Yang–Mills theories. Nonetheless, remarkably, it *is* sometimes possible to go back and forth between solutions to *all* equations of the two theories. (See, e.g., Samuel 1988, Torre 1990).

Finally, we come to the reality conditions. Let us first recall the situation in terms of the three-metric q_{ab} and the extrinsic curvature K^{ab}. If we let these fields on a real three-manifold Σ take on complex values, we obtain a phase-space formulation for complex solutions $g_{\mu\nu}$ of Einstein's equations on $\Sigma \times \mathbb{R}$. To restrict to the real solutions, we must ask that q_{ab} and K_{ab} be real; reality is then preserved by the Hamiltonian flow. In terms of the new variables, the situation is rather similar. To recover real general relativity from our complex phase-space, we have to simply impose "reality conditions": both the triad E^a_i and its time derivative, given by (2.6), must be real. Since the evolution equations are polynomial, these reality conditions are also polynomial (in fact, at worst cubic) in the new variables.[3] Again, the conditions are preserved under time-evolution. From the point of view of the classical initial value problem, the reality conditions are just additional equations imposed on the canonical pair, (A^i_a, E^a_i), and therefore play the same role as constraints. Finally, since this framework is to serve as a platform for canonical quantization, in the above discussion I focused on the Hamiltonian methods. A manifestly covariant Lagrangian formulation is also available (Samuel 1987; Jacobson and Smolin 1987, 1988a) and could serve as a point of departure for a path-integral quantization.

This finishes the summary of the classical theory. Let us now turn to the quantization program. We wish to proceed using the following steps.

(i) Introduce operator-valued distributions, $\hat{E}^a_i(x)$ and $\hat{A}_a^i(x)$, subject to canonical commutation relations.

(ii) On the algebra generated by these operators, introduce a *-operation by requiring that $\hat{E}^a_i$ be its own *-adjoint and its "dot," given by the quantum analog of (2.6), be its own *-adjoint. This incorporates the reality conditions only at the algebraic level.

(iii) Choose a representation of the algebra, or a polarization. The most convenient choice is to use for states holomorphic functions of the complex potential A_a^i, represent $\hat{A}_a^i$ as a multiplication operator and $\hat{E}^a_i$ as a differential operator, $\hbar\frac{\delta}{\delta A_a^i}$. At this stage, the *-relations are ignored. Since the incorporation of these relations requires the availability of a Hermitian inner product, and since we expect to have an unambiguous inner product only on *physical* states, it is appropriate to postpone the incorporation of these "quantum reality conditions" until *after* physical states have been extracted.

(iv) Solve the quantum constraints. Since the expressions of the classical constraints involve only A_a^i and E^a_i and *not* their complex conjugates, we can continue to ignore the *-relations in the algebra. The space of solutions is the (complex vector) space of physical states. Of interest to us will be those operators in our algebra that map this space to itself.

(v) On this space of physical states, introduce a Hermitian inner-product that now incorporates the *-relations. The operators $\hat{E}^a_i$ and its "dot" themselves will not be observables. However, the *-relations of the initial algebra induce *-relations on the space of observables, which maps the space of physical states onto itself, and *these* relations are to be faithfully reflected in the Hermitian adjointness relations by the appropriate choice of the inner product. Thus, in quantum theory, the constraints (2.4) and the reality conditions are not on the same footing; the former determine the space of physical states while the latter constrain the inner product on this space. In practice, the introduction of the inner product may well require that we isolate "time" from among the various components of $A_a{}^i$ and interpret the scalar constraint as a Schrödinger equation.

(vi) Isolate physically interesting observables and make predictions.

Steps (ii), (iv), and (vi) will require suitable regularization procedures.

These steps are to be regarded only as general guidelines. A number of variations are possible. In particular, the canonical algebra introduced in steps (i) and (ii) is not sacrosanct. For example, Rovelli and Smolin (1988, 1990a) have suggested an alternate candidate, the so-called T-algebra, which has several appealing features. Similarly, the choice of the A_a-representation is not forced upon us. The T-algebra, for example, leads one naturally to the so-called "loop-space representation." In this article, however, I will adhere to the choices made in the above sketch of the quantization program.

3. Present Status of the Program

Let me now return to the first two of the three groups of questions raised in Section 1 and report on their current status.

Because of the simplicity of the expressions of the constraint functionals, Jacobson and Smolin (1988b) were able to find a large class of solutions to the difficult Hamiltonian (quantum) constraint. To my knowledge, in the *full* theory, not a single solution was known before.

In their work, they exploited, in an essential way, the new relation between the Einstein and the Yang–Mills theories; their solutions are the gravitational analogs of Wilson loops in QCD. Rovelli and Smolin (1988, 1990a) have taken this work further. By working in the *loop-space representation*, where quantum states are expressed as suitable functionals of closed loops on the spatial three-manifold, they were able to obtain an infinite-dimensional space of solutions to *all* quantum constraints. (See the articles by Rovelli and Smolin that follow). It is still not clear if this class is "large enough" from a physical point of view. An analysis in the weak-field limit revealed that their procedure does recover the entire Fock space of states (Ashtekar et al. 1989a). However, the implications of this result for the full theory are not well-understood because the diffeomorphism group has a qualitatively different role to play in the weak field limit. (These developments are discussed in some detail in the articles by Rovelli and Smolin that follow.) The issue of solving quantum constraints is simplified also in the restricted context of quantum cosmology. In particular, Kodama (1988) for the first time has obtained some *exact* solutions to all quantum constraints in the Bianchi type IX models.[4] Recently, Husain and Smolin (1989) have analyzed a "midisuperspace" model consisting of gravitational fields with two space-like commuting Killing fields. Although the symmetry condition *is* a restriction, the model has an *infinite* number of degrees of freedom (whence the name "midisuperspace"). Again, they were able to find an infinite-dimensional space of solutions to all constraints and discover some physical observables.

The next step in the full theory is the introduction of "time," re-interpretation of the scalar constraint as the quantum Schrödinger equation, and the introduction of the scalar product on the space of wave functions satisfying the constraints. In Section 4, we shall carry out these steps in the weak field limit. Starting with the exact canonical framework, we will be able to "derive" the Schrödinger equation in the weak field limit, *without* having to introduce the notion of space-time. We shall see that to achieve this, nonlinearities are essential; the construction requires that we include the "interactions among the weak-field gravitons." While the procedure suggests strategies for addressing these problems in the full theory, it is too early to say if the constructions can in fact be extended. Work on application of these techniques to mini- and midisuperspace models mentioned above is now in progress.

Recently, there has also been considerable interest in $(2+1)$-dimensional gravity. It is well known that Einstein's theory trivializes in

$(2 + 1)$-dimensions because solutions to the vacuum field equations are *everywhere flat*. Thus, in this theory, there are no local degrees of freedom, and hence no gravitons; one has only (a finite number of) global, topological degrees. Nonetheless, since it has the same structure as the $(3 + 1)$-dimensional Einstein theory, like minisuperspaces, it is a very useful toy model. Recently, there has been a surge of activity in this area following Witten's (1988) discovery that this theory can be imbedded into the $(2 + 1)$-dimensional Chern–Simon theory using variables that are analogous to our pair $(A_a{}^i, E^a_i)$. Bengtsson (1989) has pointed out that there is a precise sense in which Witten's construction can be regarded as an application of the framework presented in Section 2 to $(2 + 1)$-dimensions, the only difference being that the internal gauge group is now $SO(2,1)$ rather than (complexified) $SO(3)$. Therefore, it is natural to ask if the quantization program sketched above can be carried out in this simple model. The answer has turned out to be in the affirmative (Ashtekar et al. 1989c). In this case, both the A_a^i and the loop space representations exist, one can introduce a scalar product canonically and identify all the observables. The issue of time is being investigated now.

The second class of problems that I mentioned above deals with the relation between canonical quantum gravity and other theories. Because of the close similarity between the present phase-space description of gravity and the standard canonical formulation of Yang–Mills theory, it is indeed possible to import into quantum gravity ideas from QCD. For example, Renteln and Smolin (1989) have carried over the Hamiltonian lattice techniques developed by Kogut and Susskind in QCD to quantum gravity. (See also Renteln 1988). Samuel (1988) has shown that the one-instanton solution in Yang–Mills theory can be taken over to gravity just by reinterpreting the variables: It becomes the four-sphere instanton. In fact, the entire analysis (in the canonical framework) of the θ-vacua and the associated CP-violation can be taken over from Yang–Mills theory to general relativity (Ashtekar et al. 1988a, 1989a). Furthermore, the use of new variables supplies some of the missing steps in the older analyses of this issue and also reveals some errors in the literature.

The availability of the loop representation for gravity suggested that one look for a similar representation also for Maxwell and Yang–Mills theories. It turned out, somewhat surprisingly, that the loop representation does exist *rigorously* for the Maxwell theory (Ashtekar and Rovelli 1989). In particular, the transform introduced by Rovelli and Smolin (1988, 1990a) as a heuristic device for quantum gravity can be taken over in the Maxwell case and does indeed provide a precise isomorphism between the photon Fock space and the loop space states. It is therefore of considerable interest now to bring in Dirac fields—there is an obvious way to introduce the required coupling—and analyze issues such as the QED phase-transition. More recently, Rovelli and Smolin (1990b) have used the T-algebra—first introduced by them as an alternative to the canonical algebra of $A_a{}^i$ and E^a_i in quantum gravity—and the loop representation to analyze (source-free) Yang–Mills theory both on a lattice and in the continuum (see also Brügman (1990)). While these developments are strictly outside quantum gravity, they do suggest that the ideas surrounding the loop-representation may provide an elegant, unified mathematical framework to describe *all* basic interactions of nature.

The introduction of these variables was motivated in part by the work of Newman and Penrose on self-dual solutions. Indeed, as we saw in Section 2, in any solution to field equations, A^i_a is a potential for the anti-self-dual part of the Weyl curvature. (From a four-dimensional perspective, the spinorial version, $A_{aA}{}^B := A^i_a - \tau_i{}^B_A$, where τ_i are the Pauli matrices, is just the pull-back to the three-manifold Σ of the four-dimensional spin connection on unprimed spinors). The framework is therefore well-suited for analysing the structure of self-dual and anti-self-dual graviational fields. To obtain the self-dual solutions, one can just set A^i_a equal to zero in the constraint equations (2.4) and the evolution equations (2.6). (Since self-dual solutions are in general complex, the reality conditions disappear. Euclidean self-dual solutions that *are* real are obtained by requiring that the triads E^a_i be real.) This simplifies the equations enormously and provides us with a new characterization of self-dual solutions (Ashtekar et al. 1988c). This characterization involves simple, geometric differential equations on triads and, unlike in the formulations introduced by Newman and Penrose, the entire description is *local* in space-time. Therefore, one can apply standard field-theoretic methods to it. In particular, it is not difficult to construct a phase-space description of these (complex) fields. The geometric form of the reduced field equations—and their similarity with self-dual Yang–Mills system—suggests that this may be a bi-Hamiltonian system with infinitely many conserved charges. Quantization of this theory would provide an interesting *four-dimensional*, field-theoretic model, which would be useful in testing ideas that one wants to use in full quantum gravity.

4. The Choice of Time

Let me now discuss the issue of time in some detail. For simplicity of presentation, in the main body of this section I will focus only on the scalar constraint, assume that Σ is topologically $\mathbb{R}^3$, and use the asymptotically flat boundary conditions. Comments on other constraints and the spatially compact case will be made at the very end.

As is clear from equations (2.4) and (2.5), in the asymptotically flat context there is a distinction between constraints and Hamiltonians, or, equivalently, "gauge" and dynamics. (For details, see Chapters II.2 and III.2 in Ashtekar et al. 1988a). Therefore, it is possible to adopt the standpoint that the overall situation is quite analogous to that in, say, non-Abelian gauge theories. (See, e.g., Ashtekar and Horowitz 1982.) To discuss quantum dynamics, one may first complete steps (i)–(v) in the above program and write down the Hamiltonian operator on the Hilbert space of physical states. Since the classical Hamiltonian weakly commutes with all constraints and is weakly real, one would expect that with an appropriate choice of ordering, the quantum Hamiltonian would be a well-defined, self-adjoint operator on the physical Hilbert space. It would therefore generate a one-parameter family of unitary transformations. One could simply call the parameter "time" in quantum theory. Thus, it is possible to argue that it is straightforward to tackle the issue of time in the asymptotically flat context.

My present view is that although it *is* consistent to adopt such an attitude, this strategy is not very useful in practice. To see this, let us recall the situation in the classical theory. Given the classical phase-space, one can pass to the reduced phase-space by factoring out the constraint surface by the orbits of the constraint vector fields. Classically, this is precisely the space of physical states. In the asymptotically-flat context, there is nontrivial dynamics on this new symplectic manifold, and its generating functional is simply the projection of the Hamiltonian from the full to the reduced phase-space. (Note that the reduced Hamiltonian depends *only* on the asymptotic values of the lapse and the shift). One can construct the Hamiltonian vector field on the reduced phase-space and call its affine parameter "time" in the classical theory. Although this procedure is "correct," without a suitable gauge-fixing prescription to tie the value of the affine parameter to some geometric field in space-time, the resulting notion of time is not very useful in practice. By itself, it has very little *direct* relation to the operational notions, such as the proper time measured by clocks that we actually use in general relativity.

In quantum theory, the notion of time as the parameter along the group generated by the Hamiltonian operator suffers from the same drawbacks. To make this notion useful, we have to make it more "concrete." We need the quantum analog of the classical gauge-fixing procedure. However, now the issue is substantially more complicated because the end result of solving all quantum equations is *not* a space-time. What we need is "internal clocks": some of the mathematical degrees of freedom of the theory are to be the clocks, with respect to which other degrees—identified as the physical modes—"evolve."

There is a second reason why the above strategy does not seem very useful in practice. Step (v) above assumes that we have identified the "correct" inner-product on physical states. Now, simple examples of constrained mechanical systems—such as the parametrized nonrelativistic particle—suggest that in practice it may be difficult to find the correct inner-product without first separating "time" from true degrees of freedom. More precisely, these examples suggest that we use the following procedure to complete our step (v). First, we should try to split $A_a{}^i$ into two parts, A^T representing "time," and A^R representing the rest—i.e., the "physical degrees of freedom" contained in $A_a{}^i$—such that the scalar constraint can be reinterpreted as a Schrödinger equation for wave functionals $\Psi(A) \equiv \Psi(A^R, A^T)$, which tells us how Ψ "evolves" with respect to A^T under the action of a Hamiltonian that acts only on its argument A^R. If this could be achieved, the situation would be similar to that in the case of the parametrized particle. The "correct" scalar product would then involve integration of $|\Psi(A^R, A^T)|^2$ *only* over the "true degrees of freedom" A^R.

These ideas are not new. They have been around for a long time now in the context of spatially compact space-times (although their relevence to the asymptotically flat case has been somewhat obscure). However, in the traditional metric representation, it has proved to be very difficult to implement them even in the weak field limit. (See, e.g., Kuchař 1970.) Let us begin by analyzing the source of the difficulty. Consider, first, the parametrized free particle with phase-space coordinates $(q^1, \ldots, q^n; p_1, \ldots, p_n)$ subject to the constraint

$$p_1 + H(q^2, \ldots, q^n, p_2, \ldots, p_n) = 0. \tag{4.1}$$

If we use the q-representation, the quantum constraint can be identified with the time-dependent Schrödinger equation $i\hbar \frac{\partial}{\partial q^1} \Psi(q^i) = \hat{H} \cdot \Psi(q^i)$, in which the role of time is played by q^1. On the other hand, in the momentum representation, where p_1 is just a multiplication operator, the quantum version of the classical constraint equation, $p_1\Psi(p_i) = -\hat{H}\cdot\Psi(p_i)$, does *not* resemble the time-dependent Schrödinger equation. Thus, although two representations may be mathematically equivalent, one may be better suited to single out "time" than another. I would like to argue that in the gravitational case the metric representation is ill-suited, much in the same way as the momentum representation is ill-suited in the above example. To see this, recall that, in terms of the traditional canonical variables, the scalar constraint can be rewritten as:

$$\mathcal{E}(q) - H_N(q,p) = 0, \tag{4.2}$$

where $\mathcal{E}(q)$ is the ADM energy, p_{ab} is the momentum canonically conjugate to the three-metric q_{ab} and $H_N(q,p)$ is the Hamiltonian corresponding to the lapse N. This constraint is very similar to (4.1), with $\mathcal{E}(q)$ playing the role of p_1. The similarity suggests that we should think of $\mathcal{E}(q)$ as the variable conjugate to "time." This idea is attractive especially because $\mathcal{E}(q)$ has the interpretation of the total energy of the system. Furthermore, the structure of (4.2) and the interpretation of $\mathcal{E}(q)$ is unaffected by inclusion of matter sources. The similarity also suggests that it would be very difficult to extract "time" in the q_{ab}-representation where $\hat{\mathcal{E}}(q)$ is just a multiplication operator.

The analysis of the parametrized particle suggests that the situation would be better in the $A_a{}^i$-representation: since $\mathcal{E}$ is now expressible as a functional of the triads $E^a{}_i$ only, $\hat{\mathcal{E}}$ would be a differential operator in the $A_a{}^i$-representation. The "part of $A_a{}^i$ conjugate to $\mathcal{E}$" in the argument of the wave functional $\Psi(A)$ is now to play the role of time.

I have been able to implement this idea only in an approximate way. Let us suppose that we have carried out the first three steps in the quantization program and have available wave functions $\Psi(A)$. The next step is to solve the quantum constraints. It is here that I would like to use an approximation scheme. Let us introduce a background configuration $(\overset{\circ}{E}{}^a_i, \overset{\circ}{A}{}^i_a = 0)$, where $\overset{\circ}{E}{}^a_i$ is a flat triad, and expand out fields that appear in the constraint equations in powers of deviations, $(\hat{E}^a{}_i - \overset{\circ}{E}{}^a_i)$ and $(\hat{A}^i_a - 0)$. Since we now have access to a background triad $\overset{\circ}{E}{}^a_i$, we can convert the internal indices to vector indices and, using the flat metric $\overset{\circ}{q}_{ab}$ constructed from the background triad, carry out the standard decomposition of symmetric tensor fields into transverse-traceless, longitudinal, and trace parts. With this machinery at hand, let us solve the constraints order by order. If we keep terms just to *first order* in deviations, we find that the only nonzero part of $\hat{A}_{ab} := \hat{A}_a^{\ i}\overset{\circ}{E}_{bi}$ is the symmetric, transverse traceless part, $\hat{A}^{TT}$. That is, all other parts of $\hat{A}_{ab}$ are of at least second order. If we impose the scalar constraint to second order, after some simplification we obtain:[5]

$$\frac{-2}{G}(\Delta\hat{E}^T)\cdot\Psi(A^{TT},A^L,A^T,A^S) \equiv \hbar\frac{\delta}{\delta\tau(x)}\cdot\Psi(A^{TT},A^L,A^T,A^S)$$
$$= G\hat{A}_{ab}^{TT}(x)(\hat{A}_{ab}^{TT}(x))^*\cdot\Psi(A^{TT},A^L,A^T,A^S), \qquad (4.3)$$

where, Δ is the Laplacian of the flat metric $\overset{\circ}{q}_{ab}$; $\hat{E}^T$ is the trace part of $\hat{E}_{\ i}^a\overset{\circ}{E}_i^b$; A^L, A^T and A^S are, respectively, the longitudinal, the trace and the skew parts of $\hat{A}_{ab}$, and where $\tau(x) = \frac{-G}{2}(\Delta)^{-1}A^T(x)$ is the field canonically conjugate to $\frac{-2}{G}(\Delta E^T)$. This equation constrains the functional dependence of $\Psi(A)$: If we know the value of $\Psi(A)$ at all configurations at which A^T is, say, zero, the equation determines $\Psi(A)$ at *all* values of $A_a^{\ i}$. In this sense, it is a quantum *constraint* equation. On the other hand, if we simply integrate the equation over Σ using the volume element provided by the background $\overset{\circ}{q}_{ab}$, we obtain:

$$\hbar\left(\int d^3x\frac{\delta}{\delta\tau(x)}\right)\cdot\Psi(A^{TT},A^L,A^T,A^S)$$
$$= \hbar\left(\int d^3k\,|k|\,A_{ab}^{TT}(\vec{k})\frac{\delta}{\delta A_{ab}^{TT}(\vec{k})}\right)\cdot\Psi(A^{TT},A^L,A^T,A^S) \qquad (4.4)$$
$$\equiv \hat{H}_{(2)}\cdot\Psi(A^{TT},A^L,A^T,A^S),$$

where $\hat{H}_{(2)}$ denotes the second-order, truncated Hamiltonian. We now note that $\hat{H}_{(2)}$ is precisely the Hamiltonian of the linearized theory. Next, since $\Psi(A)$ is *holomorphic*, we have:

$$\frac{\delta}{\delta\tau(x)}\cdot\Psi(A) = i\frac{\delta}{\delta(Im\tau(x))}\cdot\Psi(A), \qquad (4.5)$$

where $Im\tau(x)$ is the imaginary part of $\tau(x)$. Therefore, (4.4) *can* indeed be interpreted as a Schrödinger equation provided we identify $\int d^3x\frac{\delta}{\delta(Im\tau(x))}$ with the time translation operator[6] $\frac{\partial}{\partial t}$: then (4.4) just reduces to $i\hbar\frac{\partial}{\partial t}\Psi(A) = \hat{H}_{(2)}\cdot\Psi(A)$, with $\hat{H}_{(2)}$ independent of t. Thus, the scalar constraint of general relativity does contain, at least approximately, a time-dependent Schrödinger equation. Again, this feature remains intact if one includes matter sources.

Recall that in canonical quantum gravity we do *not* have access to a classical space-time. Even in the above truncation procedure, we worked only with the three-manifold Σ and expanded operators around some background fields on Σ. Thus, we did not have access to a classically defined time variable. Rather, we were able to isolate from among the mathematical variables contained in $A_a{}^i$, a preferred variable t which serves as an "internal clock" with respect to which the wave function "evolves." Put differently, by identifying time in the components of $A_a{}^i$ we have *derived* the Schrödinger equation of linearized gravity without having access to a space-time metric—or indeed, even a four-manifold. An immediate consequence is that the theory is unitary in this weak-field limit. Thus, from the viewpoint of canonical quantum gravity, the standard quantization of the spin-two field on a Minkowskian background is consistent because, to this approximation, there is a well-defined truncated theory in which one can single out an internal clock.

Note that the procedure required that we keep terms to second order and thus include "interactions" between linear gravitons. In particular, the construction breaks down if we set Newton's constant G equal to zero. The rough picture is that the linear gravitons, being transverse and traceless, themselves do not have a Coulombic degree of freedom. However, since they carry energy, they act as sources of a Coulombic gravitational field in a second-order calculation. It is the second-order field that carries information about "time" (or, more precisely, $\frac{\partial}{\partial t}$).

What is the situation with respect to other constraints? The final picture is completely analogous to that sketched above for the scalar constraint. The action of the truncated Gauss-law constraint determines the dependence of the wave function $\Psi(A^{TT}, A^L, A^T, A^S)$ on the skew part, A^S, of A_{ab} while that of the truncated vector constraint determines the dependence on the longitudinal part A^L. Thus, in the truncated theory under consideration, it is only the dependence of the wave function on the "physical modes" A^{TT} that is freely specifiable. Once we know this dependence, the dependence on all other parts of A_{ab} is *completely* determined by the truncated quantum constraints. This is the sense in which the quantum constraints "generate gauge." On the other hand, we can identify certain "displacement operators," $\frac{\partial}{\partial t}$ and its analogs for the Gauss and the vector constraints, and reinterpret these equations in space-time terms. Then, the quantum scalar constraint implies that the (truncated) Hamiltonian generates "time evolution," and the quantum vector equation implies that the (truncated) three-momentum generates "space translations."

What is the interpretation of the above "time" variable in the classical theory? In terms of the more familiar variables, $Im\, A^T$ is just the trace part, K^T, of the extrinsic curvature K_{ab}, when the Gauss law is satisfied. Thus, the "local" time function $t(x)$ of the truncated theory is given by $t(x) = \frac{-1}{2}(\Delta)^{-1}K(x)$, where Δ is the Laplacian of the background metric $\overset{\circ}{q}_{ab}$ and $K(x)$ is the trace of the extrinsic curvature. It is easy to check that the resulting t is indeed an affine parameter of the Hamiltonian vector field in linearized gravity. Thus, what we have accomplished is to tie the affine parameter to a concrete, geometric field encoded in the components of the connection $A_a{}^i$. That this is a viable choice of the time function in the weak-field limit was pointed out by Kuchař already in 1970. However, while the representation used by Kuchař exists only in the weak field limit, the $A_a{}^i$ representation exists in the full theory.[7] Therefore, we can now attempt to extend the procedure to the full theory. Whether this can be achieved in a clean fashion is not yet clear. The analysis should not be very difficult because the full constraint is at worst quartic in the canonical variables $A_a{}^i$ and E^a_i. If the answer turns out to be in the negative, the theory would predict that the notion of time — and associated unitarity of quantum evolution — is only approximate. One may then be faced with the type of problems that are raised in Ted Jacobson's article in the session on "time."

Having singled out a preferred time variable in the weak-field limit, we can attempt to introduce a scalar product on the physical degrees of freedom A_{ab}^{TT}. In the program outlined in Section 2, the choice is to be dictated by the requirement that the classical reality conditions be taken over as Hermitian adjointness relations between the quantum observables. It turns out that, in the weak-field limit under consideration, this condition suffices to pick out the "correct" inner product without having to invoke Poincaré invariance of the measure (Ashtekar and Lee 1989). Whether the reality conditions are strong enough in the full theory to pick out an inner product on the space of physical states is, however, still an open question.

Finally, let me comment briefly on the spatially compact case. In this case $\mathcal{E}$ vanishes identically and we cannot repeat the above argument as is. However, since (4.3) continues to hold in the compact case, it would seem that one should be able to extract a notion of "local time" $t(x)$ in a suitably truncated theory. However, it is probably impossible to construct a preferred *global* displacement operator $\frac{\partial}{\partial t}$.

Acknowledgments. Jim Hartle suggested that I give this talk and made up the title for me. The title in turn forced me to look at the whole program from a new perspective. The work reported in Section 4 was initially inspired by a discussion with Joohan Lee. Over the last three years, my views on the entire subject have been deeply influenced by discussions with numerous colleagues, especially Ted Jacobson, Carlo Rovelli, Joseph Samuel and Lee Smolin. To all these individuals, I am most grateful. This work was supported by NSF grants PHY86-12424, INT88-15209, and PHY 90-16733.

NOTES

[1] For bosonic strings, the divergence was established by Gross and Periwal (1988). One expects the situation to be essentially the same for superstrings. Private communications: D. Gross (1988), J. Ambjorn (1989).

[2] In this framework, the lapse is a scalar density of weight minus one, so that the integrand of the Hamiltonian is, as required, a density of weight one.

[3] In several earlier papers, it was erroneously stated that the reality conditions are polynomial in the Lorentzian case. In effect, in those papers, the reality condition was expressed in terms of *old* variables and the nonpolynomial dependence came in precisely in the passage to the old variables. Note that, in the Riemannian (i.e., +, +, +, +) case, the second reality condition is satisfied if and only if A_a^i itself is real.

[4] His analysis of the physical interpretation of these solutions is, however, incomplete. I believe that a correct treatment of the issue of appropriate inner products is feasible in this case and is likely to show that some of these exact solutions are physically interesting.

[5] (4.3) follows from the fact that, in the full theory, the scalar constraint can be re-expressed as: $-2\epsilon^{abc}\partial_a\Gamma_{ab} = G^2 (A_a^{*b} A_b^a - A^* A)$, where $\Gamma_{ab} = \Gamma_a^i e_{bi}$ and $A_a^b = A_a^i e^b_i$.

[6] The fact that "time" arises as the imaginary part of a complex variable suggests a new way of performing the Wick rotation to pass to the "Euclidean" regime. Unlike in Hawking's Euclidean program, this rotation would be only a computational tool.

[7] Also, the close relation between (4.1) and (4.2) and the role of energy $\mathcal{E}$ in the procedure seems to have been overlooked in the older treatments.

REFERENCES

Ashtekar, A. (1986a). "Self-duality and spinorial techniques." In *Quantum Concepts in Space and Time,* eds. C.J. Isham and R. Penrose, Oxford: Oxford University Press, 302–317.

Ashtekar, A. (1986b). "New variables for classical and quantum gravity." *Physical Review Letters* 57: 2244–2247.

Ashtekar, A. (1987). "A new Hamiltonian formulation of general relativity." *Physical Review D* 36: 1587–1603.

Ashtekar, A., and Horowitz, G.T. (1982). "On the canonical approach to quantum gravity." *Physical Review D* 26: 3342–3353.

Ashtekar, A., Mazur, P., and Torre, C.T. (1987). "BRST structure of general relativity in terms of new variables." *Physical Review D* 36: 2955–2962.

Ashtekar, A. (with invited contributions) (1988a). *New Perspectives on Canonical Gravity* Bibliopolis, Naples.

Ashtekar, A., Romano, J.D., and Tate, R. (1988b). "New variables for gravity: inclusion of matter." *Physical Review D* 40: 2572–2587.

Ashtekar, A., Jacobson, T., and Smolin, L. (1988c). "A new characterization of half-flat solutions to Einstein's equation." *Communications in Mathematical Physics* 115: 631–648.

Ashtekar, A. and Lee, Joohan (1989). "Weak field limit of general relativity ion terms of new variables." Syracuse University preprint.

Ashtekar, A. and Rovelli, C. (1989). "Quantum Faraday lines: Loop representation for the Maxwell field." Syracuse University preprint.

Ashtekar, A., Balachandran, A.P., and Jo, S. (1989a). "The CP-problem in quantum gravity." *Int. J. Mod. Phys. A* 4: 1293–1514.

Ashtekar, A., Rovelli, C., and Smolin, L. (1989b). "Loop representation for linearized gravity." (in preparation).

Ashtekar, A., Husain, V., Samuel, J., and Smolin, L. (1989c). "(2 + 1) quantum gravity as a toy model for the (3 + 1) theory." *Classical and Quantum Gravity* 6: L185–L193.

Bengtsson, I. (1989). "Yang–Mills theory and general relativity in three and four dimensions." *Physics Letters B* 220: 51–53.

Brügman, B. (1990). "The method of loops applied to lattice gauge theory." Syracuse University preprint.

Gross, D., and Periwal, V. (1988). "String perturbation theory diverges." *Physical Review Letters* 60: 2105–2108.

Husain, V. and Smolin, L. (1989). "Exactly solvable quantum cosmologies from two Killing field reduction of general relativity." *Nuclear Physics B* 327: 205–238.

Jacobson, T. (1988a). "New variables for supergravity." *Classical and Quantum Gravity* 5: 923–935.

Jacobson, T. (1988b). "Fermions in canonical gravity." *Classical and Quantum Gravity* 5: L143–L146.

Jacobson, T. and Smolin, L. (1987). "The left-handed connection as a variable for quantum gravity." *Physics Letters B* 196: 39–42.

Jacobson, T. and Smolin, L. (1988a). "Covariant action for Ashtekar's form of canonical gravity." *Classical and Quantum Gravity* 5: 583–594.

Jacobson, T. and Smolin, L. (1988b). "Nonperturbative quantum geometries." *Nuclear Physics B* 299: 295–345.

Ko, M., Ludvigsen, M., Newman, E.T., and Tod, K.P. (1981). "The theory of $\mathcal{H}$ space." *Physics Reports* 71: 51–139.

Kodama, H. (1988). "Specialization of Ashtekar variables for Bianchi-IX cosmologies." *Progress in Theoretical Physics* 80: 1024–1040.

Kuchař, K. (1970). "Ground state functional of the linearized gravitational field." *J. Math. Phys.* 11: 3322–3334.

Renteln, P. (1988). Ph.D. Thesis (Harvard University).

Renteln, P. and Smolin, L. (1989). "A lattice approach to spinorial quantum gravity." *Classical and Quantum Gravity* 6: 275–294.

Rovelli, C. and Smolin, L. (1990a). "Loop representation for quantum general relativity." *Nucl. Phys. B* 331: 81–152.

Rovelli, C. and Smolin, L. (1990b). "Loop representation for lattice gauge theory." Pittsburgh-Syracuse preprint.

Rovelli, C. and Smolin, L. (1988). "Knot theory and quantum gravity." *Physical Review Letters* 61: 155–158.

Samuel, J. (1987). "A Lagrangian basis for Ashtekar's reformulation of canonical gravity." *Pramana Journal of Physics* 28: L429–L431.

Samuel, J. (1988). "Gravitational instantons from the Ashtekar variables." *Classical and Quantum Gravity* 5: L123–L126.

Torre, C.G. (1990). "Perturbations of gravitational instantons." *Physical Review D* 41: 3620–3625.

Witten, E. (1988). "2 + 1-dimensional quantum gravity as an exactly soluble system." *Nucl. Phys. B* 311: 46–78.

Discussion

DEWITT: Doesn't the requirement that the connection be self-dual over-constrain it?

ASHTEKAR: No, it doesn't.

DEWITT: I would worry about that.

ASHTEKAR: That is the surprising feature about the whole thing. All the simplifications arise by requiring that the connection be self-dual, and yet, the requirement doesn't over-constrain it.

DEWITT: So the restriction to the self-dual part somehow cancels your having made the connection complex.

ASHTEKAR: That's exactly right.

DEWITT: On your transparencies, you set the shift equal to zero. How much more complicated does it get if you don't specify the shift and lapse?

ASHTEKAR: Not much more.

DEWITT: I'm thinking of computer relativity.

ASHTEKAR: Yes, I do think that this framework might help. In fact Nakamura's group in Tokyo—I don't know what progress they've made—but they were thinking of using these variables in order to do numerical classical relativity.

DEWITT: Does the fact that you don't need to raise and lower indices on E^a mean that you can evolve and get solutions of Einstein's equations running right through singular metrics?

ASHTEKAR: Yes, I think so.

DEWITT: Changing the signature, even?

ASHTEKAR: I have not looked at the question of change of signature of the *space-time* metric. But it's certainly true that you could evolve initial data through configurations where the determinant of the metric vanishes. As Chris Isham was saying a while ago, at certain points it's possible that this will actually just let you evolve through singular configurations.

YORK: To reconstruct the metric from your variables, one must use square roots of q. If you want to build back your space-time metric, numerically or however, you will have to worry about the singular metrics.

ASHTEKAR: Yes, that is correct. The metric itself will be ill-defined at points where the determinant vanishes.

JACOBSON: I wanted to make a comment about singular metrics. You don't *have* to do a (3+1) decomposition, you can just take the formalism and use the Palatini method; write the equations of motion in terms of the connection and the tetrad, and they are perfectly well-defined even when the tetrad becomes degenerate.

ASHTEKAR: That's true. But for numerical work, one would have to use the (3+1) framework.

JACOBSON: While discussing the transform between the loop-space and A-representations, you mentioned the transformation between position and momentum representations as an analogy. But there, the two spaces have the same dimension. Here the loop space is so much bigger than the space of connections.

MASON: It's smaller. Loop space is three functions of one variable, whereas the space of connections is three functions of three variables; so the loop space is considerably smaller than the space of connections.

JACOBSON: My question is: is there an inverse transform from the loop-space representation to the A-representation?

ASHTEKAR: I believe that there is an inverse transform although I do not have it. What is the general idea? It is the following: Given any loop, we've an object called the "form factor of the loop." As Chris Isham said, the "basis" I introduced in the loop-space representation of the Maxwell field seems to be over-complete in the same sense that coherent states are over-complete. One would like to introduce a complex structure on the loop space and restrict oneself only to those functions on the loop space that are holomorphic. Then, the conjecture would be: there is a 1–1 correspondence between a suitable class of holomorphic functions of an A, with holomorphicity so defined, and photon Fock states. I say "suitable class" because, as we heard from Carlo Rovelli, there are further conditions that ensure that a function on the loop space is built from holonomies for *some* connection.

HOROWITZ: There is an obvious complex structure on the loop space anyhow.

ASHTEKAR: That's not the complex structure that works in this case because the correct complex structure so-to-say knows about and distinguishes between photons, gravitons, and so on.

YORK: You usually talk about asymptotically flat cases; are there any essential differences, any interesting things that come up when you do compact spaces?

ASHTEKAR: The general program I outlined just goes through. The treatment of the issue of time doesn't, because I really used the existence of nonzero energy to give me $\delta/\delta t$. But the general program goes through in the spatially compact case without any problem at all. If the topology of the manifold is nontrivial, then you have new effects. In general, even if the topology is just $\mathbb{R}^3$, because I'm talking about wave functions of the connection, you have things like θ-vacua. These are different from the θ-vacua that Rafael Sorkin, John Friedman, Don Witt, and Jim Hartle have been talking about because they have to do with the "internal" group rather than the diffeomorphism group. What has happened is that now, because of the inclusion of the internal degree of freedom, the kinematic symmetry group has been enlarged from the three-dimensional diffeomorphism group to a semi-direct product of the three-dimensional diffeomorphism group and the group of internal rotations. There is a much richer structure topologically, which is being investigated. In fact Chris Isham and Kunstater wrote a paper quite some time ago about trying to look at the θ-vacua structure of Yang–Mills theory in the presence of nontrivial topology, and amazingly we can literally take all their results and transport them here.

KUCHAŘ: I would like to return to the question of the reality condition. It seems that the main constraints have such a nice structure, but there may be a real difficulty in imposing the reality condition in the full theory. What are the commutation relations between the reality condition and the constraints, if one tried to impose it initially?

ASHTEKAR: Poisson brackets?

KUCHAŘ: In classical theory, yes.

ASHTEKAR: The reality condition, if imposed initially, is preserved in time, that means that the commutators are just constraints.

KUCHAŘ: But aren't there structure *functions*?

ASHTEKAR: Probably, yes.

KUCHAŘ: So, if you quantize, you must somehow ensure properly ordered commutation relations between the constraints.

ASHTEKAR: Absolutely; I think that's an extremely important and difficult problem.

KUCHAŘ: How far is this realized in the loop representation that you have? How much attention is paid there to the reality condition and its commutators?

SMOLIN: Maybe I could answer that after my talk.

ASHTEKAR: I am not up to date on this but let me say what I know. Surprisingly, on the loop space, it seems that the reality condition—which I was always afraid of—has a nice form, a really simple form. It is telling you something geometrical. I'm more optimistic than I used to be. That is an enormously important problem, which has not been addressed in any real degree.

HOROWITZ: At one point, as I understood it, there was a question of how to factor order the momentum constraint. One factor ordering was preferred on general grounds, whereas another was used to solve the constraint equations. Is that still an issue?

ASHTEKAR: No. The most satisfactory statement about solutions of the constraints is in the loop space, as we will see in the Rovelli and Smolin talks. It turns out that the factor ordering that closes formally is precisely the factor ordering that is relevant in the loop space representation.

YORK: I have a question about using a lapse function with a different density weight. I mean, I realize that you can write down scalar densities all day long without their being explicitly dependent on the metric or anything. You just say it's a scalar density of weight minus one, and there's the end of it. And then, you proceed to solve your constraints and carry out your quantum program. But, on the other hand, if I want then to see how classical space-time emerges in some classical limit, then the customary interpretation of the lapse function is that it has something to do with the proper time between two time slices. And I can see, in linearized theory, how you can get around this. In the exact theory, I don't see how you can get around the problem of saying that now the proper time depends on the choice of coordinates on the slice; because no matter how you represent it, in the canonical theory you will have some slices. And if you change your three-dimensional coordinates, all your objects are perfectly well behaved, excluding the lapse function, which picks up a Jacobian in front. So the proper time between two slices then depends on the three-dimensional coordinates. I don't understand that.

ASHTEKAR: Two remarks. First of all, as you say, it is true that one can do everything consistently if the lapse has density-weight minus one. In other words the constraints close. You never have to change the density weights in the calculation of Poisson brackets. The second point is that in order to reconstruct a space-time, you really do have to do something nonpolynomial. You have to extract the scalar lapse out of the density. You really have to de-densitize N by multiplying it by the square root of the determinant of the three-metric. And so, the physical time that you are talking about is the de-densitized object. It is not the density object. That doesn't have any coordinate dependence. But one does have to do this work.

KUCHAŘ: Can we return once more to the problem of foliation fixing? In the case that you have an asymptotically flat space-time, which is also linearized, you can interpolate an almost flat slice between the data at infinity and the inside. In the full quantum theory, what is the way in which you can to fix your foliation, or does it not matter?

ASHTEKAR: If I take this canonical point of view, then strictly speaking in the quantum theory there is no four-manifold with a four-metric and so on, and so I'm not foliating anything. I have to solve the Schrödinger equation. I have to isolate what I mean by time, and if the weak field result is a good indicator, "time" may come out of the imaginary part of A^T also in the full theory. That's what I would mean by "time." Then I'll have to worry about conditions under which I can reconstruct the four-dimensional space-time. Are there some nice wave functions that can evolve dynamically, and actually correspond to four-dimensional space-times? That's an important question for me and the other people working on this subject. But it's not an abstract question about the formalism. It's a question about particular solutions to the equations.

KUCHAŘ: I would say that the state of the field is different on two slices related by a "bubble-time evolution" because there are different internal degrees of freedom included.

ASHTEKAR: You can say that only if you have a space-time.

KUCHAŘ: Well, if we come to compact spaces in this discussion, I think that this question will reappear.

ASHTEKAR: Oh yes. In fact, the prescription I was giving, at least on the face of it, doesn't work on compact spaces. Maybe one can modify it.

SORKIN: You gave this action using only the self-dual part of the connection. If you used that in a functional integral, would you obtain a solution in your self-dual representation? In fact, there is some statement of Weinberg's relating a functional integral to a wave function. I suppose it would have to be in the A-representation.

ASHTEKAR: I don't know the answer to that question. Charles Torre has a general formalism for computing propagators, Feynman path integrals, and so on, using complex coordinates on the phase-space. But it is an open problem to actually use his general formalism; to actually do the calculations, for example, to see what happens if you did the Hartle–Hawking proposal, and so on. Jim Hartle and I were just talking about it last night, and there were some conjectures, but I do not know exactly what happens.

TORRE: The four-dimensional action that Joseph Samuel et al. found is complex. I don't think anybody knows how to define a functional integral with complex actions of that sort.

ASHTEKAR: But yesterday we just heard from Jim Hartle that it was kosher for the action to be complex.

TORRE: No, no, it may be O.K. I'm just saying that I don't know anybody who has defined it directly, or a definition of what it means to integrate, because there are reality conditions on your Hilbert space.

ASHTEKAR: One could ask this question in stages. One can first ask the question in just the Euclidean domain. If you don't believe in the Euclidean domain, it is only an exercise to go to the Lorentzian domain. In the Euclidean case, everything is real, and one should be able to do this. Then, one has to get the wave function in the Lorentzian regime. The conjecture is something like the following: by doing the Euclidean integral, you get a wave function that is a function of real A. It's just $SU(2)$-valued. The "correct" Lorentzian wave function is a holomorphic function of complex A's, and it could be just an analytic continuation of the one resulting from the Euclidean integral.

MASON: At least in the linearized theory, would the anti-self-dual polarization lead to the same quantization?

ASHTEKAR: Yes, but there is a catch. If for an harmonic oscillation, for example, you fix a symplectic structure once and for all, then naively one would think that d/dz or $d/d\bar{z}$ are equally allowed polarizations. It turns out that's not the case. If you've fixed the sign of the symplectic structure, and d/dz gives a factor of $e^{-z\bar{z}}$, then $d/d\bar{z}$ will give $e^{+z\bar{z}}$. So if you have fixed your symplectic structure, that decides whether you should use d/dz or $d/d\bar{z}$ as your polarization. So, it is the choice of the overall sign of the gravitational symplectic structure that decides whether you should use the self-dual or the anti-self-dual polarization.

Loop Representation in Quantum Gravity: The Transform Approach

Carlo Rovelli

1. Introduction

In this talk, I present an approach to quantum gravity which was developed in the last months in collaboration with Lee Smolin.

The central idea of this approach is to define a "representation" for the states the the operators of quantum gravity in terms of functionals over a loop space (Rovelli and Smolin 1988, 1990). "Representation" is used here in the sense of Dirac (Dirac 1930). Quantum mechanics is given by an abstract Hilbert space with an algebra of operators. The abstract state space may be realized concretely by specific spaces of functions, that is by different choices of a basis.

The idea of the loop representation for quantum gravity is a natural development of two previous results: Ashtekar's reformulation of general relativity (Ashtekar 1986, 1987, 1988) and Jacobson and Smolin's discovery of a class of solutions of the Wheeler–DeWitt equation (reformulated in the Ashtekar variables) related to loops in three dimensions (Jacobson and Smolin 1988).

The central result obtained in the loop representation is that one can find a large class of solutions of the *full* set of quantum constraint equations. More precisely, we find the general solution of the diffeomorphism constraint and an infinite-dimensional space of solutions of all the constraints.

These solutions turn out to be classified by the knot and link classes (see, for instance, Kauffman 1983) of the three-manifold that represents the space. These are the topologically inequivalent ways in which one loop, or a set of loops, may be knotted and linked in three dimensions.

There are two completely independent ways to arrive at the definition of the loop representation. The first is the one through which the loop representation was originally found. One starts from the self-dual representation (Ashtekar 1988), which is the natural quantization in the Ashtekar variables, and defines a linear mapping from the self-dual state space into the space of the loop functionals (Rovelli 1988). We call this mapping $\mathcal{F}$.

This procedure is analogous to the definition of the momentum representation for a single quantum particle obtained by first defining the coordinate representation and then using the Fourier transform to "bring" all the observables into momentum space. The mapping $\mathcal{F}$ can be expressed as a functional transform, and can be viewed as an infinite-dimensional analog of the Fourier transform. This approach to the loop representation is physically intuitive, but is not very rigorous, particularly because we have little control over functional integrals.

The second approach is more rigorous. One starts from the classical theory (in Ashtekar formulation), chooses a suitable Poisson algebra of classical observables, and then quantizes it by finding a representation of this algebra in terms of linear operators on a linear space. The algebra of observables that leads directly to the loop representation is a graded noncanonical algebra of nonlocal observables. We call it the $\mathcal{T}$ algebra. The first approach, which makes use of the functional transform, can be considered as a heuristic device for finding the operators that quantize the $\mathcal{T}$ algebra.

This second approach to the loop representation is in the spirit of Chris Isham's approach to quantum gravity (Isham 1984). In particular, Isham has long argued that the correct quantization for general relativity should not be a Fock space quantization (Isham and Kakas 1984), and that, therefore, one has to quantize a noncanonical algebra of observables.

In this talk, I shall describe the first approach to the loop representation, the transform approach. In the next lecture, Lee Smolin will describe the $\mathcal{T}$ algebra quantization approach to the loop representation.

2. Loop Solutions

The natural "Schrödinger" quantization of the Ashtekar reformulation of general relativity is the so-called self-dual representation (Ashtekar 1988). In this representation, the unconstrained states are represented by wave functions $\Psi[A]$ that are holomorphic functionals of the Ashtekar connection $A_a(x)$. The elementary observables that are quantized are the connection $A_a(x)$ and its conjugate momentum $\tilde{\sigma}^a(x)$. The corresponding quantum operators are the multiplicative operator $A_a(x)$ and the functional derivative operator

$$\tilde{\sigma}^a(x)\Psi[A] \equiv \frac{\delta}{\delta A_a(x)}\Psi[A].$$ (1)

The dynamics is expressed by the quantum constraints

$$\mathcal{G}(x)\Psi[A] = \mathcal{D}_a \frac{\delta}{\delta A_a(x)}\Psi[A] = 0,$$ (2)

$$\mathcal{C}_a(x)\Psi[A] = Tr\left[F_{ab}(x)\frac{\delta}{\delta A_b(x)}\right]\Psi[A] = 0,$$ (3)

$$\mathcal{C}(x)\Psi[A] = Tr\left[F_{ab}(x)\frac{\delta}{\delta A_a(x)}\frac{\delta}{\delta A_b(x)}\right]\Psi[A] = 0.$$ (4)

The unexpected result that constitutes the point of departure of the approach that I am describing is the discovery by Ted Jacobson and Lee Smolin (Jacobson and Smolin 1988) of a large class of solutions of the first and third of these equations.

These solutions are related to loops in three dimensions and are obtained as follows. Given a loop γ, we define the functional of the connection

$$H[\gamma, A] = Tr\, P\, e^{\int_\gamma A},$$ (5)

where P means path ordered. H will be a fundamental object in what follows. It is the trace of the holonomy (or Wilson loop) of the connection $A_a(x)$ along the loop γ. For a multiple loop $\{\gamma\}$, namely a finite set of loops γ_i, we define

$$H[\{\gamma\}, A] = \prod_i H[\gamma_i, A].$$ (6)

Under suitable conditions on the loop (or multiple loop), H is a solution of the first and third constraint equations. The conditions are that the loop (or the multiple loop) be differentiable and without intersections.

Other solutions are given by certain linear combinations of holonomies of loops with intersections with coefficients that satisfy certain algebraic conditions (Jacobson and Smolin 1988, Husain 1988).

These results can be obtained by a formal calculation, as well as in the context of a well-defined regularization of the constraint operators.

At this point, the open problem is to solve the remaining constraint equation (3), which is the diffeomorphism constraint. The diffeomorphism constraint requires the wave functionals to be invariant under the action of the diffeomorphism group. It is easy to see that the diffeomorphism constraint generates a loop state, the action of a diffeomorphism on the loop itself. It "moves the loop around" by an infinitesimal amount. Thus, the Smolin–Jacobson loop solutions are not diffeomorphism invariant; however, they are transformed among themselves by the action of the diffeomorphism group.

We would like to define solutions of the diffeomorphism constraint by "averaging" the loop wave functionals over all the positions of the loop. But we do not know how to do this because of the difficulties of defining diffeomorphism-invariant measures on the space of the loops.

The alternative way that we may, and will, follow is to take seriously the loop solutions and try to express everything in terms of them. This is a general way of proceeding in quantum mechanics. If we know the solutions of the time-independent Schrödinger equation for an harmonic oscillator, or for the hydrogen atom, we can use them as a new basis in the Hilbert space. For the oscillator, this is the Fock basis, which diagonalizes the energy; for the Hydrogen atom, we use the basis that diagonalizes H, L^2, L_z, and so on. In these bases, the complexity of the quantum system is greatly simplified, but the intuitive content of the Schrödinger equation in the "coordinate" basis is partly lost. Similarly, for a free field, or as a starting point for a perturbation expansion, we use the Fock basis, which diagonalizes energy and momentum.

In a similar way, since the Hamiltonian constraint is in part "diagonalized" in the loop states, we want to "go to the loop basis," to see if in this basis the theory could be more tractable.

A useful analogy is given by the Klein–Gordon operator in the co-ordinate representation of a free relativistic particle. This can be seen as the quantization of the classical constraint $p^2 = m^2$, which defines the dynamics of the classical theory. Solutions of the Klein–Gordon equation are given by plane waves that satisfy a suitable condition on the momentum. We know that it is useful to go to a representation of functions of the momentum, and this can be obtained by defining an integral transform (the Fourier transform) that has the plane waves as integral kernel.

We want to do the same in the present context. By analogy, this would mean to go to a representation in which the states are expressed as functionals $A[\gamma]$ of the loops (the loops "label" the holonomies like the momenta label the plane waves). Similarly, by analogy, we want to change basis by defining an infinite-dimensional integral transform, which has the holonomy H as integral kernel:

$$\mathcal{A}[\gamma] = \int d\mu[A]\, H[\gamma, A]\, \Phi[A]. \tag{7}$$

Intuitively, a given loop functional $\mathcal{A}[\gamma]$ will express the superposition of holonomy states $H[\gamma, A]$, just as momentum wave functions are superpositions of exponentials.

The problem with equation (7) is the choice of the functional measure $\mu[A]$. However, it can be shown that for our purposes, namely, to bring the theory to loop space, the choice of the measure is largely unimportant. In Rovelli (1988) it is shown, in fact, that a redefinition of the measure can be formally matched by a redefinition of the basic self-dual observables. Here we shall follow a more direct approach, and show that the mapping from the self-dual to the loop representation can be defined without the need of a measure, just by using a topology on the state space.

In the next section, we shall formalize these ideas and define the change of basis in the quantum state space of general relativity. Two comments are worthwhile. The first one is that we do not expect, in general, that a formal change of basis in a quantum field theory could give us a mathematically equivalent structure. It is certainly possible, and also reasonable, that the structure that we shall build in the new representation is mathematically inequivalent to the one of the self-dual representation. We do not see this fact as negative. A quantum theory of gravity still doesn't exist, and our long-term aim is to find at least one complete quantum structure which has general relativity as classical limit. The shift from a naive "Schrödinger quantization" to an inequivalent structure with the same classical limit is not only allowed, but perhaps required.

The second comment is that, from the physical point of view, a change of basis in the state space of the quantum theory may be a device for the introduction of new concepts for the description of a quantum system. For instance, the Fock basis for the quantum Maxwell field provides us with the notion of photons. The hope that the loop solutions could be indications for the introduction of some of the new concepts needed to describe space-time at the Planck scale is one of the motivations for the present work.

3. The Mapping $\mathcal{F}$

Let S be a space of functionals $\Psi[A]$ of the Ashtekar connection. We do not specify this space more precisely here, but we assume that it is a well-defined topological vector space. As we said, the present considerations can be regarded as just a heuristic way to arrive at a redefinition of the theory in loop space.

A first technical step is to go to the space S^* dual to S. S^* is the space of the continuous linear functionals (distributions) Φ over S. The conjugate representation of the observables is naturally defined on S^* by the formula

$$O^+\Phi(\Psi) = \Phi(O\Psi), \tag{8}$$

which defines the operator O^+ on S^* corresponding to any observable O on S. Physically, S^* forms the bra representation, while the states in S are the kets.

Now we consider the space of the piecewise smooth loops and multiple loops in the three-manifold, and the space S of the functionals over this space. These functionals are objects like $\mathcal{A}[\{\gamma\}]$, that associate an amplitude to any set of a finite number of loops. Note the "Fock-like" structure

$$\mathcal{A}[\{\gamma\}] = (\mathcal{A}[\gamma_1], \mathcal{A}[\gamma_1, \gamma_2], \ldots). \tag{9}$$

We then define the linear mapping $\mathcal{F}$ from S^* into $\mathcal{S}$,

$$\mathcal{F}: \Phi \longrightarrow \mathcal{A}, \tag{10}$$

in the following way:

$$\mathcal{A}[\{\gamma\}] = \Phi(H_{\{\gamma\}}), \tag{11}$$

where $H_{\{\gamma\}}$ is the holonomy functional $H[\{\gamma\}, A]$ previously defined.

$\mathcal{F}$ associates to any quantum state expressed in the self-dual representation (more precisely to every bra) the same quantum state expressed in the new loop representation.

This is a bit formal. In order to reexpress this mapping in a more concrete form, it is useful to introduce a measure $\mu[A]$ on the space of the connections. $\mu[A]$ has no direct physical meaning, in particular it is not related to the physical scalar product and we do not require it to be gauge invariant or diffeomorphism invariant. For instance we may choose it to be a Gaussian measure. We disregard the problems related to the support of the measure and to the actual definition of the set of integrable functionals. Given $\mu[A]$, some of the distributions Φ in S^* (a subset that forms a dense subspace of S^*) can be expressed in terms of functionals $\Phi[A]$:

$$D(\Psi) = \int d\mu[A]\,\Psi[A]\,\Phi[A]. \tag{12}$$

Then the transform $\mathcal{F}$ may be expressed as an integral transform:

$$A[\gamma] = \int d\mu[A]\,H[\gamma, A]\,\Phi[A]. \tag{13}$$

This is the form given in equation (7).

The transform $\mathcal{F}$, in spite of being a functional integral, is less out of control than appears at first sight: in certain cases, the functional integral may be explicitly computed. Abhay Ashtekar mentioned in his talk the application of the present formalism to the linearized theory. In this case, there is a natural choice for $\mu[A]$, and the integral may be computed over all the $\Phi[A]$ that represent the states of the standard Fock basis (Ashtekar, Rovelli and Smolin 1988). Thus, at least in the linearized case, the transform has a well-defined mathematical meaning.[1]

4. Observables

The main reason for defining the transform is that we can use it to transport the observables to the loop space S. Given an operator O on S (and its equivalent O^+ on S^*), we say that an operator $\tilde{O}$ on S is equivalent to O if the transform intervenes between the two, namely, if

$$\tilde{O}\mathcal{F} = \mathcal{F}O^+. \tag{14}$$

Commutation relations are preserved by this equivalence. Therefore the equivalent operators $\tilde{O}$ form another representation of the algebra of the observables O. It is straightforward to verify that O and $\tilde{O}$ are equivalent if and only if

$$\tilde{O}H[\gamma, A] = OH[\gamma, A], \tag{15}$$

where $\tilde{O}$ acts on the loop argument, while O acts on the connection argument. In the Fourier analogy, we have that, since

$$i\frac{d}{dk}e^{-ikx} = xe^{-ikx}, \tag{16}$$

the operator id/dk in the momentum representation is equivalent to the operator x in the coordinate representation.

Not every operator O has an equivalent $\tilde{O}$ (one can actually find the condition on O for an equivalent to exist). In particular, the elementary operators A and $\frac{\delta}{\delta A}$ do not have an equivalent. However, enough have equivalents to construct a complete representation of the gauge-invariant observables. We describe here some of these.

Consider again, for a given loop γ, the expression $H[\gamma, A]$. This is a function of A; therefore, it is a classical observable. In order to stress that we are considering it as an observable, we introduce a new notation for it, which is $T[\gamma]$. $T[\gamma]$ is the first of a large class of objects, called T^n, which Lee Smolin will describe in his talk. The corresponding quantum operator in the self-dual representation is a multiplicative operator:

$$T[\gamma]\Psi[A] = H[\gamma, A]\Psi[A]. \tag{17}$$

Now we want the loop equivalent. Namely, a loop operator $\tilde{T}$ that satisfies

$$\tilde{T}[\gamma]H[\eta, A] = T[\gamma]H[\eta, A]. \tag{18}$$

We have

$$\tilde{T}[\gamma]H[\eta, A] = H[\gamma, A]H[\eta, A] = H[\{\gamma, \eta\}, A] \tag{19}$$

(remember the definition of H for multiple loops). Therefore we can define the operator $\tilde{T}[\gamma]$ as

$$\tilde{T}[\gamma]\mathcal{A}[\{\eta\}] = \mathcal{A}[\gamma \cup \{\eta\}]. \tag{20}$$

Clearly, the (Abelian) algebra of the T operators in the self-dual representation is isomorphic to the algebra of these operators, which is Abelian because the $\cup$ operation is commutative.

In order to completely define the theory in the loop space, we also need observables that contain the conjugate momentum $\tilde{\sigma}$. The operators that have an equivalent have to be gauge invariant; moreover we want to avoid the coincidence of functional derivatives at the same point because their action is ill-defined. These considerations force us to the following choice for the other observables to bring into loop space. We first consider the classical observables

$$T^a[\gamma](s) = Tr\left(P\,e^{\oint_s^s A\,d\gamma}\tilde{\sigma}^a(\gamma(s))\right), \tag{21}$$

where s is the parameter of the loop γ. $T^a[\gamma](s)$ is the trace of the holonomy of γ with a $\tilde{\sigma}$ inserted at the point $\gamma(s)$. We call this insertion a "hand" on the loop γ. The corresponding self-dual quantum operator is

$$\hat{T}^a[\gamma](s) = Tr\left(P \, e^{\oint_s^s A \, d\gamma} \frac{\delta}{\delta A_a(\gamma(s))}\right). \tag{22}$$

It is not difficult to verify that this operator has an equivalent. In fact

$$\hat{T}^a[\gamma](s)H[\eta, A] = \tilde{T}^a[\gamma](s)H[\eta, A], \tag{23}$$

where $\tilde{T}^a[\gamma](s)$ is defined by

$$\tilde{T}^a[\gamma](s)\mathcal{A}[\eta] =$$
$$= \frac{1}{2} \int dt \, \delta^3(\gamma(s), \eta(t))\dot{\eta}^a(t)(\mathcal{A}[\gamma\#\eta] - \mathcal{A}[\gamma\#\eta^{-1}]) \tag{24}$$

In this formula, $\gamma\#\eta$ is defined as the loop obtained starting from the intersection of γ and η and going first around γ and then around η (the r.h.s. is zero if the two loops do not intersect in correspondence of the hand of γ). η^{-1} is the loop η with reversed orientation.

$\tilde{T}^a[\gamma](s)$ is therefore the quantum operator in the loop representation corresponding to the classical observable $T^a[\gamma](s)$. We will not go through the straightforward calculation that gives equation (23). We just note the following two facts: first, the result is zero unless the two loops intersect at the hand because the hand is a derivative operator and the holonomy depends only on the values of A at the point on γ. Second, the way the two loops rejoin is given by the algebra of $SU(2)$ matrices.

The other observables that one can bring into the loop space are obtained by inserting more than one hand in the holonomy. The corresponding loop operators have features similar to the ones we introduced: they act by breaking and rejoining loops at the intersections. We will not describe them here, but Lee will discuss them at length in the following talk.

The main property of this set of observables are the following. They completely define the kinematics of the theory, in the sense that any (gauge-invariant) observable of the classical theory can be expressed in terms of them. They have a closed Poisson algebra, which we call a T algebra; and, in the quantum context, a closed commutator algebra.

Therefore, they completely define a quantum kinematics that reduces to the kinematic of general relativity in the classical limit. Thus, we can forget the transform $\mathcal{F}$ and *define* the kinematical framework of quantum gravity in terms of the state space $\mathcal{S}$ and this algebra of operators.

The following facts may be noted. First, the algebra $\mathcal{T}$ is not a canonical Poisson algebra. More generally, there is no obvious way to bring into the space $\mathcal{S}$ a set of conjugate operators with canonical commutation relations. We take this result as support for the idea that the canonical commutation relations do not form the correct algebra for quantum gravity.

Second, all the operators that we defined on the loop space are non-local. They are related to extended structures in the three-space: the loops. Again, we think that this is an indication that quantum gravity cannot be expressed in terms of standard local quantum field operators.

The structure suggested by the loop solutions, and obtained by the transform, defines an infinite-dimensional quantum system in terms of noncanonical and nonlocal observables. As we shall see in the next section, the physical states will essentially be defined solely in terms of topological properties of extended structures. These features are perhaps necessary for a nonperturbative quantization of a generally covariant quantum field theory.

5. Constraints and Solutions

Let us now consider the constraints that code the dynamics of the system.

First, the gauge constraints are automatically solved in going to the loop representation, which is entirely gauge invariant.

Next, we consider the diffeomorphism constraints, which are the ones that we were not able to solve in the self-dual representation. Not surprisingly, the operator equivalent to the diffeomorphism constraints turns out to be the generator of diffeomorphisms in the loop space. These are the generators of infinitesimal displacement of loops by vector fields in the three-manifold.

Now, a loop functional annihilated by such generators has to be constant on the orbits generated by the action of the diffeomorphism group on the manifold of the loops.

The orbits of the diffeomorphism group in loop space are well-known and well-studied in the mathematics literature. Given two loops, they are in the same orbit if and only if they are knotted in the same way. More generally, given a set of loops, it can be deformed in another set of loops if the two sets are knotted and linked in the same way. The inequivalent ways of knotting a loop are called knot classes of the manifold. For multiple loops, they are called the link classes of the manifold.

Since we are considering here also loops with intersections and "corners" (this is required for the operators to be well-defined), we have to generalize the definition of link classes by including linked loops with intersections and with "corners." A functional that solves the diffeomorphism constraints is one that is constant on these generalized link classes. Conversely, we have one independent solution of the diffeomorphism constraints of quantum general relativity for any generalized link class of the manifold.

Finally, let's consider the Hamiltonian constraint. The easiest way to understand why one can find solutions to it in the loop representation is to return to our analogy of the free relativistic particle. Recall that the Klein–Gordon equation becomes diagonal in the momentum representation. This means that it becomes just a statement of the support of the wave function. We expect the same thing to happen here. That is, we expect that the loop functionals that have support on the differentiable self-intersecting loops are indeed solutions of the loop equivalent of the Hamiltonian constraint. This result is in fact confirmed by a more rigorous calculation (Rovelli and Smolin 1990).

Now, the set of differentiable nonintersecting loops is closed under the action of the diffeomorphisms. Thus, this set can also be decomposed into equivalence classes under the action of the diffeomorphism group. The link classes formed by differentiable nonintersecting loops are exactly the ordinary link classes considered in knot theory. Thus, we have the following result: There is one independent solution of the full set of constraint equations for any ordinary link class of the three-space.

The set of solutions that we have found forms a sector of the space of physical states in nonperturbative quantum general relativity. This sector has a rich structure. For instance, there is a "loop number quantum number" that counts the number of loops. Links are infinite but countable; therefore, they define a discrete basis in the state space. They can be ordered by some definition of "growing complexity," and one can define a "vacuum-like" state which has support only on the no-loop.

These results indicate that the structure of the physical state space of quantum gravity is very different from that of the standard Fock space. Without a definition of the physical observables, however, it is difficult to give a physical interpretation to these exact solutions of the dynamical equations of quantum gravity.

6. Conclusion

To summarize, we have the following results:

(1) States, observables and constraints may be naturally represented in terms of functionals over the space of piecewise-smooth loops. This is a concrete representation of the abstract state space and operator algebra of quantum gravity.

(2) This representation may be formally obtained through a linear mapping $\mathcal{F}$ from the (representation conjugate to the) self-dual representation.

(3) In this representation, the constraints can be solved. The general solution of the diffeomorphism constraints is given by the functionals of the generalized link classes of the three-manifold.

(4) A set of independent solutions of the full set of constraints is in one-to-one correspondence with the ordinary link classes of the manifold.

Let me stress that these results are obtained without any additional physical input or assumptions besides general relativity and standard quantum mechanics.

The theory is still far from being a complete theory of quantum gravity since two elements are missing: the Hilbert structure on the physical states, and the definition of the physical, gauge-invariant operators.

Nevertheless, we think that these results show that it is possible to develop techniques and concepts to study the nonperturbative structure of quantum gravity; and that, therefore, in spite of the failure of the perturbation expansion, the possibility that quantum general relativity could be a viable physical theory is still open.

NOTE

[1] During the editing of these notes we learned that Witten has shown that (in our language) it is possible to explicitly compute the transform of the vacuum state $\Psi[A] = 1$ by using the diffeomorphism and gauge-invariant measure given by the exponential of the Chern–Simon form. The result is a Jones Polynomial (Witten, 1988).

REFERENCES

Ashtekar, Abhay (1986). "New Variables for Classical and Quantum Gravity." *Physical Review Letters* 57: 2244–2247.

———(1987). "New Hamiltonian Formulation of General Relativity." *Physical Review D* 36: 1587–1602.

Ashtekar, Abhay, ed. (1988). *New Perspectives in Canonical Gravity*. Naples: Bibliopolis.

Ashtekar, Abhay; Rovelli, Carlo; and Smolin, Lee (1990). *Linearized Quantum Gravity in Loop Space*. Forthcoming.

Dirac, Paul A.M. (1930). *The Principles of Quantum Mechanics*. Oxford: Clarendon Press.

Husain, V. (1988). "Intersecting Loops Solutions to the Hamiltonian Constraint of Quantum General Relativity." Yale University preprint.

Isham, Christopher J. (1984). "Topological and Global Aspects of Quantum Theory." In *Relativity, Groups and Topology II*. Bryce S. DeWitt and Raymond Stora, eds. New York: North-Holland, pp. 1059–1290.

Isham, Christopher J., and Kakas, A.C. (1984). "A Group Theoretical Approach to the Canonical Quantization of Gravity: I. Construction of the Canonical Group." *Classical and Quantum Gravity* 1: 621–632.

Jacobson, Ted, and Smolin, Lee (1988). "Nonperturbative Quantum Geometries." *Nuclear Physics B* 299: 295–345.

Kauffman, Louis H. (1983). *Formal Knot Theory*. Mathematical Notes, no. 30. Princeton: Princeton University Press.

———(1987). *On Knots*. Annals of Mathematics Studies, no. 115. Princeton: Princeton University Press.

Rovelli, Carlo (1988). "Loop Space Representation." In *New Perspectives in Canonical Gravity*. Abhay Ashtekar, ed. Naples: Bibliopolis, pp. 275–303.

Rovelli, Carlo, and Smolin, Lee (1988). "Knot Theory and Quantum Gravity." *Physical Review Letters* 61: 1155–1158.

———(1988b). "Loop Representation for Quantum General Relativity." *Nuclear Physics B* 331: 80–152 .

Witten, Edward (1988). In *Proceedings of the 9th International Conference on Mathematical Physics*. B. Simon et al., eds. Bristol: Adam Hilger.

Nonperturbative Quantum Gravity via the Loop Representation

Lee Smolin

1. Introduction

In this talk I would like to describe a new approach to the quantization of general relativity that Carlo Rovelli and I have been developing over the last year. This approach allows exact results to be obtained concerning the structure of the space of physical states of the gravitational field. It is based on a new representation for canonical quantum gravity, called *the loop representation*, which is based on functionals over sets of loops in a three-dimensional manifold Σ.

The construction of this new representation was motivated by certain problems that arose in working with the self-dual representation of Ashtekar (1986, 1987, 1988). In particular, there it was possible to find exact solutions to the Hamiltonian constraint (Jacobson and Smolin 1988), while no solutions were found to the spatial diffeomorphism constraints. The loop representation was invented to remedy this problem by constructing a representation space on which the spatial diffeomorphism group acts naturally, while preserving the simplicity of the action of the Hamiltonian constraint in the self-dual representation (Rovelli and Smolin 1988, 1990, Rovelli 1988).

Using the loop representation, it is straightforward to find the complete set of solutions to the constraints that generate diffeomorphisms of Σ. These are expressed in terms of a countable basis, the elements of which are in one-to-one correspondence with the knot and link classes of Σ. (More precisely, the elements are in one-to-one correspondence with a generalization of what the mathematicians usually call the link classes, which allows loops to intersect and be kinked.) Thus, knot theory (Kaufmann 1983, 1987a, b, c, Rolfson 1976) becomes, through the loop representation, a basic tool for studying the structure of the space of physical states of quantum general relativity at the nonperturbative level.

Further, the Hamiltonian constraint turns out to have a rather simple action on elements of the loop representation. This makes it possible to find a large set of exact solutions to the Hamiltonian constraint, including a set of states that are also annihilated by the diffeomorphism constraints. These are certainly not the most general solutions to the combined set of constraints, but they are examples of exact physical states of the gravitational field.

The derivations of these results involve a lot of technicalities, which are described in detail elsewhere. I would like to focus here on the conceptual issues that underlie and motivate the construction of the loop representation. In particular, I would like to begin by discussing some of the basic problems that any completely nonperturbative quantization of the gravitational field must confront. Following this, I will give a statement of the basic strategy or philosophy guiding the construction of the loop representation. After these preliminary remarks, I will give a schematic account of the construction of the loop representation, and the derivation of the main results, which I hope can serve as a critical or annotated introduction to the more detailed discussion of Rovelli and Smolin 1990. I will end with a discussion of open problems.

The account of the construction of the loop representation given here is meant to be complementary to the discussions of Ashtekar and Rovelli in this volume. The basic definitions of the new variables, as well as a description of the general quantization program based on those variables, is given in Ashtekar's contribution. In addition, Ashtekar discusses the application of the loop representation of free field theories, which allows us to develop our intuitions about it in a context in which we understand the whole theory.

The loop representation for quantum gravity is discussed by Rovelli from a point of view that is complementary to the one taken here. He describes an approach in which the loop representation is constructed by means of a functional transform from the self-dual representation. This approach is formal, but rather intuitive, and was, in fact, the way that we first discovered the loop representation. In my talk I will describe an approach that is to begin with less intuitive, but completely free of formal constructions, and hence on firmer ground. In this approach, the representation is constructed directly from the classical theory by means of a quantization of a certain algebra of observables on the classical phase space. We developed this approach because it was clear that the basic structure of the algebra of operators in the loop representation was perfectly well-defined in spite of the fact that the measure in the transform was not specified. The approach that I will be describing then is one that arrives at the same construction as described in Carlo's talk, after taking a somewhat longer but more rigorous route.

In the concluding section I will briefly discuss the relationship between the two approaches.

2. Problems Confronting any Nonperturbative Quantization of the Gravitational Field

There are two aspects of any nonperturbative quantization of a gravitational theory that make such a theory qualitatively different from any conventional quantum field theory, or any perturbative construction of a quantum gravity theory. These are:

(1) The absence of any background metric. In a nonperturbative quantization, the metric of space-time must be an operator. There will then be no c-number metric available to use in the construction of the quantum theory. Similarly, there is no background connection structure.

(2) The replacement of the Poincaré group by the diffeomorphism group as the governing symmetry group of the theory.

Conventional quantum field theory relies heavily on the existence of a background metric and the associated Poincaré group. The problem of making a nonperturbative quantization of a gravitational theory is to a large degree the problem of how to construct a quantum field theory in four space-time dimensions without these background structures, and in a way that realizes the diffeomorphism invariance of the classical theory. Any attempt to do this must confront a number of issues that are not encountered in conventional quantum field theory. We will discuss here five of these issues; each of them plays a crucial role in the construction of the loop representation.

Issue 1: How to Regulate the Theory

Any quantization of a nontrivial field theory requires a regularization procedure because the dynamics of the classical theory is expressed in terms of functions which contain local products of the elementary observables. In conventional quantum field theories, the background metric plays a crucial role in the regularization procedure as it is used to measure the scale at which the regularization is imposed.

However, in nonperturbative quantum gravity there is no background metric that we can use to measure (or define) the scale of the cutoff. Furthermore, since diffeomorphism invariance is a local gauge symmetry of general relativity, we must face the problem of whether the regularization of the theory can be done in a diffeomorphism-invariant manner.

ISSUE 2: WHAT IS THE SHORT-DISTANCE STRUCTURE OF THE THEORY?

There has been a lot written about what the short-distance structure of space-time should be in a quantum theory of gravity (foam, strings, conformal invariance, etc.), but comparatively little written about how that short-distance structure is to be defined in terms of the operators and states out of which the theory is to be constructed. I would like to suggest that this is a far from trivial problem.

To see why, let us begin by considering how we go about measuring the short-distance structure in a conventional quantum field theory. This is done by studying the limiting behavior of n-point functions as the points are brought close to each other (where close is measured in terms of the shortest scale of the renormalized quantum field theory). For example, the two-point function behaves as

$$< 0|\phi(x)\phi(y)|0 > = \frac{const.}{|x - y|^d} \tag{2.1}$$

for small $|x - y|$, where this distance, of course, is measured in terms of the background metric. The fact that the behavior of the n-point functions simplifies in a characteristic way as points are brought close together is what we mean by the statement that a quantum field theory has a characteristic short distance structure. The existence of such structure is reflected most generally, in the existence of the operator product expansion

$$\phi(x)\phi(y) = \sum_n \frac{O_n(x)}{|x - y|^n}. \tag{2.2}$$

In nonperturbative quantum gravity, there is no background metric in terms of which the coefficients of an operator product expansion could be defined. Thus, we are led to ask whether the operator product expansion, or something like it, could exist in a nonperturbative quantum theory of gravity, and if so what form it would take.

It is easy to see that in its naive form the operator product expansion could not exist in a nonperturbative quantum theory of gravity. Among the constraints of any such theory will be the generators of the spatial diffeomorphisms, $D_a(x)$, and in a canonical treatment the physical states will have to be annihilated by these constraints,

$$D_a(x)|\psi >= 0. \qquad (2.3)$$

Let us consider the expectation value of any product of local operators in such a state. It is clear that since any n points can be moved around to any other n points by a diffeomorphism, the expectation value could only depend on the number of points (and perhaps the topology of the space). For example, for the two-point function we must have, for $x \neq y$,

$$< \psi|\phi(x)\phi(y)|\psi > = constant. \qquad (2.4)$$

This strongly suggests that there is no operator product expansion, or at least none that is constructed in terms of local operators.

Now, as the reader may have realized, this argument is, in an important sense, bogus. This is because the operator $\phi(x)\phi(y)$ is not a physical observable as it does not commute with the diffeomorphism constraints $D_a(x)$. In order to extract information from physical states, we must use only physical observables, that is observables that are operators on the physical state space. This brings us to our next important issue, which we will have to discuss before coming back to the issue of the short-distance structure.

ISSUE 3: WHAT ARE THE PHYSICAL OBSERVABLES?

In a theory with local gauge symmetry, physical observables are those which commute with all of the constraints that generate the gauge symmetry (Dirac 1964). In general relativity there are two kinds of constraints, the (three-dimensional) diffeomorphism constraints and the Hamiltonian constraint, which generates, locally, the dynamics. Thus, it will be convenient to divide the discussion into two parts:

What are the (spatially) diffeomorphism-invariant observables? In the classical Hamiltonian theory we know of many observables that are spatially diffeomorphism-invariant. A familiar set of examples is all integrals of local densities constructed from the canonical variables, all multiple integrals over multi-densities, and so on. However, there are two problems which arise when we try to carry over these observables to the quantum theory. First, all of them involve products of the basic observables q_{ab} and $\tilde{p}^{cd}$ at the same point. As such, we do not expect them to be well-defined as operators in any representation defined by the canonical commutation relations. Indeed, all but a small set of them involve either $\sqrt{\det q}$ or q^{ab}, which are nonpolynomial in the basic observables. Thus, we do not expect any of these observables to be well defined naively; instead, they must be introduced into the quantum theory in terms of some regularization procedure.

However, as we have just discussed, it is not a trivial problem to invent regularization procedures that do not destroy the diffeomorphism covariance of these local expressions. This problem is made more difficult by two circumstances: first, we must regularize not just simple products, but nonpolynomial expressions. Second, if we regularize by point-splitting, we have to worry about how to tie up the indices of the q_{ab}'s and $\tilde{p}^{cd}$'s, which are now at different points. Without introducing any additional structure, the only way to do this in a diffeomorphism-covariant way at the classical level is to introduce parallel transporters along geodesics joining the points. However, these parallel transporters are also nonpolynomial functions of the q_{ab}, because they involve the exponential of the metric connection (!!!) and because the expression that determines which curve are the geodesics will be nonpolynomial. When we go to the quantum theory, it is not at all obvious that any of this structure can be introduced in a way that is well defined. If we try, for example, to make the parallel propagator well defined, we must regulate it, but that will destroy its transformation properties under diffeomorphisms,... and it is clear we are in an infinite regress!

Of course, we might hope that, having destroyed the transformation character of these observables by regulating them, the diffeomorphism invariance could be restored in some limit in which the regularization is removed. While we can cannot conclusively argue against this, we may note that this is not an easy thing to do in a nonperturbative context, as the regularized operators will act on the larger, unphysical, state space, which has many extra degrees of freedom not present in the physical states. In perturbative electrodynamics, where the decoupling of the physical and gauge degrees of freedom can be easily accomplished, problems like this can be handled; but it is not so clear how to do it in the present, much more complicated circumstances.

We must, then, conclude that we do not in fact know whether any of the classically diffeomorphism-invariant observables that are written in terms of local functions of the q_{ab}'s and p^{ab}'s can be made into good diffeomorphism-invariant operators in the quantum theory.

This problem will play an important role when we come to discuss the interpretation of the results from the loop representation, and we will come back to it. For now, we go on to the second difficulty with the known spatially diffeomorphism-invariant observables, which is that, while we know of an infinite number of examples, *we do not have any systematic understanding of them.*

This is illustrated by listing the things that we do not know about the local diffeomorphism-invariant observables. We do not know how many are needed to coordinatize the constraint surface of the constraints $D_a(x)$. We do not have a complete set of diffeomorphism-invariant observables that are mutually commuting in terms of the Poisson brackets. Finally, we do not have any systematic understanding of their algebra.

These problems are reflected in the metric representation of the quantum theory (DeWitt 1967, Wheeler 1968, Bergmann 1958, 1961, Bergmann and Goldberg 1955). It is often asserted that the problem of solving the spatial diffeomorphism constraints is trivial in the metric representation. Indeed, it is trivial to construct examples of such states, for example, any diffeomorphism-invariant functional of the q_{ab}'s is a diffeomorphism-invariant state in the metric representation. While this is true, in order to construct a usable quantum theory, we need much more information about the space of diffeomorphism-invariant states. For example, does the space have a countable basis? Can we exhibit at least one complete basis explicitly? We may remark that if these questions were solved, we would be able to also solve our problems concerning the diffeomorphism-invariant observables in the quantum theory as, given an explicit countable basis, the operators on the space can be expressed directly as matrices.

As far as I know, there has not been any progress on these questions using the metric representation. We will see that one of the major results of the loop representation is that it does lead to an explicit countable basis for the space of diffeomorphism-invariant states. However, as I have just hinted, the problem of their physical interpretation is not trivial.

What are the observables that also commute with the Hamiltonian constraint? As has been stressed many times, one of the unique and difficult features of quantum gravity is that one of the constraints, the Hamiltonian constraint, plays the dual role of generating both a local gauge symmetry and the local dynamics. This is because, without imposing boundary conditions such as asymptotic flatness, the canonical transformations that are generated by a change in the time coordinate must be the same as those associated with evolution. For this discussion we will restrict ourselves to the spatially compact case in which the Hamiltonian is proportional to a linear combination of the constraints. In this case, we have the circumstance that any physical observable must also be a constant of the motion.

Thus, in general relativity the problem of the physical observables is intertwined with the dynamics in a way that one does not find in more conventional theories. We do not have the freedom to find the physical observables first, and then later introduce the dynamics and study their evolution. Since the Hamiltonian is proportional to the constraints, the physical observables do not evolve. Instead, finding the full set of physical observables must be equivalent to the problem of finding the general solution to the dynamics, as they are functions of the phase-space variables that are constants of the motion for any initial data satisfying the constraints.

Given this situation, it is not surprising that for the case we are considering, the vacuum Einstein equations in the spatially compact case, *not a single physical observable is known explicitly as a function of the phase space variables.*

This said, I would like to return to the problem of the short-distance structure. It follows from our discussion of the problem of physical observables that to make a meaningful statement about the short-distance structure in quantum general relativity at the nonperturbative level, one must find an observable that is (1) physical at the classical level, (2) can be regulated in a way that does not destroy the fact that it commutes with the constraints, and (3) explores some aspect of the short-distance behavior.

These are very nontrivial requirements, and there is no reason to believe that any of the classical observables which we can write as integrals over local densities will survive this process to become observables in the quantum theory even if we restrict our attention to the class of spatially diffeomorphism-invariant observables. We will see that in the loop representation we are able to fulfill these requirements with respect to the diffeomorphism constraints, and thus give a construction for the complete set of diffeomorphism-invariant observables. Using them, we will be able to discover that the short-distance structure of quantum gravity nonperturbatively is completely different from that described by the perturbation theory. In particular, we will see that the diffeomorphism-invariant observables do not measure metrical relationships among local observables, as they do in ordinary quantum field theories. Instead, we find that all spatially diffeomorphism-invariant observables measure topological relations among nonlocal observables.

ISSUE 4: WHAT IS THE PHYSICAL INNER PRODUCT?

In the Dirac quantization procedure the physical inner product exists only on the space of physical states. Furthermore, while we may put an inner product on the whole space of unconstrained states, being in the kernel of the constraints, the physical states will have infinite norm with respect to most inner products on the larger space.[1] This means that the problem of regulating and solving the constraints has to be treated in a context in which there is no physically meaningful inner product structure. This makes the problem of solving the constraints in quantum gravity different from the problem of solving for the eigenstates of the Hamiltonian in a conventional quantum field theory.

In particular, we cannot use the physical inner product to study the convergence problems that arise when the Hamiltonian constraint equation is expressed, as it must be, in terms of a regularization procedure.

We will discuss this issue in more detail when we come to it.

In general, since the physical inner product is imposed on the space of solutions to the constraints, and since the local dynamics is coded into the Hamiltonian constraint, this means that in gravity, the choice of the inner product depends on the dynamics of the theory in a way that it does not in ordinary quantum field theories.

ISSUE 5: WHAT ARE THE CONSEQUENCES OF REPLACING POINCARÉ INVARIANCE BY DIFFEOMORPHISM-INVARIANCE?

As I mentioned above, much of the usual structure of perturbative quantum field theory is a consequence of Poincaré invariance. The existence of particle states and the Fock representation for the Hilbert space are both direct consequences of the representation theory of the Poincaré group. In a quantum theory without a background flat metric and its Poincaré group of isometries, we should not expect either structure to play a role.

In order to get an idea of how three-dimensional diffeomorphism invariance might play a role in a quantum field theory, it is helpful to look at the example of conformal field theory, in which one is constructing a quantum field theory that is invariant under the infinite-dimensional group of conformal mappings on a Riemann surface. It is very striking that much of the progress in this area depends on the fact that a lot is known about the representations (and projective representations) of Diff(S^1) and the loop groups. This knowledge makes possible, in the case of the rational conformal field theories, the construction of quantum field theories entirely by means of the representation theory (Belavin et al. 1984, Friedan et al. 1984, Friedan and Schenker 1987, Moore and Seiberg 1989).

It is thus tempting to speculate how we would approach quantum general relativity at the nonperturbative level if as much were known about the representation theory of Diff(S^3) as is known about the representation theory of Diff(S^1). Might we be able to construct the possible states and observables directly through the group theory by constructing singlets from combinations of the irreducible representations?

Unfortunately, very little is known about the representation theory of Diff(S^3). I would like to suggest that the construction of such a theory should be an important problem for those interested in quantum gravity at the nonperturbative level.[2] In the conclusion, I will suggest that the results of the loop representation strongly suggest that knot theory should play an important role in this representation theory.   ·

3. A Strategy

The problems faced by anyone who would like to quantize a gravitational theory nonperturbatively are thus very daunting. We are required to set up a representation of the theory in which the constraints may be defined, regularized and solved, without knowing what the physical observables of the theory are, either classically or quantum mechanically. Furthermore, we know that if the theory exists, the description it must give of the geometry of space-time must be very different from a classical background geometry; otherwise, we would expect perturbation theory to be reliable. But the perturbation theory of quantum general relativity is nonsense, so the only hope is that it is not reliable. Thus, in this situation, we do not want to introduce too soon any approximations based on an intuitive guess about what quantum geometry is like: since all our experience with geometry is classical, if we can guess it, it is likely to be wrong.

I would like to suggest a strategy for proceeding in this situation.

— *Follow the Dirac quantization procedure strictly.* Fix no gauges and never use a path integral unless we know a rigorous nonperturbative definition for its measure. Essentially, because we are completely ignorant about the structure of the physical state space and observable algebra, we should avoid making any ad hoc assumptions, at least until these have been found.

— *The quantization procedure must be completely regulated.* Trust no result that does not come from a completely regulated construction.

An example of such a construction is a lattice regularization. However, such a regularization destroys diffeomorphism covariance. In particular, I know of no lattice regularization such that the generators of a diffeomorphism group retain the property of having a first class algebra in their regulated form.[3] As the crucial problem is the structure of the solution space to the constraints, I do not believe this is acceptable. Thus, we should demand:

— *The representation must be regulated in such a way that the regulated observable algebra carries a representation of the spatial diffeomorphism group.* The strongest argument I know for using the loop representation is that it does have this property.

— *Make no approximations until forced to. In particular, try to proceed at least to the stage of finding the physical states before making any approximations.* As we do not know the physical inner product until after we have solved the constraints, one cannot measure the errors that result from the use of an approximation procedure before the solutions to the constraints have been found. This might seem a completely hopeless requirement, were it not for the fact that in the Ashtekar form the constraints are simplified to the point that exact solutions can be found (Jacobson and Smolin 1988).[4]

— *Only draw physical conclusions from physical operators.* Trust no assertion about the physics of the system unless it comes from the evaluation, in the physical inner product, of an expectation value of a physical observable between physical states.

As the main point of this strategy is to use the Dirac quantization procedure, I will list its steps here. They are:

1. Choice of a preferred subalgebra $\mathcal{A}$ of the classical observable to be the elementary observables of the quantum theory.

2. Choice of a linear space $\mathcal{S}$, on which there exists a completely regulated algebra of linear operators $\tilde{A}$ that is a deformation of the classical algebra of elementary observables, $\mathcal{A}$.

3. Definition of the constraints and the Hamiltonian of general relativity in terms of the elements of $\tilde{A}$.

4. Solution of the quantum constraints by finding the subspace $S_{Phys.}$ of S that is in the kernel of the regularized constraints in the limit that the regularization is removed.

5. Definition of the physical observables that constitute the operator algebra on the space $S_{Phys.}$. In this step, we are required to do two things; first, find the algebra, and second, give its elements a physical interpretation.

6. Definition of the physical inner product on $S_{Phys.}$. This choice must implement both the reality conditions of the classical theory and the physical interpretation of the physical observables in the sense that operators that correspond to classical physical observables that are real must be Hermitian with respect to the physical inner product.

The first three steps have been completed in the loop representation, and some progress has also been made concerning steps 4 and 5. In the rest of this paper, I will describe first what has been done, after which I will comment on what remains to be done.

We begin with the first step.

4. The Classical Loop Algebra

The quantization of a classical theory consists of the association of classical observables, defined as functions on the phase-space of the theory, with linear operators on some representation space such that the commutator algebra of the latter goes over into the Poisson algebra of the former in the limit $\hbar \rightarrow 0$. For reasons of both operator-ordering and regularization, in any quantization of a quantum field theory most of the classical observables will not have an unambiguous representation in terms of the operator algebra of the quantum theory. What we need to do, then, is to choose a subalgebra of classical observables that will be represented unambiguously in terms of the operator algebra in the quantum theory. These are called the elementary observables. Any other classical observable will then be represented indirectly in terms of this preferred subalgebra. Typically what will happen is that at the classical level, most observables will be expressible in terms of a limit of a sequence of elementary observables. In that case their quantum counterparts will be defined in terms of the same limit of the linear operators associated with the elementary observables in the sequence. These limits, of course, may or may not exist.

This is precisely the process of regularization of a quantum field theory. For example, in the conventional quantization of a scalar field theory, the elementary observables are taken to be the fields $\phi(x)$ and their conjugate momenta $\pi(x)$. Their algebra, $\{\phi(x), \pi(y)\} = \delta^3(y, x)$, can be taken over directly to a representation of the quantum theory, for example the Fock representation. Now, in this representation, there are no operators that correspond to local products, such as $\phi^3(x)$. Instead, one first finds at the classical level a sequence of elementary observables whose limit is the desired local product, in this case, $\lim_{y, z \to x} \phi(y)\phi(z)\phi(x)$. Thus, there will be an operator associated with ϕ^3 in this representation if this sequence, acting on states in the representation, has a well-defined limit.

Of course, in this case, the limit does not exist. This does not prevent there being a well-defined quantum field theory for a scalar field. What is important is that the Hamiltonian, which is $H = \int d^3x[\pi^2 + \phi^2]$, exist. In the Fock representation, as we know, there is an ordering such that the Hamiltonian does exist.

Now, my point is that the choice of a set of elementary observables conditions the rest of the quantization procedure. The set of elementary observables must constitute a closed algebra under Poisson brackets. The set must be small enough so that every element in its algebra can be represented in terms of a well-defined linear operator on the representation space. This is the sense in which the quantization of the elementary observables gives a regularization of the theory. However, the set of elementary observables must be large enough so that the constraints, the Hamiltonian, and a large enough[5] set of physical observables must be expressible at the classical level through limits of sequences of elementary observables. This is the sense in which the algebra of elementary observables must be complete.

Of course, having a choice of elementary observables that are both regulated and complete does not guarantee that a representation based on them will lead to a sensible quantum theory. For this to occur, the limits of sequences necessary to define the constraints, the Hamiltonian, and other physical observables must also exist in the quantum theory. There is no general procedure for discovering whether a choice of a set of elementary observables will work; one must set up the quantization and try it.

In canonical quantum gravity, the traditional choice of elementary observables has been the canonical algebra based on the three-metric, q_{ab}, and canonical momenta, p^{cd}, with their usual canonical commutation relations. This choice, together with a point-splitting regularization prescription for observables outside of this set, admits a quantization in terms of functionals of the three-metric–the metric representation. This choice is complete and regulated in the sense I have described. Unfortunately, it has not led to any definite results because no way has been found to express the Hamiltonian constraint in terms of regulated operators in this representation in any way that allows results to be obtained about their solution space.

To construct the loop representation, we choose a different set of elementary observables that are based on loops, rather than points, in the three-manifold. In order to construct these observables, we need first to introduce some notation. The quantization will be based on a three-manifold Σ, which is without any background metric or connection structure. The phase-space of general relativity will be described using the new variables A_a and E^b, as is described in Ashtekar's lecture here and in Ashtekar 1986, 1987, 1988. We will be concerned, in addition, with closed loops in Σ, which will be denoted by Greek letters α, β, γ, etc. These loops will be assumed to be *piecewise smooth and parametrized,* with *nonvanishing tangent vectors.* Given a loop γ, and two points on it given by their parameter values, s and t, we may define the parallel transport to be,

$$U_\gamma(s, t) \equiv P e^{\int_s^t du A_a(\gamma(u))\dot\gamma(u)^a}. \tag{4.1}$$

The parallel transport all around the loop, beginning and ending at a point s (also called the holonomy), will be denoted $U_\gamma(s)$.

The loop variables we are about to define form a closed algebra under Poisson brackets, which is called the classical $\mathcal{T}$ algebra. It is graded by the nonnegative integers. The zero elements are just the traces of the holonomy around loops. They are denoted

$$T^0[\gamma] \equiv Tr U_\gamma. \tag{4.2}$$

We also find it necessary to introduce observables corresponding to unordered sets of loops in Σ. Such a set is called a *multiloop,* and is denoted

$$\{\gamma\} \equiv \{\gamma_1, \gamma_2, \ldots\}. \tag{4.3}$$

Corresponding to each multiloop $\{\gamma\}$, we also have a T^0 observable, which is defined to be

$$T^0[\{\gamma\}] \equiv \prod_i Tr U_{\gamma_i}. \tag{4.4}$$

Under the Poisson bracket, these T^0 observables may be seen to form an over-complete set of commuting gauge-invariant[6] observables.

These observables are over-complete because they satisfy a set of relations that follow from properties of $SL(2, C)$ matrices and their definition in terms of parallel transport. These include:

i. Invariance under reparametrization of the loop parameter s.

ii. Invariance under interchange of a loop for its inverse, where by the inverse of a loop we mean $\gamma^{-1}(s) = \gamma(s_{max} - s)$, where s_{max} is the value of the loop parameter at the end of the loop (which is usually taken to be 1.)

iii. The spin network identity: Consider two loops, α and β, that share a common base point. Then, referring to Figure 4.1 below, we see that there are three gauge-invariant expressions that can be formed by connecting the parallel transporters around the two loops. These are

$$T^0[\alpha \circ \beta], \ T^0[\alpha \circ \beta^{-1}], \ T^0[\alpha]T^0[\beta]. \tag{4.5}$$

However, because of the basic spinor identity,

$$\delta_A^C \delta_B^D - \delta_A^D \delta_B^C = \varepsilon_{AB} \varepsilon^{CD}, \tag{4.6}$$

these are related by

$$T^0[\alpha \circ \beta] + T^0[\alpha \circ \beta^{-1}] = T^0[\alpha]T^0[\beta]. \tag{4.7}$$

iv. Identity under retracing: Consider, as in Figure 4.2, a loop α, from whose base point grows an open curve η, ending at another point. Then,

$$T^0[\eta \circ \alpha \circ \eta^{-1}] = T^0[\alpha]. \tag{4.8}$$

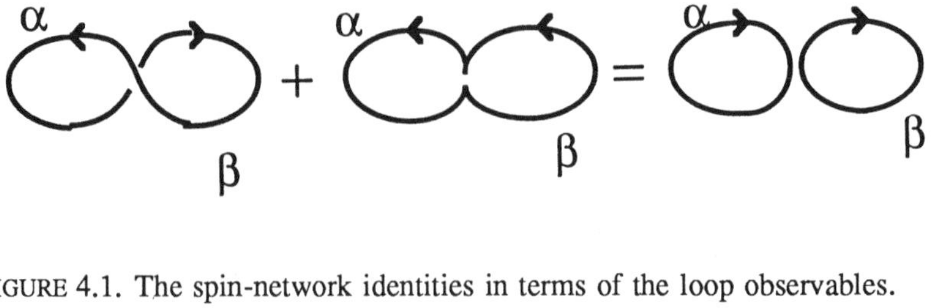

FIGURE 4.1. The spin-network identities in terms of the loop observables.

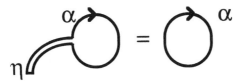

FIGURE 4.2. The retracing identities.

In order to have a complete algebra of observables, we need some observables that also depend on the conjugate E^b fields. We would like these observables to be gauge-invariant. Looking ahead to the problem of representing this algebra by a regulated set of linear operators in a construction of the quantum theory, we should require that our elementary observables not include any that involve more than one E^b at any point of Σ. We have chosen to extend the T^0 observables to a complete set that satisfies these criteria in the following way:

We define a T^1 observable by inserting a conjugate E^b field into the trace of the parallel transport around the loop at some given point s. The full definition of this observable is

$$T^1[\gamma]^a(s) \equiv Tr[U_\gamma(s)E^a(\gamma(s))]. \tag{4.9}$$

It depends on a loop and on a point of the loop where the E^a is inserted, and it carries a spatial tangent-space index at that point, corresponding to the index of the E^a.

Clearly, this definition can be extended. Given any loop γ, and any n *distinct* points $s_1 < s_2 < \ldots < s_n$ on it, we may define a T^n observable such that the parallel transport around γ is interrupted by the insertion of an E^{a_i} at each of the selected points s_i, $i = 1, \ldots, n$. For example, the T^2 observable is defined by[7]

$$T^2[\gamma]^{ab}(s,t) \equiv Tr[U_\gamma(t,s)E^a(\gamma(s))U_\gamma(s,t)E^b(\gamma(t)))]. \tag{4.10}$$

The definition of the T observables may be extended to multiloops in a natural way. We leave to the reader to fill in the details.

The T^n observables enjoy four basic properties that are basic to their use as a set of elementary observables in the construction of the quantum theory.

PROPERTY 1: THE T^n FORM A CLOSED ALGEBRA UNDER
POISSON BRACKETS

In order to understand this algebra, which we call the *classical loop algebra*, let us consider the Poisson bracket of $T^1[\alpha](s)$ with $T^0[\beta]$. An examination of their definitions shows that the Poisson bracket will be zero unless the curve β passes through the point $\alpha(s)$, at which the only conjugate E^a field in the two observables lives. A straightforward calculation shows that the result of the Poisson bracket is then

$$\{T^1[\alpha]^a(s), T^0[\beta]\} = i\Delta^a[\alpha, \beta](s)\left(T^0[\alpha \circ \beta] - T^0[\alpha \circ \beta^{-1}]\right). \quad (4.11)$$

Here the Δ^a stands for the distributional quantity,

$$\Delta^a[\alpha, \beta](s) \equiv \int dt\, \delta^3(\alpha(s), \beta(t))\dot{\beta}^a(t), \quad (4.12)$$

and the notation $\alpha \circ \beta$ stands for the curve that results from composing α with β at the point $\alpha(s)$ where, it is assumed, they intersect.

The Poisson brackets that arise in the T algebra may be visualized by using the following graphic notation and language. In general, the Poisson bracket of a T^n with a T^m will be zero unless the two curves intersect at a point where one of these has an E^a inserted. We notate these observables by drawing dots on the curves, as illustrated in Figure 4.3. These dots refer to the E^a insertions, and are called *hands*. In any given Poisson bracket, a hand will play a role if the loop associated with the other observable passes through it. If it does, we say that the hand is *active*; otherwise, it is *passive*.

If a hand is active, then it contributes a term to the Poisson bracket that is of the form of Figure 4.4. That is, the hand is removed, the two loops are broken at the point of intersection and joined to each other in the two possible ways. In one of these, an orientation will have to be reversed to produce an observable with consistent orientation, and this costs a sign.

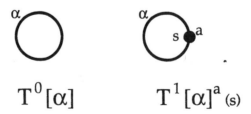

FIGURE 4.3. Graphical notation for the loop observables.

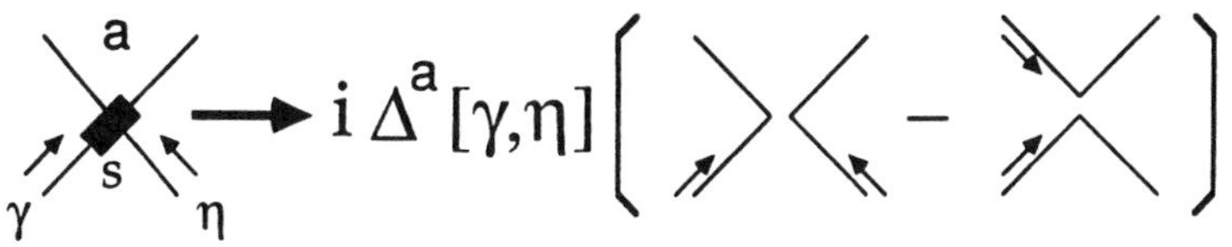

FIGURE 4.4. A grasp and its result.

This action of a hand at a point of intersection is called a *grasp*, and the two curves that result, which have one less hand, are called the *results of the grasp*.

With this graphical notation, we indicate the Poisson bracket (4.11) as in Figure 4.5.

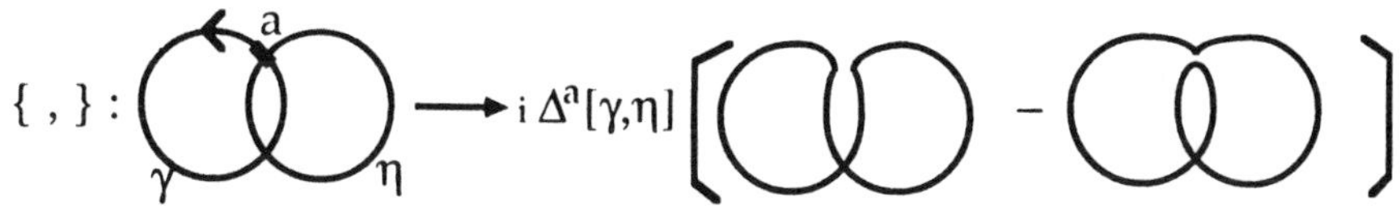

FIGURE 4.5. The Poisson bracket $\{T^1[\alpha]^a(s), T^0[\beta]\}$.

With this nomenclature, the general Poisson bracket in the classical T algebra may be seen to be of the form

$$\{T^n[\alpha], T^m[\beta]\} = i \sum_{\text{grasps}} \Delta[\alpha, \beta] T^{n+m-1}[\text{result of the grasp}]. \quad (4.13)$$

PROPERTY 2: COMPLETENESS ON THE GAUGE-INVARIANT OBSERVABLES

As I stressed above, most observables on the classical phase space will not be elements of the subalgebra of elementary observables, which are the only ones that will be represented directly in terms of a regulated algebra in the construction of the quantum theory. All other observables are to be represented indirectly in terms of limits of sequences of the elementary observables. We must then insure that a given set of elementary observables is large enough so that: (1) they coordinatize either the whole classical phase-space, or a constraint surface of that space (in which case the corresponding constraint will be realized identically in the quantization), and (2) a large class of observables, including the constraints and Hamiltonian, should be expressible as limits of sequences of the elementary observables.

These two conditions are fulfilled by the elements of the classical T algebra. As shown in Jacobson and Smolin 1988, the T^0 observables coordinatize the gauge-invariant and holomorphic functions of the connection A_a. The remaining ones are enough to coordinatize the gauge-invariant functions of both E^a and A_a, as discussed in Rovelli and Smolin 1990.

As to the second requirement, one can show that any gauge-invariant and local function of F_{ab} and E^a may be constructed in terms of limits of sequences of T observables. This result follows from the fact that, as is well known, F_{ab} may be expressed in terms of parallel transport in the limit of small loops. In particular, if we have a small loop, based at a point x and bounding a square of area δ^2 in the coordinate $\hat{a}\hat{b}$ plane, we have the identity

$$U_\gamma(s) - U_{\gamma^{-1}}(s) = \delta^2 F_{ab} + O(\delta^3). \qquad (4.14)$$

Thus, by inserting appropriate combinations of small loops and their inverses into elements of the T algebra, one can get local functions that converge to powers of F_{ab}.

Since, in Ashtekar's formalism, the constraints of general relativity are polynomial, local, and gauge invariant functions of F_{ab} and E^a, they can be expressed in terms of sequences of elements of the T algebra.

PROPERTY 3: THE LOOP ALGEBRA IS CLOSED UNDER THE ACTION OF THE SPATIAL DIFFEOMORPHISMS

In order to establish this property, one may first show that acting on gauge-invariant observables, the diffeomorphism constraints of Ashtekar's formalism indeed generate the action of spatial diffeomorphisms (Ashtekar 1988). One may then construct the action of a spatial diffeomorphism on elements of the T algebra by exponentiating the action generated by the diffeomorphism constraints, and the result is that the action of a diffeomorphism on a T^n observable is to take it into a linear combination of other T^n observables in the natural way, according to the action of the diffeomorphism group on loops. For example, for the T^0 observables one has, for $\phi \in \text{Diff}(\Sigma)$,

$$\phi \circ T^0[\alpha] = T^0[\phi \circ \alpha]. \qquad (4.15)$$

For the higher $T^n[\alpha](s_1, \ldots, s_n)$, the action of the diffeomorphisms is slightly more complicated to take into account the fact that they must transform as vector densities at each of the points $\alpha(s_i)$.

PROPERTY 4: THE DISTRIBUTIONAL SINGULARITIES APPEARING IN THE LOOP ALGEBRA MAY BE REMOVED BY AN APPROPRIATE SMEARING PROCEDURE

In scalar field theory, the distributional singularities appearing in the canonical Poisson brackets, $\{\phi(x), \pi(y)\} = \delta^3(x, y)$, are removed by introducing smooth test functions on the manifold and constructing the algebra of observables, $\phi(f) \equiv \int d^3x f(x)\phi(x)$ (and similarly, for $\pi(g)$). The Poisson algebra of these smeared observables is then completely free of singularities.

This can be done also for the elements of the loop algebra. The space of test functions is in this case more complicated. For each $T^n[\gamma]$, one constructs a $2n$-parameter congruence of curves containing γ, and then constructs smeared operators by integrating over the $2n$-dimensional parameter space with smooth test functions. The details of this construction are given in Rovelli and Smolin 1990; the result is that, just as in the case of the canonical algebra, there is a closed algebra of smeared observables that is completely nonsingular.

We are now ready to quantize the theory by constructing a representation for the $\mathcal{T}$ algebra. This is done in the next section.

5. Construction of the Loop Representation

To give a regularized quantization of the kinematics of general relativity, we now construct an algebra of linear operators acting on a suitable space of functions that is a deformation of the Poisson algebra of the classical $\mathcal{T}$ observables. Since those observables are parametrized by loops in Σ, it is natural to choose a representation space composed of functionals on the loop space of Σ. This, however, is not exactly what we will do here. For reasons that will be clear shortly, it turns out to be better to construct a representation on a set of functionals of the *multiple loops*.

Thus, we define our representation space $\mathcal{S}$ to consist of all complex-valued functionals of multiple loops, $\{\gamma\}$, that satisfy certain conditions, which will be given below. That is, given any unordered set of loops (defined as in the previous section) and an element $\mathcal{A}$ of $\mathcal{S}$, we have the complex amplitude $\mathcal{A}[\{\gamma\}]$.

Thus, the space $\mathcal{S}$ has a kind of "Fock-like" structure. Any $\mathcal{A}$ in $\mathcal{S}$ can be decomposed as

$$\mathcal{A} = \{\mathcal{A}^1[\alpha], \mathcal{A}^2[\alpha, \beta], \ldots\}, \tag{5.1}$$

where $\mathcal{A}^n$ is an amplitude on n unordered loops.

The space of such functionals has the structure of a linear vector space. However, as I mentioned, not every multiloop functional will be an element of our representation space $\mathcal{S}$. We impose four linear conditions on $\mathcal{S}$, which are necessary in order that it be a good representation space for quantum gravity. These conditions are:

i'. *Reparametrization invariance.* If $f(s)$ is a monotonic function of the loop parameter s with nonvanishing derivative, we require

$$\mathcal{A}[\gamma(f(s)), \ldots] = \mathcal{A}[\gamma(s), \ldots], \tag{5.2}$$

where the ",..." indicates any other loops that may be present.

ii'. *Invariance under replacement of a loop by its inverse*:

$$\mathcal{A}[\gamma^{-1}, \ldots] = \mathcal{A}[\gamma, \ldots]. \tag{5.3}$$

iii'. *The spin network identity.* Let α and β be two loops that share a common basepoint, as in Figure 4.1. Then we require

$$\mathcal{A}[\alpha \circ \beta, \ldots] + \mathcal{A}[\alpha \circ \beta^{-1}, \ldots] - \mathcal{A}[\alpha, \beta, \ldots] = 0. \tag{5.4}$$

Note that this condition, while nonlinear in the loop arguments, is linear in the functionals. This is a basic reason why it is more convenient to construct the loop representation in terms of functionals of multiloops. If we had tried to construct a representation in terms of functionals of single loops, the analogous conditions would have been nonlinear, of the form $a[\alpha \circ \beta] + a[\alpha \circ \beta^{-1}] = a[\alpha]a[\beta]$, making it unsuitable for a representation space for a quantum theory.

iv'. *The retracing identity.* Under the conditions shown in Figure 4.2, we require

$$\mathcal{A}[\eta \circ \alpha \circ \eta^{-1}, \ldots] = \mathcal{A}[\alpha, \ldots]. \tag{5.5}$$

$\mathcal{S}$ is then defined as the space of complex functionals over the multiloops that satisfy these four conditions. As these conditions are linear, $\mathcal{S}$ retains a linear structure and is thus suitable for a representation space for a quantum field theory.

The four conditions that we impose on $\mathcal{S}$ were, as the reader may have guessed, invented to correspond with the conditions (i) to (iv) that, as we saw in the previous section, are satisfied by the loop variables T^0. We shall see the reason for this in a moment.

Having defined the representation space $\mathcal{S}$, our next task is to define a suitable algebra of linear operators acting on it. We begin with a set of operators, parametrized by the loops, which we will call $\tilde{T}^0$.[8]

These are defined as[9]

$$(\tilde{T}^0[\alpha]\mathcal{A})[\{\gamma\}] \equiv \mathcal{A}[\alpha \cup \{\gamma\}]. \tag{5.6}$$

These operators are lowering operators; that is, if a loop functional $\mathcal{A}$ has support only on sets of n loops, the functional $\tilde{T}^0[\alpha]\mathcal{A}$ has support only on sets of $n-1$ loops.

It is clear that all of the $\tilde{T}^0$ commute with each other. Furthermore, given the conditions (i') to (iv') above, one may show that the $\tilde{T}^0$ operators satisfy the conditions (i) to (iv) that are satisfied by the classical loop observables T^0 by virtue of their definition in terms of parallel transport by an $SL(2,C)$ connection.

We may also define linear operators on $\mathcal{S}$ that are in correspondence with the T^n for $n > 0$. For example, we define a "one-handed" loop operator, $\tilde{T}^1[\gamma](s)$, which depends on a loop γ and a loop parameter s, by the condition

$$(\tilde{T}^1[\alpha]^a(s)\mathcal{A})[\beta] \equiv \hbar\Delta^a[\alpha,\beta](s)\big(\mathcal{A}[\alpha \circ \beta] - \mathcal{A}[\alpha \circ \beta^{-1}]\big). \tag{5.7}$$

The action of this operator is illustrated in Figure 5.1.

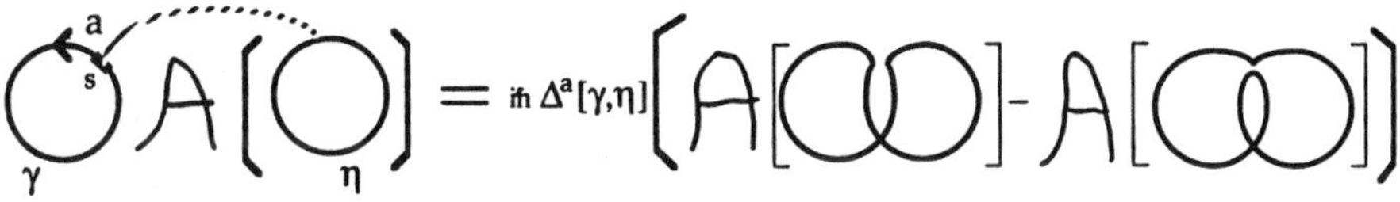

FIGURE 5.1. The action of $\tilde{T}^{-1}[\alpha]^a(s)$.

It is straightforward to show that the commutation relations of these observables with themselves and with the $\tilde{T}^0$'s are isomorphic to the Poisson bracket algebra of the classical observables of the same names times $\iota\hbar$.

The representatives of T^n, for $n > 1$, are defined analogously, with one important difference. In the classical Poisson algebra of loop observables, one hand acts at a time; the Poisson bracket is equal to the sum over all of the grasps of the active hands. For example, consider the Poisson bracket of a T^2 with a T^0, which is illustrated in Figure 5.2. Here, the loops intersect at two points, and we have a hand at each of the intersection points, but each term of the Poisson bracket only comes from an action at one of the intersection points.

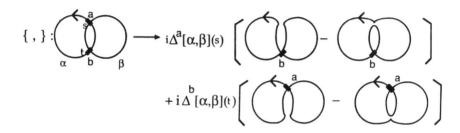

FIGURE 5.2. The Poisson bracket $\{T^2[\alpha]^{ab}(s,t), T^0[\beta]\}$.

In the corresponding action of the quantum operator, the hands act simultaneously. Let us then consider the action

$$(\tilde{T}^2[\alpha]^{ab}(s,t)\mathcal{A})[\beta], \tag{5.8}$$

which is illustrated in Figure 5.3. As in the classical case, the $\tilde{T}^2$ depends

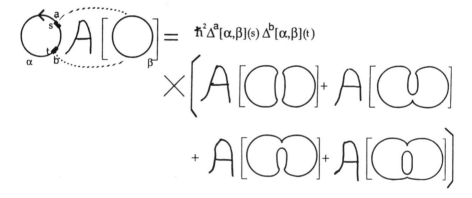

FIGURE 5.3. The action of $\tilde{T}^2[\alpha]^{ab}(s,t)$ on a loop functional $\mathcal{A}$ evaluated at the loop β.

on two preferred points on its loop, s and t, and on a tangent space index at each point. These preferred points must be distinct. The action is defined to be zero unless both hands are active, i.e., the loop β in the argument of the loop functional passes through both of them. In that case, the action is equal to the product of the Δ distributions at each hand times the sum over the results of both hands grasping simultaneously. There is also a factor of $\hbar^2$, one for each of the hands that has acted.

The difference between equation (5.27) and the classical Poisson bracket then shows up in the fact that the four terms in Figure 5.2 are different from the four terms in Figure 5.3.

One sees from Figure 5.3 why the preferred points s and t, at which the hands reside, must be distinct. The action is only well defined if the simultaneous actions of breaking and joining the loops occur at topologically distinct points on the loops. In addition, if the points coincided we would have a product of distributional singularities coming from the Δ's. But, it is important to emphasize, this is only a reflection of the fact that we cannot even write a well-defined action for both hands acting simultaneously if they coincide.

There is also a sign factor in front of each term that is defined in the following way. Each loop may be oriented initially. In each of the resulting terms, one has to assign an orientation to one or more new loops that have been created by the breaking and joining at each intersection. Sometimes one has to reverse the orientation along a segment of the curves to do this. The sign factor is defined to be (-1) raised to the number of times the orientations must be reversed along these segments to get a consistent orientation on the curves resulting from the grasps.

One may then compute the commutator of the operators $\tilde{T}^2[\alpha]^{ab}(s,t)$ and $\tilde{T}^0[\beta]$. (This is not actually that hard; the reader who wants to understand in detail what is going on is advised to work this example as an exercise.) The result, shown in Figure 5.4, is a sum of two kinds of terms. First we find the classical terms, shown in Figure 5.2, multiplied by $\hbar$. These result from terms in which one of the hands of the $\tilde{T}^2$ acted inside the commutator, leaving one over to act on states to the right. Thus, these terms are proportional to a $\tilde{T}^1$ as one hand is left over.

But there are also the quantum corrections, in which both hands of the $\tilde{T}^2$ acted on the $\tilde{T}^0$, and none are left over to act on states. The result is then proportional to a $\tilde{T}^0$. Further, as both hands have been used up, the result must be proportional to $\hbar^2$; these are then the quantum corrections to the commutator.

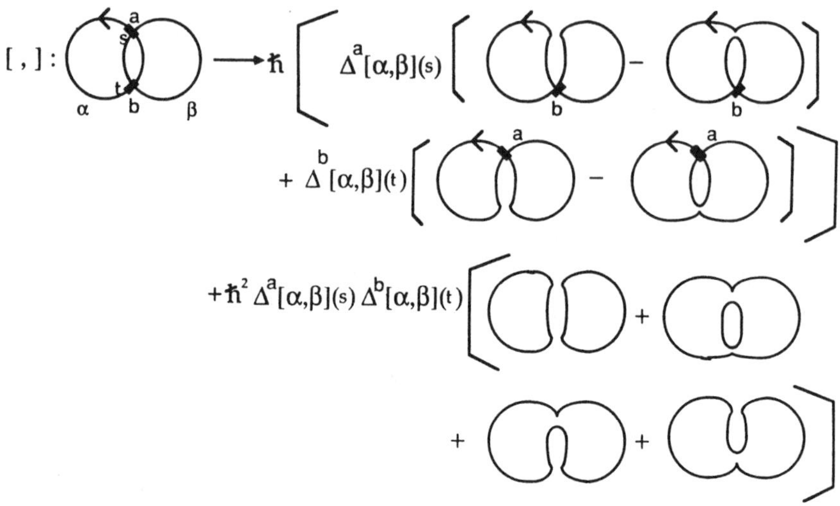

FIGURE 5.4. Commutator of $\tilde{T}^2[\alpha]^{ab}(s,t)$ with $\tilde{T}^0[\beta]$.

The general action of a $\tilde{T}^n$, for all $n > 2$, is defined completely analogously, the action being zero unless all hands can act simultaneously on the argument of the loop functional. The details are in Rovelli and Smolin (1990). Once the general loop operator is defined, it is straightforward to work out the general commutator. It is then of the form, for $n \geq m$,

$$\left[\tilde{T}^n, \tilde{T}^m\right] = \hbar\Delta\tilde{T}^{n+m-1} + \hbar^2\Delta\Delta\tilde{T}^n + \cdots + \hbar^n\Delta\ldots\Delta\tilde{T}^m. \quad (5.9)$$

Comparing this with (4.13), we see that the quantum algebra of the $\tilde{T}^n$'s is exactly a deformation of the classical Poisson algebra of the T^n's. This, together with the fact that both sets of observables satisfy the relations (i) to (iv) given in the last sections, makes it possible for us to identify the loop operators as a representation in terms of a set of linear operators acting on a representation space of the classical loop observables on the phase-space of general relativity.

The quantum loop algebra has two additional properties that are important to mention.

First, the algebra is completely regularized by the requirement that the hands on each $\tilde{T}^n$ must be at distinct points. All of the distributional singularities may be eliminated by introducing a smearing procedure in which the action we have just described is extended to an action defined in terms of a congruence of loops. The congruence has two parameters for each hand, and the action of congruence is defined by integrating the action defined above over the $2n$-parameter space of the congruences with smooth $2n$-parameter test functions. When this is done, no distributional singularities appear, and the result of any commutator is completely finite (Rovelli and Smolin 1990).[10]

Second, by the natural action of the spatial diffeomorphisms on the loop space, the quantum loop algebra is completely covariant under the action of the spatial diffeomorphisms. The loop functionals carry a linear representation of the spatial diffeomorphism group, which allows us to define linear operators that act as generators of the diffeomorphisms. The commutation relations of these operators with the quantum loop operators are found to be isomorphic to the Poisson brackets of the classical loop observables with the classical diffeomorphism constraints. This is described in the next section.

The result is that *the quantum loop algebra gives a complete, regularized, and diffeomorphism-covariant quantization of the unconstrained observable algebra of general relativity.*

6. Diffeomorphism-Invariant States

I now turn to one of the basic results of the loop representation, which is the general solution to the problem of constructing diffeomorphism-invariant states and operators. We begin by constructing the operators on S that represent the diffeomorphism constraints.

The natural action of Diff(Σ) on the loop space induces a linear representation of that group on S. For $\phi \in$ Diff(Σ), this is given by

$$U(\phi)\mathcal{A}[\{\gamma\}] \equiv \mathcal{A}[\phi^{-1} \circ \{\gamma\}]. \tag{6.1}$$

If ϕ_t is a one-parameter group of diffeomorphisms generated by a vector field v^a, we can define the representative of the diffeomorphism constraint to be the linear operator $D(v)$ such that

$$D(v)\mathcal{A}[\{\gamma\}] \equiv \frac{d}{dt}(U(\phi_t)\mathcal{A}[\{\gamma\}]). \tag{6.2}$$

We need first to worry about whether this operator will be well-defined, as we have not put any differentiability conditions on $\mathcal{A}$. This is analogous to the question of whether $p = \iota d/dx$ is well-defined on the space of L^2 functions, as not every L^2 function is differentiable at every point. The answer to this problem is that there is a dense subset of functions that are differentiable, and that is enough to define the action of p. Similarly, in the case of the loop functionals, there is a dense subset that are differentiable under the action of a vector field on the loop space, which generates the action of a diffeomorphism on the manifold. The $D(v)$ are defined in terms of this dense subset.

We may note that it is important to make the distinction between functionals on the loop space that are differentiable under the action of vector fields on the manifold, and the much stronger condition of differentiability with respect to any vector field on the loop space. We will see that the diffeomorphism-invariant states are differentiable (in fact constant) with respect to the diffeomorphisms of the manifold, but, in general, they will not even be continuous on the loop space.

On the dense subset on which they are defined, we may compute the commutator algebra of the $D(v)$. We find

$$[D(v), D(w)] = D([v, w]), \qquad (6.3)$$

so that the representation on the loop functionals of the Lie algebra of diffeomorphisms is faithful.

We must also compute the commutator of the $D(v)$ with the basic operators $\tilde{T}^n$. If we define

$$\phi \circ \tilde{T}^n \equiv U(\phi)\tilde{T}^n U(\phi^{-1}), \qquad (6.4)$$

we find, after a short calculation,

$$\phi \circ \tilde{T}^0[\gamma] = \tilde{T}^0[\phi^{-1} \circ \gamma]. \qquad (6.5)$$

A similar result is found for the general $\tilde{T}^n$, although there one has to take into account the density character of the operators and the index structure. These, we may recall, are built into the definition of the quantum loop operators through the presence of the distributional functions Δ^a in the definitions (4.12) and (5.7). The result is that one may show generally that

$$[D(v), \tilde{T}^n] = \iota\hbar\{\widetilde{C(v)}, T^n\}, \qquad (6.6)$$

where $C(v) = \int v^a Tr(F_{ab}E^b)$ is the classical diffeomorphism constraint, and the tilde over the whole right hand side means to use the one-to-one correspondence between the T^n and the $\tilde{T}^n$.

This last result, together with equation (6.3), tells us that we should identify $D(v)$ as the representative of $C(v)$ in the loop representation. We are now ready to find the solution space of the diffeomorphism constraint in the loop representation.

We want to find the general solution to the quantum constraint equation,

$$D(v)\mathcal{A}[\{\gamma\}] = 0. \tag{6.7}$$

By (6.1) and (6.2), we know that, if this is true, $\mathcal{A}$ is constant on the orbits of the identity-connected component of the diffeomorphisms,

$$\mathcal{A}[\{\gamma\}] = \mathcal{A}[\{\phi_t^{-1} \circ \gamma\}], \tag{6.8}$$

for any $\phi_t \in \text{Diff}(\Sigma)$. We can state this result in the following way. Let us define $L(\{\gamma\})$ to be the equivalence class, under $\text{Diff}(\Sigma)$, of the multiloop $\{\gamma\}$. These equivalence classes are generalizations of the link classes that knot theorists study in that, in addition to knotting and linking, these loops can intersect and overlap. We will call them the *generalized link classes*. Then the general solution to the diffeomorphism constraints can be written

$$\mathcal{A}[\{\gamma\}] = \mathcal{A}[L(\{\gamma\})]. \tag{6.9}$$

This result gets its power from two circumstances. First, it can be shown that the generalized link classes are, like the ordinary link classes, countable. Thus, we have not only the general solution to the quantum diffeomorphism constraint, but we have a countable basis for the space of solutions as well. This basis consists of loop functionals that are constant on the generalized link classes and are nonvanishing only on one particular class. Second, a great deal is known about invariants on knot and link classes, and much of this work points to connections with two-dimensional statistical mechanics and conformal field theory. Thus, this result means that tools developed in those areas may have a role to play in quantum gravity.

Since we have a countable basis for the diffeomorphism-invariant states, we know all diffeomorphism-invariant operators. Any diffeomorphism-invariant operator in the loop representation must have a representation as a matrix in the countable basis we have just described. That is, it must be a linear operator on the vector space of functions over the generalized link classes. Conversely, every such linear operator is a diffeomorphism-invariant observable in the loop representation of quantum gravity.

This result raises a very serious problem of interpretation. In the classical theory, we can write down an infinite number of diffeomorphism-invariant functions on the phase-space; for example, those that involve integrals over scalars constructed from the phase space variables. Quantum field theory being what it is, we expect that most of them will not survive the quantization procedure, as all except a finite number involve nonpolynomial expressions in q that may not survive a regularization procedure and the application of the constraints. But, at least a countably infinite number of diffeomorphism-invariant observables must survive the quantization procedure if that procedure is to be judged adequate.

Now, we have exactly a countable set of diffeomorphism-invariant operators in the loop representation. The problem is that we do not know how to make a correspondence between any of them and the classical diffeomorphism-invariant observables. The reason for this is two-fold. First, none of the elementary observables on which the quantization is based are diffeomorphism invariant. So, on both the classical and the quantum sides, the diffeomorphism-invariant objects are constructed indirectly in terms of the objects whose correspondence we know. Second, we know by construction a great deal about the algebra of diffeomorphism-invariant operators in the loop representation; for example, we know that the whole algebra is $GL(\infty, C)$, and that many of the elements can be interpreted in terms of simple operations that take us between link classes. However, essentially nothing is known about the Poisson algebra of diffeomorphism-invariant observables in the classical phase space. Thus, we cannot make a correspondence between them by recognizing relations between the classical and quantum algebras.

The problem of identifying quantum operators on the kernel of the diffeomorphism constraints with classical diffeomorphism-invariant observables is a basic issue for the interpretation of the theory. That we face this issue is the price we pay for having proceeded completely nonperturbatively. We have uncovered a new structure by solving the quantum constraints exactly. But we have done it in such a way that the interpretation of the resulting operators in terms of classical observables does not come from the construction. It is perhaps ironic that the solution to this problem, which arises from a nonperturbative quantization, may lie in a much deeper understanding of the algebra of diffeomorphism-invariant observables *in the classical theory!*

7. The Hamiltonian Constraint and its Solutions

As mentioned in Rovelli's talk, the construction of the loop representation was originally motivated by the fact that it had proved possible, in the self-dual representation, to find an infinite set of solutions to the Hamiltonian constraint. These solutions are functionals of Ashtekar's connection A_a that have the form of products of traces of holonomies around a set of loops. When the loops are smooth and nonintersecting, these states are in the kernel of the Hamiltonian constraint.

In the self-dual representation it was not, however, clear how to proceed to construct solutions that are invariant under spatial diffeomorphisms.[11] As we have just seen, in the loop representation we can give a complete solution to this problem. Having done this, we would like to know if we can also recover the existence of exact solutions to the Hamiltonian constraint in this representation. The answer is that we can; further, when one does this, certain technical issues raised in Jacobson and Smolin 1988 connected with the regularization of the Hamiltonian constraint can be cleaned up.

Before describing the construction of these solutions, I would like to make some general remarks about the problem of solving the Hamiltonian constraint. The Hamiltonian constraint carries all of the local dynamics of general relativity. Thus, it is not surprising that, as is true of the Hamiltonian of all known quantum field theories, the Hamiltonian constraint requires a regularization procedure if it is to be well defined as a quantum-mechanical operator. This can be seen from its form in terms of the new variables,

$$C(x) \equiv Tr\big(F_{ab}(x)E^a(x)E^b(x)\big). \tag{7.1}$$

As this expression involves a product of two canonical momenta E^a at a point, it is very unlikely that there is any complete representation of the quantum theory in which it can be expressed directly, without a regularization procedure.

This means that the Hamiltonian constraint will be represented in the quantum theory by a sequence of operators, $\tilde{C}^\delta(x)$ that have the property that, in the classical phase space,

$$\lim_{\delta \to 0} C^\delta(x) = C(x). \tag{7.2}$$

Now, for $\delta > 0$, each of these operators $\tilde{C}^\delta$ is completely well defined on S, the representation space of the unconstrained theory. At the same time, while the limit $\delta \to 0$ must exist in the classical theory, it is unlikely that, on the whole of $\mathcal{S}$, the $\lim_{\delta \to 0} \tilde{C}^\delta(x)$ will exist.

It is important to realize that this does not prevent there being a well-defined notion of the kernel of the regulated constraint. We will *define* this kernel to consist of the states $|\Psi\rangle$ in the unconstrained state space that satisfy

$$\lim_{\delta \to 0} \left(\tilde{C}^\delta(x) | \Psi > \right) = 0. \tag{7.3}$$

We will show that, in the loop representation, an infinite-dimensional subspace of $\mathcal{S}$ exists consisting of states that satisfy this condition. On all other states in $\mathcal{S}$, $\lim_{\delta \to 0} \left(\tilde{C}^\delta(x) | \Psi \rangle \right)$ will be undefined. But this is all right, as those are, by definition, not physical states; there is no problem if the Hamiltonian constraint is not well defined when acting on them in the limit when the regulator is removed. What is required is only that this limit exist for physical states, which is to say it exists on space of states defined by (7.3). There the limit clearly does exist; it is represented there by the operator multiplication by zero.

There are two points that must be stressed. First, no renormalization is needed to define, and find the kernel of the regulated Hamiltonian constraint. There can exist solutions to (7.3) without any infinite scaling of coupling constants in (7.1). Indeed, since the Hamiltonian constraint is a homogeneous equation, there are no coupling constants that might be scaled, aside from an overall factor multiplying the whole equation. But this is fixed by the classical condition (7.2).

Second, to ask that a regulated constraint have solutions in the sense of (7.3) is a much weaker requirement than asking that a Hamiltonian in a quantum field theory have a full spectrum of eigenstates with finite energy. In the latter case, we have a regulated Hamiltonian, which typically consists of a sum of terms

$$H^\delta \equiv \sum_i Z(\delta)_i H_i^\delta, \tag{7.4}$$

where each term H_i^δ is regulated and multiplied by a renormalization factor $Z(\delta)_i$. The requirement for this Hamiltonian to be solvable in a quantum field theory is that there exists an infinite set of states which satisfy

$$\lim_{\delta \to 0} H^\delta |\psi> = E_\Psi |\Psi>. \tag{7.5}$$

These states are required by the Hermiticity of H to be normalizable under the inner product, and to be a basis for the full representation space of the theory.

Thus, $\lim_{\delta \to 0} H^\delta$ is required to exist on the whole representation space of the quantum field theory. This is a much stronger requirement than that $\tilde{C}^\delta$ must satisfy; there the corresponding limit is allowed not to exist, except on that subspace that is the kernel in the sense of (7.3).

Of course, general relativity also has a Hamiltonian, which is proportional to a boundary integral and hence nonvanishing in the asymptotically-flat case. Whether there exists, on the physical subspace, a complete set of eigenstates to this Hamiltonian with finite energy and finite norm under the physical inner product is very much an open problem.

Let us return to the Hamiltonian constraint. I will give here a schematic description of how its solutions are found; the details are in Rovelli and Smolin 1990.

We begin by asking how the classical Hamiltonian constraint can be expressed in terms of our elementary observables, the T^n. The answer is that it must be expressed in terms of a sequence of regulated operators, C^δ, each of which is, to incorporate the two E^a's, made up of T^2's. This is done in the following way.

We first of all must break the diffeomorphism invariance by picking a particular coordinate system in terms of which the regulated operators are to be described. The use of these coordinates is only that they are a convenient way to pick out a preferred set of loops, out of which the regulated operators are built. These loops are called $\gamma^\delta_{\hat{a}\hat{b}}(x)$ for $\hat{a} > \hat{b}$. They are shown in Figure 7.1. Each of them is base-pointed at x, lies in the $\hat{a}\hat{b}$ plane, and is a coordinate circle of radius δ. It is parametrized by a parameter s that runs between 0 and 2π.

Then, if the connection is slowly varying on the of δ, we may expand the parallel transport around the loop as

$$U^{AB}_{\gamma^\delta_{\hat{a}\hat{b}}(x)} = \varepsilon^{AB} + \delta^2 F^{AB}_{\hat{a}\hat{b}}(x) + O(\delta^3). \tag{7.6}$$

We note that the symmetric part (in the spinor indices AB of $U_{\gamma^\delta_{\hat{a}\hat{b}}(x)}$) then has a leading term that is proportional to $\delta^2 F_{\hat{a}\hat{b}}$. We may use this fact to construct a regulated form of the Hamiltonian constraint using loop observables. There are several ways to do this; for the purposes of the solutions that have so far been found, it doesn't matter which we choose. The simplest is to construct the three observables $T^2[\gamma^\delta_{\hat{a}\hat{b}}(x)]^{\hat{a}\hat{b}}(0, \delta^2)$. It is straightforward to show that

$$Tr\left[F_{\hat{a}\hat{b}}E^{\hat{a}}E^{\hat{b}}\right] = \lim_{\delta \to 0} \frac{1}{\delta^2} T^2\left[\gamma^\delta_{\hat{a}\hat{b}}(x)\right]^{\hat{a}\hat{b}}(0, \delta^2). \tag{7.7}$$

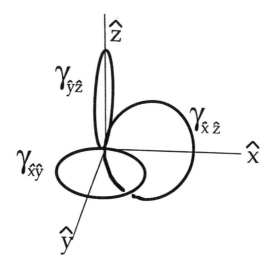

FIGURE 7.1. Construction of the regulated Hamiltonian constraint.

Note that the second point is put at a parameter value δ^2 in order to make subleading a term that would otherwise mar this result.[12] We may then define,

$$C^\delta(x) \equiv \frac{1}{\delta^2} \sum_{\hat{a} > \hat{b}} T^2 \left[\gamma_{\hat{a}\hat{b}}^\delta(x) \right]^{\hat{a}\hat{b}} (0, \delta^2). \qquad (7.8)$$

It then follows that classically $\lim_{\delta \to 0} C^\delta(x) = C(x)$. Note that, while the limit is a scalar density under diffeomorphisms, the regulated operators transform in a more complicated way under diffeomorphisms because of their dependence on a particular set of loops.

We then define the regulated quantum-mechanical constraint operator $\tilde{C}^\delta$ by the same formula as the classical regulated constraint, with the corresponding tilded T^2 operators.

One may now prove the following result: Let $\mathcal{A}$ be an element of $\mathcal{S}$ that is bounded and has support only on multiloops that are smooth and nonintersecting. Then,

$$\lim_{\delta \to 0} \left(C^\delta(x) \mathcal{A} \right) = 0. \qquad (7.9)$$

In order to give a complete proof of this result, one must work in the fully-smeared picture I mentioned before, in which each loop becomes a congruence. Here I will avoid these complications, and try to explain why the result is true. Consider Figure 7.2, where the curve α is a curve which might appear in the argument of $\mathcal{A}$.

Now, let there be two points s and t along α at which it intersects the loop $\gamma_{\hat{a}\hat{b}}^{\delta}(x)$ coming from the C^{δ}. Let us assume also that the two hand of the T^2 are at those two points; if they are not, then the action is zero. (Actually, in the fully smeared case we will be averaging over a space of congruences, one element of which will fulfill these conditions).

We may begin by noting that, by the assumption that $\mathcal{A}$ only has support on smooth multiloops, we know that the right-hand side of (7.9) vanishes unless α is smooth. This is because, if α is not smooth, the curves that appear in the argument of the right-hand side of (7.9), which are made up of segments of α and of the $\gamma_{\hat{a}\hat{b}}^{\delta}(x)$, will not be smooth. Thus, we may assume that α is smooth.

Now, recall that the action of a $\tilde{T}^2[\ldots]^{\hat{a}\hat{b}}$ is proportional to $\Delta[\ldots]^{\hat{a}}$ $\Delta[\ldots]^{\hat{b}}$, where these are the distributional functions defined in (4.12). The indices in the $\Delta[\ldots]^{\hat{a}}$'s are proportional to tangent vector of the "grasped" curve, in this case $\dot{\alpha}^{\hat{a}}$. In the fully-smeared form, the delta functions inside of the Δ's are eliminated, but this dependence on the tangent vectors remains.

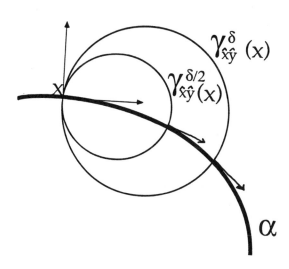

FIGURE 7.2. Action of the Hamiltonian constraint on a loop functional.

Since the regulated C^δ involves the skew part of a $\tilde{T}^2$, the result of the action (7.9) is, when it is nonvanishing, proportional to $\dot{\alpha}^{[\hat{a}}(s)\dot{\alpha}^{\hat{b}]}(t)$. Since the curve is smooth, we may write, when s and t are brought together by the limit $\delta \to 0$,

$$\dot{\alpha}^{[\hat{a}}(s)\dot{\alpha}^{\hat{b}]}(t) = K \times \delta\dot{\alpha}^{[\hat{a}}(s)\ddot{\alpha}^{\hat{b}]}(s), \qquad (7.10)$$

where K is a coordinate-dependent but finite constant. By (7.10), we then have, for bounded $\mathcal{A}$'s, with support only on smooth, nonintersecting loops,

$$\tilde{C}^\delta \mathcal{A} \leq \delta K', \qquad (7.11)$$

where K' now depends on the coordinate system and the state. The result (7.9) then follows.

To put it simply, the operator C^δ, being built of $\tilde{T}^2$'s, when acting on a loop functional $\mathcal{A}[\{\alpha\}$, measures the tangent vectors of the arguments of the loop functional at the two hands of the $\tilde{T}^2$'s. More precisely, because of the antisymmetry in the F_{ab} in the classical constraint, it is constructed to measure the *skew part* of the tangent vectors, $\dot{\alpha}^{[\hat{a}}(s)\dot{\alpha}^{\hat{b}]}(t)$. When we take the limit $\delta \to 0$, which removes the regularization, the two tangent vectors are brought together. If the curve in the argument of the loop functional is smooth at the point where the operator is acting, the skew part of the tangent vectors vanishes.

One then has to make sure that there are no subleading terms that might spoil the result. This was a problem with the point-splitting regularization described in Jacobson and Smolin 1988. There are no subleading terms here, because the regulated operator has been constructed in such a way as to make the offending term of high enough order in δ that it vanishes in the limit $\delta \to 0$. This, however, is a sufficiently technical point that I refer the reader to the paper Rovelli and Smolin 1990, rather than try to describe the details here.

8. Conclusions and Prospects

I would like to split my concluding remarks into three parts. First, I will recall what has so far been done with the loop representation. Equally important, I will follow that with a list of what has not (yet) been done. Finally, I will end with a discussion of open problems and issues that are the subject of current work.

WHAT HAS BEEN DONE?

Let us summarize what has been accomplished so far with the loop representation.

(1) The loop representation is, at the kinematical level, a complete quantization of the phase space of general relativity that is completely regulated and diffeomorphism-covariant in the sense that the algebra of regulated operators carries a representation of the spatial diffeomorphism group.

(2) The diffeomorphism and Hamiltonian constraints may be expressed in the loop representation, the former by their natural geometrical action on the loop space, the latter in regulated form.

(3) The general solution to the diffeomorphism constraint is found in the loop representation, and expressed in terms of a countable basis. This countable basis is in one-to-one correspondence with the generalized link classes of the manifold.

(4) The general diffeomorphism-invariant operator then has the form of a matrix in this basis.

(5) An infinite, but not complete, set of states that are in the kernel of the Hamiltonian constraint is also found. These states consist of all loop functionals with support on loops that are smooth and nonintersecting.

(6) The Hamiltonian and diffeomorphism constraints are compatible, in the sense that an infinite set of physical states that are in their simultaneous kernel may be constructed. This space has a countable basis, which is in one-to-one correspondence with the ordinary link classes.

For completeness, I will mention here results discussed in the talks of Ashtekar and Rovelli.

(7) A functional transform taking states in the self-dual representation to states in the loop representation may be constructed formally, as described in Rovelli's talk.

(8) For free field theories this transform may be explicitly constructed, and gives a construction of a loop representation for the Fock space of free photons and free gravitons (Ashtekar and Rovelli 1990, Rovelli and Smolin 1990).

WHAT HAS NOT BEEN DONE?

In Section 3 above, I listed the steps to give a nonperturbative quantization of general relativity via the canonical formalism. Let me stress the things on that list that have not been accomplished.

(1) *Completeness of the solution space of the constraints.* The set of solutions to the Hamiltonian constraint described here is almost certainly not a complete set. We know this for two reasons. First, a large set of additional solutions has been found in the self-dual representation associated with intersecting loops (Jacobson and Smolin 1988, Husain 1989). We expect that those solutions will exist in the loop representation as well, although the work has not yet been done to show that they exist, with all the details of regularization and smearing taken into account. The solutions that have so far been found suggest that there is a very intricate structure of intersecting solutions, with solutions existing for any number of loops intersecting at a point. It is tempting to suspect that much of the dynamics of the theory is coded in what happens at intersection points.

Second, we might expect that the solutions found so far, including the intersection solutions I've just mentioned, are not a complete set because their construction uses only the antisymmetry of the indices on the two hands in the operator $\tilde{C}^\delta$. Thus, it is possible to construct other operators, whose continuum limit classically is not the Hamiltonian constraint that also annihilate these states. Examples are known for the solutions described here; whether this is also the case for all of the intersecting solutions is still an open problem. Now this does not necessarily mean that the space of solutions is incomplete; for example, the complete set of solutions to $\Box \Phi = 0$ are also annihilated by the operators $\Box\Box$ and $v^a \partial_a \Box$, for any vector field v^a, and it is possible that the additional operators that annihilate the states we have found are all of this kind. But it is worrying, and this is an issue that must be settled.

2. *Physical interpretation of the physical operator algebra.* Given a description of the solution space to the constraints in terms of a countable basis, one immediately knows how to construct the general operator acting on that space. Thus, given our results about states, we have the general diffeomorphism-invariant operator, and a large class of completely physical operators. What we do not have, however, is any correspondence between these operators and diffeomorphism-invariant or physical observables in classical general relativity.

3. *The physical inner product.* The Ashtekar quantization is, to begin with, based on the complexified Einstein equations. This constraints and evolution equations, as we have been using them, are supplemented in the classical theory by reality conditions.[13] In a quantum theory, reality conditions must be expressed in terms of an inner product because the classical restriction that an observer be real is expressed in the quantum theory by the restriction that the corresponding operator be Hermitian.

Thus, the choice of an inner product is connected with the expression of the reality conditions. This means that, if someone proposes an inner product on the space of physical states, one must be able to check that any operator on the physical states whose classical limit is real when the reality conditions are imposed is Hermitian. However, this condition requires that we have a correspondence between classical and quantum physical observers, which, as we have just mentioned, we do not have. Thus, at present, while we have candidates for the physical inner product—for example the L^2 norm given the countable basis of link classes—we have not been able to check whether any of them correctly express the reality conditions.

CURRENT OPEN PROBLEMS

I will close by giving a list of open problems.

(1) *Construction of the asymptotically-flat theory.* In the asymptotically-flat context we know many explicit physical observables, such as the energy, momentum, angular momentum, multipole moments, and so on. Moreover, there is a Hamiltonian; so the conceptual difficulties of quantum cosmology do not confront us, and we might hope to construct a conventional quantum mechanics with a Hermitian Hamiltonian and a conserved inner product. Thus, it is very important to extend the loop representation to the asymptotically-flat case to see if these things can be done. I believe that a fair test of the whole approach will be whether an extension to the asymptotically-flat case exists such that it can be shown that there is a complete set of physical states that have both finite energy and finite norm in the physical inner product. This work is currently in progress, and I will return to it below in point 6.

(2) *Existence of the transform.* Given the construction of the loop representation I have presented here, which does not depend on the existence of the self-dual representation, the transform discussed in Ashtekar's and Rovelli's talks can be seen in a new light. Recall that the formula for the transform is

$$\mathcal{A}[\{\gamma\}] = \int d\mu(A) H(\{\gamma\}, A) \Psi(A), \qquad (8.1)$$

where $H(\{\gamma\}, A) \equiv T^0[\{\gamma\}]$ and $\Psi(A)$ is a state in the self-dual representation. We know that, at the kinematical level, both the loop representation and the self-dual representation exist. We do not know whether they are equivalent. For example, we do not know whether there are any states in the self-dual representation that are in the kernel of both the gauge and the spatial diffeomorphism constraints. Certainly, it is a very nontrivial problem to write explicitly an infinite set of such functionals (I know of two classes: the constants, and any function of the integral of the Chern–Simon form).

Thus, the question we would like to answer is: does there exist a measure $d\mu(A)$ and a domain of the space of gauge-invariant functionals of A such that the transform exists?

The recent work of Witten on knot theory (Witten 1989) is extremely interesting from this point of view. He shows that the functional

$$J[\{\gamma\}] = \int dA e^{\iota b \int Y_{\text{Chern-Simon}}} H(\{\gamma\}, A) \qquad (8.2)$$

is, for $\{\gamma\}$ smooth and nonintersecting, computable exactly, and is equal to, the Jones polynomial (Jones 1983) of the link $\{\gamma\}$. He does this in an extremely clever way that exploits the Hamiltonian quantization of the Chern–Simon theory. He does not compute (8.2) by integrating a function over a measure. Instead he uses a $(2 + 1)$-decomposition, together with the fact that the Chern–Simon field theory, having no local degrees of freedom, has a finite dimensional phase space, to solve the problem by means of certain techniques in conformal field theory. It would be very interesting to know if Witten's method is equivalent to the existence of a measure $d\mu(A)$ on the three-dimensional space.

Witten's work also raises some questions about the classical theory associated with the integral (8.2). Since the Chern–Simon theory has no local degrees of freedom, one might think that the integral (8.2) is determined by a sum over its critical points. These critical points are the connections that are flat everywhere except on the loops $\{\gamma\}$. On the loops, they solve the equation[14]

$$\iota b \varepsilon^{abc} F_{bc}^k(x) = \sum_{\gamma \in \{\gamma\}} \int dt \delta^2(x, \gamma(s)) \dot{\gamma}(s) \frac{Tr \, \tau^k U_\gamma(s)}{Tr \, U_\gamma}. \qquad (8.3)$$

This is a nonlinear coupled equation, whose solutions are invariants of the link $\{\gamma\}$. It is interesting to note that each solution to this equation is characterized by a map from π^1 of the complement of the knot into $SU(2)$. Furthermore, each solution is also characterized by the values of $H(\{\gamma\}, A) \equiv \prod Tr \, U_\gamma$. For each link, I would like to call the possible values of this quantity the *spectrum* of the knot. I have been studying these equations for a variety of links, and I find that the spectrum does contain information about the linking of the loops in the knot (Smolin 1989). It is interesting to speculate that the information contained in the classical solutions to (8.3) is realted to the knot polynomials in the sense that Witten's formula (8.2) could be interpreted as giving an expression for the Jones polynomial in terms of a sum over solutions to (8.3), with each solution weighted by its value of $H(\{\gamma\}, A)$.

(3) *Study of degenerate solutions.* As Ashtekar has noted, Einstein's equations, when written in terms of the new variables, are completely polynomial. In particular, they are well-defined in cases for which $\det(\tilde{\tilde{q}}^{bc}) = 0$, where the ordinary Einstein equations do not make sense. Thus, Ashtekar's formalism is actually an extension of general relativity, and is only equivalent to it when the additional condition $\det(\tilde{\tilde{q}}^{bc}) \neq 0$ is satisfied.

Thus, there are classical solutions to the constraints and equations of motion in Ashtekar's formalism that are not solutions to Einstein's equations. Examples of such solutions in the spherically-symmetric case have been given in Bengtsson 1988. It is important to understand whether the existence of these extra solutions plays an important role in the classical and quantum mechanics of the theory.

On the classical side, it is interesting to note that there is a class of singularities in general relativity, in which g_{ab} becomes singular or degenerate while the Weyl curvature remains finite. Ashtekar has speculated that this may make it possible to study the evolution through certain kinds of singularities using the new variables.

On the quantum side, there are reasons to believe that the solutions to the Hamiltonian constraint that were found in the self-dual representation [Jacobson and Smolin 1988] are connected with classical solutions involving degenerate E^a's. An argument that this might be the case was sketched in there. I would like to complete that argument here by exhibiting a large class of solutions to Ashtekar's equations that are static and smooth, but involve degenerate E^a's.

The constraint and evolution equations are given in Ashtekar's contribution to this volume and in Ashtekar 1988. Let us begin by making an Ansatz that is suggested in some recent work of Samuel (1988) and Ashtekar and Renteln (Ashtekar 1988):

$$F_{ab}^k(x) = m^2 \varepsilon_{abc} E^{ck}(x), \qquad (8.4)$$

where m^2 is a dimensional constant. One may observe that the Gauss law constraint is satisfied as a consequence of the Bianchi identity, and that the spatial diffeomorphism constraints are identically satisfied:

$$D_a = F_{ab}^k E^{bk} = m^2 \varepsilon_{abc} E^{ck} E^{bk} = 0. \tag{8.5}$$

Now, let us plug our Ansatz into the Hamiltonian constraint. Samuel, Ashtekar and Renteln were interested in the case in which there is a cosmological constant,

$$C = \varepsilon_{ijk} F_{ab}^k E^{a\,i} E^{b\,j} - \Lambda \varepsilon_{ijk} \varepsilon_{abc} E^{ai} E^{bj} E^{ck} = (m^2 - \Lambda) \det E^{ak}. \tag{8.6}$$

In the case that there is a cosmological constant, they find that there are solutions if we set $m^2 = \Lambda$. However, let us consider the case without a cosmological constant. In that case we find,

$$\det E^{ak} = 0. \tag{8.7}$$

Thus, any choice of A_a and E^a that satisfies the Ansatz and the condition that E^a is degenerate is a solution to the constraints (actually it is a solution for any value of the cosmological constant!). We can construct a large class of static solutions by doing the following. Let us assume that the Ansatz is taken as the definition of E^a, and let us consider A_a that satisfy a smoothed version of (8.3). This is an equation that arises by smoothing the loop functionals $H[\{\gamma\}, A]$ by introducing a smooth congruence of curves. (This is actually necessary to evaluate (8.2); it introduces what is known as a framing of the links.) Let us, to be specific, consider a smooth, two-parameter congruence of curves $\gamma_\sigma(s)$ that fills some region R of a three-manifold Σ. Then, through each point $x \in \Sigma$ there is a unique curve, which we will call γ_x, with tangent vector $\dot{\gamma}^a(x)$. Let us also have a smooth test function on the space of congruences, $f(\sigma)$. Given the congruence, this induces a smooth function $f(x)$ on Σ. For convenience, we may work in the coordinate system in which $x = (\sigma, s)$. Then, we may consider the "smoothed Chern–Simon equation,"

$$\iota b \varepsilon^{abc} F_{bc}^k(x) = f(x) \dot{\gamma}_x^a(s(x)) \frac{Tr\, \tau^k U_{\gamma_x}(s(x))}{Tr\, U_{\gamma_x}}. \tag{8.8}$$

For a given framed link, one would then like to investigate whether this equation has solutions. It is possible to show that solutions exist for some classes of links; the general problem is still under study.

Given a solution to (8.8), the E^a defined by the Ansatz (8.4) is then degenerate, so that the Hamiltonian constraint is satisfied. Furthermore, one may verify that the equations of motion, equations (2.6) of Ashtekar's contribution, are also satisfied, with $\dot{A}_a = E^a = 0$; so that these solutions are static.

Thus, we have a large class of exact solutions to Einstein's equations with smooth but degenerate E^a's that are associated with congruences of curves. It is interesting to note that these solutions are associated with the critical points of the integral (8.3), which Witten shows is proportional to the Jones polynomial. It is also very tempting to suspect that these solutions are somehow connected to the exact solutions to the constraints that we have discovered in the loop representation.

The existence of large classes of degenerate solutions to Ashtekar's equations raises one last question that is crucial for the quantum theory. We know that perturbation theory for general relativity is non-renormalizable when we expand around a background which is a *non-degenerate* solution to Einstein's equations. It is, however, possible that the perturbation theory in which one expands around one of these degenerate solutions might be renormalizable. This ideas was suggested by Witten in his work on $(2+1)$-dimensional gravity (Witten 1988b), where there exists a renormalizable perturbation expansion around a degenerate solution. It is then very interesting to ask whether a renormalizable perturbation theory might be developed around one of these degenerate solutions. Work on this is in progress.

(4) *Study of model theories.* In order to get a better understanding of the issues raised by the use of the loop representation, it would be useful to study a model system in which these same issues arise. Recently, Kodama has shown that the dynamical equations of the homogeneous cosmological models are very much simplified by the use of the Ashtekar variables. He has quantized these models in the self-dual representation, and shown that, for the Bianchi type IX, one may find exact solutions to the Hamiltonian constraint (Kodama 1988). Many of the issues raised by the existence of exact physical states in the loop representation may be studied here, including the questions of the physical observables and the physical inner product.

For a more precise investigation of the issues raised by the loop representation, the homogeneous models are probably too simple. One needs a model in which not all of the gauge and spatial-diffeomorphism symmetry has been fixed. The simplest such models are probably the Gowdy cosmologies, which have two space-like Killing vectors and a

Diff(S^1) symmetry. These models are currently under investigation by V. Husain (Husain and Smolin 1989). It is interesting to speculate that Diff(S^1) symmetry will make it possible to introduce techniques from conformal field theory into the study of these models.

Another class of model systems that can be very profitably studied from the present point of view if $(2+1)$-dimensional gravity. Recently, Witten has attacked this problem from the point of view of his program of topological quantum field theory (Witten 1988), and there has been interesting work as well on this problem by Deser, Jackiw and 't Hooft (Deser and Jackiw 1984, 1989, Deser, Jackiw, and 't Hooft 1984). Two-plus-one general relativity can also be quantized directly in loop representation (Ashtekar, Husain, Rovelli, Samuel, and Smolin 1989). In this case, the loop representation becomes a linear space of functionals over π^1 of the two-dimensional spatial manifold, modulo the relations satisfied by the loop observables.

(5) *Knot theory and the representation theory of* Diff(S^3). Of the various results I have presented, perhaps the most startling is that all spatially diffeomorphism-invariant observables may be expressed as matrices in the basis of generalized link classes. This means that any diffeomorphism-invariant observable measures the topological relations among nonlocal operators. All physically relevant notions of short-distance structure must then be constructed from such operators.

I mentioned in the first section that, if we had a theory of the representation theory of Diff(S^3), we could construct a classification of diffeomorphism-invariant observables from the rules of how singlets arise from products of representations, as is usual in applications of group theory to quantum mechanics. We do not have such a theory, but we do have a classification of the diffeomorphism-invariant observables. We would then like to ask what this classification tells us about the existence of irreducible representations of Diff(S^3). While I cannot make a rigorous argument, I believe that what it tells us is that there is a class of irreducible representations that are in one-to-one correspondence with the generalized link classes of S^3.[15]

(6) *Relationship with conformal field theory.* There are cases in which theories with a diffeomorphism-like symmetry group are completely solved in terms of the infinite-dimensional group theory. These are the rational conformal field theories (Belavin et al. 1984, Friedan et al. 1984, Friedan and Schenker 1987, re and Seiberg 1989). Further, these theories have something to do with the invariants of knot theory, as there is a large class of rational conformal field theories that are associated with

link invariants (Alverez-Gaume et al. 1988, Kauffmann 1987a,b, Akutsu and Wadati 1987a, b, Kuniba, Akutsu, and Wadati 1986, Frohlich 1987). It is hard to believe that the occurrence of link invariants in that work and in the work described here can be completely coincidental. Just as a certain amount of information about the representation of compact Lie groups can be understood in terms of $SU(2)$ representations, because they all contain nontrivial $SU(2)$ subgroups, the fact that Diff(S^1) is a subgroup of Diff(S^3) must imply certain relations among their representation theories; and these relations should imply relations between the state space of $(3 + 1)$-dimensional quantum gravity and conformal field theories.

I would like to suggest that it may be extremely fruitful to look for points of contact between $(3 + 1)$-dimensional quantum gravity and conformal field theory. I have already mentioned two places where such contact may be made: the Gowdy models, and the Witten analysis of the loop transform. I would like to make one more suggestion along these lines.

Let us consider the quantization of the gravitational field in a three-manifold Σ, which has a boundary $\partial\Sigma$ (see Figure 8.1). A certain structure may be fixed on $\partial\Sigma$, in which case the diffeomorphism group that is of interest is that subgroup of Diff(Σ) that fixes that structure. I will call this group Diff(Σ)$_0$. Now, that structure also determines a subgroup of Diff($\partial\Sigma$), which I will call Diff($\partial\Sigma$)$_0$. The asymptotically-flat case is a special case of this, in which $\partial\Sigma$ is a two-sphere on which a metric and certain other structures have been fixed.

Now let us reconsider the argument I made in Section 2 about n-point functions. Let $|\Psi>$ be a physical quantum state of $(3 + 1)$-dimensional quantum gravity, which is then invariant under the action of Diff(Σ)$_0$. Now, let us consider, as before, an n-point function $<\Psi|\Phi_1(x_1)\ldots\Phi_n(x_n)|\Psi>$ for $x_i \in \Sigma$. In general, this is not the expectation value of any physical observable. But now let us consider taking the limit in which the points x_i are brought to the boundary. Define, for n points $s_i \in \partial\Sigma$,

$$G^n(s_1,\ldots,s_n) \equiv \lim_{x_i \to s_i \in \partial\Sigma} <\Psi|\Phi_1(x_1)\ldots\Phi_n(x_n)|\Psi> . \qquad (8.9)$$

This G^n may have physical meaning because of the restriction of the diffeomorphism group on the boundary. In particular, we may ask the following question. Given the full set of G^n defined in this way, does there exist a quantum field theory on the two-surface $\partial\Sigma$ that has a vacuum state $\langle 0|$ and a set of fields $\xi_i(s)$ such that

$$G^n(s_1, \ldots, s_n) = \langle 0|\xi_i(s_1) \ldots \xi_n(s_n)|0\rangle \ ? \qquad (8.10)$$

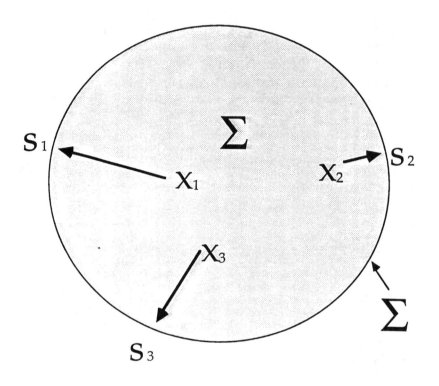

FIGURE 8.1. Does a diffeomorphism invariant quantum state on a bounded region induce a quantum field theory on the boundary?

Indeed, if the conditions of the standard reconstruction theorems from Euclidean quantum field theory are satisfied by the G^n, we know that such a theory must exist. If this can be established, then we have a correspondence between a *quantum state of* $(3+1)$*-dimensional quantum gravity* and a *quantum field theory in two Euclidean dimensions.*[16]

Acknowledgments. I would first of all like to thank my collaborators in the work described here, Abhay Ashtekar, Ted Jacobson and Carlo Rovelli, for conversations, ideas, encouragement and inspiration. My views concerning nonperturbative quantum gravity were formed in conversation with them, as well as through interaction, both spoken and written, with Bryce DeWitt, Chris Isham and Karel Kuchař. I have learned most of what I know about the mathematical issues raised by the work described here from many stimulating conversations with Louis Crane. I would also like to thank Viqar Husain, Paul Renteln, Steve Shenker, and Richard Woodard for critical conversations about this work.

This work was supported by NSF grant number PHY86-12424 to Syracuse University.

NOTES

[1] Very recently, Ashtekar has suggested that, at least in simple examples, this piece of conventional wisdom may be false if one uses a holomorphic polarization. If this is true for quantum gravity, it would be a welcome development.

[2] This point has been emphasized by Chris Isham (Isham 1984, Isham and Kakas (1984).

[3] A lattice approach to the self-dual representation was constructed in collaboration with Paul Renteln (Renteln and Smolin 1989). Detailed calculations by Renteln show that there exists no ordering for the lattice diffeomorphism constraints in three spatial dimensions for which the algebra remains first class (Renteln 1988). (Although it does work in two spatial dimensions.) There is also a Hamiltonian formulation of Regge calculus by Friedman and Jack in which the algebra of the regulated diffeomorphism constraints also fails to be first class (Friedman and Jack 1986).

[4] As the idea of being able to find exact solutions to the constraints in quantum general relativity is a little hard to get used to, let me mention that the fact that this is possible is very likely a reflection of the fact that there are sectors of the classical theory where the general solution can be found. The most important of these is the self-dual sector. One of Ashtekar's motivations for looking for a new Hamiltonian formalism was that there ought to exist coordinates on the phase space of general relativity that would allow one to understand why the self-dual sector could be exactly solved by the twistor and H-space methods (Ashtekar 1988). Indeed, the self-dual equations take a very simple form using the Ashtekar variables (Ashtekar, Jacobson, and Smolin 1988), and there is some hope that further work will show that this sector is integrable.

[5] The meaning of large enough is large enough to distinguish between all of the physical states of the theory.

[6] I use the following nomenclature: "gauge" refers to the local $SU(2)$ gauge rotations of the frame fields, "diffeomorphism" refers to the three-dimensional diffeomorphisms of the spatial manifold Σ, and "physical" refers to things that are invariant under, or commute with, all of the constraints of the theory, gauge,

diffeomorphism and Hamiltonian.

[7] Clearly, the numerical index n on the T^n is redundant; it counts the number of insertions of E^a, and hence the number of indices. Hence it can be dropped, and will be from time to time.

[8] The linear operators in the loop representation are labeled by tildes to distinguish them from the corresponding classical observables.

[9] Again, there is an obvious generalization to operators depending on multiloops. This occurs throughout, although we usually shall not mention it here.

[10] This construction is necessary in order to address the problem of what happens when one takes the commutator of two operators, whose hands happen to be at the same point of space. The smearing solves this problem because these coincidences turn out to be of measure zero in terms of the integrals over the parameter spaces of the congruences.

[11] Some attempts to do this were described in Smolin 1988.

[12] There are several ways of dealing with this term; another one is to eliminate it by taking the symmetric part by hand, using $U_\gamma^{(AB)} = U_\gamma^{AB} - U_{\gamma-1}^{AB}$.

[13] These reality conditions are actually polynomial, as was realized by Ashtekar recently (Ashtekar, Romano and Tate 1988).

[14] Note that I am using here the triad notation described in Ashtekar's talk and in Ashtekar 1988.

[15] Some related work is in Ismagilov 1975, 1976; Gelfand et al. 1977, Vershik et al. 1975).

[16] Work is in progress on implementing this idea in the loop representation in asymptotically-flat general relativity.

REFERENCES

Akutsu, Yasuhiro, and Wadati, Miki (1987a). "Knot Invariants and the Critical Statistical Systems." *Journal of the Physical Society of Japan* 56: 839–842.

———(1987b). "Exactly Solvable Models and the New Link Polynomials: I. *N*-State Vertex Models." *Journal of the Physical Society of Japan* 56: 3039–3051.

Alverez-Gaume, L.; Gomez, G.; and Siertra, G. (1988). "Hidden Quantum Symmetries in Rational Conformal Field Theories." CERN preprint, CERN-TJ-5192/88 (UGVA-DPT 071583/88).

Ashtekar, Abhay (1986). "New Variables for Classical and Quantum Gravity." *Physical Review Letters* 57: 2244–2247.

———(1987). "New Hamiltonian Formulation of General Relativity." *Physical Review D* 36: 1587–1602.

Ashtekar Abhay, ed. (1988). *New Perspectives in Canonical Gravity*. Naples: Bibliopolis.

Ashtekar, Abhay; Husain, Viqar; Rovelli, Carlo; Samuel, Joesph; and Smolin, Lee (1989). "2+1 Quantum Gravity as a Toy Model for the 3+1 Theory." *Classical and Quantum Gravity* 6: L185–L193.

Ashtekar, Abhay; Jacobson, Ted; and Smolin Lee (1988). "A New Characterization of Half-Flat Solutions to Einstein's Equation." *Communications in Mathematical Physics* 115: 631–648.

Ashtekar, Abhay; Romano, J.; and Tate, R. (1988). "New Variables for Quantum Gravity: Inclusion of Matter." *Physical Review D* 40: 2572–2587.

Ashtekar, Abhay and Rovelli, Carlo (1990). "Loop Representation for Quantum Maxwell Fields." In preparation.

Ashtekar, Abhay; Rovelli, Carlo; and Smolin, Lee (1990). "Loop Representation for Linearized Quantum Gravity." In preparation.

Belavin, A.A.; Polyakov, A.M.; and Zamolodchikov, A.B. (1984). "Infinite Conformal Symmetry in Two-Dimensional Quantum Field Theory." *Nuclear Physics B* 241: 333–380.

Bengtsson, Igemar (1988). "Note on Ashtekar's Variables in the Spherically Symmetric Case." *Classical and Quantum Gravity* 5: L139–L142.

Bergmann, Peter (1958). "Conservation Laws in General Relativity as Generators of Coordinate Transformations." *Physical Review* 112: 287–289.

———(1961). " 'Gauge-Invariant' Variables in General Relativity." *Physical Review* 124: 274–278.

Bergmann, Peter and Goldberg, Irwin (1955). "Dirac Bracket Transformations in Phase Space." *Physical Review* 98: 531.

Deser, S., and Jackiw, R. (1984). "Three Dimensional Cosmological Gravity: Dynamics of Constant Curvature." *Annals of Physics* 153: 405–416.

———(1989). "Classical and Quantum Scattering on a Cone." *Communications in Mathematical Physics* 118: 495–509.

Deser, S.; Jackiw, R.; and 't Hooft, G. (1984). "Three-Dimensional Einstein Gravity: Dynamics of Flat Space." *Annals of Physics* 152: 220–235.

DeWitt, Bryce S. (1967). "Quantum Theory of Gravity. I. The Canonical Theory." *Physical Review* 160: 1113–1148.

Dirac, Paul A.M. (1964). *Lectures on Quantum Mechanics*. Belfer Graduate School of Science Monographs, no. 2. New York: Belfer Graduate School of Science, Yeshiva University.

Friedan, Daniel, and Schenker, Stephen (1987). "The Analytic Geometry of Two-Dimensional Conformal Field Theory." *Nuclear Physics B* 281: 509–545.

Friedan, Daniel; Qiu, Zongan; and Schenker, Stephen (1984). "Conformal Invariance, Unitarity, and Critical Exponents in Two Dimensions." *Physical Review Letters* 52: 1575–1578.

Friedman, John L., and Jack, Ian (1986). "3+1 Regge Calculus with Conserved Momentum and Hamiltonian Constraints." *Journal of Mathematical Physics* 27: 2973–2986.

Frohlich, J. (1987). "Statistics of Fields, the Yang–Baxter Equation, and the Theory of Knots and Links." To appear in *Nonperturbative Quantum Field Theory*, Cargèse Lectures, New York: Plenum.

Gelfand, I.M.; Graev, M.I.; and Vershik, A.M. (1977). "Representations of the Group of Smooth Mappings of a Manifold X into a Compact Lie Group." *Compositio Mathematica* 35: 299–334.

Gerbert, P. de Sousa, and Jackiw, R. (1988). "Classical and Quantum Scattering on a Spinning Cone." MIT preprint, CTP 1594.

Husain, Viqar (1989). "Intersecting Loop Solutions to the Hamiltonian Constraint of Quantum General Relativity." *Nuclear Physics B* 313: 711–724.

Husain, Viqar and Smolin, Lee (1989). "Exactly Solvable Quantum Cosmologies from Two Killing Field Reductions of General Relativity." *Nuclear Physics B* 327: 205–238.

Isham, Christopher J. (1984). "Topological and Global Aspects of Quantum Field Theory." In *Relativity, Groups, and Topology II*. Bryce S. DeWitt and Raymond Stora eds. Amsterdam and Oxford: North-Holland, pp. 1059–1290.

Isham, Christopher J., and Kakas, A.C. (1984). "A Group Theoretical Approach to the Canonical Quantisation of Gravity: I. Construction of the Canonical Group." *Classical and Quantum Gravity* 1: 621–632.

Ismagilov, R.S. (1975). "Unitary Representations of Group of Diffeomorphisms of Space R^n, $n \geq 2$." *Functional Analysis and Its Applications* 9: 154–155.

———(1976). "Ob Unitarynykh Pryedstavlyehiyakh Gruppy C_0^∞ (X,G), $G = SU_e$." *Matematicheskiy Sbornik* 100 No. 1: 117–131.

Jacobson, Ted, and Smolin, Lee (1988). "Nonperturbative Quantum Geometries." *Nuclear Physics B* 299: 295–345.

Jones, V.F.R. (1983). "Index for Subfactors." *Inventions Mathematicae* 72: 1–25.

———(1985). "A Polynomial Invariant for Knots via Von Neumann Algebras." *American Mathematical Society. Bulletin* 12: 103–111.

Kauffman, Louis H. (1983). *Formal Knot Theory.* Mathematical Notes, no. 30. Princeton: Princeton University preprint.

———(1987a). *On Knots.* Annals of Mathematical Studies, no. 115. Princeton: Princeton University Press.

———(1987b). "Statistical Mechanics and the Jones Polynomial." University of Illinois at Chicago preprint.

———(1987c). "States Models and the Jones Polynomial." *Topology* 26: 395–407.

Kodama, H. (1988). "Specialization of Ashtekar's Formalism to Bianchi Cosmology." *Progress of Theoretical Physics* 80: 1024.

Kuchař, Karel V. (1981). "Canonical Methods of Quantization." In *Quantum Gravity 2: A Second Oxford Symposium*. Christopher J. Isham, Roger Penrose, and Dennis W. Sciama, eds. Oxford: Clarendon Press, pp. 329–376.

Kuniba, Atsuo; Akutsu, Yasuhiro; and Wadati, Miki (1986). "Virasoro Algebra, von Neumann Algebra and the Critical Eight Vertex SOS Models." *Journal of the Physical Society of Japan* 55: 3285–3288.

Moore, G., and Seiberg, N. (1989). "Classical and Quantum Conformal Field Theory." *Communications in Mathematical Physics* 123: 177.

Renteln, P. (1988). Ph.D. dissertation. Harvard University.

Renteln, P., and Smolin, Lee (1989). "A Lattice Approach to Spinorial Quantum Gravity." *Classical and Quantum Gravity* 6: 275–294.

Rolfson, D. (1976). *Knots and Links*. Boston: Publish or Perish.

Rovelli, Carlo (1988). "Loop Representation." In *New Perspectives in Canonical Gravity*. Abhay Ashtekar, ed. Naples: Bibliopolis, pp. 275–303.

Rovelli, Carlo, and Smolin, Lee (1988). "Knot Theory and Quantum Gravity." *Physical Review Letters* 61: 1155–1158.

———(1990). "Loop Representation for Quantum General Relativity." *Nuclear Physics B*. 331: 81–152.

Samuel, Joseph (1988). "Gravitational Instantons from the Ashtekar Variables." *Classical and Quantum Gravity* 5: L123–L125.

Smolin, Lee (1988). "Knot Theory, Loop Space and the Diffeomorphism Group." In *New Perspectives in Canonical Gravity*. Abhay Ashtekar, ed. Naples: Bibliopolis, pp. 245–266.

———(1989). "Invariants of Links and Critical Points of the Chern–Simon Path Integrals." *Modern Physics Letters A* 4: 1091–1112.

't Hooft, G. (1988). "Non-Perturbative 2 Particle Scattering Amplitudes in 2+1 Dimensional Quantum Gravity." *Communications in Mathematical Physics* 117: 685–700.

Vershik, A.M.; Gel'fand, I.M.; and Graev, M.I. (1975). "Representations of the Group of Diffeomorphisms." *Russian Mathematical Surveys* 30 (no. 6): 1–50.

Wheeler, John A. (1968). "Superspace and the Nature of Quantum Geometrodynamics." In *Battelle Rencontres: 1967 Lectures in Mathematics and Physics*. Cécile DeWitt and John A. Wheeler, eds. New York: Benjamin, pp. 242–357.

Witten, Edward (1988). "Quantum Gravity in 2 + 1 Dimensions." *Nuclear Physics B* 311: 46–78.

———(1989). "Quantum Field Theory and the Jones Polynomial." *Communications in Mathematical Physics* 121: 351.

Formal Commutators of the Gravitational Constraints Are Not Well-Defined

J. L. Friedman and Ian Jack

1. Summary

Ashtekar's factor ordering of the gravitational constraint equations formally implies a certain symmetric factor ordering of the Hamiltonian constraint in the Schrödinger (metric) representation. This ordering is among those for which the constraints were believed not to close, and a straightforward formal computation of the commutator fails to give closure. However, with an alternative formal computation (equivalent, in the Schrödinger representation to Ashtekar's spinorial computation) the constraints close. More generally, there is no well-defined formal factor ordering problem in quantum gravity: different formal computations of the same commutator yield different results. If the constraints are regularized by point-splitting, using a fixed flat background metric, their commutators have, in general, no well-defined coincidence limit. An alternative, covariant point-splitting prescription that uses the exponential map is available, but adopting it implies that all orderings of the momentum constraint are formally equivalent; all orderings of the Hamiltonian constraint would then lead to a formally closed commutator algebra.

2. Introduction

This talk summarizes work reported in a recent paper by I. Jack and myself (1988). Many of the same conclusions were reached independently by Tsamis and Woodard (1987), and the main claim of ambiguity appears in Woodard's earlier PhD thesis (1984). Factor orderings of the gravitational constraints for which the commutator algebra appears formally to close were proposed by Schwinger (1963), Komar (1979b), Christodoulakis and Zanelli (1987), and Ashtekar (1986, 1987). Ashtekar's ordering involves a new set of variables, but it can be simply rephrased in terms of a triad formalism. The first constraint has the meaning that state vectors are invariant under triad rotations, are functionals only of the three-metric; and the triad representation (or the σ-representation in which state vectors are functionals of Ashtekar's variable σ) is thereby equivalent to a metric representation. The remaining constraints are then particular factor orderings of the usual Hamiltonian and momentum constraints, and the Hamiltonian constraint has a unique translation to the operators h_{ab} (the three-metric) and conjugate momentum π^{ab}.

As noted in the summary, the ordering of the Hamiltonian constraint that we obtain is not one for which the gravitational constraints were thought to close (Anderson 1963, Komar 1979b). The difference between the formal computation, for which the constraints do not close, and Ashtekar's computation, for which they do, corresponds to two different ways of regularizing the constraint by point-splitting. In particular, we find that formal evaluation of the commutator of two Hamiltonian constraints does not yield an expression independent of the computation unless all orderings of the momentum constraint are taken to be equivalent. This may not be so unreasonable as it seems because any point-splitting computation that uses the exponential map (Riemann-normal coordinates) to relate vectors at nearby points gives a vanishing coincidence limit for the difference between two differently ordered momentum constraints.

3. From Ashtekar's Variables to the Schrödinger Representation

3.1. TRIAD FORMALISM

A brief recapitulation of the triad form of Ashtekar's formalism will be given here, in part to fix the notation. If one regards the action,

$$I = \int d\tau \, ^4R,$$

as a function of the tensor density,

$$\tilde{e}_i{}^a = h^{1/2} e_i{}^a,$$

where $\{\mathbf{e}_i\}$ is a triad, then the conjugate momentum is the tensor $2K^i{}_a$,

$$K^i{}_a = e^{ib}\,K_{bc},$$

with K_{ab} the extrinsic curvature. (Indices $a, b, \dots$ are abstract. Indices $i, j, \dots$ are concrete or, if the reader prefers, may be regarded as indices in the Lie algebra $so(3)$.) In the corresponding Hamiltonian formalism, one has the constraints

$$\mathcal{G}_i = 2\varepsilon_{ijk}\,\tilde{e}^{ja}K^k{}_a = 0,$$
$$\mathcal{H}_a = -2\nabla_b(K^b{}_a - \delta^b{}_a K) = 0,$$
$$\mathcal{H} = -h^{1/2}(R - K^i{}_a K_i{}^a + K^2) = 0.$$

3.2. A COMPLEX CANONICAL TRANSFORMATION

Ashtekar's change of variables corresponds to a complex canonical transformation of phase-space, $\{q, p\}$, of the form

$$q_\alpha \to q_\alpha,$$
$$p_\alpha \to p_\alpha - \partial_\alpha F(q).$$

Let

$$\Gamma^i{}_a = \frac{1}{2}\varepsilon^{ijk}\,\Gamma_{jka},$$

where $\Gamma^i{}_{ja}$ is the connection form given by

$$\nabla_a e_j{}^b = \Gamma^i{}_{ja} e_i{}^b.$$

Let

$$F(\tilde{e}) = 2i \int \tilde{e}_i{}^a\,\Gamma^i{}_a.$$

One then finds the relation,

$$\frac{1}{2}\frac{\delta}{\delta\tilde{e}_i{}^a}F = i\Gamma^i{}_a,$$

and the corresponding transformation has the form

$$\tilde{e}_i{}^a \to \tilde{e}_i{}^a,$$
$$K^i{}_a \to K^i{}_a - i\Gamma^i{}_a.$$

Thus, $\frac{1}{2}i\cdot$(new conjugate momentum) is the quantity

$$A^i{}_a = \Gamma^i{}_a + i\,K^i{}_a.$$

The transformation is canonical because of the identity

$$\{P_\alpha, P_\beta\} \longrightarrow \{P_\alpha - \partial_\alpha F, P_\beta - \partial_\beta F\}$$
$$= (\partial_\beta \partial_\alpha - \partial_\alpha \partial_\beta)F$$
$$= 0.$$

Here,

$$\{A^i{}_a(x), A^j{}_b(y)\} = -\frac{1}{4}\left[\frac{\delta}{\delta \tilde{e}_i{}^a(x)}\frac{\delta}{\delta \tilde{e}_j{}^b(y)} - \frac{\delta}{\delta \tilde{e}_j{}^b(y)}\frac{\delta}{\delta \tilde{e}_i{}^a(x)}\right]F$$
$$= 0.$$

Equivalently,

$$\{K^i{}_a(x), \Gamma^j{}_b(y)\} - \{K^j{}_b(y), \Gamma^i{}_a(x)\} = 0.$$

One can regard $A^{ij}{}_a$ and $\Gamma^{ij}{}_a$ as one-forms valued in the Lie algebra $so(3)$. The gauge-covariant derivative of a Lie-algebra-valued tensor $\xi^i{}_{a...b}$ is

$$\mathcal{D}_a \xi^i{}_{b...c} = \nabla_a \xi^i{}_{b...c} + A^i{}_{ja}\,\xi^j{}_{b...c},$$

and the curvature of A is

$$F = dA + A \wedge A = 2\,\partial_{[a}A^{ij}{}_{b]} + 2\,A^i{}_{\ell[a}A^{\ell j}{}_{b]}.$$

In terms of $\tilde{e}$, A the constraints take the form

$$\mathcal{G}_i = -2i\mathcal{D}_a\,\tilde{e}_i{}^a = 0,$$

$$\mathcal{H}_a + K^i{}_a\mathcal{G}_i = \frac{1}{i}\varepsilon^{ijk}\,\tilde{e}_k{}^b\,F_{ijab} = 0,$$

$$\mathcal{H} + i\nabla_a\mathcal{G}^a = -\tilde{e}^{ia}\,e^{jb}\,F_{ijab} = 0.$$

3.3 QUANTUM CONSTRAINTS

The simplest way to translate the constraints with Ashtekar's ordering to the Schrödinger representation is to begin in the $\tilde{e}$ representation with variables K, $\tilde{e}$ where:

$$K = \frac{1}{2i}\frac{\delta}{\delta\tilde{e}}.$$

(The formal relation to the A, $\tilde{e}$ representation is given in Friedman and Jack 1988.)

Frame rotation constraint:

Given rotations $\mathcal{R}(\phi) = \exp(\phi R)$, write

$$R_{ij} = \varepsilon_{ijk}\, R^k,$$

$$\mathcal{G}_R = \int dx\, R^i(x)\mathcal{G}_i(x) = \frac{1}{i}\int dx\, R_{jk}\,\tilde{e}^{ja}\frac{\delta}{\delta\tilde{e}_k{}^a}.$$

Then,

$$R\Psi(\tilde{e}) = i\frac{d}{d\phi}\Psi(\mathcal{R}(\phi)e) = 0$$

$$\Rightarrow \Psi \text{ invariant under frame rotations,}$$

or $\Psi = \Psi(h)$, where h_{ab} is the metric constructed from $\tilde{e}$.

Hamiltonian Constraint:

Claim: The equality

$$\tilde{e}_i{}^a\, \tilde{e}_j{}^b\, F^{ij}{}_{ab}\Psi(h) = h^{1/2}(R - \pi^{ab}\, h^{-1}\, G_{abcd}\, \pi^{cd})\Psi$$

follows from formal manipulation of the operators. The result is unambiguous as a relation in a formal module of operators at a point.

Proof: Let $\phi^I(x)$, $\pi_I(x)$ be operator-valued distributions satisfying

$$[\phi^I(x), \pi_J(y)] = i\delta^I{}_J(x, y).$$

Define a module generated by sums of

(1) operator-valued distributions of x,

(2) formal products of expressions of the form

$$f(\phi(x)), \quad \pi_I(x), \quad \delta(x, x),$$

subject to the commutation relations

$$[f(\phi(x)), \pi_I(x)] = i\delta(x, x)\partial_I f,$$
$$[\delta(x, x), f(\phi(x))] = 0, \quad [\delta(x, x), \pi_I(x)] = 0.$$

Here,

$$\Phi^I = (\tilde{e}_i{}^a, h_{ab}), \quad \pi_I = (2K^i{}_a, \pi^{ab}).$$

From the definition of F, we have

$$\tilde{e}_i{}^a e_j{}^b F^{ij}{}_{ab} \Psi(h) = [h^{1/2}R + 2\tilde{e}_i{}^a e_j{}^b K^i{}_{[a}K^j{}_{b]}]\Psi \equiv -\Psi.$$

Now,

$$\frac{\partial h_{bc}}{\partial \tilde{e}_i{}^a} = -2h^{-1/2} G^i{}_{abc},$$

where $G_{abcd} = \frac{1}{2}(h_{ac}h_{bd} + h_{ad}h_{bc} - h_{ab}h_{cd})$, and it follows that for any differentiable function of three-metrics, $\Phi = \Phi(h)$,

$$K^i{}_a \Phi = -h^{-1/2} G^i{}_{abc} \pi^{bc} \Phi.$$

This identity and the commutation relations

$$[\tilde{e}_i{}^a(x), K^j{}_b(y)] = \delta^j{}_i \, \delta^a{}_b(x,y)$$

give the claimed result,

$$\mathcal{H}\Psi = -h^{1/2}(R - \pi^{ab} h^{-1} G_{abcd} \pi^{cd})\Psi. \tag{1}$$

But there is a well-known formal computation of the commutators:

$$[\mathcal{H}(x), \mathcal{H}(y)], \quad [\mathcal{H}(x), \mathcal{H}_a(y)], \quad [\mathcal{H}_a(x), \mathcal{H}_b(y)];$$

and, for $\mathcal{H}$ of the form (1), this computation fails to give closure.

3.4. Ambiguity of the Formal Algebra

For

$$\mathcal{H} = -h^{1/2}R + h^{-\beta-1/2} \pi^{ab} h^{\beta} G_{abcd} \pi^{cd},$$

the standard computation of the commutator $[\mathcal{H}, \mathcal{H}]$ gives

$$[H_M, H_N] = \frac{1}{i} \int dx \, u_a [h^{(-3/2)-\beta} \nabla_b(\pi^{ab} h^{(3/2)+\beta}) + \nabla_b \pi^{ab}], \tag{2}$$

where $H_M = \int dx \, M$ and $u_a = M\partial_a N - N\partial_a M$. Only for $\beta = -\frac{3}{2}$ or $+1$ does the right-hand side of (2) have the natural ordering of the momentum constraint: For $\beta = -\frac{3}{2}$,

$$[H_M, H_N] = i \int dx \, u_a \, h^{ab} \mathcal{H}_b,$$

where $\mathcal{H}_a = -2h_{ac}\nabla_b \pi^{bc}$; and for $\beta = 1$, $\mathcal{H}_b$ is replaced by $\mathcal{H}_b + \mathcal{H}_b^\dagger$. $\mathcal{H}_a$ has the natural (Higgs) ordering as a generator of diffeos: If

$$P_V = \int dx\, V^a \mathcal{H}_a$$

$$[P_V, h_{ab}] = i\mathcal{L}_V h_{ab}, \qquad [P_V, \pi^{ab}] = i\mathcal{L}_V \pi^{ab},$$
$$\Rightarrow [P_V, H_M] = iH_{\mathcal{L}_V M}, \qquad [P_V, P_W] = iP_{\mathcal{L}_V W}. \tag{3}$$

The translation of the Ashtekar ordering, however, has $\beta = -1$, and the constraints fail to close under the formal computation that gives closure in equation (3). What has gone wrong?

In formal manipulation of "operators" at a point, one sets to zero expressions of the form

$$[h^{-1}(x) - h^{-1}(x)]\nabla_b \pi^{ab}(x)\, h(x).$$

But the operator-valued distribution

$$Q^a(x, y) = [h^{-1}(y) - h^{-1}(x)][\nabla_b \pi^{ab}(x)\, h(y) - h(y)\nabla_b \pi^{ab}(x)]$$
$$= -\frac{1}{2}\delta(x, y)[h^{ab}\, \partial_b \log h(x) + h^{ab}\, \partial_b \log h(y)]$$

then has vanishing coincidence form; and the difference between the $\mathcal{H}_a$'s produced by $[\mathcal{H}, \mathcal{H}]$ for differently ordered $\mathcal{H}$'s is of the form Q^a!

For example, consider the expression

$$h^\alpha h_{ac}\nabla_b(\pi^{bc} h^{-\alpha}) - h^\gamma h_{ac}\nabla_b(\pi^{bc} h^{-\gamma}).$$

By point-splitting or formal computation (i.e., just set $x = y$), we have

$$h^\alpha(x)\, h_{ac}(x)\nabla_b \pi^{bc}(y)\, h^{-\alpha}(x) - h^\gamma{}_{ac}(x)\, h_{ac}(x)\nabla_b \pi^{bc}(y)\, h^{-\gamma}(x)$$
$$= \frac{i}{2}(\gamma - \alpha)\,\delta(x, y)[\partial_a \log h(x) + \partial_a \log h(y)]$$
$$= (\alpha - \gamma)\, h_{ab}\, Q^b(x, y). \tag{4}$$

Therefore, one has no consistent set of formal rules governing formal manipulation of operators at a point. "Formal" closure has meaning only with respect to a particular computation, and there is some formal computation for any ordering that gives closure.

3.5. A COVARIANT POINT-SPLITTING PRESCRIPTION

To evaluate $\pi^{ab}(x)\Psi$, $h_{ab}(x)\Psi$ at a metric g_{ab}, one may regard Ψ as a functional of metrics on $T_x M$, using the exponential map of g_{ab} to pull back to $\tilde{h}_{ab}$ on $T_x M$ each metric h_{ab} defined in a neighborhood of x. Then, the difference,

$$\partial_a \ln h \equiv \tilde{C}^m_{am} = \tilde{\nabla}_a - \partial_a,$$

between the connection of $\tilde{h}$ and the flat connection ∂_a of $T_x M$ is $O(r)$. If the flat parallel transport of $T_x M$ is used to compare tensors at $\tilde{y}$ with tensors at $\tilde{z} \in T_x M$, the difference (3) between differently ordered versions of the momentum constraint will then be $O(r)$. In the coincidence limit, all orderings of the momentum constraint would then be equivalent, and all orderings of the constraints would formally close.

We, of course, do not mean to imply by these statements about formal coincidence limits that it is possible to find a well-defined algebra of quantum operators that includes all the constraints of quantum gravity, and in which the constraints close. Even if there were a well-defined algebra that included the constraints, it would not in general include the set of operators that arise in a formal manipulation leading to closure.

Acknowledgment. This research of one of us (J. L. Friedman) is supported in part by NSF Grant No. PHY 8603173.

Note. The translation from spinors to triads is simple and was found independently by ourselves (Friedman and Jack 1988), by Joohan Lee, by J. N. Goldberg (1988), and by M. Henneaux et al. (1989).

REFERENCES

Anderson, James. L. (1963). "*Q*-Number Coordinate Transformations and the Ordering Problem in General Relativity." In *Proceedings of the Eastern Theoretical Physics Conference October 26–27, 1962*. M. E. Rose, ed. New York: Gordon and Breach, pp. 387–392.

Ashtekar, Abhay (1986). "New Variables for Classical and Quantum Gravity." *Physical Review Letters* 57: 2244–2247.

———(1987). "New Hamiltonian Formulation of General Relativity." *Physical Review D* 36: 1587–1602.

Christodoulakis, T. and Zanelli, J. (1987). "Canonical Approach to Quantum Gravity." *Classical and Quantum Gravity* 4: 851–867.

DeWitt, Bryce S. (1967). "Quantum Theory of Gravity. I. The Canonical Theory." *Physical Review* 160: 1113–1148.

Friedman, J. L., and Jack, I. (1988). "Formal Commutators of the Gravitational Constraints Are Not Well-Defined: A Translation of Ashtekar's Ordering to the Schrödinger Representation." *Physical Review D* 37: 3495–3504.

Goldberg, J.N. (1988). "Triad Approach to the Hamiltonian of General Relativity." *Physical Review D* 37: 2116–2120.

Henneaux, M., Nelson, J.E. and Schomblond, C. (1989). "Derivation of Ashtekar Variables from Tetrad Gravity. *Physical Review D* 39: 434–437.

Komar, Arthur (1979a). "Constraints, Hermiticity, and Correspondence." *Physical Review D* 19: 2908–2912.

———(1979b). "Consistent Factor Ordering of General-Relativistic Constraints." *Physical Review D* 20: 830–833.

Schwinger, Julian (1963). "Quantized Gravitational Field. II." *Physical Review* 132: 1317–1321.

Tsamis, N. C., and Woodard, R. P. (1987). "The Factor-Ordering Problem Must be Regulated." *Physical Review D* 36: 3641–3650.

Woodard, R. P. (1984). "Invariant Formulation of Radiative Corrections in Quantum Gravity." Ph.D. dissertation. Harvard University.

Insights from Twistor Theory

L. J. Mason

1. Introduction

Twistor theory was originally conceived as a possible primary geometric framework in which to present a quantum theory of gravity (Penrose 1968). This point of view was given considerable support by Penrose's discovery of the nonlinear graviton (Penrose 1976a, 1976b). The nonlinear graviton construction encodes the information of an anti-self-dual solution of the Einstein vacuum equations into the complex structure of a deformed twistor space. Unfortunately, there is an obstacle to the realization of the original aims: it appears to be essential for the correct description of physics to find a twistor space description of general vacuum space-times. At present, although many promising ideas have been presented, it is not yet clear that this can be done in sufficiently elegant a fashion that one would feel compelled to regard the twistor space as the primary geometry from which space-time is derived. Part of this program has become known as the googly problem, which is the problem of encoding the information of a self-dual gravitational field into a curved twistor space. Nevertheless, twistor constructions do exist for general curved space-times, and it is reasonable to hope that they will provide important mathematical methods as part of a conventional scheme for quantizing gravity.

The purpose of this article is to review briefly the existing twistor constructions that can be applied to general space-times, and to discuss how they might yield insights into conceptual problems arising in quantum gravity.

The point of view I shall take is that the twistor constructions provide new coordinates on the gravitational phase space. This is analogous to the way one views Ashtekar's new variables (Ashtekar 1988). An important difference is that Ashtekar's variables are local functionals of the standard ones, whereas the twistor variables are nonlocal. The transformation to the twistor variables is more like a nonlinear analogue of a Fourier transform. This has the disadvantage that it is less easy to work with the twistor variables, but the advantage is that there is the possibility of deeper insights, especially into nonlocal concepts such as positive frequency.

I wish to discuss three ways the twistor constructions can contribute. The first is that the hypersurface twistor-space construction appears to lead to a type of polarization condition on initial data sets that is similar to the requirement that it be of positive frequency, a definition that is simply expressed in terms of the twistor variables (this definition is a generalization of one given in Penrose 1976b, but it differs from the positive-frequency condition in linearized theory). The second is that the new variables supplied by the twistor constructions provide the mathematical background necessary to realize (at least in principle) a suggestion in Penrose 1975 that one should quantize space-time points rather than null directions. The third is that the natural "gauge group" arising in the use of twistor variables is unrelated to the diffeomorphism group of space or space-time, but is related to motions of regions in twistor space. This provides a separation of gauge and dynamics since the dynamics are not associated with motions of twistor space but with motions of space-time.

So far as far as the physics is concerned, these ideas are somewhat speculative. However they have already led to interesting mathematics (see, for instance, Penrose 1983 and references therein).

2. The Twistor Constructions

There are two constructions that are applicable to general curved space-times. The construction with the most potential for application is the hypersurface twistor-space construction. This is tied to a choice of hypersurface in space-time, and encodes the information of a gravitational initial-data set into a complex three-manifold together with a pair of cohomology classes. The second is the ambitwistor space construction. This encodes the information of an entire space-time (up to conformal equivalence) into a curved five-dimensional complex manifold, the space of complex null geodesics in the complexified space-time.

These constructions are most easily formulated in terms of a complexified space-time. We shall therefore assume that our original (Lorentzian) space-time is analytic so that we can extend the patching functions and the metric to complex values of the coordinates. This leads to a complex four-manifold, M^4, with complex analytic metric, g_{ab}. Such a complex thickening of a real space-time is only assumed to be defined in a small neighborhood of the "real slice." (There are limiting cases of both constructions when the space-time is not analytic, but these will not concern us here).

I shall concentrate on the hypersurface twistor construction because of its greater potential. An account of spaces of complex null geodesics can be found in Baston and Mason 1987; the conjectures in that paper have, however, recently been proved.

3. A Brief Review of Hypersurface Twistor Space

Consider a complexified initial-data set $(\Sigma^3, h_{ij}, k_{ij})$, consisting of a complex three-manifold Σ^3, a complex analytic three-metric h_{ij}, and extrinsic curvature k_{ij}. The associated (projective) hypersurface twistor space, PT^3, is the three-complex-dimensional space of certain null curves in Σ^3. These null curves can be defined as follows.

Let $(\Sigma^3, h_{ij}, k_{ij})$ be a hypersurface together with its first and second fundamental forms, embedded in a complexified space-time M^4. A hypersurface twistor curve in Σ^3 is a null curve Γ, whose tangent vector $V(\pi)^a$ satisfies the following propagation law. The tangent vector $V(\pi)^a$ can be thought of as a four-vector that happens to be tangent to Σ^3, so that it has a spinor decomposition $V(\pi)^a = \pi^{A'} \eta^A$ for some $\pi^{A'}$ and η^A (spinor conventions, etc., follow those of Penrose and Rindler 1986). The spinor $\pi_{A'}$ is propagated along Γ according to the rule

$$V(\pi)^a \nabla_a \pi_{B'} = 0,$$

where ∇_a is the restriction of the spin connection from M^4. (This connection has become known as the Sen connection, and the connection coefficients are Ashtekar's new variables for canonical gravity, Ashtekar 1988.) The propagation of η^A is determined by that of $\pi^{A'}$ since $V(\pi)^a = \pi^{A'} \eta^A$ is tangent to Σ^3 iff η^A is proportional to $T^{AA'} \pi_{A'}$, where $T^{AA'} = T^a$ is normal to Σ^3 in M^4.

Remarks

(1) In a space-time M^4 with anti-self-dual Weyl curvature, twistor space can be defined as the space of totally null self-dual two-surfaces, α-surfaces. This definition guarantees that when such α-surfaces exist in

M^4, the hypersurface twistor curves are the intersections of Σ^3 with the α-surfaces. This has the advantage that, in flat space-time or a space-time with anti-self-dual Weyl curvature, the hypersurface twistor space is independent of the choice of hypersurface Σ^3 since it can always be identified with the space of α-surfaces.

(2) When $k_{ij} = 0$, hypersurface twistor space is the space of null geodesics in Σ^3 of the metric h_{ij}. When $k_{ij} \neq 0$, hypersurface twistor curves are the null geodesics of a connection on Σ^3 with torsion $\tau^i{}_{jk} = i\varepsilon^l{}_{jk}k^i{}_l$, where $\varepsilon_{ijk} = \varepsilon_{[ijk]}$, etc., so that the hypersurface twistor space encodes k_{ij}, as well as h_{ij}.

(3) As a real manifold, PT^3 has topology $S^2 \times \mathbf{R}^4$ when Σ^3 is topologically trivial. When Σ^3 is the complex thickening of some real topologically nontrivial three-manifold $\Sigma^3_{\mathbf{R}}$, PT has topology $S^2 \times \Sigma^3_{\mathbf{R}} \times \mathbf{R}$.

(4) It is convenient also to introduce the nonprojective hypersurface twistor space T. This is the space of pairs $(\Gamma, \pi_{A\prime})$, where $\pi_{A\prime}$ is the spinor field on Γ that is covariantly constant on Γ, as above, and aligned along Γ. This space is four-complex-dimensional, and is the total space of a complex line bundle, denoted $\mathcal{O}(-1)$, over PT. We can then define the line bundles: $\mathcal{O}(n) = \mathcal{O}(-1)^{-n}$.

(5) Cohomology classes on PT correspond to initial data sets satisfying appropriate constraint equations for linear massless fields on Σ^3; $H^1(PT, \mathcal{O}(-n))$ corresponds to initial data for massless fields with helicity $(n - 2)$.

3.1. Reconstruction of Initial Data from the Twistor Data

In order to be able to regard the twistor data as new coordinates on the gravitational phase-space, we must be able to reconstruct the initial data set from the twistor data. The information about PT^3 on its own does not determine the initial data set. We must also include two extra cohomology classes, ι and σ, on the hypersurface twistor space PT^3. One, ι, is straightforward to understand. The definitions above are conformally invariant so that if two initial data sets are conformally equivalent, they will give rise to the same hypersurface twistor space. We therefore require an additional structure on PT^3 that encodes the information about the conformal factor. In order to understand the reason for the requirement of the second structure, σ, we must go into some detail concerning the reconstruction procedure. First I shall state a proposition:

Proposition. *An initial data set, $(\Sigma^3, N, h_{ij}, k_{ij})$, where the function N is some choice of lapse function, determines and is determined by the twistor data (PT^3, ι, σ), where ι and σ are elements of $H^1(PT, \mathcal{O}(-2))$. This correspondence is stable under small deformations of both the initial data set and the twistor data.*

Sketch of the Proof: First of all, we must construct the three-manifold Σ^3. A point $p \in \Sigma^3$ is represented in PT^3 by the set L_p of α-curves through p. The set L_p is one-dimensional as a complex manifold, and is a Riemann sphere, $\mathbf{C} \cup \infty$. There is a rigidity to compact holomorphic submanifolds of a complex manifold, so that the parameter space M^4 of such holomorphically embedded Riemann spheres is only four-dimensional. The space M^4 contains Σ^3, but we do not have the information on the location of Σ^3 inside M^4.

Let us assume for the moment that we have the information on the location of Σ^3 inside M^4. Then, if we can construct a conformal structure on M^4, this will induce a conformal equivalence class of initial data on Σ^3. The conformal structure on M^4 can be constructed by defining two points p, $p' \in M^4$ to be connected by a null geodesic if the corresponding pair of Riemann spheres L_p and $L_{p'}$ in PT^3 intersect. It can be shown that this definition is unambiguous (cf. Penrose 1976a).

Each point $Z \in PT^3$ leads to an α-surface in M^4, consisting of those p for which $Z \in L_p$. This implies that the Weyl curvature of the conformal structure is self-dual (Penrose 1976a). It was remarked above that in this situation the hypersurface twistor curves are the intersection of the α-surfaces in M^4 with Σ^3 and, given M^4, PT^3 is independent of the location of Σ^3 in M^4.

The extra structures we require, then, are the information of the location of Σ^3 in M^4 and the conformal factor. Both the conformal factor and the location of Σ^3 can be encoded by means of certain cohomology classes, ι and σ respectively, in $H^1(PT^3, \mathcal{O}(-2))$; elements of $H^1(PT^3, \mathcal{O}(-2))$ correspond to solutions of the conformally invariant wave equation, so we can take σ to correspond to the solution that is zero on Σ^3 and whose normal derivative is given by the lapse function, N. Under the conformal rescaling $g_{ab} \rightarrow \Omega^2 g_{ab}$, a solution ϕ of the conformally invariant wave equation transforms as $\phi \rightarrow \phi/\Omega$. We can take ι to correspond to the solution of the wave equation that equals one when g_{ab} has the correct conformal factor. $\square$

3.2. COORDINATE EXPRESSION OF THE TWISTOR DATA

In terms of coordinates, the result of this transform can be understood as follows. The data for PT^3 is just that of a complex manifold. In order to express the data for the manifold in concrete terms, recall that a complex manifold can be covered by a collection of coordinate neighbourhoods $\{U_I\}$, each with holomorphic coordinates z_I^i, with $i = 1,\ldots,3$. In the case of PT^3, only two sets are required, $I = 0, 1$. On the overlap, $U_{IJ} = U_I \cap U_J$, we therefore have that $z_I^i = f_{IJ}^i(z_J^j)$, where the f_{IJ}'s are the patching functions. The collection of patching functions $\{f_{IJ}^i(z^j)\}$ determines the manifold. For PT^3 we just require $f_{01}^i(z^j)$.

The patching functions are also subject to a coordinate or "gauge" freedom. Let g_I^i be a change of the coordinates z_I^i on U_I. Then, in the new coordinates, the patching functions are $\{g_I \circ f_{IJ} \circ g_J^{-1}\}$, where "∘" denotes functional composition. The patching functions $\{f_{IJ}\}$, therefore, determine the same manifold as $\{g_I \circ f_{IJ} \circ g_J^{-1}\}$. (When there are three sets or more, the patching functions f_{IJ} are not freely prescribable, but must satisfy the compatibility condition on triple overlaps: $f_{IJ} \circ f_{JK} = f_{IK}$. However, for topologically trivial Σ^3, PT can be covered by just two open sets.)

The cohomology classes ι and σ can be represented similarly by holomorphic sections of $\mathcal{O}(-2)$ on U_{01}. The line bundle $\mathcal{O}(-2)$ can be trivialized on U_{01} so that ι and σ can be represented by holomorphic functions $\iota(z)$ and $\sigma(z)$ respectively, defined on the overlap of U_0 with U_1. The functions $f^i(z)$, $\iota(z)$ and $\sigma(z)$ on U_{01} are effectively freely prescribable.

Remarks

(1) I have so far not discussed the constraint and evolution equations. These have been articulated in hypersurface-twistorial terms (Mason 1985). However, the hypersurface twistor spaces were presented there from a rather different point of view, and the results cannot be directly interpreted in terms of the structures introduced above. The constraints had a remarkably simple form. This leads me to hope that, in terms of the structures introduced above, the constraints would also be elegant. However, the required work remains to be done.

(2) Hypersurface twistor spaces can also be defined for null hypersurfaces; however, the above proposition does not apply. One loses the information on the self-dual part of the initial data.

(3) The data can be expressed in "real" terms. The complex structure on PT can be represented by means of the $\bar{\partial}$-operator, which maps from $(0, p)$-forms to antiholomorphic $(0, p + 1)$-forms, where a (p, q)-form is an expression of the form

$$\alpha_{i_1 \cdots i_p j_1 \cdots j_q} dz^{i_1} \wedge \cdots \wedge dz^{i_p} \wedge d\bar{z}^{j_1} \wedge \cdots d\bar{z}^{j_q}$$

in local coordinates. The cohomology classes ι and σ can then be represented by equivalence classes of $(0, 1)\bar{\partial}$-forms.

4. A Polarization Condition for Initial Data Sets

The canonical quantization approach to quantum gravity has yet to incorporate the notion of positive frequency which is so crucial to the development of ordinary quantum field theories. One of the reasons for this is the fact that it is not clear what one wants from a concept of positive and negative frequencies in the full nonlinear regime. From the point of view of geometric quantization, what is required is a splitting of a general perturbation of a given background into its positive and negative frequency parts. Unfortunately, there are several, possibly terminal, technical difficulties with this type of definition for quantum gravity. (It is still possible that this kind of approach will work). Here, I will present a generalization of Penrose's proposal (Penrose 1976b).

The proposal is that one should introduce the notion of positive frequency for full solutions of the vacuum equations. Such a positive frequency solution of the vacuum equations, then, is a "nonlinear graviton"; a graviton carries a finite amount of curvature. The definition presented here generalizes that in Penrose 1976b in various respects. It is ambidextrous, in that the graviton is not required to be in an eigenstate of helicity. However, this definition has the defect that it doesn't agree with the standard definition in linearized theory. Instead, in linearized theory, the anti-self-dual part of the field satisfying this polarization condition is positive frequency, but the self-dual part is negative frequency. Nevertheless, this definition is natural, at least in contexts where a spacelike hypersurface has been chosen, and may perhaps have some utility in the canonical quantization program.

The polarization condition is motivated from the definition of positive frequency in the context of zero-rest-mass fields on flat space-time. The hypersurface twistor spaces defined above are generalizations of projective twistor space $\mathbf{PT}$, which is $\mathbf{CP}^3$ for flat space-time $\mathbf{M}$. $\mathbf{PT}$

is divided into three regions, **PT$^+$**, **PN**, and **PT$^-$** on which a certain pseudo-Hermitian form $Z \cdot \bar{Z}$ of signature $(2, 2)$ is positive, zero, and negative, respectively. The real five-dimensional submanifold **PN** is the common boundary of **PT$^+$** and **PT$^-$**. Cohomology classes on regions in **PT** correspond to massless fields on space-time. It turns out that a massless field on space-time is positive frequency if and only if its corresponding twistor cohomology class extends over **PT$^+$**.

In the context of hypersurface twistor spaces, we shall say that an initial-data set satisfies the polarization condition if the hypersurface twistor space has a region in it analogous to **PT$^+$**.

Let us consider the case in which the hypersurface Σ^3 in space-time is the complexification of a real three manifold $\Sigma_{\mathbf{R}}^3$. (For the purposes of the definition, the data h_{ij} and k_{ij} will in general be complex on $\Sigma_{\mathbf{R}}^3$). The hypersurface $\Sigma_{\mathbf{R}}^3$ will be taken to have the topology of S^3, so that we are working either with a spatially compact universe, or a one-point compactification of the asymptotically-flat case (in this latter case, the fields will be singular at the point $i_0 \in \Sigma_{\mathbf{R}}^3$, where i_0 is the point at infinity). Then, it is possible to define regions, PT^+, PN, and PT^- in hypersurface twistor space PT. The five-real-dimensional submanifold PN is defined as the set of hypersurface twistor curves that intersect $\Sigma_{\mathbf{R}}^3$; this has topology $S^2 \times \Sigma_{\mathbf{R}}^3$. This real hypersurface divides PT into two halves, PT^+ and PT^-, each with topology $S^2 \times S^3 \times \mathbf{R}$. The manifolds $PT^\pm$ have PN as a common boundary, but also, in general, each has a natural boundary (i.e., a boundary, past which it is impossible to analytically continue the manifold). See Figure 1.

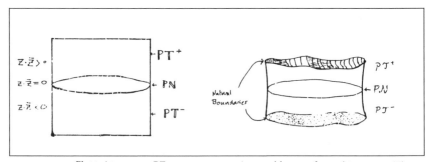

Flat twistor space, **PT**   A general hypersurface twistor space, $P\mathcal{T}$

FIGURE 1.

Definition. An initial data set, $(\Sigma_{\mathbf{R}}^3, h_{ij}, k_{ij})$, in which $\Sigma_{\mathbf{R}}^3$ has topology S^3 as in the previous paragraph, will be said to satisfy the polarization condition when the only boundary component of the region PT^+ in the hypersurface twistor space PT is PN. This means that there is no natural boundary, and the $S^3 \times \mathbf{R}$ factor in the topological decomposition of PT^+ can be filled in to become an open ball in $\mathbf{R}^4$.

Remarks

(1) This polarization condition does *not* agree with the positive frequency condition in the linearized limit. Nevertheless, at least when the hypersurface is (conformally) flat, it has a conformally-invariant interpretation. A field satisfies the above polarization condition if the anti-self-dual part is positive frequency, and if the self-dual part is negative frequency. When the hypersurface is general (but the field still linear), this condition still requires that the anti-self-dual part of the field be of positive frequency, but the self-dual part may have more complicated behavior. When the hypersurface is null, this definition agrees with positive frequency. However, one loses the information on the self-dual part of the field. It is perhaps worth noting that it is possible to have real solutions of the vacuum equations satisfying this polarization condition.

(2) There will perhaps be classes of data for which the $S^3 \times \mathbf{R}$ factor of PT^+ can be filled in, but only with some nonstandard topology, such as that obtained from taking a topologically general real four-manifold, and then taking out an open ball. This would then be a gravitational analogue of the four-dimensional twistor conformal field theory ideas (cf. Hodges, Penrose and Singer 1989).

(3) Many more ideas will be required to build a quantum theory of gravity using these ideas (cf. Penrose 1976a, b).

5. Quantization of Space-Time Points Versus Null Directions

In the usual approach to canonical quantization, components of the metric are subject to commutation relations, and therefore to quantum fluctuations and uncertainty relations. As a result, null directions through a point are no longer well-defined in the quantum theory. However, space-time points remain clearly defined. Penrose (1975) has argued that it is perhaps more appropriate to have clearly defined null directions in the theory, but a notion of space-time point that is subject to quantum fluctuations and uncertainties. If we assume for the moment that it is possible to achieve this, one can see that the latter approach is desirable both in order to define spinors and to define massless particles on a quantized background; massless particles must travel along null directions, and the definition of spinors is also tied to the light cone.

The use of the hypersurface twistor variables can be thought of as realizing this, at least heuristically, in the following sense. If the twistor variables are used, then it is the twistor variables which become subject to commutation relations and therefore to quantum fluctuations and uncertainties. The points of twistor space retain their identity. (That this is the case is perhaps clearer when the twistor data is presented in terms of the $\bar{\partial}$-operator, rather than patching functions). When it comes to finding the space-time points, one must locate the global holomorphic S^2's in PT. However, if the complex structure (and ι and σ) are subject to uncertainties, the location of such global holomorphic S^2's will not be well-defined. The heuristic space-time picture one might have is that of well-defined hypersurface twistor curves with smeared intersections so that they do not focus down clearly onto space-time events. The space-time events therefore become smeared as the price one pays in order that the null directions should retain their identity. See Figure 2.

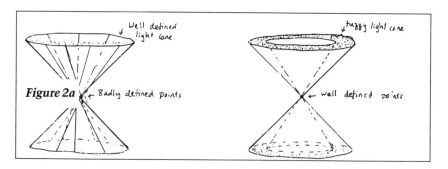

Twistor based quantization scheme   Space-time based quantization scheme

FIGURE 2.

6. Separation of Gauge and Dynamics

The last point I wish to make is that the gauge freedom in the twistor data is, as described towards the end of Section 1, the group of holomorphic coordinate transformations on the open sets covering the twistor space. (One could perhaps also include transformations from data obtained with respect to one cover of twistor space to data obtained from another, but this is not essential.) Alternatively, if one expresses the data using the $\bar{\partial}$-operator, etc., the gauge group is the real diffeomorphism group of PT. In both cases, the gauge groups are unrelated to space-time diffeomorphisms. If one were to attempt to quantize using the twistor variables, the diffeomorphism group of Σ^3 would no longer have any role to play.

The twistor gauge groups present less difficulties than the space-time ones. Firstly, the Einstein constraint equations generate motions of space-time, not of twistor space, so the problem of eliminating the constraints is not tied to the problem of handling the gauge freedom; this makes it easier to work geometrically and invariantly. Secondly, the dynamics corresponds to motions of space-time; these motions are not now confused with the gauge freedom in the description of the geometry.

7. Conclusions and Outlook

Many problems remain to be solved before hypersurface twistors can become an effective mathematical tool for quantum gravity. In particular, it will be essential to obtain the canonical formalism and constraint equations in a tractable form in terms of the new variables. Nevertheless, it can be seen that they hold much promise.

Acknowledgments. I should like to thank Roger Penrose and Abhay Ashtekar for important discussions and contributions. I would also like to thank Abhay Ashtekar and the physics department at Syracuse University for their hospitality while I was thinking about some of these issues. This research was supported in part by an Andrew Mellon Fellowship, a Fulbright Scholarship and the NSF Grant PHY80023.

REFERENCES

Ashtekar, Abhay, ed. (1988). *New Perspectives in Canonical Gravity.* Naples: Bibliopolis.

Baston, R., and Mason, L. J. (1987). "Conformal Gravity, the Einstein Equations and Spaces of Complex Null Geodesics." *Classical and Quantum Gravity* 4: 815–826.

Hodges, A.P., Penrose, Roger and Singer, M.A. (1989). "A twistor conformal field theory for four space-time dimensions." *Phys. Lett. B* 216: 48–52.

LeBrun, Claude (1983). "Spaces of Complex Null Geodesics in Complex-Riemannian Geometry." *American Mathematical Society. Transactions* 278: 209–231.

Mason, L. J. (1985). Twistors in Curved Space-time. Ph.D. Dissertation. Oxford University.

Penrose, Roger (1968). "Twistors Quantisation and Curved Space-time." *International Journal of Theoretical Physics* 1: 61–99.

———(1975). "Twistor Theory. Its Aims and Achievements." In *Quantum Gravity: An Oxford Symposium.* Christopher J. Isham, Roger Penrose, and Dennis W. Sciama, eds. Oxford: Oxford University of Press, 268–407.

———(1976a). "Nonlinear Gravitons and Curved Twistor Theory." *General Relativity and Gravitation* 7: 31–52.

———(1976b). "The Nonlinear Graviton." *General Relativity and Gravitation* 7: 171–176.

———(1983). "Physical Space-time and Nonrealizable CR Structures." In *The Mathematical Heritage of Henri Poincaré.* Proceedings of Symposia in Pure Mathematics, vol. 39, part 1. Felix E. Browder, ed. Providence: American Mathematical Society, 401–422.

Penrose, Roger, and Rindler, Wolfgang (1986). *Spinors and Space-time.* Cambridge and New York: Cambridge University Press.

Sparling, G. A. J. (1983). "Twistor Theory and the Characterisation of Fefferman Conformal Structures." University of Pittsburgh preprint.

Discussion

SMOLIN: Those null curves have nothing to do with null curves in conformal theory. It's just a complexification of the three-metric?

MASON: This is just a complexification of the three-metric. When the four-dimensional space-time is anti-self-dual, these are the intersections of the space-like three-hypersurface with the totally null two-planes, the twistor planes, that you have in anti-self-dual space-times.

JACOBSON: Is there any natural way of including the condition of the nondegeneracy of the metric? How is that translated into a condition on the twistor space or space of complex null geodesics?

MASON: Well, if the metric is singular, the null geodesics will become badly behaved in a neighborhood of that singularity. So, I suppose, you might expect some peculiar behaviour in the quadrics; their normal bundles will jump in some way. It won't be easy to read off from the patching functions.

JACOBSON: So, if you have a bona fide complex manifold, this implies that the metric is nonsingular?

MASON: No, because the quadrics can still "jump" even though the complex manifold is perfectly regular.

JACOBSON: What information do you put on the space of complex null geodesics in order to reconstruct the conformal scale of the metric?

MASON: There's a naturally defined vector bundle on the space of complex null geodesics; global sections of that vector bundle provide conformal factors for which the metric is vacuum if such a global section exists. It generalizes the earlier nonlinear gravitonconstruction.

ROVELLI: Is it possible to articulate the definition of a positive frequency initial data set in space-time without using twistor space?

MASON: I haven't really worked it out, but I expect that the translation is of the form of having a large analytic extension of the three-manifold and initial data into the complex, in certain directions.

Nonlinear Sigma Models in 4 Dimensions as Toy Models for Quantum Gravity

Bryce DeWitt

1. Introduction

Conventional quantum gravity is not a perturbatively renormalizable theory. For this reason many theorists have taken the view that it cannot stand on its own feet, and have turned to string theory as the only theory capable of providing an ultimate foundation for quantum-gravitational effects. While string theory may indeed provide such a foundation, it is by no means a proven fact that lack of perturbative renormalizability implies that a theory cannot be renormalized at all, or cannot be meaningful in its own right in a fundamental sense. No one knows whether conventional quantum gravity exists or not; the question constitutes unfinished business for the theorist.

Repeatedly in the history of physics, even after interest has turned elsewhere, one has had to come back and settle unfinished business. No way is known of extracting meaning out of a theory that is not perturbatively renormalizable, or for that matter, out of a perturbatively renormalizable theory in the strong coupling regime, except by computing the Feynman functional integral for a lattice simulation of the theory and attempting to determine an asymptotic behavior in the continuum limit. A small group at the University of Texas (Jorge de Lyra, See Kit Foong, Timothy Gallivan, and the author) has undertaken a program to study lattice quantum gravity. For many reasons, this will be a very difficult study and, by requiring some very sophisticated techniques, will stretch the capacities of current supercomputers to their limits. In order to get their bearings on the subject of lattice simulations of functional integrals and also to get some preliminary computer experience, the Texas group has decided to look first at a simpler system, which, however, has enough similarities to the gravitational field to make it interesting—namely, the nonlinear sigma model in four dimensions.

There is really an infinity of nonlinear sigma models, each characterized by a group coset space constituting a dynamical configuration space. I shall report here on preliminary results obtained for the $O(2)$, $O(3)$, and $O(1,2)$-models, the configuration spaces being $O(2)$, $O(3)/O(2)$, and $O(1,2)/O(2)$, respectively. It is useful to compare the classical action functional for these models with that of the gravitational field, the latter being given by

$$
\begin{aligned}
S = -\frac{1}{2}\mu^2 \int (-g)^{1/2} \big[& g^{\mu\nu}(g^{\sigma\rho}g^{\tau\lambda} - g^{\sigma\tau}g^{\rho\lambda})g_{\sigma\tau,\mu}g_{\rho\lambda,\nu} \\
& + 2g^{\mu\nu}g^{\sigma\tau}g^{\rho\lambda}(g_{\sigma\tau,\mu}g_{\nu\rho,\lambda} - g_{\mu\sigma,\rho}g_{\nu\lambda,\tau})\big] d^4x,
\end{aligned}
\tag{1}
$$

where $g_{\mu\nu}$ is the metric tensor, $g^{\mu\nu}$ its inverse, g its determinant, and μ a scale parameter known as the *Planck mass* (in units for which $\hbar = c = 1$). Apart from numerical factors, μ^2 is the reciprocal of the gravity constant.

The prototypical form for the classical action of a nonlinear sigma model is

$$
S = -\frac{1}{2}\mu^2 \int G_{ab}(\phi)\phi_{a,\mu}\phi_{b,}{}^{\mu} d^4x,
\tag{2}
$$

where the fields ϕ_a are coordinates in the configurate space; $G_{ab}(\phi)$ is the group-invariant metric on this coset space; commas followed by Greek indices denote differentiation with respect to the space-time coordinates; the Minkowski metric, which raises and lowers the Greek indices, has signature $-++++$; and μ is a scale parameter having the dimensions of mass.

The following similarities between the gravitational and sigma fields may immediately be noted: (1) Both $g_{\mu\nu}$ and ϕ_a are dimensionless. (2) Both fields possess only a single adjustable scale parameter μ. (3) Except for the $O(2)$-model, corresponding Feynman graphs in perturbation theory are equally divergent. These similarities are enough to make nonlinear sigma models of immediate interest as toy models for quantum gravity.

There are also some important differences between the two kinds of fields, which will engender significant differences of detail in the computer approaches to their respective theories, and which may in the end prove to be more important than their similarities.

(1) Gravity has a local symmetry group ($Diff(M)$, where M is the space-time manifold), whereas the sigma models have global symmetry

groups. The existence of the local symmetry group for gravity creates the complication of having to introduce ghosts into the theory. The existence of the global symmetry group for a sigma model leads to the complication of spontaneous symmetry breaking.

(2) Even if a sigma model proves to have no sensible or nontrivial continuum limit, this will not mean that the same is true for quantum gravity. A quantized gravitational field implies a quantized light cone, and it has long been believed (Klein 1955; Landau 1955; Pauli 1956; Deser 1957; DeWitt 1964; Isham, Salam and Strathdee 1971, 1972) that a quantum-smeared light cone may constitute a possible lattice-independent regulator for quantum gravity, so that not only may a nontrivial continuum limit be found to exist for the theory, but both the bare and renormalized Planck masses may remain *finite* in the limit. Such a mechanism is unavailable for sigma models since the light cone for them remains constantly that of Minkowski space-time.

(3) To evaluate the functional integral for any field on a computer, one must pass to the Euclidean version of the theory, in which a Wick rotation is performed on the time variable, and contour rotations of some of the functional integration variables may also have to be carried out. In the case of the sigma model, the simple Wick rotation suffices to yield a *Euclidean action* that is bounded from below. This is not true for the gravitational field. There, not only must one rotate the g_{i4} variables ($i = 1, 2, 3$), but one must also perform a rotation on the so-called *conformal factor,* with complications that depend on the existence and number of negative eigenvalues possessed by the so-called *conformally covariant Laplacian operator.* Since our concern in this paper is with the nonlinear sigma models, I shall not describe these complications here.

2. The $O(2)$ Sigma Model

The Euclidean action for this model has the form

$$S_E = \frac{1}{2}\mu^2 \int \phi_{i,\mu}\phi_{i,}{}^{\mu} d^4x \qquad i = 1, 2, \tag{3}$$

with the field variables subjected to the following constraint:

$$\phi_i \phi_i = 1. \tag{4}$$

The configuration space of this model is the circle S^1, which is intrinsically flat. Perturbatively, the $O(2)$-model is linear and hence trivial.

Only the nontrivial topology of the configuration space makes this model nontrivial.

In the computer simulations of all the models that I shall describe, the "universe" consists of a hypercubical lattice of $(N + 1)^4$ points. The lattice spacing is a and the dimensions of the hypercube are $L = (N + 1)a$ on a side. The field variables in the continuum limit depend on four Euclidean coordinates, a dependence that is expressed in the form $\phi_i(x^1, x^2, x^3, x^4)$. If the origin of these coordinates is placed at a corner of the hypercube, then the lattice variables may be expressed in the form

$$\phi_{i\alpha\beta\gamma\delta} = \phi_i(\alpha a, \beta a, \gamma a, \delta a), \tag{5}$$

where α, β, γ, and δ are integers that range from 0 to $N + 1$.

Taking note of the fact that the volume element d^4x for the lattice is a^4, one obtains the following expression for the lattice approximation to the action (3):

$$
\begin{aligned}
S_E &= \frac{1}{2}\mu^2 a^2 \sum_{\alpha,\beta,\gamma,\delta=0}^{N} \left[(\phi_{i\alpha+1\,\beta\gamma\delta} - \phi_{i\alpha\beta\gamma\delta})^2 + (\phi_{i\alpha\beta+1\,\gamma\delta} - \phi_{i\alpha\beta\gamma\delta})^2 \right. \\
&\quad \left. + (\phi_{i\alpha\beta\gamma+1\,\delta} - \phi_{i\alpha\beta\gamma\delta})^2 + (\phi_{i\alpha\beta\gamma\delta+1} - \phi_{i\alpha\beta\gamma\delta})^2 \right] \\
&= \beta \sum_{\alpha,\beta,\gamma,\delta=0}^{N} \left[4 - \phi_{i\alpha\beta\gamma\delta}(\phi_{i\alpha+1\,\beta\gamma\delta} + \phi_{i\alpha\beta+1\,\gamma\delta} \right. \\
&\quad \left. + \phi_{i\alpha\beta\gamma+1\,\delta} + \phi_{i\alpha\beta\gamma\delta+1}) \right] \\
&= 4(N+1)^4\beta - \beta \sum_{\text{links}} \phi_i(\text{one end of link})\, \phi_i(\text{other end of link}),
\end{aligned}
\tag{6}
$$

where the constraint (4) has been used to simplify the sum and where

$$\beta \equiv \mu^2 a^2. \tag{7}$$

The coefficient β is a dimensionless number. It and the integer N constitute the only adjustable constants available to us in the study of sigma models.

When β is very large, S_E is large, and fluctuations of the ϕ_i about some "frozen" configuration are suppressed. Because of the constraint (4), the frozen configuration corresponds to some point on the configuration-space circle. Since all points on the circle are physically equivalent, it is obvious that the $O(2)$-model has a degenerate ground state and that the frozen configuration corresponds to a spontaneously broken symmetry. In order to prevent the frozen configuration from wandering in the Monte Carlo simulation of the lattice functional integral (see below), we hold the field fixed at "infinity," i.e., at the edge of the universe, in this case at the surface of the hypercube. The fixed value is chosen to be $\phi_1 = 0$, $\phi_2 = 1$. In terms of computer variables this corresponds to setting

$$
\begin{aligned}
\phi_{1\,\alpha\beta\gamma\delta} &= 0 \quad \text{if any of the } \alpha, \beta, \gamma, \delta \text{ is equal to 0 or } N+1 \\
\phi_{2\,\alpha\beta\gamma\delta} &= 1 \quad \text{if any of the } \alpha, \beta, \gamma, \delta \text{ is equal to 0 or } N+1.
\end{aligned}
\tag{8}
$$

This boundary condition may be referred to as "seeded" symmetry breaking.

3. Monte Carlo Simulation of the Functional Integral

In the Monte Carlo simulation, the field is randomly changed at successive lattice sites. If the amount of the change at a given site is equal to $\Delta\phi_i$, the corresponding change in the Euclidean action is

$$
\Delta S_E = -\beta \sum_{\substack{\text{neighboring} \\ \text{sites}}} \Delta\phi_i(\text{site}) \, \phi_i(\text{neighboring site}).
\tag{9}
$$

The new ϕs at the given site are chosen randomly with uniform weight around the configuration-space circle. (This is equivalent to choosing θ randomly with uniform weight in the interval $0 \leq \theta \leq 2\pi$ and setting $\phi_1 = \cos\theta$, $\phi_2 = \sin\theta$, although the actual numerical technique used is somewhat different from this). Then, a random number ξ between zero and one is chosen. The new field configuration is kept if $\exp(-\Delta S_E) > \xi$. Otherwise, the field at the given site is left unchanged. This process is then repeated at the next site, and so on.

At the end of each sweep over the lattice, one arrives at a field $\phi_{i\alpha\beta\gamma\delta}(\text{sweep})$. One repeats this process until many thousands ($\sim 80,000$) of sweeps have been made. It can be shown that the functional-integral average of any functional $A[\phi]$ of the field is then well approximated by

$$
\langle A[\phi] \rangle = \frac{\sum_{\text{sweeps}} A[\phi(\text{sweep})]}{\text{number of sweeps}}.
\tag{10}
$$

Furthermore, error bars on this approximation can be determined by dividing the sweeps into groups of 2,000, computing the average for each group, and then computing the variance of these averages around the overall average.

4. Spontaneous Symmetry Breaking: "Magnetization"

For each choice of the dimensionless parameter β, the extent of the spontaneous symmetry breaking, or "freezing," of the field can be measured by the following field average, called the *magnetization:*

$$M \equiv \frac{1}{N^4} \left\langle \sum_{\alpha,\beta,\gamma,\delta=1}^{N} \phi_{2\,\alpha\beta\gamma\delta} \right\rangle. \tag{11}$$

As we have already remarked, when β is large the field is "frozen," which, with the boundary conditions (8), corresponds to $M \approx 1$. As β decreases, a phase transition is found to occur. The critical value β_c at which this takes place may be estimated roughly by the following reasoning: The phase transition will occur when the ϕ_i at each site begin to be free to range over the whole of configuration space, thus restoring the $O(2)$ symmetry. This will happen, roughly, when $\exp(-\Delta S_E) \approx \frac{1}{2}$ where $\Delta S_E \approx 2\beta$, corresponding to a change $\Delta\phi_i$ from one side of configuration space to the other. The critical β is therefore approximately

$$\beta_c \approx \frac{1}{2}\ell n2 = .35. \tag{12}$$

Actual computer runs for the $O(2)$, $O(3)$, and $O(4)$ models yield the following numbers:

$$\left\{ \begin{array}{l} O(2):\quad \beta_c = .30 \\ O(3):\quad \beta_c = .45 \\ O(4):\quad \beta_c = .55 \end{array} \right\}. \tag{13}$$

The qualitative dependence of M on β and N is shown in the following figure:

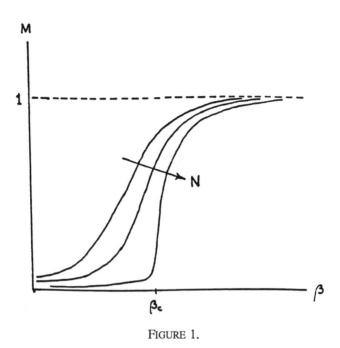

FIGURE 1.

5. Two-Point Functions (Propagators)

Of prime importance in assessing the existence of a nontrivial continuum limit for a quantum field theory, whether perturbatively renormalizable or not, is a study of the two-point function. This function, also known as the propagator, yields information both on field renormalizations and on physical mass parameters. It is most conveniently studied in momentum space.

"Momentum space" usually means "Fourier transform." When box boundary conditions are imposed, as in the computations reported here, "Fourier transform" must be replaced by "box-mode-function transform." The box-mode-function transforms of the fields ϕ_i are

$$\tilde{\phi}_{iklmn} \equiv \frac{4\pi}{(N+1)^3} \sum_{\alpha,\beta,\gamma,\delta=1}^{N} \phi_{i\alpha\beta\gamma\delta} \sin \frac{\pi k\alpha}{N+1} \sin \frac{\pi l\beta}{N+1} \cdot$$

$$\cdot \sin \frac{\pi m\gamma}{N+1} \sin \frac{\pi n\delta}{N+1}, \tag{14}$$

where k, l, m, n are positive integers and where the normalization has been chosen to simplify the expression for the two-point function. When ϕ_i is taken to be a massless free field (with configuration space $\mathbb{R}$ instead of S^1), one finds, both theoretically and very accurately on the computer,

$$\langle \tilde{\phi}_{1\ klmn}^{\ \ 2} \rangle = \frac{1}{\beta} \frac{1}{k^2 + l^2 + m^2 + n^2} \quad \text{independent of } N, \quad (15)$$

showing the usual $1/p^2$ behavior of the propagator of a massless field in four dimensions. [The momentum here is given by $p^2 = (\pi/L)^2(k^2 + l^2 + m^2 + n^2)$.]

The two-point function for the $O(2)$-model is found to be well approximated by an expression of the form

$$\langle \tilde{\phi}_{i\ klmn}^{\ \ 2} \rangle = \frac{1}{\beta_R} \frac{1}{k^2 + l^2 + m^2 + n^2 + \Delta^2}. \quad (16)$$

$1/\beta_R$ represents the "renormalized" residue at the pole of the propagator and Δ is a kind of mass parameter. β_R and Δ^2 are obtained on the computer by a least-squares best-fit adjustment and are found to have the qualitative behavior shown in Figures 2 and 3. These figures, which, for $N \geq 6$, are practically N-independent, can be understood as follows. If $\beta > \beta_c$, the field fluctuations do not probe the whole of configuration space, and the theory rapidly becomes that of a massless free field. Hence, in this region $\beta_R \to \beta$ and $\Delta \to 0$. If $\beta < \beta_c$, then the field fluctuations feel the constraints $-1 \leq \phi_1 \leq 1$, $-1 \leq \phi_2 \leq 1$, and the propagator gets depressed compared to that of a truly free field. In principle, this can show up either as $\beta_R > \beta$ or $\Delta > 0$, or both. Surprisingly, except for an anomalous bump in β_R near β_c, it shows up largely as $\Delta > 0$. Moreover, Δ tends to infinity so rapidly as the continuum limit $\beta \to 0$ is approached that it is very unlikely that a continuum limit (with a finite propagator mass) exists for this model.

6. The $O(3)$ Sigma Model

The computer analysis of the $O(3)$-model is very similar to that of the $O(2)$-model. The only difference is that the indices i in equations (3)–(9) now range over the values 1, 2, 3, and the Monte Carlo weighting is chosen

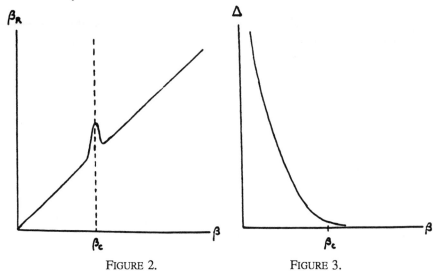

FIGURE 2.   FIGURE 3.

differently because the configuration space is now the two-sphere S^2 instead of the circle S^1. To determine the weighting, one first shows that if ϕ_3 is expressed in terms of ϕ_1 and ϕ_2, the Euclidean action may be written in the form

$$S_E = \frac{1}{2}\mu^2 \int G_{ab}(\phi)\phi_{a,\mu}\phi_{b,\mu}\, d^4x \tag{17}$$

(cf. eq. (2)), where the metric $G_{ab}(\phi)$ is given by

$$G_{ab} = \delta_{ab} + \frac{\phi_a\phi_b}{1 - \phi_c\phi_c}, \tag{18}$$

the indices a, b ranging over the values $1, 2$. One then computes

$$G \equiv \det(G_{ab}) = \frac{1}{1 - \phi_c\phi_c} \tag{19}$$

and writes the functional-integration measure in the form

$$G^{1/2}d\phi_1 d\phi_2 = d\phi_3\, d\theta, \tag{20}$$

where

$$\phi_1 = \sqrt{1 - \phi_3^2}\cos\theta, \qquad \phi_2 = \sqrt{1 - \phi_3^2}\sin\theta. \tag{21}$$

It is evident that, in the Monte Carlo simulation, one must effectively choose θ and ϕ_3 randomly with uniform weights in the respective intervals $0 \le \theta \le 2\pi$, $-1 \le \phi_3 \le 1$, and then compute ϕ_1 and ϕ_2 from equations (21).

The boundary conditions for "seeded" symmetry breaking in the case of the $O(3)$-model, are conveniently taken to be $\phi_1 = 0$, $\phi_2 = 0$, $\phi_3 = 1$ at the edge of the "universe." The magnetization curves for the $O(3)$-model are found to resemble very much those of the $O(2)$-model. The only difference is that β_c is now equal to .45 and, in the definition of the magnetization, the subscript 2 in equation (11) is replaced by 3.

There are now three two-point functions. Two of them are found to be equal (as a priori symmetry arguments in any case require) and to be well approximated by

$$\langle \tilde{\phi}_{1\,klmn}{}^2 \rangle = \langle \tilde{\phi}_{2\,klmn}{}^2 \rangle = \frac{1}{\beta_R} \frac{1}{k^2 + l^2 + m^2 + n^2 + \Delta^2}. \tag{22}$$

The third is found to have a value consistent with zero (as symmetry requires):

$$\langle \tilde{\phi}_{1\,klmn}\, \tilde{\phi}_{2\,klmn} \rangle = 0. \tag{23}$$

The curves for the functions β_R and Δ, which, for $N \ge 6$, are again found to be practically N-independent, now have the following forms:

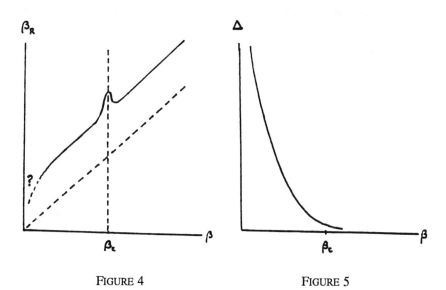

FIGURE 4   FIGURE 5

The chief difference between these curves and those for the $O(2)$-model are that $\beta_R > \beta$ for all β. This inequality can be understood as follows: for a given field configuration, the Euclidean action (17), with G_{ab} given by equation (18), is always greater than that of the corresponding linear free field with $G_{ab} = \delta_{ab}$. The two-point function is therefore depressed compared to that of the corresponding free field, whence $\beta_R > \beta$.

On the other hand, when β is large, the field fluctuations about $\phi_1 = \phi_2 = 0$ are small, so that the field does not "feel" the curvature of the configuration space and behaves like a free field. This implies

$$\lim_{\beta \to \infty} \frac{\beta_R}{\beta} = 1, \tag{24}$$

which agrees with the plotted behavior showing the β_R curve paralleling the 45° dotted line $\beta_R = \beta$ for large β.

Once again, because of the behavior of Δ as $\beta \to 0$, this model probably does not have a sensible continuum limit. For this reason, no attempt has been made to get good values for β_R and Δ as β gets small. The computer effort has instead concentrated on the $O(1,2)$-model.

7. The $O(1,2)$ Sigma Model

The configuration space $O(1,2)/O(2)$ of this model is topologically $\mathbb{R}^2$ and metrically equivalent to a space-like hyperboloid in a Minkowski space of $1+2$ dimensions. The Euclidean action is given by

$$S_E = \frac{1}{2}\mu^2 \int \eta_{ij}\phi_{i,\mu}\phi_{j,\mu}\, d^4x, \qquad \eta_{ij}\phi_i\phi_j = -1, \tag{25}$$

$$(\eta_{ij}) = \mathrm{diag}(1,1,-1), \tag{26}$$

that, upon elimination of ϕ_3, takes the form (17) with

$$G_{ab} = \delta_{ab} - \frac{\phi_a\phi_b}{1+\phi_c\phi_c}, \qquad G = \det(G_{ab}) = \frac{1}{1+\phi_c\phi_c}. \tag{27}$$

The functional-integration measure is

$$G^{1/2}\,d\phi_1 d\phi_2 = d\phi_3 d\theta, \tag{28}$$

where

$$\phi_1 = \sqrt{\phi_3^2 - 1}\cos\theta, \qquad \phi_2 = \sqrt{\phi_3^2 - 1}\sin\theta. \qquad (29)$$

ϕ_3 now ranges from 1 to ∞. Since random numbers cannot be chosen with uniform weight in an infinite interval, one must alter the Monte Carlo simulation for this model. What one does is (effectively) choose θ randomly with uniform weight in the interval $0 \le \theta \le 2\pi$, set $\phi_3 = 1 + |\zeta|$ where ζ is chosen by a random walk starting from zero, and compute ϕ_1 and ϕ_2 from equations (29). The displacements $\Delta\zeta$ in the random walk are chosen with uniform weight in the interval $-\alpha \le \Delta\zeta \le \alpha$, where, for each β, α is selected so as to obtain approximately a 50% rejection rate in the field trials. Note that equations (6) and (9) must now be replaced by

$$S_E = -4(N+1)^4\beta$$
$$- \beta \sum_{\text{links}} \eta_{ij}\phi_i(\text{one end of link})\phi_j(\text{other end of link}), \quad (30)$$
$$\Delta S_E = -\beta \sum_{\substack{\text{neighboring} \\ \text{sites}}} \eta_{ij}\Delta\phi_i(\text{site})\phi_j(\text{neighboring site}), \qquad (31)$$

respectively.

Since the configuration space of the $O(1,2)$-model is noncompact, it is never fully probed by the field at a given site. The field is therefore *always* partly frozen, and the $O(1,2)$-model does not have a phase transition. This is confirmed by computations of the magnetization curve, which is found to be insensitive to N and to have the form:

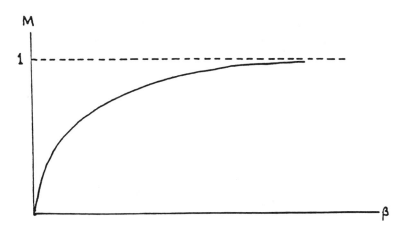

FIGURE 6.

In this case, the magnetization is defined by

$$M \equiv N^4 \left\langle \sum_{\alpha,\beta,\gamma,\delta=1}^{N} \phi_{3\,\alpha\beta\gamma\delta} \right\rangle^{-1} \tag{32}$$

and the boundary conditions at the edge of the "universe" are $\phi_1 = \phi_2 = 0$, $\phi_3 = 1$. Note that an alternative way of saying that this model does not have a phase transition is simply to say $\beta_c = 0$.

It should be noted that the gravitational field too has a noncompact configuration space. The standard boundary condition of asymptotic flatness corresponds to a "seeded" broken symmetry ($g_{\mu\nu} = \delta_{\mu\nu}$) at the edge of the "universe." The $O(1,2)$-model therefore has more relevance for quantum gravity than the other sigma models and is being studied more intensively than the others at the present time.

8. Two-Point Function for the $O(1,2)$ Sigma Model: Renormalization

Equation (23) is found to hold, within the error bars, for the $O(1,2)$-model. Equation (22) is also reasonably well satisfied, although Δ is found to remain small (nearly consistent with zero) for all values of β. The curve for β_R as a function of β now has the form

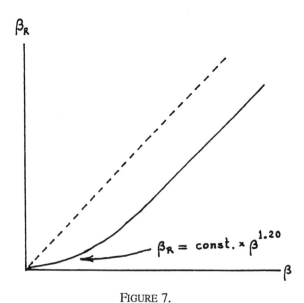

FIGURE 7.

and is fairly insensitive to N. In this case, we find $\beta_R < \beta$ for all β because now, for a given field configuration, the Euclidean action (17), with G_{ab} given by (27), is always less than that of the corresponding free field. The two-point function is therefore less depressed than that of the corresponding free field.

Just as in the case of the $O(3)$-model, when β is large, the field fluctuations around $\phi_1 = \phi_2 = 0$ are small, so that the field does not "feel" the curvature of the configuration space and behaves like a free field. Once again, therefore,

$$\lim_{\beta \to \infty} \frac{\beta_R}{\beta} = 1, \tag{33}$$

which agrees with the figure. For small β, on the other hand, β_R is found very accurately to obey a power law formula of the form

$$\beta_R \xrightarrow[\beta \to \infty]{} \text{const.} \times \beta^{1.20}. \tag{34}$$

This formula has important implications for the continuum limit.

If expression (22), with $\Delta = 0$, is truly an accurate representation of the two-point function, then, in lowest approximation, the *effective* action for the $O(1,2)$-model must have the form

$$\Gamma[\phi] \approx \frac{1}{2}\mu_R^2 \int G_{ab}(\phi)\phi_{a,\mu}\phi_{b,\mu}d^4x \,, \tag{35}$$

where, because of the $O(1,2)$-invariance of the model, the G_{ab} appearing in the integrand is identical with (27) and where

$$\mu_R = \sqrt{\beta_R}/a. \tag{36}$$

μ_R is the *renormalized* scale parameter or "Planck mass" of the theory. If μ_R is to be held fixed, then the continuum limit $a \to 0$ requires us to take $\beta_R \to 0$ and hence $\beta \to 0$. This is the hardest region to reach on the computer. (The error bars get big). However, if we can trust the computer results, then

$$\frac{\mu_R^2}{\mu^2} = \frac{\beta_R}{\beta} \xrightarrow[\beta \to 0]{} \text{const.} \times \beta^{.20} = \text{const.} \times \beta_R^{.167} \xrightarrow[\substack{\beta_R \to 0 \\ a \to 0}]{} 0, \tag{37}$$

which implies that the *bare* scale parameter μ must diverge in the continuum limit:

$$\lim_{a \to \infty} \mu^2 = \infty \quad \text{(for fixed } \mu_R\text{)}. \tag{38}$$

That a bare parameter should diverge is not surprising. The same thing happens with perturbatively renormalizable theories. However, in order to conclude that the theory can make nontrivial sense in the continuum limit, one must check that expression (35) is, in fact, a good approximation to the effective action, at least for long-wavelength field disturbances. A description of how this can be done is given in the next section. First, however, let us examine a little more closely just how the continuum limit is best approached.

The renormalized scale parameter μ_R defines the energy scale at which quantum processes begin to cause the dynamics of the field to depart wildly from the dynamics of the corresponding classical field. In order to probe this energy scale accurately, one must choose the lattice spacing small compared to μ_R^{-1}. On the other hand, the "universe" should be chosen big enough so that the classical realm (long wavelengths) can be reached. This requires L to be chosen large compared to μ_R^{-1}. Thus, we want

$$a \ll \mu_R^{-1} \ll L. \tag{39}$$

Since $L = (N + 1)a$, the best that we can do is to get

$$a = \frac{\mu_R^{-1}}{\sqrt{N + 1}}, \qquad L\sqrt{N + 1}\mu_R^{-1}, \tag{40}$$

$$\beta_R = \mu_R^2 a^2 = \frac{1}{N + 1}. \tag{41}$$

Although other approaches to the continuum limit are possible, equation (41) represents the optimum.

9. The Four-Point Function

Only the quadratic (in ϕ) part of the effective action contributes to the two-point function. In order to check whether expression (35) is a good approximation to Γ in the long-wavelength limit, and hence whether a sensible classical realm exists for the theory, one must examine also the four-point function, and hence the nonlinearities, or field interactions, to which expression (35) leads. Theoretically, the full (renormalized) four-point function is given by

$$\Gamma_{ab'c''d'''} = -\int d^4y \int d^4y' \int d^4y'' \int d^4y''' \Gamma_{ae}(x, y).$$

$$\cdot \Gamma_{bf}(x',y')\Gamma_{cg}(x'',y'')\Gamma_{dh}(x''',y''')$$

$$\times \frac{\delta^4\Gamma}{\delta\phi_e(y)\delta\phi_f(y')\delta\phi_g(y'')\delta\phi_h(y''')}, \tag{42}$$

where $\Gamma_{ae}(x,y)$ is the two-point function. Using (35), one finds, for the functional derivative in the integrand,

$$\frac{\delta^4\Gamma}{\delta\phi_e(y)\delta\phi_f(y')\delta\phi_g(y'')\delta\phi_h(y''')} \approx \mu_R^2\big[\delta_{ef}\delta_{gh}\delta(y,y')\delta(y'',y''')\delta_{,\mu\mu}(y,y'')$$
$$+ \delta_{eg}\delta_{fh}\delta(y,y'')\delta(y',y''')\delta_{,\mu\mu}(y,y') \tag{43}$$
$$+ \delta_{eh}\delta_{fg}\delta(y,y''')\delta(y',y'')\delta_{,\mu\mu}(y,y')\big].$$

Taking the box-mode-function transform of eq. (42), approximating the two-point functions in the integrand by expression (22) with $\Delta = 0$, and remembering that

$$\Gamma_{ab'c''d'''} = \langle\phi_a(x)\phi_b(x')\phi_c(x'')\phi_d(x''')\rangle$$
$$- \langle\phi_a(x)\phi_b(x')\rangle\langle\phi_c(x'')\phi_d(x''')\rangle$$
$$- \langle\phi_a(x)\phi_c(x'')\rangle\langle\phi_b(x')\phi_d(x''')\rangle \tag{44}$$
$$- \langle\phi_a(x)\phi_d(x''')\rangle\langle\phi_b(x')\phi_c(x'')\rangle,$$

one obtains, after some calculation,

$$\frac{\langle\tilde{\phi}_{1\,1111}^{\,4}\rangle - 3\langle\tilde{\phi}_{1\,1111}^{\,2}\rangle^2}{\langle\tilde{\phi}_{1\,1111}^{\,2}\rangle^2} = \frac{\langle\tilde{\phi}_{2\,1111}^{\,4}\rangle - 3\langle\tilde{\phi}_{2\,1111}^{\,2}\rangle^2}{\langle\tilde{\phi}_{2\,1111}^{\,2}\rangle^2}$$
$$= 3\frac{\langle\tilde{\phi}_{1\,1111}^{\,2}\tilde{\phi}_{2\,1111}^{\,2}\rangle - \langle\tilde{\phi}_{1\,1111}^{\,2}\rangle\langle\tilde{\phi}_{2\,1111}^{\,2}\rangle^2}{\langle\tilde{\phi}_{1\,1111}^{\,2}\rangle\langle\tilde{\phi}_{2\,1111}^{\,2}\rangle}$$
$$= \frac{81}{(4\pi)^2(N+1)^2}\frac{1}{\beta_R}. \tag{45}$$

The reason for choosing here the box-mode-function transforms with $k = l = m = n = 1$ is to test expression (35) at the longest wavelengths available.

The computer results for the above averages are readily obtained although the error bars increase in size for small β_R. The curves that one gets have the forms shown in Figure 8. Not only do the computer curves show negative rather than positive values for the four-point function, but also the magnitudes of these values are much smaller than that of expression (45). Clearly, therefore, expression (35) yields inconsistent results.

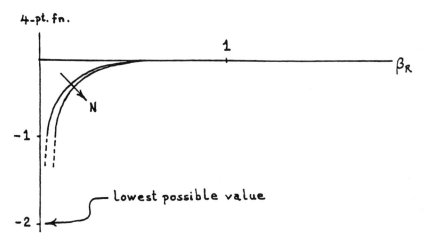

FIGURE 8.

This does not yet mean that a consistent continuum quantum theory does not exist because an assumption has been made in the above analysis that is not valid. *The long-wavelength, or low-momentum, region cannot actually be reached even with modern supercomputers.* In the present context, "low-momentum" means

$$p^2 \ll \mu_R^2 \quad \text{or} \quad \frac{\pi^2}{L^2} \ll \frac{\beta_R}{a^2} \quad \text{or} \quad \frac{\pi^2}{(N+1)^2} \ll \beta_R. \qquad (46)$$

With β_R at the optimum value of $1/(N+1)$ [see eq. (41)], this implies

$$\frac{\pi^2}{N+1} \ll 1. \qquad (47)$$

But, as a practical matter, N cannot be chosen appreciably greater than 10. We are therefore forced to work directly in the critical energy region $p^2 \sim \mu_R^2$.

Plans are currently underway to replace the *Ansatz* (22) for the two-point function by

$$\langle \tilde{\phi}_{1\, klmn}^2 \rangle$$
$$= \frac{1}{\beta_R} \frac{1}{k^2 + l^2 + m^2 + n^2 + \alpha \frac{\pi^2}{(N+1)^2 \beta_R}(k^2 + l^2 + m^2 + n^2)^2}, \qquad (48)$$

and to determine β_R and α by a least-squares best fit. The *Ansatz* (48) corresponds to giving the two-point function a momentum dependence of the form

$$\frac{1}{\mu_R^2}\frac{1}{p^2 + \alpha\mu_R^{-2}p^4},\tag{49}$$

and to correcting the approximation (35) for the effective action by the addition of a fourth-order (in derivatives) term. $O(1,2)$ invariance implies that the most general fourth-order effective action consistent with (49) must have the form

$$\begin{aligned}
\Gamma \approx{}& \frac{1}{2}\mu_R^2 \int G_{ab}(\phi)\phi_{a,\mu}\phi_{b,\mu}\,d^4x \\
&+ \frac{1}{2}\alpha \int G_{ab}(\phi)\big[\phi_{a,\mu\mu} + \Gamma^a{}_{cd}(\phi)\phi_{c,\mu}\phi_{d,\mu}\big] \\
&\quad\times \big[\phi_{b,\nu\nu} + \Gamma^b{}_{ef}(\phi)\phi_{e,\nu}\phi_{f,\nu}\big]\,d^4x \\
&+ \frac{1}{8}\gamma \int \big[G_{ab}(\phi)\phi_{a,\mu}\phi_{b,\mu}\big]^2 d^4x \\
&+ \frac{1}{4}\delta \int G_{ab}(\phi)\phi_{a,\mu}\phi_{b,\nu}G_{cd}(\phi)\phi_{c,\mu}\phi_{d,\nu}\,d^4x,
\end{aligned}\tag{50}$$

where $\Gamma^a{}_{bc}$ is the Riemannian connection corresponding to the metric G_{ab}.

With the coefficient α determined by the least-squares best fit to the two-point function, one will compare the computer-generated four-point function with the four-point function yielded by expression (50), and see if consistency can be achieved. There are two remaining parameters, the coefficients γ and δ, that can be adjusted in order to secure consistency. Since there are two adjustable parameters, a good consistency check will require an examination of the momentum dependence of the four-point function.

If consistency can be achieved, this will constitute strong evidence (the first to date) that a theory that is not perturbatively renormalizable can nevertheless exist and have a sensible and nontrivial classical regime. The results of this extended analysis will be announced when they emerge.

Acknowledgment. This work is supported in part by grants from the U.S. National Science Foundation and the Pittsburgh supercomputer center.

REFERENCES

Deser, S. (1957). "General Relativity and the Divergence Problem in Quantum Field Theory." *Reviews of Modern Physics* 29: 417–423.

DeWitt, Bryce S. (1964). "Gravity: A Universal Regulator?" *Physical Review Letters* 13: 114–118.

Isham, Christopher J., Salam Abdus, and Strathdee, J. (1971). "Infinity Suppression in Gravity-Modified Quantum Electrodynamics." *Physical Review D* 3: 1805–1817.

———(1972). "Infinity Suppression in Gravity-Modified Electrodynamics. II." *Physical Review D* 5: 2548–2565.

Klein, Oskar (1955). "Quantum Theory and Relativity." In *Niels Bohr and the Development of Physics*. Wolfgang Pauli, Leon Rosenfeld, and Victor Weisskopf, eds. London: Pergamon, 96–117.

Landau, Lev D. (1955). "On the Quantum Theory of Fields." In *Niels Bohr and the Development of Physics*. Wolfgang Pauli, Leon Rosenfeld, and Victor Weisskopf, eds. London: Pergamon, 52–69.

Pauli, Wolfgang (1956). Discussion following: Oskar Klein "Generalizations of Einstein's Theory of Gravitation Considered from the Point of View of Quantum Field Theory." In *Fünfzig Jahre Relativitätstheorie. Bern, 11–16. Juli 1955. (Helvetica Physica Acta. Supplementum* 4.) André Mercier and Michel Kervaire, eds. Basel: Birkhauser, 69–75.

Discussion

ISHAM: Bryce, can you tell whether the results are independent of coordinates? Can you actually do that calculation in a coordinate-independent fashion?

DEWITT: I fear that what we've done here is definitely coordinate dependent. I may redo this calculation using Riemann-normal coordinates. But then, suppose I do an $O(1, 2)$-transformation. I could transform my state, which means that I could transform my boundary conditions. And I'm not quite sure about the structure of the effective action under those circumstances, with the given boundary conditions.

ROVELLI: In Yang–Mills theory, one can explicitly integrate over part of this freedom. I imagine that this procedure doesn't work here. Did you try it explicitly?

DEWITT: We have not tried anything but what I've already spoken about. I'm very open to suggestions. I've tried to speak to the experts on this subject. I had a nice talk with Kogut, who is trying to determine whether there is a phase transition in ordinary QED. He says that the field is wide open. He has no words of wisdom. He doesn't believe that just because a theory is perturbatively non-renormalizable, it doesn't exist. Now we've no evidence yet that the theory doesn't exist. We obtain simple behavior, apparently, as β goes to zero. If such behavior continues—and we've no reason to suspect that it won't—into the non-linear region, nothing weird should happen. The four-point function is not yet consistent. If we can make it consistent, then I would say: "It looks like we have a consistent nonlinear σ model in four dimensions." And there would be all the more reason to hope that one might have a consistent quantum gravity. If this doesn't pan out, well then I would say: "In that case I have to appeal back to the light cone smearing." And maybe that will help out.

SMOLIN: Just two technical questions. Do you have enough accuracy to measure the momentum dependence of the four-point function?

DEWITT: Yes. George DeLyra had already done that a bit. But the four-point function he was using in his thesis is not really the right one. I'm not even sure we are using the right one, i.e., the right variables. That is Chris Isham's question.

SMOLIN: The second question: Could you make a high-temperature expansion just by hand to check the results?

DEWITT: Not here, because the internal space is noncompact, so you can't make the expansion that is sometimes made for the $O(3)$-model.

GELL-MANN: Can you treat this theory as a restricted theory? I mean, say you have two scalar fields, and the sum of the squares is one.

DEWITT: That's the way it started.

GELL-MANN: Right. So, what if you integrate over the two directions and construct the normals to the areas, but now you do it with a computer procedure that puts in this restriction. Are there any theorems that appear, or anything like that, that relate the restricted theory to the unrestricted one?

DEWITT: You can prove that it's identical, i.e., identical on the computer, to what we're doing.

GELL-MANN: No, no, I didn't mean that. What I mean is, is there a theorem that would compare the nonlinear theory of the restricted variety to the unrestricted one, i.e., to the plain free theory, because one of them is simply the same integral with a different range?

DEWITT: The only theorems I know of so far are just the simple-minded ones that β_R is either greater than β or smaller than β, depending on comparison of the action for a given field to the other.

GELL-MANN: The hope that these theories might make sense, contrary to the contemporary picture, is based on the idea that there would be a restricted version of the theory that did make sense. And since all you are doing is restricting your range, that is nothing horrible.

DEWITT: In other words, put a "tight" potential around that range.

GELL-MANN: I don't know if it leads anywhere or not, but we conjectured that as a possibility back in 1959, and I think somebody explored whether it made sense or not.

DESER: Of course, Murray Gell-Mann is one of the people guilty of starting this.

GELL-MANN: Yes.

ANDERSON: I've a couple of remarks about the relation between path integrals, perturbation theory, and the numerical results that you're actually exploring. If you did a continuum path integral in the semi-classical limit, you've chosen boundary conditions so that you're doing a configuration evolution between spins, all lined up. The vacuum state for your full theory need not necessarily have only one maximum.

DEWITT: That is a good point.

ANDERSON: In addition to having selected particular configurations, you have the possibility of more than one evolution connecting your configurations. Moreover, you've chosen boundary conditions that compactify back, so that there are clearly two evolutions. Now the question arises: when I go to the discrete version, where I clearly can go continuously between the two because there is not a homotopic classification in the discrete model, can you really have preserved that with your Metropolis method of choosing the configurations? Because the configurations in the other homotopy class will, in some sense, be a long way away over a very high barrier that increases as the numbers of points in your lattice goes up.

DEWITT: If it is high barrier, the computer certainly isn't going to pick it up.

ANDERSON: So you may have put yourself in the region of the one evolution. This is the trivial evolution that has all spins lined up initially, which is good if you're comparing to perturbation theory around the trivial evolution. But if you believe in nonperturbative effects, you want contributions...

DEWITT: Well, hold on, I'm not convinced that including other topological possibilities is necessary in order to get away from perturbative analysis. You've already gotten away by probing the internal space.

ANDERSON: I'm not convinced that the structure that one is probing in perturbation theory is the nonperturbative structure, as seen by the effect of the higher potential terms. If I do a Riemann-normal coordinate expansion, I have a free theory plus some potential, instead of doing it as you've done. Then I see the other evolutions most strongly, in a semiclassical approximation, as being other histories that get enormously large contributions. And I don't see them as small perturbations around the trivial evolution. So one would hope that you're trying to include this. But if your Metropolis method constrains you too strongly to the trivial evolution, you may only see perturbation theory. And you won't see the modification of perturbation theory that you did in numerical relativity.

DEWITT: Well, already we don't see perturbation theory here because we get finite results on the computer.

ASHTEKAR: Let me go back to this question of what we can say about gravity now, from here. Let me just understand, at a very naive level, what the analogy is. Here you said that $O(2, 1)$ was noncompact;

by that did you mean that the configuration space was noncompact?

DEWITT: That's right.

ASHTEKAR: And the analogy is that, in gravity, the configuration space is noncompact. You're comparing the metric to the Wheeler–DeWitt metric in the gravity case when you say that...

DEWITT: Actually, there isn't a metric in the gravity case, is there? That's too strong a statement, but there's something like a curved internal space.

ASHTEKAR: That's my question. Then what is curved?

DEWITT: Well, there is also a factor that multiplies the two spatial derivatives, and that's not part of your internal space. You take the time–time part, then... alright, you know that...

ASHTEKAR: Can you do something like what can be done in gravity? There is a strong coupling expansion that one can work with. Is there anything analogous to that? In other words, you start out doing this strong coupling expansion of the field itself.

DEWTT: Well, I don't see any easy way, but maybe there is.

ASHTEKAR: In gravity, the first few things can be almost trivially worked out analytically. So the question is: what happens in this case?

DEWITT: Well of course this doesn't break neatly. There is no Hamiltonian constraint. It doesn't break neatly into a kinetic part and a curvature part, so we don't have those things.

DESER: You gave a table of analogies and presumably all this is partly to get...

DEWITT: Just to get experience, for one thing...

DESER: ...with a non-renormalizable model.

ZUREK: Two related questions: How do you start the system? How do you choose the initial configuration?

DEWITT: Just start it, all lined up: $\phi_1 = 0$, $\phi_2 = 0$, $\phi_3 = 1$ over the whole lattice. Just start rolling the dice and pretty soon it's away from that. Anyway, we throw away the first 2,000 or so sweeps.

ZUREK: This was a question that was raised earlier. If you try starting it in some awkward configuration; maybe you could end up in different...

DEWITT: Yes, if you can give me a map, an actual explicit map from S^4 to S^2, I'll try to mimic it on the computer. We can then see what happens.

ZUREK: The other, related question: if you pull the system away from where it's fluctuating, how many sweeps does it take to get back to where it was? If you just brute-force it away? Sort of a relaxation time.

DEWITT: Well, typically the relaxation time is pretty fast here; even after the first 2,000 sweeps, the quantities don't differ that much from the succeeding one, but we throw them away anyway.

ZUREK: What about if you approach the critical temperature?

DEWITT: There is no critical temperature in this model. It's at infinity, or $\beta = 0$.

JACOBSON: What was the reason you gave, or did you give it, as to why you think the $O(3)$ and the $O(2)$-models don't agree?

DEWITT: I didn't discuss this, I don't even have it prepared. I forget the details. You can just see that the $O(3)$-model is trivial in the chaotic limit. Your action is essentially zero in the continuum limit. The alternative is to drive everything to the critical point.

HARTLE: A few years ago, there were some ideas by Weinberg that some theories might be asymptotically safe, even if they were non-renormalizable. Does this fall into that class?

DESER: That was $R + R^2$.

DEWITT: Well, no. He was really saying that you need an infinity of counter terms and an infinity of β functions, and maybe they could all somehow conspire. I don't think I want to try facing...

DESER: That was a last hope that a lot of us just waved our hands with.

ANDERSON: If the theory were not to have existed, what would you have expected to see?

DEWITT: It may still not exist, if we cannot get consistency between the two-point and four-point functions. I would say that something's wrong here. I'm not quite sure what.

PART V

Role of Topology and Black Holes in Quantum Gravity

Space-Time Topology and Quantum Gravity

John L. Friedman

1. Summary

Characteristic features are discussed of a theory of quantum gravity that allows space-time with a non-Euclidean topology. The review begins with a summary of the manifolds that can occur as classical vacuum space-times and as space-times with positive energy. Local structures with non-Euclidean topology—topological geons—collapse, and one may conjecture that in asymptotically-flat space-times non-Euclidean topology is hidden from view. In the quantum theory, large diffeos (diffeos not isotopic to the identity) can act nontrivially on the space of states, leading to state vectors that transform as representations of the corresponding symmetry group, $\pi_0(\text{Diff})$. In particular, in a quantum theory that, at energies $E < E_{\text{Planck}}$, is a theory of the metric alone, there appear to be ground states with half-integral spin and, in higher-dimensional gravity, with the kinematical quantum numbers of fundamental fermions. The possibility of topology change is considered in the context of a path integral approach; requiring a Lorentz metric on an n-dimensional space-time is shown to provide no additional restrictions on the possible transitions from one three-manifold to another. A path integral can lead to nontrivial representations of $\pi_0(\text{Diff})$ only if diffeos of the final three-surface are not extendible to diffeos of the space-time, and inextendibility is shown for space-times that correspond to the creation of pairs of topological geons. Finally, some objections are outlined to a theory that allows changing topology.

2. Introduction

> "The Riemann point of view allows, also for real space, topological conditions entirely different from those realized by Euclidean space."[1]— H. Weyl

There is little reason to believe that on either the largest scale or the smallest, the topology of space-time is Euclidean. On the largest scale, countably many closed three-manifolds are consistent with the isotropy and homogeneity of either the hyperbolic or the spherical FRW models and with deSitter (Ellis 1971)[2], and without a constraint of global homogeneity, we now know that there are also no constraints on the spatial topology of either vacuum space-times or space-times with everywhere positive energy (Witt 1986a). On scales of the order of the Planck length, if one can speak of a space-time manifold at all, the path-integral approach to quantum gravity suggests that the microscopic topology of space-time, as well as its geometry, will fluctuate (Wheeler 1962, 1964, 1968; Hawking 1978), because smooth Lorentzian (and Euclidean) four-geometries interpolate between space-like three-geometries with different spatial topologies. Curiously, if ours is a child universe, born from the inflation of a quantum fluctuation in an older space-time[3] (or if, as in the Hartle–Hawking (1983) picture, our past is simply a compact space-time), then our large-scale topology preserves an initial choice made by the microscopic child.

It is particularly difficult to dismiss the possibility of microscopic non-Euclidean topology when the most attractive way to unify gravity and the fundamental gauge bosons is to suppose that our universe has small "internal" dimensions. Again their topology—and the corresponding symmetry of their geometry—presumably reflect early choices. Although most of the presentation is in a four-dimensional context, generalizations to higher-dimensional space-times (with Kaluza–Klein asymptotics) will be mentioned.

It will be helpful before talking about quantum gravity to begin with a discussion of topology in the context of the classical theory and with a brief characterization of three-manifolds.

3. Topology in Classical Relativity

3.1. WHAT MANIFOLDS DOES GRAVITY ALLOW?

If one requires a space-time M to have a Cauchy surface, a compact, space-like hypersurface whose domain of dependence is M itself, then a straightforward proof by Geroch (1967) shows that M must be a product manifold of the form $S \times R$, (space $\times$ time) where S is a three-manifold. Even without the requirement of causality, Tipler (1977) has shown for almost all space-times satisfying a positive-energy condition ($R_{ab}u^a u^b \geq 0$, all null u^a) that $M \simeq S \times R$. It is not difficult to extend the theorems to higher-dimensional gravity (Tipler 1985), or to asymptotically-flat space-times.

Then for classical, vacuum space-times, or space-times with sources whose energy density is locally positive, one would like to know what three-manifolds S occur. As Witt (1986b) has recently shown, the answer is that they all do: there are vacuum space-times (and space-times with everywhere-positive energy) that have arbitrary spatial topology. Proving the results for asymptotically-flat space-times is more difficult than for the spatially closed case. For closed space-times, the result follows quickly from an earlier theorem of Kazden and Warner (1975), and I shall recapitulate its proof for vacuum space-times. Because every vacuum initial data set has a finite time evolution, the problem is to show that one can satisfy the vacuum constraint equations (Hawking and Ellis 1973).

Theorem (Witt 1986a, b). *Every closed three-manifold has vacuum initial data.*

Proof. Let Σ be a closed three-manifold. A vacuum initial data set on Σ is a pair (g_{ab}, K_{ab}), with g_{ab} a positive-definite three-metric, and K_{ab} a symmetric tensor satisfying the momentum constraint:

$$\nabla_b(K^{ab} - g^{ab}K) = 0, \qquad (3.1)$$

and the Hamiltonian constraint:

$$R - K_{ab}K^{ab} + K^2 = 0. \qquad (3.2)$$

(K_{ab} will be the extrinsic curvature of Σ in the space-time evolved from the initial data; $K = K^c{}_c$). Given any function R on Σ that is not everywhere positive, the Kazden–Warner theorem (1975) states that there is a three-metric g_{ab} whose curvature scalar is R. Pick g_{ab} with $R = -6$, and set $K_{ab} = g_{ab}$. The constraint equations immediately follow. $\square$

In the asymptotically-flat case, one considers three-manifolds obtained by removing a point from a closed manifold (the point represents spatial infinity). The theorem establishes the existence of asympototically-flat, vacuum initial data on any such space (Witt 1986b). Previous work (Schoen and Yau 1979a, b; Yau 1980; Geroch 1978; Friedman and Mayer 1982) had shown the existence of vacuum initial data on the comparatively small set of asympototically-flat space-times whose prime factors are spherical spaces or handles (the terms are defined in (c) below); and Brill (1983) had given an example of an asympototically-flat space-time with positive energy whose topology is $T^3 - \{P\}$, where $T^3 \simeq S^1 \times S^1 \times S^1$ is the three-torus.

3.2. A BRIEF DISCUSSION OF THREE-MANIFOLDS

Although no complete classification of three-manifolds yet exists (Seifert and Threlfall 1980), there is a fairly simple description of a large class (Thurston 1982), and a less transparent construction that yields all three-manifolds (but does not tell how to count each inequivalent manifold once).

   With the recent activity in string theory, two-manifolds are now much more familiar to physicists, and the geometrical construction of two-manifolds is also a key to the much more interesting three-dimensional case. Recall that every closed two-manifold is a sphere with handles and crosscaps attached—and every orientable two-manifold is a sphere with handles. The attachment of a handle is an example of a connected sum. A connected sum of two n-manifolds M_1 and M_2 is a manifold $M_1 \# M_2$ obtained by removing from each an n-dimensional ball and gluing the resulting manifolds-with-boundary along their bounding spherical surfaces. A plane with two handles is the connected sum $R^2 \# S^1 \# S^1$ of the plane with two copies of the torus. The terms in the sum are called factors of the topology. A closed manifold M is prime if $M \simeq M_1 \# M_2$ implies M_1 or M_2 is an n-sphere.

   In the late 19th century, work by Poincaré, Schwarz, Fuchs and Klein led to the result that every orientable two-manifold except the torus and the sphere is obtained as a quotient space of the hyperbolic plane H by a discrete group of isometries (Milnor 1982). H is the hyperboloid of unit time-like vectors in three-dimensional Minkowski space. The metric induced on H has the familiar form

$$ds^2 = d\chi^2 + \sinh^2 \chi \, d\phi^2,$$

and its isometry group is the three-dimensional Lorentz group. Thus, every orientable two-manifold except S^2 and T^2 is diffeomorphic to a quotient H/G, where G is a discrete subgroup of the three-dimensional Lorentz group. The torus is simpler, being a quotient space of the Euclidean plane E by translations in two orthogonal directions.

One forms the quotient space of a manifold M by a group G that acts on M by identifying all points in an orbit: a point of M/G is an orbit of G. If G is a discrete group acting freely on M, there is a unit cell C, whose images under G tile the space M. Each orbit of G contains either one interior point of C or two or more boundary points of C. Thus one can construct M/G from C by identifying boundary points in the same orbit. For the torus, G is generated by unit translations in the x or y direction. The unit cell is a square and the group action maps the left boundary to the right boundary, the bottom to the top. The torus is then presented in the familiar form of a square with edges identified as in Figure 1. The second diagram gives the analogous instructions for constructing a surface of genus two from an octagon by sewing together edges labeled by the same letter.

In three-dimensions there are a countably infinite number of prime manifolds in contrast to the two (crosscap and handle) prime surfaces. Nevertheless, with a natural way of counting, most three-manifolds can be built in an analogous way by identifying faces of a polyhedron serving as the unit cell for a group of isometries of a symmetric space.

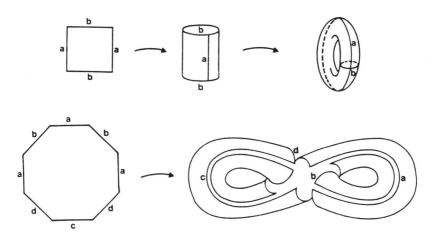

FIGURE 1.

First, it is worth mentioning that there is a single, somewhat different construction that produces every three-manifold. The discovery that one gets all three-manifolds is due to Lickorish (1962, 1963), and the construction to Dehn. One simply removes solid tori from S^3 and then sews them back in. The subtlety is in the sewing instructions: one must sew some closed curve on the surface of the inner tube to a homotopically different curve on the surface of the outer tube. A diffeo of the two-torus to itself that takes a closed curve to a homotopically inequivalent closed curve is not deformable to the identity. If we call two diffeos equivalent if one is deformable to the other, then the equivalence classes of diffeos form a group, the zero-th homotopy group of the diffeos. Here one gets inequivalent diffeos by cutting the torus vertically or horizontally, twisting the resulting cylinder n times and sewing it back together. There is one inequivalent diffeo for every relatively prime pair of numbers (m, n). (To get nonorientable three-manifolds, one must end the construction by taking the connected sum with a nonorientable handle $S^1 \times S^2$.)

The difficulty in classifying three-manifolds is that the tori removed from S^3 can be knotted. What is remarkable is that, regardless of the knotting or of how many tori you remove, for all but a finite number of the sewing instructions, the resulting three-manifold can, as in the two-dimensional case, be obtained as a quotient H/G of hyperbolic space by a discrete group of Lorentz transformations (Thurston 1982). Thus, in this way of counting, almost all closed three-manifolds are hyperbolic.

In addition to these hyperbolic spaces, there are ten flat spaces, obtained by discrete subgroups of the Euclidean group acting on the flat plane, and the spherical spaces, obtained by finite subgroups of the four-dimensional rotation group acting on S^3, To summarize the terminology:

H/G	$G \subset$ Lorentz Group	Hyperbolic Spaces
E/G	$G \subset$ Euclidean Group	Flat Spaces
S^3/G	$G \subset SO(4)$	Spherical Spaces

The unit cells of a discrete group action in three-dimensions are polyhedra, and each corresponding three-manifold can be constructed from a polyhedron by identifying faces. Figure 2 shows some examples of spherical spaces with $G \subset SU(2) \subset SO(4)$. ($SU(2)$ acts naturally on $S^3 \simeq SU(2)$ by left multiplication.)

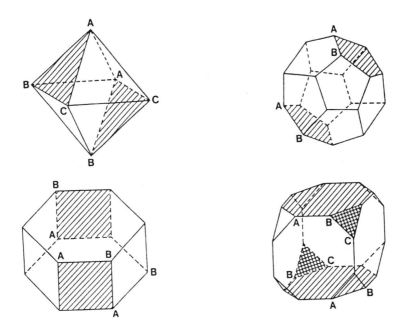

FIGURE 2.

Finally, there are circle bundles over any two-manifold and two-manifold bundles over a circle: the best known three-manifolds, the handles, are of this type, and they again are obtained from a homogeneous covering space by a finite group of isometries.

Witt's theorems, together with previous results of Gromov and Lawson (1983) (see also earlier work of Shoen and Yau (1979b)) imply that, if one uses Thurston's counting, almost no manifolds admit metrics with a maximal slice (a space-like hypersurface Σ for which $K = 0$). That is, the Gromov–Lawson work shows that no hyperbolic space admits a metric with $R \geq 0$, while the Hamiltonian constraint with $K = 0$ and positive energy implies $R \geq 0$. Related, recent work of Eardley and Witt (1990) shows the existence of space-times with no $K = \text{constant}$ slices. Here, however, it is only the particular metrics that fail to have space-like hypersurfaces with constant K; each manifold of the form $\Sigma \times \mathbb{R}$ (or $[\Sigma - \{i_0\}] \times \mathbb{R}$), with Σ compact, admits a Lorentz metric having $K = \text{constant}$ slices.

3.3. GANNON'S THEOREM AND A
TOPOLOGICAL CENSORSHIP CONJECTURE

If all three-manifolds occur as solutions to the vacuum field equations, one might ask why spatial structures with non-Euclidean topology are not a part of our ordinary experience. The reason is that on macroscopic scales topological structures collapse, and this is the content of a theorem due to Dennis Gannon. Only topological structures comparable in size to the visible universe (or small enough that quantum effects play a crucial role) can survive from the big bang to the present. As will be discussed in Section 5, one can conceive of stabilizing local regions with non-Euclidean topology by using a quantum field to violate the positive-energy condition, but large structures require large energies.

In 1975, Gannon used techniques developed by Penrose, Hawking and Geroch to give a relatively simple proof that every asympototically-flat three-geometry with non-Euclidean topology has a singular time evolution.[4] Let $(M, g_{\alpha\beta})$ be a space-time that satisfies the weak energy condition and has a nonsimply-connected Cauchy surface that is asympototically-flat at spatial infinity. Then M is geodesically incomplete.

The term "topological geon" will be used for a localized configuration with non-Euclidean topology (Wheeler's term "geon" (1962, 1964, 1968), short for "geometrical entity," is used for topological configurations by Finkelstein and Misner (1959). Then Gannon's theorem can be interpreted as showing that in the classical theory, there can be no stationary topological geons. If one assumes cosmic censorship, the theorem implies that in classical relativity, topological geons collapse, forming black holes; and black holes are thus the only stationary geons of classical gravity.

I am aware of only a few examples of asympototically-flat space-times, or initial data sets with positive energy, on three-manifolds other than $\mathbb{R}^3$. In these examples, the topology is shrouded: an experimenter cannot determine that the manifold has non-Euclidean topology and report her results to a distant observer. This suggests that a form of topological censorship may conceivably exist, forbidding null or time-like curves that start and end in asymptopia and traverse a homotopically nontrivial path. To state a precise conjecture, let $(M, g_{\alpha\beta})$ be an asympototically-flat space-time and $\eta_{\alpha\beta}$ a flat metric that agrees asymptotically with $g_{\alpha\beta}$. Call N a constant neighborhood of spatial infinity in M if N is a region of the form $r > R$, where r is a radical coordinate of the flat metric. One may then rephrase as follows a

Topological censorship conjecture: *Let* $(M, g_{\alpha\beta})$ *be a space-time that satisfies the weak energy condition and has a nonsimply-connected Cauchy surface with asympototically-flat initial data. Then there is a constant neighborhood N of spatial infinity in M, for which every time-like or null curve that starts and ends in N is homotopic to 1 rel N.*

A related conjecture, likely to be simpler to decide, is that apparent horizons always isolate regions with non-Euclidean topology on hyper-surfaces with asympototically-flat initial data.

3.4. LARGE-SCALE TOPOLOGY

"It is true that locally hyperbolic space is not necessarily open. It is possible to construct with such space polytrops, i.e. Klein's forms of finite volume. This is true even for Euclidean space."
— G. Lemaître

The remarkable result that most three-manifolds allow homogeneous, iso-tropic metrics means that the observed matter distribution is consistent with most global topologies, even under the assumption that we are in a typical region of the full universe. In particular, as Ellis emphasized (1971), there is no reason to conclude that a matter density below closing density means an infinite space, or that the radius of a finite universe need be comparable to the radius of curvature. A small spherical or hyperbolic universe would mean that one is seeing the same galaxies over and over again at successively earlier times. It turns out to be surprisingly difficult even to decide whether the universe is larger than 1/10 its apparent radius: Gott (1980)[5] places an observational limit $R_H > 400 \ (H_0/50 \, \text{km/s})^{-1}$ Mpc from the absence of multiple images of nearby galaxies, and possible observational tests have been mentioned recently by a number of authors (including Ellis and Schreiber (1986), Fang and Sato (1985), Bernstein and Shvartsmann (1980)).[6] The best hope for verifying a small universe would be the observation of a repeated large feature, such as a void or a filamentary structure of galaxies.

4. Quantum Kinematics: Small and Large Diffeos

4.1. THE MEANING OF THE MOMENTUM CONSTRAINT

In the canonical approach to quantum gravity, invariance under spatial diffeomorphisms is enforced by the momentum constraint, (3.1), which can be written in the form

$$\nabla_b \pi^{ab} = 0, \tag{4.1}$$

where $\pi^{ab} = -\frac{1}{16\pi} g^{1/2}(K^{ab} - g^{ab}K)$ is the momentum conjugate to g_{ab}.

In classical gravity, two metrics are physically equivalent if they are related by a diffeomorphism. The classical momentum constraint can be regarded as a function on phase-space, with g_{ab} and π_{ab} as configuration and momentum space variables, respectively. Any function on phase-space generates canonical transformations, and the transformations associated with the momentum constraint are spatial diffeomorphisms. The quantum version of the constraint is analogous (Higgs 1958), but not identical.

For a family of diffeos χ_λ, whose orbits $\lambda \to \chi_\lambda(p)$ are tangent to a family ξ_λ of vector fields, (4.1) implies that for any vector ξ^a that vanishes at spatial infinity, the Dirac-ADM momentum associated with ξ vanishes:

$$\mathbb{P}_\xi = \int d\tau \, \mathcal{L}_\xi g_{bc} \pi^{bc} = 0. \tag{4.2}$$

The operator $\mathbb{P}_\xi$ is the generator of the diffeos χ_λ. That is, the canonical commutation relation,

$$[g_{ab}, \pi^{cd}] = i\delta^c_{(a}\delta^d_{b)},$$

implies that

$$[\mathbb{P}_\xi, \pi^{ab}] = \frac{1}{i}\mathcal{L}_\xi \pi^{ab} = \frac{1}{i}\frac{d}{d\lambda}\chi_\lambda \pi^{ab},$$

$$[\mathbb{P}_\xi, g_{ab}] = \frac{1}{i}\mathcal{L}_\xi g_{ab} = \frac{1}{i}\frac{d}{d\lambda}\chi_\lambda g_{ab}.$$

In particular, for a wave function $\Psi(g)$,

$$\mathbb{P}_\xi \Psi(g) = \frac{1}{i}\frac{d}{d\lambda}\Psi(\chi_\lambda, g),$$

and $\mathbf{P}_\xi = 0$ implies that the state space is invariant under asymptotically trivial diffeos that are isotopic to the identity (deformable to the identity through a continuous family of diffeos).

Diffeos that do not leave the state vectors invariant thus belong to two classes: those that asymptotically correspond to elements of the symmetry group at spatial infinity (e.g., rotations and translations), and those that, while trivial at infinity, are not isotopic to the identity. The change in a state vector under diffeos in the first class determines the momentum and angular momentum of the state. Elements in the second class give rise to an analogue in quantum gravity of the Yang–Mills θ-states, states that transform as irreducible representations of the group of gauge transformations that are not deformable to the identity (Friedman and Sorkin 1980, 1982; Isham 1981, 1982).

The θ-states of gravity are representations of the group $\pi_0(\mathrm{Diff})$ of disconnected components of the group Diff of asymptotically trivial diffeos. Diffeos that approach the identity at infinity can be regarded as fixing (at least) a point and a frame of the compact manifold obtained by adding a point at infinity. The amount of structure that is to be fixed depends on the precise definition of asymptotic flatness at spatial infinity, but because the space Diff_F of diffeos that fix a frame at x has the same homotopy type as the space of diffeos that fix an entire disk about the point x, the group of inequivalent diffeos is isomorphic to $\pi_0(\mathrm{Diff}_F)$, independent of details of asymptotic flatness.

4.2. LARGE DIFFEOS

The geometrical meaning of large diffeos—diffeos that are not isotopic to the identity—is often easy to picture if one uses the construction, outlined in Part 3.2 above, of three-manifolds from polyhedra. It is again helpful to begin with two-manifolds. A handle on a plane can be constructed by removing a square from the plane and identifying opposite edges as indicated in Figures 3a and 3b. The dashed line depicts a curve γ that is not homotopic to the identity in $\pi_1(M, i_0)$, where i_0 is again the point at spatial infinity. A diffeo χ that exchanges the two ends of the handle is shown in Figures 4a and 4b. Because the image $\chi \circ \gamma$ of the curve is not homotopic to γ, χ itself is not isotopic to the identity. If we regard γ as a geodesic of a metric g_{ab} on M, then $\chi \circ \gamma$ is a geodesic of the inequivalent metric $\chi\, g_{ab}$. When the manifold is represented, as in Figures 3a and 4a by identifications of a removed polygon, symmetries of the polygon correspond to large diffeos—here, diffeos that rotate the square by an angle $n\pi/2$, and which rotate points a distance r away from the square by an angle that goes to zero as $r \to \infty$. In particular, a rotation by 2π corresponds to a nontrivial diffeo, and the large diffeos thus include the covering group in $\overline{SO(2)}$ of the symmetries of the square, regarded as elements of $SO(2)$.

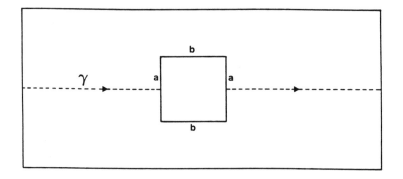

FIGURE 3a

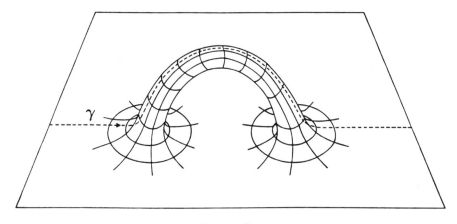

FIGURE 3b

Diffeos of prime three-manifolds commonly have a similar character: The group of inequivalent diffeos is a double covering of symmetries of the polyhedron from which the manifold is constructed (Friedman and Witt 1986; Witt 1986b). The group is a double covering because the diffeo $\mathcal{R}$ corresponding to a 2π rotation is not isotopic to the identity: Figure 5 shows the result of the diffeo $\mathcal{R}$ on the asympototically-flat Poincaré dodecahedron space. The diffeo $\mathcal{R}$ is associated with the ADM-Dirac angular momentum operators by $\exp(2\pi i J_z)\Psi(g) = \Psi(\mathcal{R}g)$. When $\mathcal{R}$ is not isotopic to the identity, the space of state vectors satisfying the momentum constraint includes states with half-integral angular momentum. In other words, the kinematics of quantum gravity allows one to construct topological geons with half-integral spin. I shall review in Section 5 the arguments that such states can be expected to occur dynamically as well. Finkelstein and Mismer had suggested as early as (1959) that topological geons might have half-integral spin, although it arises in a different way than the one they were considering.

The diffeo corresponding to exchange of identical topological geons is also not deformable to the identity, and states representing geons with

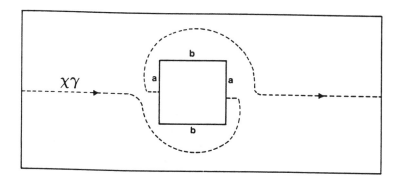

FIGURE 4a

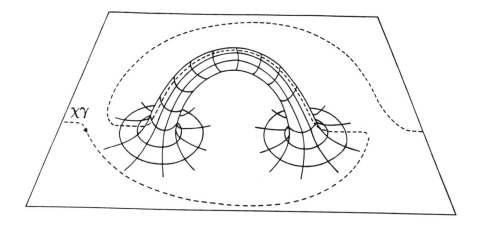

FIGURE 4b

odd statistics are again part of the space of states satisfying the momentum constraint. However, no relation between spin and statistics of geons holds on a purely kinematical level; the Finkelstein–Rubinstein (1968) proof of a spin-statistics relation for kinks (Skyrmions) uses paths of fields involving kind creation. The precise proof certainly does not go through for geons, but it may be that natural assumptions about a path-integral formulation for space-times with non-Euclidean topology allow one to construct an analogous argument (Sorkin 1988).

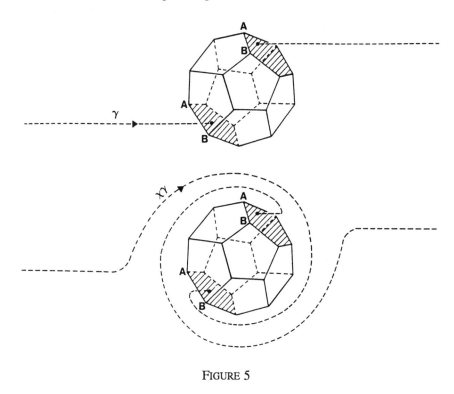

FIGURE 5

A classification of $\pi_0(\text{Diff}_F)$ for all spherical spaces is given by Witt (1986b). For each of the spaces shown in Figure 2, $\pi_0(\text{Diff}_F)$ is isomorphic to crystallographic symmetry groups to the double coverings in $SU(2)$ of the symmetry groups of the polyhedra.

State vectors with half-integral spin constructed in a similar way from the metric alone have recently been considered in a Kaluza–Klein framework—that is, in higher-dimensional gravity with Kaluza–Klein asymptotic conditions. One can easily show that if $\mathcal{R}$ is not homotopic to the identity on a four-dimensional manifold X, it is not homotopic to the identity on a higher-dimensional manifold with topology $X \times Y$; state vectors in higher-dimensional gravity with half-integral angular momentum thus arise in the same way as in four dimensions (Friedman and Higuchi 1990). In addition, Sorkin (1984) showed that the Kaluza–Klein monopole he discovered gives rise to states of half-integral angular momentum in essentially the same way that the Jackiw–Rebbi (1976)–Hasenfratz–'t Hooft (1976) construction gives spin-1/2 from isospin for a Yang–Mills system with an isospinor field. Here the isospinor is replaced by a state in which the vertical component of the metric differs from one by a function δg_{55} of x_5 with behavior $\exp(ix_5)$, representing an excitation of minimum charge. This mechanism differs somewhat from that of the four-dimensional case: each diffeo that corresponds to a rotation includes an asymptotic gauge transformation, and for the metric behavior mentioned, a 2π rotation corresponds to the element -1 of the gauge group. There are two additional constructions, also different from the two already discussed and from each other. J.G. Williams (1985) constructs a kink in the light-cone structure of the internal space. The space of $(n+4)$-dimensional Lorentzian metrics is not simply connected, and a 2π-rotation path in that space is not deformable to the identity. S. Carlip (1986) also considers the construction of kinks, this time from the spatial metric alone on a manifold whose four-dimensional space-time is topologically Euclidean, but for which the internal space has topology $S^1 \times S^2$. In the construction of Sorkin, one does not need to change the topology in order to create half-integral spin geons, and the spin-statistics proofs for the Yang–Mills solitons (Friedman and Sorkin 1983) and kinks (Finkelstein and Rubinstein 1968), respectively, should imply a spin-statistics relation for the types of geons they consider. This may also be true for William's construction, but because it involves the four-metric in an essential way, and the proofs are presently understood only in the canonical framework, the situation is less clear.

The space of metrics modulo diffeos, $\mathcal{S} := \mathrm{Riem}(M)/\mathrm{Diff}(M)$, is called superspace (Fischer 1980); for asympotically-flat space-times, $\mathcal{S} = \mathrm{Riem}(M)/\mathrm{Diff}_F(M)$. In the Schrödinger representation, a state vector ψ is a complex function on $\mathrm{Riem}(M)$, and the momentum constraint implies that ψ can be regarded locally as a function on $\mathcal{S}$. Globally, because $\pi_1(\mathcal{S}) = \pi_0(\mathrm{Diff})$, is not simply connected, and ψ is a cross section of a bundle whose fibers are n copies of C, where $n = \dim[\pi_0(\mathrm{Diff})]$. This equivalent way of treating gravity's θ-states was initially used by Isham (1981, 1982) (see Imbo and Sudarshan (1988) for a more recent treatment), and is a special case of the quantization of systems on nonsimply-connected space-times considered by Laidlaw and C.M. DeWitt (1971).

5. Comments on Dynamics

5.1. TOPOLOGY CHANGE VIA PATH INTEGRALS

In the path-integral approach to quantum gravity, the amplitude for a transition from one three-geometry to another is a sum over four-geometries of the form

$$\int dg \, e^{iS(g)}, \tag{5.1}$$

where S is the classical action. One expects a nonzero amplitude for transitions from one spatial topology to another because any two spaces can be joined by a smooth space-time with a Lorentzian metric. Two $(n-1)$-manifolds Σ_1 and Σ_2 are called *cobordant* if there is an n-manifold M, whose boundary is the disjoint union of Σ_1 and Σ_2: $\partial M = \Sigma_1 \cup \Sigma_2$. We shall call Σ_1 and Σ_2 *Lorentz cobordant* if, in addition, there is a Lorentzian metric on M for which Σ_1 and Σ_2 are space-like. Sorkin (1986a, b) shows that imposing Lorentz cobordism does not restrict the possible transitions from one $(n-1)$-manifold to another:

Theorem. *If two $(n-1)$-manifolds are cobordant, they are Lorentz cobordant.*

Because Sorkin presents an explicit proof only for n odd, a sketch of the full proof will be given here. First, however, note that the theorem implies an earlier result (Reinhart 1963) for four-dimensional space-times.

Corollary. *Any two three-manifolds Σ_1 and Σ_2 are Lorentz cobordant.*

The corollary is a consequence of Lickorish's theorem (1963) that any three-manifolds are cobordant.

To prove the theorem, one uses the fact that a manifold admits a Lorentz metric iff it admits a nonvanishing direction field $\pm t^{\alpha}$. (A direction field is a vector field defined only up to sign, like the Killing field perpendicular to the boundary of a Möbius strip.) That is, one can always choose a Euclidean signature metric, $e_{\alpha\beta}$, and from it construct a Lorentzian metric by writing

$$g_{\alpha\beta} = e_{\alpha\beta} - t_\alpha t_\beta,$$

where $e_{\alpha\beta}$ has been used to lower and normalize $\pm t^\alpha$.

Let M be a cobordism, $\partial M = \Sigma_1 \cup \Sigma_2$. One can always construct on M a vector field with isolated zeroes that is transverse to Σ_1 and Σ_2, and which has index (winding number) ± 1 at each zero. If $n = \dim M$ is odd, then a field pointing radially inward has index -1, while one pointing radially outward has (for any n) index $+1$. Cut out a ball around each zero and glue in a copy of RP^n. The radial vector field can then be extended inward to a smooth direction field (for $n = 2$, RP^2-[a disk] is the Mobius strip and the direction field is the Killing field perpendicular to the circular boundary of the strip).

If n is even, a vector field with index $+1$ can again be represented by a radial field and extended inward by gluing in a copy of RP^n. To extend a field with index -1, however, one must replace a ball containing the zero by a copy of a manifold N with Euler number $\chi(N) = 3$, so that the Euler number of $M \# N$ is $\chi(M \# N) = \chi(M) + \chi(N) - 2 = \chi(M) + 1$. One example is $N = CP^2 \times RP^2 \times \cdots \times RP^2$. A more concise, less constructive, version of the proof for n even is this: an even dimensional manifold M' has a nonvanishing vector field transverse to $\partial M'$ iff $\chi(M') = 0$, and $\chi(M') = 0$ for either $M' = M \# N \# \ldots \# N$ or $M' = M \# RP^n \# \ldots \# RP^n$.

There are no restrictions on what topological geons can occur because one can always find a Lorentz cobordism relating two manifolds that differ by a pair of oppositely oriented geons (Sorkin 1986a,b; Yodzis 1972, 1973; Friedman and Higuchi 1990). If n is even (e.g., in four dimensions), one can use what can naturally be called a pair-creation manifold. The history of a three-dimensional geon in a Lorentzian four-manifold can be described as the interior of a tube with time-like boundary; the tube has topology $\Sigma \times I$, where Σ is the topological geon, a manifold with non-Euclidean topology. If one bends the tube into a U-shape so that both its ends Σ_a and Σ_b are in the final hypersurface (the result, is a four-manifold whose initial boundary has trivial topology and whose final boundary is the connected sum of two oppositely oriented geons). An explicit Lorentz metric on a creation-manifold of this type is given in Friedman et al. 1988. If the dimension of M is odd, it does not admit a Lorentz metric, but one can always modify M to produce a manifold N that does by gluing in copies of $\mathbb{R}P^n$: $N \simeq M \# RP^n \# \cdots \# RP^n$.

One pays for a change of topology by a loss of causality. Geroch proved that if Σ_1 and Σ_2 are not homeomorphic (or if M is not a product of the form $\Sigma_1 \times I$), then any Lorentz metric $g_{\alpha\beta}$ that makes Σ_1 and Σ_2 space-like must have closed time-like curves (as stated in Geroch's paper, the space-time $M, g_{\alpha,\beta}$, must either have closed time-like curves or be time-nonorientable; but by going to covering space, it is easy to show that there must be closed time-like curves in all but one case; the creation of a universe from nothing. See Section 5.2 below, for an example, a time-nonorientable space-time with one boundary, Σ_2, and no closed time-like curves). How evil this is, is not at all clear at present; certainly, without knowing whether a path integral over metrics makes sense in the first place, one cannot really decide whether including four-geometries with closed time-like curves makes matters worse. One can, however, ask whether a path integral makes sense for quantum fields on a fixed background space-time with closed time-like curves or with changing topology. Some preliminary work relevant to these simpler issues will be discussed in Section 6, below.

The probability of quantum fluctuations in topology can be estimated from the path integral (5.1) (Wheeler 1962, 1964, 1968). To change lengths by order unity in a region of characteristic size ℓ, one needs a radius of curvature $< \ell$. Then $R \approx \ell^{-2}$ implies, in Planck units, $S = \int dV_4 R \approx \ell^2$, and the corresponding probability is of order $\exp(-\ell^2)$. For ℓ of the order of the Planck scale or smaller, one expects random fluctuations in space-time topology, and the probability of correlated fluctuations rapidly falls to zero for lengths larger than ℓ_{Planck}. The precise form of the gravitational Lagrangian is not crucial if S is constructed from G, c and the metric. Because the metric can supply only quantities with dimensions $(\text{length})^n = \ell^n$, and the only quantity constructed from G, c and ℓ that has units of action is $c^3 G^{-1} \ell^2$, one has, on purely dimensional grounds, the relation

$$\frac{S}{\hbar} \approx \frac{c^3}{G\hbar} = \frac{\ell^2}{\ell_{\text{Planck}}^2}.$$

In the framework of two-dimensional gravity, the only contribution to the gravitational action is the Euler characteristic: there are no dynamics. In the Euclidean framework, contributions to the path integral from manifolds with handles are exponentially suppressed as the number of handles increases.[7]

5.2. CHILD UNIVERSES

> "Dr. Oppenheimer — Are there many universes? I believe that
> you think there are.
>
> Dr. Klein — Yes, I think so."[8]

In gravitational collapse, the fields in a collapsing region are trapped
behind a horizon, and it seems very unlikely that the subsequent black
hole evaporation can restore to the exterior space-time the information
that characterizes the interior state. If one assumes that in the real,
quantum-mechanical universe, the evolution inside the black hole is nonsingular, then the fact that most of the information about the interior
region remains forever inaccessible, suggests that the interior evolves to
a causally disconnected universe (Hawking 1987). Bardeen's earlier calculation (1968) of collapsing shells of charged dust provides an example
of collapsing matter that re-expands to form its own universe; the dust
itself expands into the second sheet of the Riessner–Nordström solution
without encountering a singularity. (The space-time is singular, and in
a perturbed collapse the dust would itself encounter a space-like singularity). If a quantum theory allows branching universes, one could ask
whether ours might be a child of a larger space-time.[3] In the inflationary
variant, the precursor of our universe would be a small bubble of false
vacuum, which again inflates as a causally disconnected region, and a
calculation by Sato et al. (1981) and Blau et al. (1987) gives a classical
solution for a bubble that expands into the left branch of the extended
Schwarzschild space-time.

In the context of Euclidean path integrals, one can represent amplitudes for changing the number of disconnected universes by sums of
metrics on manifolds whose topology is that of a four-sphere with several
balls removed to leave a set of boundary three-spheres. The boundaries
correspond to disconnected universes, some of which are assigned to
the past, some to the future. Hartle and Hawking (1983) used a topology of this kind to estimate the probability for bifurcating universes,
and a number of recent papers have treated the problems by calculating
Euclidean instantons.[9] There is a widespread misconception that bifurcating universes are inconsistent with Lorentz metrics. Although the
simplest manifolds, S^4- (copies of B^4), do not have Lorentz metrics,
Reinhart's theorem (the Corollary in Section 5.1, above) already guarantees Lorentzian space-times whose initial and final space-like hyper-

surfaces are disjoint unions of an arbitrary number of three-spheres. It is not difficult to find appropriate manifolds for such space-times. An example of a Lorentzian space-time that describes the branching off of a single child universe can be obtained from the manifold $\mathbb{C}P^2 - 3$ balls. For the space $\mathbb{C}P^2$, $\chi = 3$, and each time a ball (a four-cell) is removed, the Euler characteristic (by its definition) is decreased by one. Thus $\chi(\mathbb{C}P^2 - 3 \text{ balls}) = 0$, and one can find a time-like vector field pointing out of the three-spheres that represent the past boundary, and into the one that represents the future.

In the alternative, attractive cosmology of Hartle and Hawking (1983), the Universe has no past boundary. Again, one can construct Lorentzian examples, including examples of topology change without closed time-like curves. The simplest is RP^4- ball (Sorkin 1986a, b): Because $\chi(RP^4) = 1$, $\chi(RP^4 - \text{ball}) = 0$. The two-dimensional analogue is the Möbius strip mentioned above. Its single circular boundary is the final hypersurface of the space-time and the nonvanishing time-like direction field is the field normal to the boundary.

5.3. Dynamics of Gravity's θ-States

If the path integral approach allows one to create topological geons, one can ask whether the θ-states that were allowed by the momentum constraint can arise dynamically—whether, for example, a creation manifold will lead to a finite amplitude for the creation of geons with half-integral spin. If one assumes that the action and measure of the path integral are diffeomorphism invariant, an amplitude of the form (5.1) will be invariant under a diffeo χ of the final three-geometry if χ can be extended to a diffeo $\bar{\chi}$ of the four-manifold. Hartle and Witt (1988) exhibit a simple example of such an inextendible diffeo for a space-time with a single boundary, namely CP^2 — a ball. The boundary is then S^3, and the diffeo χ is parity, the orientation-reversing diffeo of S^3.

For a creation manifold M, one can similarly show that if χ is a diffeo of one of the geons that is not homotopic to the identity on the final hypersurface, then χ is not extendible to a diffeo of M. When χ is not extendible, then different sets of four-geometries contribute to final metrics which differ by χ, and the resulting final amplitude will in general be a superposition of representations of the group D/D_0 of inequivalent diffeos. Then by measuring, say, the spin of a geon, an observer would pick out a branch of the wave function in which the geon has either integral or half-integral spin, and there will in general be branches corresponding to other representations of $\pi_0(\text{Diff})$.[10]

Although the path-integral approach thus apparently predicts that general representations of $\pi_0(\text{Diff})$ will arise dynamically, one could change the prescription to sum not over metrics ending with a particular metric, but over four-geometries ending with a particular three-geometry: that is, one could compute the final state by summing over all diffeomorphically equivalent metrics, with the same phase for each large diffeo (Hartle and Witt 1988). This would consistently project out the trivial representation of $\pi_0(\text{Diff})$ and leave one with theory that is not as rich, but which is equally permissible.

5.4. GROUND STATES WITH NON-EUCLIDEAN TOPOLOGY

If, as the path-integral approach suggests, topology change is a key feature of quantum gravity, a natural question is whether microscopic non-Euclidean topology arises only as a quantum fluctuation or can persist as a gravitational soliton—whether one can expect to find stable topological geons.

Peter Hajicek (1981), following earlier suggestions by Salam and Strathdee (1976) and Gibbons and Perry (1980), argued that in the Einstein–Maxwell theory, charged black holes with $e = m$ (or more generally $e^2 + g^2 = m^2$) could be regarded as gravitational solitons in the semi-classical theory. In pure Einstein–Maxwell theory, the charge could not be radiated, and Hawking evaporation would imply an evolution to an extreme black hole for which the temperature is zero. Inside the black hole, the geometry would not be static, but within the semiclassical framework, only the exterior is accessible to an observer outside the hole. There is a similar set of black holes in Kaluza–Klein theory (and, of course, for the Einstein–Yang–Mills equations), which, together with Sorkin's (1984) globally nonsingular monopole can be regarded as stable endpoints of semi-classical black hole evaporation.

However, once one allows the creation of pairs of Planck-size black holes, then on phase-space grounds, larger black holes should decay to smaller ones, so that only objects of Planck-size or smaller remain. But at the Planck scale, the classical notion of event horizon does not survive fluctuations in the metric, and the term "black hole" is no longer appropriate. One nevertheless is entitled to expect, in a quantum theory of pure gravity, a ground state with half-integral angular momentum, and in higher-dimensional gravity, a ground state belonging to the fundamental representation of $\bar{G}$, where G is the "internal" part of the asymptotic symmetry group.

Consider first the four-dimensional case. We shall assume that there is a well-defined quantum theory that at low energy ($E < E_{\mathrm{Planck}}$) is a theory of the four-dimensional space-time metric alone. (Thus, other fields are excluded, but space-time is not assumed to make sense at distances short compared to the Planck scale). Then, since gravitational radiation can only carry off integral spin, states with half-integral angular momentum cannot radiate it away, and presumably must settle down to ground states with half-integral spin. Any state vector with half-integral spin must have support on manifolds with non-Euclidean topology. Thus, one expects the ground states to look like collections of isolated topological geons, and that there should be a lowest energy, stable geon with half-integral spin (plausibly spin 1/2).

The higher-dimensional case is also intriguing. Suppose as before that there exists a quantum theory that for $E < E_{\mathrm{Planck}}$ is a theory solely of an $(n+4)$-dimensional metric, and that the theory has solutions that represent stable space-times with Kaluza–Klein asymptotics with an internal symmetry group G. Then there are stable ground states that have half-integral spin, and which belong to a faithful representation of the group $\bar{G}$. The argument is essentially the same as in the four-dimensional case (Friedman and Higuchi 1990). It suggests that there are stable topological geons with the kinematic quantum numbers of leptons and quarks, since the lowest energy state with half-integral spin and belonging to a faithful representation of G is likely to be a state with the lowest quantum numbers—a state with spin 1/2 belonging to the fundamental representation of $\bar{G}$.

There is, of course, no reason to believe that the states will have masses substantially different from the Planck mass; and with ordinary Kaluza–Klein boundary conditions one cannot have chiral fermions (mirror fermions always occur) (Witten's proof (1987) for point fermions does not apply to geons with dimensions comparable to the size of the internal space, but chiral fermions are still forbidden). However, our ignorance of the dynamics of quantum gravity is large enough that zero rest-mass geons cannot be excluded, and if one relaxes the asymptotic conditions, a nonzero vacuum expectation value of a Higgs-like vector constructed from the metric can break chiral symmetry without changing the lowest-order Einstein–Yang–Mills form of the gravitational Lagrangian.

Even in the semi-classical context, it seems to difficult to decide the simplest question here: is there a spherically symmetric geon supported by its vacuum Casimir energy? The Candelas–Weinberg[11] solution is an example of a stationary, semi-classical (unstable?) state, in which the Casimir energy of a compact internal space supports it against collapse. While the analogous self-consistent spherical symmetric solution may still be too difficult to find or rule out, Morris, Thorne and Yurtsever (1988)[12] find a spherically symmetric geon supported by Casimir energy when one includes a plate surrounding the geon. (They consider a wormhole, with plates on either side, but by identifying corresponding points in the two asymptotic regions, one obtains a spherically symmetric geon with one asymptotic region; the topology is $O(3) \simeq RP^3$ with a point removed.) Their computation deliberately ignores the detailed structure of the plate, and it may happen that a more realistic plate will change the optimistic conclusion.

6. Objections to Topology Change

> "To assume spacetime is a manifold, one must disregard the possibility ... that a microscope of ever greater magnification would reveal ever new topological complications, *ad infinitum.*
> — H. Weyl[1]

An obvious difficulty with topology change in a Lorentzian framework is that one must construct a path integral that includes space-times that are either time nonorientable or have closed time-like curves.[13] One can argue that, if a sum over such a class of metrics makes sense, then the sum over, say, scalar fields on a fixed background space-time should also make sense; that is, a Lorentzian description of topology change is in trouble if one cannot at least define a quantum field on a space-time with closed time-like curves. This is not entirely obvious, since a sum over metrics might smooth out singular behavior from a particular metric, but it at least provides a class of more accessible problems.

In two dimensions, topology change is inconsistent with a Lorentzian metric. Anderson and DeWitt (1986) considered a background space-time representing a bifurcating universe, with the trousers topology $S^2 - (3$ disks), and a Lorentzian metric that was singular at a point, and flat elsewhere. The singularity in this example is severe, and they found no way to define a consistent field theory (see also Manogue et al. 1988). In two dimensions, however, the difficulty is not the presence of closed time-like curves, but a necessarily singular metric.

Morris, Thorne, and Yurtsever (1988) observed that the wormhole geometry they constructed could be converted into a space-time that has closed time-like curves by accelerating one end of the wormhole, while leaving the other end fixed. The construction is particularly interesting for two reasons. First, the past of the region with closed time-like curves together with its boundary, the Cauchy horizon, appears to be stable to small perturbations. Light that travels in a future direction along nearly closed null geodesics traverses the wormhole many times, and each time it is blue-shifted. This is the kind of effect that one would expect to make the Cauchy horizon unstable, but the optics are divergent, and the spreading of the light is rapid enough that the energy decreases with each trip. Second, while there are many closed time-like curves, the set of closed time-like geodesics appears to be a set of measure zero. As a result, initial data for dust to the past of the Cauchy horizon has a unique time evolution for almost all dust particles. It is at least plausible that, for arbitrary smooth initial data for a free scalar field, there is a unique solution in L^2 on the entire space-time. More generally, it appears likely that there is a large class of space-times with closed time-like curves, for which the closed time-like geodesics form a set of measure zero; and it may be that for a subclass of these there is a well-defined initial value problem in the classical case, and well-defined free quantum fields on such background space-times.

An additional objection due to DeWitt should be mentioned. In a path integral that involves a sum over topologically different four-manifolds, one could assign a free phase to each topologically distinct manifold. A combining rule slightly restricts the freedom (Kandrup and Mazur 1988), but does not remove it.

Acknowledgments. I am indebted to Rafael Sorkin and Donald Witt for several helpful discussions.

This research is supported in part by NSF Grant No. PHY 8603173.

NOTES

[1] Weyl 1949, p. 91.

[2] These Clifford–Klein space forms were known to mathematicians by the late 1800s. See Milnor 1982 for a brief history. The first mention of them that I know of in the context of cosmology is in a talk by Lemaître (1958).

[3] Among the early references to a universe born from quantum fluctuations

are Albrow 1973; Tyron 1973; Brout, Englert, and Gunzig 1978; Mel'nikov 1979; Zel'dovich 1981; Sato et al. 1982; Atkatz and Pagels 1982. Later references can be found in Blau, Guendelman and Guth 1987; Konstantinov and Mel'nikov 1986; see also Vilenkin 1983.

[4] Gannon 1975, Prop. 1.2. The actual statement of Gannon's theorem uses a weaker condition than asymptotic flatness—that, for a family of spheres near infinity, the inward directed null geodesics converge ($\theta > 0$) on the Cauchy surface. The proof of a subsequent result in this paper is incorrect: see Lee 1976; Frankel and Galloway 1982a, b; Galloway 1983.

[5] See also Sokolov and Shvartsman 1974.

[6] Intervening references are listed in these articles.

[7] Alvarez 1983. The two references disagree on sign; Alvarez uses the usual convention, implying suppression of high genus manifolds.

[8] Klein 1958.

[9] Kandrup and Mazur 1988; Myers 1988; Giddings and Strominger 1988; Coleman 1988; Klebanov, Susskind and Banks 1988.

[10] Hartle and Witt note that if the final state is an eigenstate corresponding to a particular representation of π_0 (Diff), it must generically be the trivial representation, because there will be manifolds contributing to the amplitude for which any diffeo will be extendible. But it seems unlikely that, in a generic evolution, the wave function would be an eigenstate corresponding to a particular representation of π_0 (Diff) for a particular three-manifold.

[11] Candelas and Weinberg 1984. See also Shen and Sobczyk 1987, and references therein.

[12] See also Morris and Thorne 1989.

[13] An alternative would be to allow the metric to degenerate at single points. See, e.g., Sorkin 1986a,b; Brill 1972.

References

Albrow, M.G. (1973). "CPT Conservation and the Oscillary Model of the Universe." *Nature-Physical Science* 241: 56–57.

Alvarez, Orlando (1983). "Theory of Strings with Boundaries: Fluctuation, Topology, and Quantum Geometry." *Nuclear Physics B* 216: 125–184.

Anderson, Arlen, and DeWitt, Bryce S. (1986). "Does the Topology of Space Fluctuate?" *Foundations of Physics* 16: 91–105.

Atkatz, David, and Pagels, Heinz (1982). "Origin of the Universe as a Quantum Tunneling Event." *Physical Review D* 25: 2065–2073.

Bardeen, James M. (1968). "General Relativistic Collapses of Charged Dust." *American Physical Society. Bulletin* 13: 41–41.

Bernstein, I.N., and Shvartsman, V.F. (1980). "On the Connection between the Size of the Universe and Its Curvature." *Soviet Physics JETP* 52: 814–820.

Blau, Steven K., Guendelman, E.I., and Guth, Alan H. (1987). "Dynamics of False-Vacuum Bubbles." *Physical Review D* 35: 1747–1766.

Brill, Dieter R. (1972). "Thoughts on Topology Change." In *Magic Without Magic: John Archibald Wheeler, A Collection of Essays in Honor of His Sixtieth Birthday.* John R. Klauder, ed. San Francisco: W.H. Freeman, pp. 309–316.

———(1983). "On Spacetimes Without Maximal Surfaces." In *Proceedings of the Third Marcel Grossmann Meeting on General Relativity, 30 August–3 September, 1982, Shanghai, China.* Hu Ning, ed. Amsterdam and New York: North-Holland.

Brout, R., Englert, F., and Gunzig, E. (1978). "The Creation of the Universe as a Quantum Phenomenon." *Annals of Physics* 115: 78–106.

Candelas, Philip, and Weinberg, Steven (1984). "Calculation of Gauge Coupling and Compact Circumferences from Self-Consistent Dimensional Reduction." *Nuclear Physics B* 237: 397–441.

Carlip, Steven (1986). "Topological Solitons in Kaluza–Klein Theories." *Physical Review D* 33: 1638–1642.

Coleman, S. (1988). "Why there is Nothing Rather than Something: A Theory of the Cosmological Constant." *Nuclear Physics B.* 310: 643–668.

Eardley, D., and Witt, D.M. (1990). "Spacetimes With no Slices of Constant Mean Curvature." Preprint.

Ellis, George F.R. (1971). "Topology and Cosmology." *General Relativity and Gravitation* 2: 7–21.

Ellis, George F.R., and Schreiber, G. (1986). "Observational and Dynamical Properties of Small Universes." *Physics Letters A* 115: 97–107.

Fang, Li Zhi, and Sato, Humitaka (1985). "Is the Periodicity in the Distribution of Quasar Red Shifts an Evidence of Multiply Connected Universes." *General Relativity and Gravitation* 17: 1117–1120.

Finkelstein, David, and Misner, Charles W. (1959). "Some New Conservation Laws." *Annals of Physics* 6: 230–243.

Finkelstein, David, and Rubinstein, J. (1968). "Connection Between Spin Statistics and Kinks." *Journal of Mathematical Physics* 9: 1762–1779.

Fischer, Arthur Elliot (1980). "The Theory of Superspace." In *Relativity: Proceedings of the Relativity Conference in the Midwest, Cincinnati, Ohio, June 2–6, 1969.* Moshe Carmeli, Stuart I. Fickler, and Louis Witten, eds. New York and London: Plenum, pp. 303–357.

Frankel, Theodore, and Galloway, Gregory J. (1982a). "Stable Minimal Surfaces and Spatial Topology in General Relativity." *Mathematische Zeitschrift* 181: 395–406.

———(1982b). "Correction to: Stable Minimal Surfaces and Spatial Topology in General Relativity." *Mathematische Zeitschrift* 182: 575–576.

Friedman, John L., and Higuchi, A. (1990). "State Vectors in Higher-Dimensional Gravity with Kinematic Quantum Numbers of Quarks and Leptons." *Nuclear*

Physics B 339: 491–515.

Friedman, John L., and Mayer, Steven (1982). "Vacuum Handles Carrying Angular Momentum: Electrovac Handles Carrying Net Charge." *Journal of Mathematical Physics* 23: 109–115.

Friedman, John L., and Sorkin, Rafael D. (1980). "Spin 1/2 from Gravity." *Physical Review Letters* 44: 1100–1103.

———(1982). "Half-Integral Spin from Quantum Gravity." *General Relativity and Gravitation* 14: 615–620.

———(1983). "Statistics of Yang–Mills Solitons." *Communications in Mathematical Physics* 89: 501–521.

Friedman, John L., and Witt, Donald M. (1986). "Homotopy is not Isotopy for Homeomorphisms of 3-Manifolds." *Topology* 25: 35–44.

Friedman, John L.; Papastamatiou, Nicolas; Parker, Leonard, and Huai, Zhang (1988). "Non-orientable Foam and an Effective Planck Mass for Point-Like Fermions," *Nuclear Physics B.* 309: 533

Galloway, Gregory J. (1983). "Minimal Surfaces, Spatial Topology, and Singularities in Space-Time." *Journal of Physics A* 16: 1435–1439.

Gannon, Dennis (1975). "Singularities in Nonsimply Connected Space-Times." *Journal of Mathematical Physics* 16: 2364–2367.

Geroch, Robert P. (1967). "Topology in General Relativity." *Journal of Mathematical Physics* 8: 782–786.

———(1978). "The Positive Mass Conjecture." In *Theoretical Principles in Astrophysics and Relativity.* Norman R. Lebowitz, William H. Reid, and Peter O. Vandervoort, eds. Chicago: University of Chicago Press, pp. 245–252.

Gibbons, G.W., and Perry, Malcolm J. (1980). "New Gravitational Instantons and their Interactions." *Physical Review D* 22: 313–321.

Giddings, Steven B., and Strominger, Andrew (1988). "Baby Universes, Third Quantization, and the Cosmological Constant." Harvard University preprint, HUTP-88/A036.

Gott, J. Richard, III (1980). "Chaotic Cosmologies and the Topology of the Universe." *Royal Astronomical Society, Monthly Notices* 193: 153–169.

Gromov, Mikhael, and Lawson, H. Blaine, Jr. (1983). "Positive Scalar Curvature and the Dirac Operator on Complete Riemannian Manifolds." *Institut des Hautes Études Scientifiques. Publications Mathématiques* 58: 83–196.

Hajicek, P. (1981). "Quantum Wormholes. I. Choice of the Classical Solution." *Nuclear Physics B* 185: 254–268.

Hartle, James B., and Hawking, Stephen W. (1983). "Wave Function of the Universe." *Physical Review D* 28: 2960–2975.

Hartle, James B., and Witt, Donald M. (1988). "Gravitational θ-States and the Wave Function of the Universe." *Physical Review D* 37: 2833–2837.

Hasenfratz, P., and 't Hooft, G. (1976). "Fermion-Boson Puzzle in a Gauge Theory." *Physical Review Letters* 36: 1119–1122.

Hawking, Stephen W. (1978). "Spacetime Foam." *Nuclear Physics B* 144: 349–362.

———(1987). "Quantum Coherence down the Wormhole." *Physics Letters B* 195: 337–343.

Hawking, Stephen W., and Ellis, George F.R. (1973). *The Large-Scale Structure of Spacetime*. Cambridge: Cambridge University Press.

Higgs, Peter W. (1958). "Integration of Secondary Constraints in Quantized General Relativity." *Physical Review Letters* 1: 373–375.

Imbo, Tom D., and Sudarshan, E.C.G. (1988). "Inequivalent Quantizations and Fundamentally Perfect Spaces." *Physical Review Letters* 60: 481–483.

Isham, Christopher J. (1981). "Topological θ Sectors in Canonically Quantized Gravity." *Physics Letters B* 106: 188–192.

———(1982). "θ-States Induced by the Diffeomorphism Group in Canonically Quantized Gravity." In *Quantum Structure of Space and Time*. Proceedings of the Nuffield Workshop, Imperial College London, 3–21 August 1981. M.J. Duff and Christopher J. Isham, eds. Cambridge: Cambridge University Press, pp. 37–52.

Jackiw, R., and Rebbi, C. (1976). "Spin from Isospin in a Gauge Theory." *Physical Review Letters* 36: 1116–1119.

Kandrup, H.E., and Mazur, P.O. (1988). "Particle Creation and Topology Change in Quantum Cosmology." University of Florida preprint, UFTP-88-10.

Kazden, Jerry L., and Warner, F.W. (1975). "Scalar Curvature and Conformal Deformation of Riemannian Structure." *Journal of Differential Geometry* 10: 113–134.

Klebanov, I., Susskind, L., and Banks, T. (1988). "Wormholes and the Cosmological Constant." SLAC preprint; SLAC-PUB-4705.

Klein, Oskar (1958). "Some Considerations regarding the Earlier Development of the System of Galaxies." In *La Structure et l'Évolution de l'Univers*. Institut International de Physique Solvay. Onzième Conseil de Physique tenu à l'Université de Bruxelles du 9 au 13 juin 1958. R. Stoops, ed. Brussels: Stoops, pp. 33–47.

Konstantinov, M.Yu., and Melnikov, V.N. (1986). "Topological Transitions in the Theory of Spacetime." *Classical and Quantum Gravity* 3: 401–416.

Laidlaw, Michael G.G., and DeWitt, Cécile M. (1971). "Feynman Functional Integrals for Systems of Indistinguishable Particles." *Physical Review D* 3: 1375–1378.

Lee, C.W. (1976). "A Restriction on the Topology of Cauchy Surfaces in General Relativity." *Communications in Mathematical Physics* 51: 157–162.

Lemaître, Georges (1958). "The Primaeval Atom Hypothesis and the Problem of the Clusters of Galazies." In *La Structure et l'Évolution de l'Univers*. Institut International de Physique Solvay. Onzième Conseil de Physique tenu à l'Université de Bruxelles du 9 au 13 juin 1958. R. Stoops, ed. Brussels: Stoops, pp. 1–25.

Lickorish, W.B.R. (1962). "A Representation of Orientable Combinatorial 3-Manifolds." *Annals of Mathematics* 76: 531–540.

———(1963). "Homeomorphisms of Non-Orientable Two-Manifolds." *Cambridge Philosophical Society. Proceedings.* 59: 307–317.

Manogue, C.A., Copeland, E., and Dray, T. (1988). "The Trousers Problem Revisited." *Pramana Journal of Physics* 30: 279–292.

Mel'nikov, V.N. (1979). "On Quantum Effects in Cosmology." *Doklady Academii Nauk* 246: 1351–1355.

Milnor, John (1982). "Hyperbolic Geometry: The First 150 Years." *American Mathematical Society. Bulletin* 6: 9–24.

Morris, Michael S., and Thorne, Kip S. (1989). "Wormholes in Spacetime and Their Use for Interstellar Travel: A Tool for Teaching General Relativity." *American Journal of Physics* 54: 395–412.

Morris, Michael S., Thorne, Kip S., and Yurtsever, Ulvi (1989). "Wormholes, Time Machines, and the Weak Energy Condition." *Physical Review Letters* 61: 1446–1449.

Myers, R.C. (1988). "New Axionic Instantons in Quantum Gravity." *Physical Review D* 38: 1327–1330.

Rajeev, S.G. (1982). "Exact Solution of Quantum Gravity in 1+1 Dimensions." *Physics Letters B* 113: 146–150.

Reinhart, Bruce L. (1963). "Cobordism and the Ruler Number." *Topology* 2: 173–177.

Salam, Abdus, and Strathdee, J. (1976). "Black Holes as Solitons." *Physics Letters B* 61: 375–376.

Sato, K.; Kodama, H.; Sasaki, M.; and Maeda, K. (1981). " Creation of Wormholes by a First-Order Phase Transition in the Early Universe." *Progress in Theoretical Physics* 65: 1443–1446.

Sato, K.; Kodama, H.; Sasaki, M.; and Maeda, K. (1981). "Multi-Production of Universes by First-Order Phase Transition of a Vacuum." *Physics Letters B* 108: 103–107.

Schoen, Richard, and Yau, Shing-Tung (1979a). "Positivity of the Total Mass of a General Space-Time." *Physical Review Letters* 43: 1457–1459.

———(1979b). "Existence of Incompressible Minimal Surfaces and the Topology of Three Dimensional Manifolds with Non-Negative Scalar Curvature." *Annals of Mathematics* 110: 127–142.

Seifert, H., and Threlfall, W. (1980). *A Textbook of Topology.* Michael A. Goldman, trans. Pure and Applied Mathematics: A Series of Monographs and Textbooks, vol. 89. Samuel Eilenberg and Hyman Bass, eds. New York: Academic Press.

Shen, T.C., and Sobczyk, J. (1987). "Higher-Dimensional Self-Consistent Solution with Deformed Internal Space." *Physical Review D* 36: 397–411.

Sokolov, D.D., and Shvartsman, V.F. (1974). "An Estimate of the Size of the Universe from a Topological Point of View." *Soviet Physics JETP* 39: 196–200.

Sorkin, Rafael D. (1984). "A 5-Dimensional Monopole." In *Monopole '83*. James L. Stone, ed. New York: Plenum, pp. 39–45.

———(1986a). "Topology Change and Monopole Creation." *Physical Review D* 33: 978–982.

———(1986b). "Non-Time-Orientable Lorentzian Cobordism Allows for Pair Creation." *International Journal of Theoretical Physics* 25: 877–881.

———(1988). "Classical Topology and Quantum Phases: Quantum Geons." In *Geometrical Aspects of Nonlinear Field Theories*. G. Marmo, ed. North Holland.

Thurston, William P. (1982). "Three Dimensional Manifolds, Kleinian Groups and Hyperbolic Geometry." *American Mathematical Society. Bulletin* 6: 357–381.

Tipler, Frank J. (1977). "Singularities and Causality Violation." *Annals of Physics* 108: 1–36.

———(1985). "Topology Change in Kaluza–Klein and Superstring Theories." *Physics Letters B* 165: 67–70.

Tyron, Edward P. (1973). "Is the Universe a Vacuum Fluctuation?" *Nature* 246: 396–397.

Vilenkin, Alexander (1983). "Birth of Inflationary Universe." *Physical Review D* 27: 2848–2855.

Weyl, Hermann (1949). *Philosophy of Mathematics and Natural Sciences*. Olaf Helmer, trans. Princeton: Princeton University Press.

Wheeler, John A. (1962). *Geometrodynamics*. New York: Academic Press.

———(1964). "Geometrodynamics and the Issue of the Final State." In *Relativity, Groups and Topology*. Lectures Delivered at les Houches during the 1963 Session of the Summer School of Theoretical Physics, University of Grenoble. Cécile DeWitt and Bryce S. DeWitt, eds. New York and London: Gordon and Breach, pp. 316–520.

———(1968). *Einstein's Vision*. Berlin: Springer-Verlag.

Williams, J.G. (1985). "Topological Spin and $(n + p)$-Dimensional Manifolds." *Lettere al Nuovo Cimento* 43: 282–284.

Witt, Donald M. (1986a). "Vacuum Space-Times that Admit No Maximal Slice." *Physical Review Letters* 57: 1386–1389.

———(1986b). 'Topological Aspects of Classical and Quantum Gravity." Ph.D. Dissertation. University of Wisconsin-Milwaukee.

———(1986c). "Symmetry Groups of State Vectors in Canonical Quantum Gravity." *Journal of Mathematical Physics* 27: 573–592.

———(1990). "Topological Obstructions to Maximal Slices." submitted to *Journal of Mathematical Physics*. Forthcoming.

——— "New Asymptotically Flat Spacetimes." Forthcoming.

Witten, Edward (1987). "Fermion Quantum Numbers in Kaluza Klein Theory." In *Modern Kaluza–Klein Theories*. Thomas W. Appelquist, Alan Chodos, and Peter G.O. Freund, eds. Menlo-Park: Addison-Wesley, pp. 438–511.

Yau, Shing-Tung (1980). "Space-Time." In *The Chern Symposium*. Proceedings of the International Symposium on Differential Geometry in Honor of S.-S. Chern, Held in Berkeley, California, June 1979. W.-Y. Hsiang, et al., eds. New York, Heidelburg, and Berlin: Springer-Verlag, pp. 255–259.

Yodzis, P. (1972). "Lorentz Cobordism." *Communications in Mathematical Physics* 26: 39–52.

———(1973). "Lorentz Cobordism. II." *General Relativity and Gravitation* 4: 299–307.

Zel'dovich, Y.B. (1981). "Birth of the Closed Universe and the Anthropogenic Principle." *Soviet Astronomical Letters* 7: 322. Trans. from *Pis'ma Astron Zh* 7 (1981): 579–581.

Discussion

DEWITT: To construct topological spaces, you looked at the action of a group of boosts acting on a hyperbolic manifold. Why do you have to discuss the metric just to discuss topology? You used a term like "boost," and "hyperbolic"—all this implies a metric.

FRIEDMAN: You need not, but it turns out to be very useful in classifying the topologies to talk about the group of isometries that you use to produce the manifold; e.g., the spherical spaces are obtained by S^3 modulo the group of isometries. It turns out that π_0 of the diffeos is related to this group of isometries. More generally, in three dimensions, topology is related to natural geometries of three-manifolds.

HOROWITZ: I would like to make a comment. There is a very interesting conclusion that comes from combining the result you just mentioned with another result. You can argue that these singularities, which we all expect to be inside black holes, must be very unusual and cannot be the standard Schwarzschild-type singularities. And the argument goes as follows: there was a student of Bartnik's who proved that an asympototically-flat space-time with a singularity of the Schwarzschild type always has a maximal slice, a slice on which the trace of the extrinsic curvature is zero. So, asympototically-flat space-times with ordinary signature are expected to have maximal slices. But then you can show that on any surface like these with nontrivial topology, you can't have a maximal slice, because the maximal slice would require a positive scalar curvature. And one can prove that there are no such metrics.

FRIEDMAN: Right, that's the Gromov–Lawson result and Schoen and Yau have a . . .

HOROWITZ: So none of these space-times that you construct by evolving these topologically nontrivial data have maximal slices; therefore, they must violate the Bartnik theorem and the only way they can do it is by having very strange, weak singularities, it seems. I haven't understood it completely. It seems to be a huge class of space-times.

FRIEDMAN: Yes, that's right. The statement about maximal slices is what Don Witt emphasized in the article where he showed the existence of asympototically-flat space-times with positive energy and arbitrary spatial topology (Witt 1986a).

GELL-MANN: So black holes are dead?

HOROWITZ: No, no. It will still have an event horizon. It will still be a black hole. It's just a question of what happens when you get right inside, near the singularity.

ASHTEKAR: John, since this discussion of angular momentum has been around for quite some time, one might worry about technical issues like the following one. In classical theory, there are those super-translation ambiguities. You really need to sensitize boundary conditions at infinity to actually eliminate them from the asymptotic symmetry group. Is this quantum analogy insensitive to this super-translation ambiguity? In this sense, I really don't know which rotation subgroup to choose.

FRIEDMAN: Yes, to say how large the symmetry group is at spatial infinity, one must say how strong you want to make your asymptotic conditions. Which diffeos are regarded as trivial at spatial infinity depends on how much structure you have at spatial infinity — how much structure the diffeos must leave invariant. However, the group π_0 of the diffeos that are trivial at spatial infinity turns out to be insensitive to this structure: π_0 (Diff) is the same whether one fixes simply a frame at spatial infinity or an entire disk in a neighborhood of spatial infinity. Of course, spatial infinity must be regular enough to contain a rotational symmetry group.

ASHTEKAR: Irrespective of which one it is?

FRIEDMAN: Yes.

DEWITT: By the way, the term "geon" was invented by "Geon" Wheeler (laughter).

YORK: What's a handle?

FRIEDMAN: You cut out two balls and identify the surfaces. It is just a 3-dimensional wormhole, one that stays in our own universe.

SMOLIN: Does Geroch's theorem on absence of topology change use either field equations or energy conditions.

FRIEDMAN: It assumes that the manifold is nonsingular, it does not use the field equations or energy conditions.

SORKIN: I want to make a comment: There is another possible mechanism that can lead to topology change if the metric has a very mild form of singularity, e.g., if the metric vanishes at isolated points. This thing turns out to be unlike the mechanism for topology change that can exist on some manifolds, but is forbidden on some other manifolds. This other one turns out to be able to live on every co-bordism. So every topology change is topologically allowed.

FRIEDMAN: Yes, I think that's right, and in part the mechanism you're going to use seems to me philosophically dependent on whether you think that, on the same scale that topology changes—or something close to that scale—the idea of a manifold is going to break down as well, or the idea of a smooth metric is going to break down as well. So I am really taking a conservative point of view here and asking: "what can you do if you suppose that you still have a smooth Lorentzian metric and a manifold, that these objects make sense at sizes on the order of the Planck length?" Now I don't think that is outlandish. One could imagine two different scales, one for which topology is non-Eulidean, but space-time retains its character as an (n+4)-dimensional manifold with Lorentz metric; at this Planck scale, the fields are unified with a symmetry group determined by the space-time geometry. Then one could imagine a smaller scale, for which the notion of manifold is no longer meaningful. I am not averse to the suggestion that on scales small compared to the Planck length we must abandon manifold and metric.

DESER: Let me make a small propaganda remark. A few years ago Jackiw and I worked with (2+1)-dimensional gravity, which is very nice lab not merely to change topology, but also for other things. We looked at solutions with and without sources.

Black Holes and Partition Functions

James W. York, Jr.

1. Summary

The train of developments leading from the early viewpoint on black hole thermodynamics to the present understanding in terms of statistical thermodynamics of the gravitational field in black hole topologies is reviewed. The key idea, which is elementary but has many consequences, is to take into strict account the boundary conditions, that is, the information supposed known to a given observer, in constructing appropriate ensembles. Statistical mechanics of gravitational fields describing the black hole topological sector, and the correspondence to thermodynamics, are then discussed in the canonical ensemble. The Euclidean action is evaluated on the constraint hypersurface and a measure is obtained from a heuristic argument, resulting in a path-integral form of the canonical partition function. From this one obtains the usual black-hole entropy supplemented by statistical corrections when the temperature and size of the system are appropriate. Under other boundary conditions, it follows that a phase transition (change of topology) must occur.

2. Introduction

When it was found on the basis of particle creation effects that a black hole can be assigned a physical temperature (Hawking 1975), Bekenstein's profound speculations on black hole thermodynamics were put on a firm basis (Bekenstein 1973). There were reasons to hope that these demonstrated relations among gravitation, quantum theory, and statistical thermodynamics might provide important hints towards an understanding of quantum gravity. Indeed, suggestive early success was achieved by Gibbons and Hawking (1977), who described black holes by exploiting the properties of the Euclidean action and its relation to the partition function of a canonical ensemble (Feynman and Hibbs 1965). Further efforts in this direction were presented by Hawking (1979).

It was evident, however, that serious difficulties had been encountered in the above approach. The essence of the difficulties lay in the way the famous black hole temperature formula seems to have been interpreted. In the case of spherical uncharged holes, to which I shall confine attention here, we have

$$T = \frac{\hbar}{8\pi GM} \tag{1}$$

in units where $c = k = 1$. Superfically, this equation tells us that a black hole of mass M has negative heat capacity C. Though it is now known that such an assertion requires careful qualification (York 1986), let us proceed initially as if C were unequivocally negative. Because negative heat capacity cannot occur in the canonical ensemble, the hope of performing a Euclidean functional integral calculation of the canonical partition function and deducing unambiguously from it the established thermal properties of black holes, as well as new results, was effectively demolished. This was a serious blow to the whole program because the black hole combines in some sense both the "hydrogen atom" and the "black-body radiation" problems of quantum gravity.

For example, if the heat capacity C is negative, then we see from

$$\langle (\Delta E)^2 \rangle = CT^2 \tag{2}$$

that the root-mean-square energy fluctuations in the canonical ensemble are imaginary and thus meaningless. Furthermore, using the Euclidean action of a Schwarzchild black hole in infinite space with the temperature (1), one cannot establish without ad hoc modifications a relation between the partition function and the density of states, as is evident from the problems that turned up in the original treatment of this problem (Hawking 1979). However, an appropriate density of states should exist and, one hopes, should be related in the usual way to the black hole entropy

$$S_{BH} = \frac{4\pi GM^2}{\hbar}, \tag{3}$$

which was established independently of these deeper issues, on the basis of the so-called laws of black hole mechanics, as a consequence of the work of Bekenstein (1973) and Hawking (1975). One has today even more confidence in the validity of (3) as entropy in the standard sense of statistical thermodynamics by virtue of the work of Zurek and Thorne

(1985) and of Martinez and York (1989). In the latter reference, it is shown that the entropies of black holes and of ordinary matter are additive when they are in thermal equilibrium.

A further consequence of negative heat capacity and the consequent thermal instability entailed by (1) was noted by Gross et al. (1982) in their study of the canonical ensemble. Differing radically from Gibbons and Hawking (1977), they interpreted the Euclidean black hole as an instability of "hot flat space" (flat space filled with linear thermalized gravitons), whereby the latter would spontaneously nucleate black holes at a certain rate. According to these authors, the black hole that would form by nucleation is the self-same one that represents the unstable configuration of the hot gravitational field. However, this interpretation resulted in a paradox, whereby there occurs a spontaneous *increase* of free energy in the process of nucleation. Hence, their interpretation required modification (York 1986).

These consequences of negative heat capacity, could they not be suitably repaired, would be especially discouraging because, despite the fact that quantum gravity played no explicit part in the original derivation of (1), the quantum radiance of black holes necessarily brings out quantum properties of space-time. This is not just for the obvious reason that the radiation of black holes is thermal, and must therefore include "gravitons." It has also been shown in the reaction of a black hole to the emission of thermal radiation that, even for the coldest black hole in "empty" space $(M \to \infty)$, the horizon structure cannot have a classical configuration (York 1984). The implication is that zero-point fluctuations of the *metric* are an essential piece of the puzzle. A detailed model of these fluctuations and their effects has been proposed (York 1983). One can only conclude that black hole radiance and quantum gravity are inseparable, though the final nature of their union is still being sought. In this paper I shall give an overview of the key ideas that have been established beyond the "old black hole thermodynamics" (York 1987), and also report new results found by Bernard F. Whiting and the author (Whiting and York 1988).

3. Temperature and Stability

It is misleading to regard (1) as the temperature of a black hole. No single temperature characterizes a black hole in thermal equilibrium, unlike the case when gravitation can be ignored. According to the analysis by

Tolman (1934), the principle of equivalence requires that the temperature measured locally by a static observer in a static gravitational field be blue-shifted with respect to a temperature like (1) that is determined at asymptotically flat spatial infinity. Thus, for a Schwarzchild black hole, we have at Schwarzchild radial coordinate locus r_B

$$T(r_B) = T_\infty |g_{oo}(r_B)|^{-1/2} = \frac{\hbar}{8\pi GM} \left(1 - \frac{2GM}{r_B}\right)^{-1/2}, \quad (4)$$

where T_∞ denotes the temperature given by (1). In the following it will generally prove more convenient to use the inverse temperature $\beta \equiv T^{-1}(r_B)$.

The shift in point of view that occurs when one characterizes a black hole by the inverse temperature β that would be observed at r_B, rather than β_∞, is significant. A calculation of the heat capacity at a fixed value of r_B yields (York 1986)

$$C = -\beta \frac{\partial S_{BH}}{\partial \beta} = \frac{8\pi GM^2}{\hbar} \left(1 - \frac{2GM}{r_B}\right) \left(\frac{3GM}{r_B} - 1\right)^{-1}. \quad (5)$$

As no single temperature characterizes a black hole, no unique value of the heat capacity does either. Thus, suppose in a canonical ensemble that the idealized fixed perfectly heat-conducting surface of a spherical cavity has the temperature (4) at a given radius r_B. Then, according to whether r_B is less or greater than $3M$, the heat capacity is positive or negative, respectively, as we see from (5). Hence, for $2M \leq r_B < 3M$, the energy fluctuations ΔE in (2) are real. Combining (2), (4), and (5) we find

$$\langle (\Delta E)^2 \rangle = \frac{\hbar}{8\pi G} \left(\frac{3GM}{r_B} - 1\right)^{-1}, \quad (6)$$

so that as $r_B \to 2GM$, $\Delta E \to M_p(4\pi)^{-1/2}$ independently of the mass of the black hole ($\hbar/G = M_p^2 =$ the square of the Planck mass).

Above I stated that the principal difficulties in making progress in black hole thermodynamics and its implications lay in the interpretation people seemed to give to the relation between T_∞ and M. One such point has been raised above, namely, one should use (4) instead of (1). The other is that, in the canonical ensemble, the given data are β and r_B so that one should regard M as the function of r_B and β that follows from (4). (More precisely, one wants to find the ensemble average of the "total internal energy," $\langle E \rangle$, as a function of r_B and β, as shall be treated later.)

The implications of (4) are interesting. If and only if

$$r_B T \geq \frac{3\sqrt{3}}{8\pi} \hbar, \tag{7}$$

one finds two real positive values for M. Otherwise, there are no real positive values. These values are equal only when equality holds in (7), in which case $r_B = 3GM$. The larger of the two values of M corresponds to a positive heat capacity, and therefore to a thermally locally-stable black hole. The larger black hole has also been shown to be locally gravitationally stable—it has the least action—among spherical geometries satisfying the canonical boundary conditions (fixed β and r_B) and the Einstein constraint equations (Whiting and York 1988). From the corresponding Euclidean action I_s (s = stable), one can obtain a "zero-loop" partition function $Z_0 (\beta, r_B) = \exp(-I_s/\hbar)$. From Z_0 one can deduce the entropy (3), which, as one sees, is independent of r_B. The details and other consequences of using Z_0 are given in York (1986) and in Braden et al. (1987). In the latter reference a satisfactory relationship is established between Z_0 and a corresponding density of states. Here I shall only recall explicitly the result for the mean energy $\langle E \rangle$. One finds that it is related to M by

$$M = \langle E \rangle - \frac{1}{2} \frac{G \langle E \rangle^2}{r_B}, \tag{8}$$

so that there is a physically sensible relation between the mass M, the thermal energy $\langle E \rangle$, and the gravitational binding energy (the *negative* final term).

Another important property of $\langle E \rangle$, besides being the thermodynamic internal energy, is that it can be inferred from a quasi-localized canonical ("Arnowitt–Deser–Misner") action, in Lorentzian space-time, for the static region between $r = 2GM$ and $r = r_B$ (York 1991). Therefore, the internal energy can be shown to have, for the static space-times that are candidates for equilibrium, a definite status as "mechanical energy" independent of the thermodynamic interpretation of the Euclidean action where it was first encountered. However, the canonical action alluded to above contains *no* counterpart of the entropy term in the Euclidean action that was displayed in York 1986, and which will be treated more generally below. Therefore, elucidation of the relationship between the Euclidean and canonical actions must be considered to be a central problem in the development of quantum gravity.

If (7) does not hold, there are no real positive solutions for M. It is significant that this condition is independent of the gravitational constant G. I have found a quite similar condition that is necessary for the nucleation of black holes to occur in hot flat space, a process first considered by Gross et al. (1982); however, as a consequence one sees readily that it is the *larger* hole that can nucleate, contrary to their original claim. (Indeed, this turns out to remove the paradox of spontaneously increasing free energy that was mentioned earlier). The action of the smaller unstable hole acts as a barrier that controls the rate of nucleation from hot flat space. The basic condition that permits nucleation of a black hole is (York 1986)

$$r_B T > \frac{27}{32\pi} \hbar. \tag{9}$$

This condition will receive further attention in the discussion of the partition function below.

Let us contrast the condition that permits nucleation of a black hole with an estimate of a corresponding condition necessary for the collapse of a ball of massless radiation under its own weight, that is, a relativistic analog of the Jeans instability (Jeans 1928). Suppose the radiation is a massless boson field. If the temperature is held constant, the energy E_{rad} increases as the cube of the radius. At a certain value of r_B, the radiation will be inside its own gravitational radius, which is of the order of GE_{rad}. Then we imagine that gravity switches on and the radiation collapses. The "Jeans" criterion is then roughly

$$r_B T^2 \gtrsim \left(\frac{\hbar^3}{G}\right)^{1/2}. \tag{10}$$

The contrast between nucleation and "ordinary collapse" is clear from a comparison of (9) and (10). The two processes are physically distinct, which suggests that they will have different rates. It appears that the rate of nucleation is typically greater than that of collapse in "hot flat space." Furthermore, I have found that the region of the $r_B\beta$ plane from which stable holes can nucleate is larger than that from which they can form by collapse. The criterion (necessary condition) for the onset of nucleation is independent of G and therefore has no classical counterpart, in contrast to the Jeans instability. The latter point was asserted, though before relations (7) and (9) were established it could not have been proved, by Gross et al. (1982).

4. Reduced Euclidean Action

The results described in this and subsequent sections were obtained in collaboration with Bernard F. Whiting (Whiting and York 1988).

Because thermodynamics can be derived from statistical mechanics in the rest of physics, one expects the same to hold true for black hole thermodynamics and the statistical mechanics of the relevant gravitational fields. In turn, one expects to learn something about quantum gravity if one is able, for example, to obtain the partition function from a functional integral over geometries. Here we use the action appropriate for Einstein's theory, but we find much that we would expect to be common to any framework for a theory of quantum gravity. In particular, one expects black holes to be present in all such theories.

In this study we go beyond previous works, which all relied explicitly on properties of the classical black hole geometries. We consider all regular spherically symmetric space-times of suitable topology in a certain class specified in detail below. Most of these geometries are not classical in that they do not satisfy all of the Einstein equations, but only the constraints. We shall include their effects in constructing a partition function for the fixed "single black hole topological sector." The ensemble is specified by the area $4\pi r_B^2$ of the metrically spherical cavity with a black hole at the center and by the constant inverse temperature β at $r = r_B$. In the Euclidean description adopted here, which is suited to the thermal equilibrium state (Gibbons and Perry 1976; Hartle and Hawking 1976; Israel 1976) the four-geometries have topology $R^2 \times S^2$, boundaries $S^1 \times S^2$, and Euler number $\chi = 2$. We construct a "reduced" Euclidean action I_* for these geometries by eliminating the constraints in the Euclidean action. The remarkable form assumed by I_* indicates heuristically a measure and enables us to formulate the partition function Z effectively as a functional integral. This particular Z can be evaluated as an ordinary integral for the geometries we consider even though the number of such geometries being summed is roughly as large as the number of smooth single-valued functions with given boundary values on a disk in R^2. From the result we obtain quantum corrections to the usual black hole entropy formula (3). We also demonstrate the existence of a phase transition accompanied by a change of topology as the local minimum of I_* rises towards zero. The local minimum of I_* actually equals zero when a definite relation holds between size and temperature, namely, $r_B = (27/32\pi)\beta\hbar$, as in (9).

We adopt for the black hole geometries metrics of the form

$$ds^2 = U(y)d\tau^2 + V^{-1}(y)dy^2 + r^2(y)d\Omega^2, \tag{11}$$

where the radial coordinate $y \in [0,1]$ and the Euclidean time τ has a period 2π. U has a prescribed constant value U_B at the boundary $y = 1$, as does $r(1) \equiv r_B$. The three-geometry of the boundary is thus fixed and related to the boundary conditions of the canonical ensemble by

$$\beta\hbar = \int_0^{2\pi} U_B^{1/2}d\tau = 2\pi U_B^{1/2}, \tag{12}$$

where β is the inverse temperature at the boundary of prescribed area $A_B = 4\pi r_B^2$. The "center" of the geometry at $r(0) \equiv r_+$ is required to be regular. We avoid a "conical" singularity at this center by imposing

$$\left[V^{1/2}(U^{1/2})'\right]_{y=0} = 1, \tag{13}$$

where a prime ($'$) denotes differentiation with respect to y. In addition, to distinguish in (11) "hot flat space" ($S^1 \times R^3$) from the black hole sector, we examine the formula for the Euler number of a four-geometry, which in the present case yields

$$2\left[V^{1/2}(U^{1/2})'(1 - \hat{V})\right]_{y=0} = 2[1 - \hat{V}(0)]. \tag{14}$$

The equality in (14) follows from (13) and we have defined $\hat{V} = V(r')^2$. Black holes have Euler number $\chi = 2$, from which (14) yields the requirement $\hat{V}(0) = 0$ in the black hole sector.

With the three-geometry of the boundary ∂M fixed and a regular center, the appropriate Euclidean action is

$$I = -\frac{1}{16\pi G}\int_M Rg^{1/2}d^4x + \frac{1}{8\pi G}\int_{\partial M}(K - K^0)\gamma^{1/2}d^3x, \tag{15}$$

where γ_{ij} is the metric induced on ∂M, K is the mean extrinsic curvature of ∂M, and K^0 is a constant chosen to make the action of flat space with the given boundary ∂M equal to zero. If we evaluate the action (15) for metrics of the form (11), we can extract a first order Lagrangian density given by

$$L = (UV)^{1/2}(r')^2 + r\left(\frac{V}{U}\right)^{1/2} r'U' + \left(\frac{U}{V}\right)^{1/2} + \text{constant} \qquad (16)$$

apart from an irrelevant multiplicative factor. (Bardeen (1973) considered a similar Lagrangian for a different purpose, in treating black holes in the conventional Lorentzian space-time picture. There, r_+ defines an inner boundary while in the Euclidean description it does not.) The Euler–Lagrange equations of L for the metric functions U, V, and r yield, as desired, the complete and correct set of nonvacuous Einstein equations for the metric (11), namely,

$$G_0^0 = \frac{1}{r^2 r'}\left[r(\hat{V} - 1)\right]' = 0, \qquad (17)$$

$$G_1^1 = G_0^0 - \frac{U}{rr'}\left(\frac{\hat{V}}{U}\right)' = 0, \qquad (18)$$

$$G_2^2 = G_3^3 = \frac{1}{2r^2 r'}\left(\frac{r^2\hat{V}U'}{Ur'}\right)' - \frac{1}{4}\frac{U}{(r')^2}\left(\frac{\hat{V}}{U}\right)'$$
$$+ \frac{1}{2}(G_0^0 - G_1^1) = 0. \qquad (19)$$

Thus, the simplification of the action by the symmetry of the metric has led to no spurious features in the variation problem, which is something one always has to check. Another relation useful for later work is an expression for the scalar curvature in the form

$$R = -\frac{1}{g^{1/2}}\left(g^{1/2}\frac{VU'}{U}\right)' - 2G_0^0. \qquad (20)$$

In performing a path integral based on the action (15), one wishes to enforce the constraints because these generate the invariances of the theory and reveal, at least implicity, the physical degrees of freedom. There are, of course, formal methods by which this can be accomplished. However, the comparative simplicity of the metrics (11) enables us to confront the question directly by simply solving and eliminating the constraints. It is this process that leads to what we call the "reduced action."

The form of (11) is adapted for analysis of the constraints on $\tau = $ constant foliations, on the leaves of which the induced three-geometries are manifestly independent of τ. This choice seems very natural for a problem involving equilibrium, particularly because periodicity is defined with respect to this variable. (Strictly speaking, the foliations degenerate at the centers $r = r_+$, where in each possible four-geometry, the three-dimensional leaves intersect on a two-sphere).

In the present case, the momentum constraints for $\tau = $ constant slices are trivially satisfied. The Hamiltonian constraint is, from (17),

$$\frac{1}{r^2 r'}[r(\hat{V}-1)]' = 0, \tag{21}$$

with solution $\hat{V} = 1 - Dr^{-1}$. The integration constant D is identified as $r(0) = r_+$ from the previously noted requirement $\chi = 2$. The parameter r_+ is the only remaining degree of freedom. This variable indicates the size of the horizon of the black hole, for example, $r_+ = 2GM$ for a Schwarzchild black hole. Because r_+ emerges in solving the Hamiltonian constraint, it can be regarded naturally as a functional of the three-geometry on $\tau = $ constant slices. Thus, r_+ can be defined by the invariant integral

$$\int_0^1 \frac{dy}{V^{1/2}} = r_B \left(1 - \frac{r_+}{r_B}\right)^{1/2}$$
$$+ \frac{1}{2} r_+ \ln\left[1 + \left(1 - \frac{r_+}{r_B}\right)^{1/2}\right]\left[1 - \left(1 - \frac{r_+}{r_B}\right)^{1/2}\right]^{-1}, \tag{22}$$

where $r^2 = g_{\theta\theta}$ and $V = g^{yy}$ are metric variables of the $\tau = $ constant hypersurfaces.

The metrics are now reduced to the form

$$ds^2 = U(y)d\tau^2 + \left(1 - \frac{r_+}{r}\right)^{-1}(r')^2 dy^2 + r^2(y)d\Omega^2, \tag{23}$$

in which radial gauge invariance, i.e., an essentially arbitrary relation between y and r, is manifest. A point of particular interest is that $U = $ (lapse function)^2 is an arbitrary positive function of y, except that it must take the fixed boundary value U_B at $y = 1$ and that its behavior near the origin is constrained by regularity conditions, as we infer from (13) and from the form of $\hat{V}$. Near $y = 0$ or $r = r_+$, we can write

$$U \underset{y \to 0}{\longrightarrow} 4r_+^2\left[\left(1 - \frac{r_+}{r}\right) + c_1\left(1 - \frac{r_+}{r}\right)^2 + c_2\left(1 - \frac{r_+}{r}\right)^3 + \cdots\right], \tag{24}$$

where c_1 and c_2 are constants. A computation of the Einstein tensor using (18) and (19) shows that its only nonzero components in the limit $y \to 0$ are

$$G^\theta_\theta = G^\phi_\phi = \frac{3c_1}{4r_+^2}. \tag{25}$$

We interpret this result as representing quantum tangential stresses within the surface of the black hole. These stresses vanish for the two "classical" Schwarzschild black holes that are possible if (and only if) the inequality (7) holds. Recall that $G^0_0 = 0$ everywhere, by construction, and note that $G^1_1 \to 0$ as $y \to 0$, meaning that G^0_0 and G^1_1 are equal at $r = r_+$, as is necessary for regularity there (Page 1982).

We shall now calculate the reduced action I_*. Recall that the boundary term in (15) as applied to metrics (23) of topology $R^2 \times S^2$ refers only to the hypersurface $y = 1$. The locus $y = 0$ is a regular "point," actually a two-surface, and not a boundary, in the Euclidean manifold that represents the black hole in equilibrium. None of the $\tau = $ constant hypersurfaces is a boundary surface because the manifold is closed in the τ-direction by the periodicity condition.

The mean extrinsic curvature of a $y = $ constant hypersurface is

$$K = -\frac{1}{g^{1/2}} \frac{\partial}{\partial x^\mu} \left(g^{1/2} n^\mu \right) = -\frac{1}{g^{1/2}} \left(g^{1/2} V^{1/2} \right)', \tag{26}$$

where n^μ is the outward-pointing unit normal vector field on $y = $ constant. The square roots of the determinants of the four-metric $g_{\mu\nu}$ and of the three-metric γ_{ij} on $y = $ constant are, respectively,

$$g^{1/2} = \left(\frac{U}{V} \right)^{1/2} r^2 \sin \theta, \tag{27}$$

$$\gamma^{1/2} = U^{1/2} r^2 \sin \theta, \tag{28}$$

where

$$V^{-1} = g_{yy} = (r')^{-2}(1 - r_+/r). \tag{29}$$

The subtraction term K^0 in (16) is obtained by taking flat four-space with topology $S^1 \times R^3$ and embedding in it a boundary surface $S^1 \times S^2$ with the same intrinsic three-geometry as has $y = 1$ in the physical four-spaces. Flat four-space is therefore conveniently described by

$$ds^2 = U_B d\tau^2 + (r')^2 dy^2 + r^2 d\Omega^2 \tag{30}$$

and we easily obtain

$$K^0 = -\left(\frac{2}{r}r'\right)_{y=1}. \qquad (31)$$

This method of constructing a subtraction term is designed to establish flat space with topology $S^1 \times R^3$ as having zero Euclidean action and serving thus as a reference point. The process is somewhat similar to discarding the zero-point energy terms from the Hamiltonian of an assembly of electromagnetic oscillator modes in a black-body cavity prior to calculating the free energy, entropy, and other such quantities. The entropy is unaffected by this subtraction, both in the black hole problem and for the electromagnetic field.

Combining the results of the previous paragraph with (13), (20), and (23) yields, in an elementary but highly instructive exercise,

$$I_* = (\beta\hbar r_B/G)\left[1 - (1 - r_+/r_B)^{1/2}\right] - \pi r_+^2/G, \qquad (32)$$

or

$$I_* = (\beta E - S_{BH})\hbar, \qquad (33)$$

where we have made the identification

$$E = (r_B/G)\left[1 - (1 - r_+/r_B)^{1/2}\right]. \qquad (34)$$

Note that the expression for E is equivalent to (8) if we put $r_+ = 2GM$. We have here dropped the angular brackets on E, as no ensemble average is supposed to have been performed yet. In (34), we have also made the identification of S_{BH} as in (3), but with $r_+ = 2GM$. We see that the reduced action does not have the form "βE" that one would expect for a single static state of energy E. The entropy term reveals the presence of a large number of degrees of freedom associated with any value of r_+. Arguments in which S_{BH} is obtained as the logarithm of the number of quantum states accessible in building a black hole (Zurek and Thorne 1985), or as the logarithm of the density of states (Braden et al. 1987), have been given for black holes that satisfy the complete set of Einstein equations. Here, of course, only the constraint equations have been used, but a counting argument such as that proposed by Zurek and Thorne (1985) should be insensitive to this, given that the entropy is associated with the surface of the black hole and its immediate neighborhood, a feature which they and others, such as Davies (1981) have observed. For example, in York 1986, it was shown using the on-shell action that the entropy (3) is obtained independently of the radius r_B of the cavity, that is, even in the limit in which r_B approaches r_+.

A further difference between I_* and the Euclidean action obtained directly from the (stable!) Schwarzschild solution is that β, r_+, and r_B are independent variables in the former, while in the latter they are necessarily related by the equilibrium temperature formula in the form (4), with $\beta = T^{-1}$ and $r_+ = 2GM$. The stationary points of I_* give a *derivation* of the Hawking temperature.

If we divide (33) by β, we obtain the "generalized free energy" posited on physical grounds by York (1986). Here, that result has been shown to follow precisely from the action evaluated on the constraint hypersurface of the gravitational phase space. One finds that if $r_B \geq (3\sqrt{3}/8\pi)\beta\hbar$, I_* has two real positive stationary points with respect to variation of r_+, or of the dimensionless energy $\xi = GE/r_B$, and at both of these stationary points the equilibrium temperature formula (4) holds. One stationary point corresponds to unstable equilibrium, while the other corresponds to equilibrium that is locally stable (see Figure 1). In previous work, these stationary points were tacitly identified with the unstable and stable Schwarszchild solutions, respectively, but the result is actually more general. Each stationary point represents an infinite number of equilibrium space-times in which the temperature formula (4) holds, but which are all foliated differently because of the freedom in choosing the lapse function. Only *one* such choice of the lapse at each stationary point reproduces a Schwarzschild solution, but such a choice is in no way singled out by the reduced action, which is essentially independent of the lapse function. This wildness in the time-slicing expresses strongly that the absence of order in gravitational thermal equilibrium is more radical than that corresponding to ordinary thermal equilibrium. In the latter case, equilibrium is thought of as occurring *in* space-time. In the present case, the concept of *an* equilibrium space-time is not physically correct.

If $r_B = (27/32\pi)\beta\hbar$, the value of I_* at the locally stable stationary point vanishes, a fact that will be of great interest below. If $r_B < (3\sqrt{3}/8\pi)\beta\hbar$, then I_* has no real positive stationary points. In Figure 1, we see that the reduced Euclidean action is not always positive. However, this does not cause a problem in the convergence of the functional integral we shall obtain. We are only considering configurations that satisfy the constraints. There is, as a result, *no* analog of the rapidly varying conformal factors that could force the Euclidean action associated with a finite region to become arbitrarily negative, as considered in Gibbons et al. (1978). Such deformations of the geometry would force it away from the constraint hypersurface. The corresponding question of "negative conformal modes" in the second variation of the Euclidean action consequently cannot arise when one stays on the constraint hypersurface. The relation of the present resolution of this issue of unphysical indications of instability to previous attempts by Gross et al. (1982) and by Allen (1984) has been analyzed in detail by B. Allen, G. Cook, and B. Whiting, in an unpublished work. Their conclusions agree with the discussion presented here.

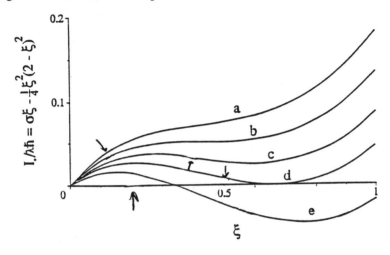

FIGURE 1. The reduced action I_* is plotted as a function for the dimensionless energy $\xi = GE/r_B$, with $\lambda = 4\pi r_B^2/r_p^2$, for five values of the dimensionless inverse temperature $\sigma = \beta \hbar/4\pi r_B$. Curve a ($\sigma = 0.4333$) has no stationary points. Curve b ($\sigma = 2/3\sqrt{3} = 0.385$) has two coincident stationary points, while curve c ($\sigma = 0.337$) has a distinct positive local minimum. For these three curves, the global minimum at $\xi = 0$ dominates the partition function. Curve d ($\sigma = 8/27 = 0.296$) has a local minimum with zero action. For curve e ($\sigma = 0.232$), the local minimum is negative and it dominates the partition function.

On physical grounds the Euclidean action for gravity should not be expected always to be positive. Consider the box of massless bosons discussed earlier in connection with the relativistic Jeans instability. The free energy F of this radiation is negative, and so therefore is its effective Euclidean action βF. When the radiation collapses, the free energy and Euclidean action of the black hole it forms cannot be positive; these quantities must remain at least as negative as they were before collapse. (This is essentially equivalent to the second law of thermodynamics). The classical Euclidean action for any gravity theory must therefore always have negative values in its range, at least for topologies suitable to the description of black holes.

We end this section by expressing I_* in terms of variables convenient for subsequent analysis of its implications. We define the dimensionless variables

$$\lambda = 4\pi r_B^2 / r_p^2, \tag{35}$$

$$\sigma = \beta\hbar / 4\pi r_B, \tag{36}$$

$$\xi = GE / r_B, \tag{37}$$

that measure, respectively, the size (surface area) of the system, its inverse temperature, and its thermodynamic energy. The positive root is chosen in (34) so that the range of ξ is $[0, 1]$, meaning that the black hole is not bigger than the box. We obtain for the reduced action (32)

$$I_* = \lambda\hbar \left[\sigma\xi - \frac{1}{4}\xi^2(2 - \xi)^2 \right]. \tag{38}$$

The reduced action, as shown in Figure 1, has local extrema with respect to ξ if and only if $\sigma \leq (2/3\sqrt{3})$, which corresponds to (7). In this case one finds

$$\frac{\partial I_*}{\partial \xi} = 0 \Leftrightarrow \sigma = \xi(1 - \xi)(2 - \xi), \tag{39}$$

which is equivalent to the equilibrium temperature relation (4).

5. Density of States

In accord with the semiclassical limit of the properties of the entropy as discussed in Zurek and Thorne (1985), Braden et al. (1987), and also in light of the result of Martinez and York (1989) on the additivity of the entropies of black holes and ordinary matter, we can take the number of energy levels between E and $E + dE$ for the gravitational field in the black hole topological sector to be given approximately by

$$dN(E) = d[\exp S_{BH}(E)], \tag{40}$$

where

$$S_{BH}(E) = \frac{1}{4}\lambda\xi^2(2 - \xi)^2. \tag{41}$$

The corresponding expression for the density of states is then

$$\frac{dN}{dE} = \frac{4\pi G r_B}{r_p^2}\xi(1-\xi)(2-\xi)\exp\frac{\pi r_B^2}{r_p^2}\xi^2(2-\xi)^2. \qquad (42)$$

This quantity rises rapidly from $\xi = 0$ and, after reaching a maximum, drops very steeply to zero at $\xi = 1$ ($GE = r_B$), where it is naturally cut off, as the density of states cannot be negative (and the black hole cannot be bigger than the box).

The maximum of dN/dE is readily shown to be given by

$$\xi_{\text{max}} = 1 - \lambda^{-1/2} - \frac{35}{2}(\lambda^{-1/2})^2 + O(\lambda^{-3/2}). \qquad (43)$$

When $r_B \gg r_p$, we need to keep only the first two terms in the expression for ξ_{max}. We then obtain, in terms of the original variables,

$$\frac{r_B}{G} - E_{\text{max}} = \frac{M_p}{(4\pi)^{1/2}}. \qquad (44)$$

The result is independent of both the energy and the size of the box. From (6) we observe that the right-hand side of (44) is equal to the root-mean-square energy fluctuation of a Schwarzschild black hole in the limit $r_B \to 2GM$, or $r_B \to G\langle E\rangle$. Thus, a Schwarzschild black hole of energy E_{max} would nearly fill the cavity, with just enough "room" left for its characteristic quantum fluctuations.

From (40), it is easy to obtain the total number of states. We have

$$N_{\text{total}} = \int_0^{r_B/G} (dN/dE)dE = \int_0^{\lambda/4}(\exp S_{BH})dS_{BH} \qquad (45)$$
$$= (\exp \pi r_B^2/r_p^2) - 1.$$

We see that $\log N_{\text{total}}$ increases with area (r_B^2), that is, more slowly than the increase with volume (r_B^3) that would apply to a conventional system. This may suggest the physical relevance of lower-dimensional quantum gravity in extending these results. (See, for example, Brown, (1988).)

6. Partition Function

We now wish to obtain an expression for the partition function as a path integral. The partition function is

$$Z = \sum_i e^{-\beta E_i} = \sum_E g(E)e^{-\beta E} = \int e^{-\beta E} dN(E), \qquad (46)$$

where i labels states and $g(E)$ denotes the multiplicity of energy levels E. In the last equality we have assumed a continuous energy spectrum with measure $dN(E)$ in energy-space. The key step is to obtain an expression for this measure. For this we shall interpret the results of the previous section as giving a heuristic indication that we adopt (40) as the measure by using $dN = (dN/dE)dE$. Then (46) becomes

$$Z = \int \left[\exp -(\beta E - S_{BH}) \right] dS_{BH} \tag{47}$$

which, because of the remarkable form (33) of the reduced action for any of geometries (23) in the black hole topological sector, becomes

$$Z = \int \exp(-I_*/\hbar) dS_{BH}. \tag{48}$$

The partition function is now expressed as a Euclidean functional integral with measure $S_{BH} = \pi r_+^2 / r_p^2$ determined by the metric only at the "center" point of the manifold, that is, at $y = 0$ where $r(0) = r_+$, so that the functional integral has been reduced to a single ordinary integral in this case. Writing (48) in terms of the dimensionless variables in (35), (36), and (37), we obtain

$$Z = \lambda \int_0^1 d\xi \, \xi(1 - \xi)(2 - \xi) \exp -\lambda \left[\sigma \xi - (1/4)\xi^2(2 - \xi)^2 \right]. \tag{49}$$

From the partition function we can obtain the entropy S_{GF} of the gravitational field in the black hole sector by evaluating

$$S_{GF} = \ln Z - \sigma \frac{\partial}{\partial \sigma} \ln Z. \tag{50}$$

Whenever $\sigma < 8/27$, we find that a stable stationary point ξ_0 dominates Z. At such a point, $\sigma = \xi_0(1 - \xi_0)(2 - \xi_0)$ as in (39), $1 \geq \xi_0 > 2/3$, and $I_*(\xi_0) < 0$. Writing $\xi = \xi_0 + w$ and expanding through quadratic order in w, we evaluate Z in terms of ξ_0 and corrections obtained from the Gaussian integral in w. The result for the entropy contains statistical corrections to S_{BH} evaluated at ξ_0, namely,

$$S_{GF} = S_{BH}(\xi_0) + \ln(2\pi\lambda)^{1/2} \frac{\xi_0(1 - \xi_0)(2 - \xi_0)}{[3\xi_0(2 - \xi_0) - 2]^{1/2}}, \tag{51}$$

$$S_{BH}(\xi_0) = \frac{1}{4}\lambda\xi_0^2(2-\xi_0)^2. \tag{52}$$

We have displayed only the dominant correction to $S_{BH}(\xi_0)$ in the case in which r_B and β are greater than the corresponding Planck values. In this case $S_{GF} > S_{BH}$ as might have been expected. Qualitatively similar results would hold for any nonexponential measure $\mu(r_+)$. Below we obtain a result more particularly dependent on the choice $\mu = S_{BH}$.

We turn to the cases in which Z is *not* dominated by the local minima of I_*. If $\sigma > (2/3\sqrt{3})$, then I_* has no real positive stationary points, and points near the origin in Figure 1 dominate Z. But these points do not describe thermal equilibrium as it is normally understood, precisely because they are not classical stationary points of the action or of the free energy $F = -\beta^{-1}\ln Z$. In fact, we expect that the correct physics will be described in another topological sector, as we discuss further below.

More interesting are the cases in which local minima of I_* exist but are not negative. These occur when $(3\sqrt{3}/8\pi)\beta\hbar \leq r_B \leq (27/32\pi)\beta\hbar$, or $(2/3\sqrt{3}) \geq \sigma \geq (8/27)$. Despite the continued existence of locally stable equilibria in this range, we will see that in the black hole sector, Z already becomes dominated by points near the origin—tiny nonclassical black holes surrounded by "quantum geometry"—when the local minimum of I_* becomes slightly positive. The case of interest is when, at its local minimum, I_* is zero, the established reference for flat space; this occurs for $\xi_0 = 2/3$ and $\sigma = 8/27$. To explore a neighborhood of this point, we set $\xi_0 = 2/3 + \delta$, and thus $\sigma = 8/27 - (2/3)\delta$, for some very small fixed δ. Then the contribution to Z from a stationary point in this region, denoted $Z_{SP}(\delta)$, can be estimated accurately by reducing (49) to a suitable Gaussian. We find

$$Z_{SP}(\delta) \simeq (8/27)(3\pi\lambda)^{1/2}\exp(4/9)\lambda\delta. \tag{53}$$

On the other hand, the contribution to Z from points near the origin $\xi = 0$ is easily seen to be

$$Z_{\text{origin}} \simeq 2/\lambda\sigma^2 = 8\pi/\beta^2 M_p^2 \tag{54}$$

for any σ not close to zero. That (54) does not depend on the size of the box seems physically correct because this case deals with black holes for which $r_+ \ll r_p$ ($\ll r_B$). That it is independent of r_B is a consequence of using the measure $\mu = S_{BH}$.

The contributions (53) and (54) are equal when $\delta = \delta_0 \simeq (9/4)\lambda^{-1} \ln(C/\lambda^{3/2})$, where $C \approx 25$. Taking as an example $r_B = 1.5\,\text{km}$, we have

$\lambda \sim 10^{77}$ and $\delta_0 \sim 10^{-76}$. A dramatic change in Z as the minimum of I_* passes through zero can be seen by noting that $Z_{SP}(0) \sim \lambda^{1/2}$, $Z_{SP}(\delta_0) \sim \lambda^{-1}$, and $Z_{SP}(2\delta_0) \sim \lambda^{-5/2}$. Therefore, Z rapidly becomes well-estimated by (54) as σ increases through 8/27. At the same time, the expectation value of the energy undergoes an enormous change, from $\langle E \rangle_{SP} \cong (2/3)(r_B/G)$ to $\langle E \rangle_{\text{origin}} \cong 2\beta^{-1} \ll M_p$. Another significant feature that follows when (54) holds can be seen by computing the entropy. We find $S_{GF} \approx -2\ln(\beta M_P)$, which is negative. This results from the fact that when $\sigma > 8/27$, the ensemble-averaged number of stationary point states is actually smaller than one.

The feature we have just described indicates that for $\sigma > 8/27$ a phase transition must occur. This should be a transition to a different topological sector, so that Z_{origin} calculated above would not actually describe a physical situation for which the entropy would remain positive. One possibility, already previously discussed (York 1986), is that the transition occurs to the $\chi = 0$ topology corresponding to a box filled with "gravitons," that is, to the hot flat space sector dressed with gravitational fluctuations for which the "one-loop" action is known to be negative. (We do not consider here "one-loop" perturbative calculations of the effect of small nonspherical fluctuations in the black hole sector. See, however, York 1985 where the effect of radiation in the box on the geometry of the black hole is treated). But there exists at present no proof that hot flat space is dominant among all available topological sectors.

Descriptions of thermodynamic ensembles appropriate to other black hole geometries have also been given using the methods we have described. Charged black holes have been considered by Braden et al. (1990). Schwarzschild–de Sitter black holes have also been successfully studied.

Acknowledgments. I am pleased to thank H. W. Braden, J. D. Brown, E. A. Martinez, and especially B. F. Whiting for their contributions as collaborators on various aspects of this work. Valuable criticism has been received from D.N. Page and J. Louko. Research support from the National Science Foundation is gratefully acknowledged.

REFERENCES

Allen, Bruce (1984). "Euclidean Schwarzschild Negative Mode." *Physical Review D* 30: 1153–1157.

Bardeen, J.M. (1973). "Rapidly Rotating Stars, Disks, and Black Holes." In *Black Holes*. Cecile DeWitt and Bryce S. DeWitt, eds. New York: Gordon and Breach, pp. 241–290.

Bekenstein, Jacob D. (1973). "Black Holes and Entropy." *Physical Review D* 7: 2333–2346.

Braden, Harry W., Whiting, Bernard F., and York, James W., Jr. (1987). "Density of States for the Gravitational Field in black hole Topologies." *Physical Review D* 36: 3614–3625.

———(1990). "Charged Black Holes in the Grand Canonical Ensemble." *Physical Review D* 42: 3376–3385.

Brown, John D. (1988). *Lower Dimensional Gravity*. Singapore: World Scientific.

Davies, P.C.W. (1981). "Is Thermodynamic Gravity a Route to Quantum Gravity?" In *Quantum Gravity 2: A Second Oxford Symposium*. Christopher J. Isham, Roger Penrose, and Dennis W. Sciana, eds. Oxford: Clarendon, 183–209.

Feynman, Richard P., and Hibbs, A.R. (1965). *Quantum Mechanics and Path Integrals*. New York: McGraw-Hill.

Gibbons, G.W., and Hawking, Stephen W. (1977). "Action Integrals and Partition Functions in Quantum Gravity." *Physical Review D* 15: 2752–2756.

Gibbons, G.W., and Perry, Malcolm J. (1976). "Black Holes in Thermal Equilibrium." *Physical Review Letters* 36: 985–987.

Gibbons, G.W., Hawking, Stephen W., and Perry, Malcolm J. (1978). "Path Integrals and the Indefiniteness of the Gravitational Action." *Nuclear Physics B* 138: 141–150.

Gross, David J., Perry, Malcolm J., and Yaffe, Laurence G. (1982). "Instability of Flat Space at Finite Temperature." *Physical Review D* 25: 330–355.

Hartle, James B., and Hawking, Stephen W. (1976). "Path-integral Derivation of black hole Radiance." *Physical Review D* 13: 2188–2203.

Hawking, Stephen W. (1975). "Particle Creation by Black Holes." *Communications in Mathematical Physics* 43: 199–220.

———(1979). "The Path-integral Approach to Quantum Gravity," In *General Relativity: An Einstein Centenary Survey*. Stephen W. Hawking and W. Israel, eds. Cambridge and New York: Cambridge University Press, 746–789.

Israel, W. (1976). "Thermo-Field Dynamics of Black Holes." *Physics Letters A* 57: 107–110.

Jeans, James H. (1928). *Astronomy and Cosmogony*. Cambridge: Cambridge University Press.

Martinez, E.A., and York, James W., Jr. (1989). "Additivity of the Entropies of Black Holes and Matter." *Physical Review D* 40: 2124–2127.

Page, Don N. (1982). "Thermal Stress Tensors in Static Einstein Spaces." *Physical Review D* 25: 1499–1509.

Tolman, Richard C. (1934). *Relativity, Thermodynamics, and Cosmology*. Oxford: Clarendon.

Whiting, Bernard F. (1990). "Black Holes and Gravitational Thermodynamics." *Classical and Quantum Gravity* 7: 15–18.

Whiting, Bernard F., and York, James W., Jr. (1988). "Action Principle and Partition Function for the Gravitational Field in black hole Topologies." *Physical Review Letters* 61: 1336–1339.

York, James W., Jr. (1983). "Dynamical Origin of black hole Radiance." *Physical Review D* 28: 2929–2945.

———(1984). "What Happens to the Horizon When a Black Hole Radiates?" In *Quantum Theory of Gravity: Essays in Honor of the 60th Birthday of Bryce S. DeWitt*. Steven M. Christensen, ed. Bristol: Adam Hilger, 135–147.

———(1985). "Black Hole in Thermal Equilibrium with a Scalar Field: The Back-Reaction." *Physical Review D* 31: 775–784.

———(1986). "Black Hole Thermodynamics and the Euclidean Einstein Action." *Physical Review D* 33: 2092–2099.

———(1987). "Black Holes and Partition Functions." Lecture notes, University of Texas at Austin.

———(1991). "Canonical Action Principle for Finite Gravitational Systems." Forthcoming.

Zurek, Wojciech, H., and Thorne, Kip S. (1985). "Statistical Mechanical Origin of the Entropy of a Rotating, Charged Black Hole." *Physical Review Letters* 54: 2171–2175.

Discussion

JACOBSON: You have the number of states going as the radius of the box squared. Somehow, that's independent of whether there happens to be a black hole and is just a statement about the thermodynamics of this section. What if you couple this minigravitational field to some other fields still within this sector?

YORK: Yes, we've done the Maxwell fields. The result is similar. Do you remember we were speaking of the case when black holes dominate this sector statistically?

JACOBSON: Then is the number of states in a Planck-length-sized box still zero or one? Is it still true that it scales as the square of the box radius?

YORK: Yes, for all of the cases we have been able to do, namely coupling to the cosmological constant or to electric charge; those are the two fundamental cases we've done. We have not done the Kerr black hole yet. There are various reasons why that's going to be hard. They aren't just technical.

JACOBSON: By charge, do you mean just one parameter?

YORK: I mean a charged black hole or Reissner-Nordström black hole that's in equilibrium with the right chemical potential at the wall. This is the grand ensemble where we'll have to sum over all charge values for the final result. In other words, I have to add all the appropriate chemical potentials. When I put the electric field charge in there, I have to add the Maxwell action. And I have to fix the inverse temperature on the boundary of the box. I also have to fix the difference in potential between the hole and the side of the box. And then I have to blue-shift it down to where I am on the box in accordance with the principle of equivalence. Then and only then does it work. You work out the whole program, solve the Einstein–Maxwell constraints, and everything goes through just like the case I described in my talk, where charge was not considered.

JACOBSON: This is, so to say, a philosophical question. How do you regard the basis of these results? It looks like you took the Euclidean form of the action and you said: Let me think of this as giving the partition function of a statistical-mechanical system and just fool around with it as if it were that and compute the thermodynamic quantities. Is that really the bottom line, or do you have more of a feeling about what this really represents?

YORK: Well, the argument that you can represent the partition function as a particular path integral is, of course, not due to me. It goes back at least to Feynman, and you can find it in his book on path integrals. My argument would be just the same as everyone else's. All these static black holes do have a natural time-like Killing vector in the Lorentzian section, for which the prescription $t \to it$ and periodic identification is just as sound, as far as I am able to tell, as it is in flat space-time, for which it does work. In short, what we're talking about is quantum gravity using a path integral. The Hawking effect makes quantum gravity and thermodynamics inseparable. In stationary cases, I'm not sure how to do it. In fact, I'm sure that it's quite difficult. But in the static case, the only part of this argument that isn't based on things that are known to work in many cases is the argument for the measure, which is a physical argument. It came from eliminating the Hamiltonian constraint first, before trying to do a sum over paths. Then you look at the form of the action and consider the meaning of entropy to get a measure.

SMOLIN: I have a question about exactly that point. In the perturbation derivation of the path integral, we solve the Hamiltonian constraint, just the linear part, which is just spatial scalar curvature equal to zero, and do perturbations around that. There's always been an issue about whether we've got a measure that is good for nonperturbative integrations. Can you do the analogue of that in this case and see if you get the same measure or a different measure?

YORK: I don't have a full answer yet. I have looked at it and the only thing I can say with certainty of course goes back to the papers of Karel Kuchař, Jim Hartle, and their collaborators. I can comment on one well-known part of the issue. The way I approach the problem, there isn't any problem with conformal rotations. In fact, if you do the problem the way I suggest, there is no freedom to make conformal rotations. In fact, with the usual machinery, it's not possible to pose the boundary-value problem that you need for the black hole at all. So, in other words, the tricks that have been used to get rid of the problem of so-called negative conformal modes fail completely unless the radius of the box is at infinity. And that is a singular point of the problem. Bernard Whiting has completely solved that problem. There is an infinite number of negative conformal modes and one interesting mode. If you study the perturbation problem in exactly the same way by imposing the Hamiltonian constraint, you either get one mode or none, exactly in accordance with my previous arguments that in one case it was stable and in one case it was unstable. So I would say that there is a very definite difference in the usual lore of perturbing the action and looking for negative modes. In the case of gravity and the problems in gravity, at least for black holes, it is completely eliminated by solving the Hamiltonian constraint explicitly.

Black Hole Thermodynamics and the Space-Time Discontinuum

Ted Jacobson

A few years ago quantum gravity was a really exciting field because people felt as if we were on the verge of some breakthrough when it turned out that formulating quantum field theory in curved spaces gave a picture of black hole thermodynamics that was remarkably consistent in a way that really couldn't have been anticipated. Since quantum physics itself was initially discovered as a result of trying to understand the thermodynamics of a box of radiation, it was felt that perhaps the key to quantum gravity had been found. Now, in the past several years that feeling seems to have died down completely. I think the reason is not so much that minds have changed about the significance of the results, but rather that, apart from a few ongoing lines of inquiry, people did not know what to do with them. I would like to suggest something that can perhaps be done, something that has not been tried. Unfortunately it is a rather vague suggestion, but I hope it is better than none; so I want to make it.

What happened when quantum physics was discovered was that, in order to apply Boltzmann's concept of entropy, Planck figured he needed to be able to count. Even though there wasn't anything to count in classical physics, he introduced something he could count just for the sake of defining entropy, and then that thing he was counting turned out to have physical significance. I want to suggest that we should be trying to count something now or, put another way, we should be trying to formulate physics in a discontinuum rather than in a continuum. This is thus an expression of an idea that has been mentioned just a bit during the conference, in particular in John Stachel's talk on Einstein's view of the quantum and in Chris Isham's talk on "group quantization" of gravity. What I am suggesting is, therefore, not aimed at a standard statistical understanding of black-hole entropy, on which significant progress has been made[1], but rather at the idea that black hole thermodynamics be used for guidance in the search for a fundamental discontinuum formulation of physics.

Rather than asserting anything specific about the structure of the space-time discontinuum, I want to state a principle in a quantum language. The principle is that within any finite-space volume, the state space of any physical system is *finite dimensional*. The idea is that this would be a quantum feature of any model of space-time which abandons the continuum. Now, a plausible consequence of this principle is that the dimension of the state space will depend upon the space-time metric simply because the volume of a region is determined by the metric. Conversely, if there is some discontinuum theory that underlies the continuum theory, then the space-time metric must actually be an aspect of this discrete state space. The heuristic I want to suggest is that this underlying finite theory must be consistent with the classical dynamics we already know for the space-time metric, and also with what seems to be true about quantum field theory in curved space-times, in particular the generalized second law of thermodynamics.[2] Of course, any new fundamental theory must be consistent with what we know, but let me try to give one example where this consistency might help us to constrain the possibilities we confront in trying to construct a finite theory.

Suppose we accept the idea that the state space is finite dimensional, and that it depends on the space-time volume. Since the space-time volume is dynamical in general relativity, it follows that the fundamental discrete theory will not be formulated in a Hilbert space fixed *a priori,* but rather the state space itself will be dynamical. One immediately confronts a question: suppose there is given a state in some finite state space, and then another dimension is added, another basis vector. What kind of evolution, what kind of law of nature would determine the new state in the larger space with an added dimension? Would the new state be a single vector in the larger state space, or would it be a *mixed* state because of the arbitrariness inherent in extending the original state into the larger space? One can invent examples of either sort of rule for evolution, but here is where a clue from black hole thermodynamics can be used. In particular, the (likely) fact that pure states go to mixed states when quantum field theory is coupled to curved space-time manifolds suggests that, at the fundamental discrete level, there is evolution from pure to mixed states when dimensions are added to (or taken away from) the state space.

This idea might be explored using as a model a field theory with a short-distance cutoff, coupled to classical dynamical gravity. The idea that the dynamics of the metric leads to a dynamics of the state space can be incorporated by specifying the cutoff in terms of a short *proper* distance rather than a coordinate distance. Thus, the state space itself is dynamical in this model, and a choice must therefore be made about how to generalize the usual description of evolution to this situation. The interplay between this choice and the generalized second law of thermodynamics could then be explored, in an attempt to gain insight into possible foundations for a finite theory.

Acknowledgment. This work is supported by NSF grant PHY 88-04561.

NOTES

[1] See for example: Zurek, W.H. and Thorne, K.S., (1985): *Phys. Rev. Lett.* 54, 2171; Thorne, K.S., Zurek, W.H. and Price, R.H. (1986): "The Thermal Atmosphere of a Black Hole" in *Black Holes: The Membrane Paradigm*, ed. K.S. Thorne, R.H. Price and D.A. Macdonald, New Haven: Yale University Press; Whiting, B.F. and York, J.W., Jr. (1988): *Phys. Rev. Lett.* 61 , 1336; Martinez, E.A. and York, J.W., Jr., (1989) *Phys. Rev. D* 40: 2124.

[2] See for example: Unruh, W.G. and Wald, R.M. (1982): *Phys. Rev. D* 25: 942; (1983): *Phys. Rev. D* 27: 2271.

Contributors

Abhay Ashtekar, Department of Physics, Syracuse University, Syracuse, NY 13244, USA.

Bryce DeWitt, Department of Physics, University of Texas, Austin, TX 78712, USA.

John L. Friedman, Department of Physics, University of Wisconsin, Milwaukee, WI 53211, USA.

Jonathan J. Halliwell, Department of Physics, Massachusetts Institute of Technology, Cambridge, MA 02139, USA.

James B. Hartle, Department of Physics, University of California, Santa Barbara, CA 93106, USA.

Gary T. Horowitz, Department of Physics, University of California, Santa Barbara, CA 93106, USA.

C. J. Isham, Blackett Laboratory, Imperial College, London, SW7 2BZ, United Kingdom.

Ted Jacobson, Department of Physics, University of Maryland, College Park, MD 20742, USA.

Karel V. Kuchař, Department of Physics, University of Utah, Salt Lake City, UT 84112, USA.

Lionel J. Mason, New College, Oxford OX1 3BN, United Kingdom.

Don N. Page, Department of Physics, University of Alberta, 412 Avadh Bhatia Physics Lab, Edmonton, Alberta, Canada T6G 2J1.

Carlo Rovelli, Department of Physics and Astronomy, University of Pittsburgh, Pittsburgh, PA 15260, USA.

Abner Shimony, Department of Physics, Boston University, 590 Commonwealth Avenue, Boston, MA 02215, USA.

Lee Smolin, Department of Physics, Syracuse University, Syracuse, NY 13244, USA.

Rafael D. Sorkin, Department of Physics, Syracuse University, Syracuse, NY 13244, USA.

John Stachel, Department of Physics, Boston University, 590 Commonwealth Avenue, Boston, MA 02215, USA.

C. G. Torre, Department of Physics, Syracuse University, Syracuse, NY 13244, USA.

W. G. Unruh, Cosmology Program, CIAR, Department of Physics, University of British Columbia, Vancouver, B.C. V6T 2A6, Canada.

Robert M. Wald, Enrico Fermi Institute and Department of Physics, The University of Chicago, Chicago, IL 60637, USA.

James W. York, Institute of Fields Physics, Department of Physics and Astronomy, University of North Carolina, Chapel Hill, NC 27599, USA.

Wojciech H. Zurek, Los Alamos National Laboratory, Los Alamos, NM 87545, USA.

Participants

Arlen Anderson
Abhay Ashtekar
David G. Boulware
Joy Christian
Sidney R. Coleman
Stanley Deser
Bryce DeWitt
John L. Friedman
Murray Gell-Mann
Jonathan J. Halliwell
James B. Hartle
Gary T. Horowitz
C. J. Isham
Ted Jacobson
Karel V. Kuchař
Lionel J. Mason
Ezra T. Newman

Don N. Page
Paul Renteln
Joseph D. Romano
Carlo Rovelli
Muhammed Samiullah
Kristin Schleich
Abner Shimony
Lee Smolin
Rafael D. Sorkin
John Stachel
Ranjeet R. Tate
C.G. Torre
W. G. Unruh
Robert M. Wald
Richard Woodard
James W. York
Wojciech H. Zurek